Quantum Electrodynamics and Quantum Optics

NATO ASI Series

Advanced Science Institutes Series

A series presenting the results of activities sponsored by the NATO Science Committee, which aims at the dissemination of advanced scientific and technological knowledge, with a view to strengthening links between scientific communities.

The series is published by an international board of publishers in conjunction with the NATO Scientific Affairs Division

A Life Sciences **B Physics**	Plenum Publishing Corporation New York and London
C Mathematical and Physical Sciences	D. Reidel Publishing Company Fordrecht Boston, and Lancaster
D Behavioral and Social Sciences **E Engineering and Materials Sciences**	Martinus Nijhoff Publishers The Hague, Boston, and Lancaster
F Computer and Systems Sciences **G Ecological Sciences**	Springer-Verlag Berlin, Heidelberg, New York, and Tokyo

Recent Volumes in this Series

Volume 104—Short-Distance Phenomena in Nuclear Physics
edited by David H. Boal and Richard M. Woloshyn

Volume 105—Laser Applications in Chemistry
edited by K. L. Kompa and J. Wanner

Volume 106—Multicritical Phenomena
edited by R. Pynn and A. Skjeltorp

Volume 107—Positron Scattering in Gases
edited by John W. Humberston and M. R. C. McDowell

Volume 108—Polarons and Excitons in Polar Semiconductors and Ionic Crystals
edited by J. T. Devreese and F. Peeters

Volume 109—Percolation, Localization, and Superconductivity
edited by Allen M. Goldman and Stuart A. Wolf

Volume 110—Quantum Electrodynamics and Quantum Optics
edited by A. O. Barut

Series B: Physics

Quantum Electrodynamics and Quantum Optics

Edited by

A. O. Barut

University of Colorado
Boulder, Colorado

Plenum Press
New York and London
Published in cooperation with NATO Scientific Affairs Division

Proceedings of the NATO Advanced Study Institute on
Quantum Electrodynamics and Quantum Optics,
held May 27–June 8, 1983,
in Boulder, Colorado

Library of Congress Cataloging in Publication Data

NATO Advanced Study Institute on Quantum Electrodynamics and Quantum Optics (1983: Boulder, Colo.)
Quantum electrodynamics and quantum optics.

(NATO ASI series. Series B, Physics; v. 110)
"Proceedings of the NATO Advanced Study Institute on Quantum Electrodynamics and Quantum Optics, held May 27–June 8, 1983, in Boulder, Colorado"—T.p. verso.
"Published in cooperation with NATO Scientific Affairs Division."
Includes bibliographical references and index.
1. Quantum electrodynamics—Congresses. 2. Quantum optics—Congresses. 3. Gauge invariance—Congresses. I. Barut, A. O. (Asim Orhan), date- . II. North Atlantic Treaty Organization. Scientific Affairs Division. III. Title. IV. Series.
QC679.N36 1983 537.6 84-9937

ISBN-13: 978-1-4612-9717-8 e-ISBN-13: 978-1-4613-2783-7
DOI: 10.1007/978-1-4613-2783-7

Softcover reprint of the hardcover 1st edition 1984

A Division of Plenum Publishing Corporation
233 Spring Street, New York, N.Y. 10013

PREFACE

The borderline of quantum electrodynamics and quantum optics offer spectacular results and problems concerning the foundations of radiation theory. Perhaps the major new viewpoint that has emerged from recent investigations is that one can now work inside a time-dependent quantum process, whereas up to now all elementary quantum processes were either stationary, or one worked with asymptotic in-and out-states, i.e. an S-matrix approach.

In the first part of this volume, the Quantum Electrodynamics, the present status of the main approaches to this most accurate of all physical theories are discussed: the Hamiltonian approach, the Green's function approach with particular emphasis to bound state problems, and the newer, nonperturbative approach. The latest numerical results on radiative corrections, Lamb shifts and anomalous magnetic moments are reviewed with new results for high Z atoms. Also discussed are different theoretical interpretations of the radiative phenomena as due to quantized field vacuum fluctuations or due to self energy.

A small group of contributions are devoted to the physics and mathematical description of decaying or unstable states in quantum theory. This remarkable phenomenon of quantum theory still needs complete clarification, it is a time-dependent phenomenon, which can be described also by asymptotic S-matrix methods, but with complex energies.

The second main part of the book is devoted to new developments in quantum optics, in particular to time-dependent processes. Strong field autoionization phenomena provide instances of studying time development of coherence, spectral observations and change of population. A controversy (which turns out to be a practical controversy rather than a fundamental one) of using the interaction $\vec{A} \cdot \vec{p}$ versus $\vec{E} \cdot \vec{r}$ in radiative calculations is also discussed extensively by several authors.

A volume on QED and Quantum optics would be incomplete without the discussion of basic quantum theory of measurement. Indeed some discussions are devoted to the new aspects of the foundations of quantum theory with emphasis on the experimental tests of Bell's inequalities.

The contributions collected here have been originally presented at the NATO Advanced Study Institute, held in Boulder, Colorado, between May 26 and June 8, 1983. I should like to thank NATO Scientific Affairs Division and the Department of Physics of the University of Colorado for their support of the Institute, and the contributors and participants for their enthusiasm which made the meeting a very lively, and fruitful experience.

A. O. Barut
Boulder, Colorado
February 1984

CONTENTS

I. QUANTUM ELECTRODYNAMICS

II. QUANTUM OPTICS

III. DECAYING STATES

IV. GAUGE INVARIANCE

V. QUANTUM CORRELATIONS

FINITE BOUNDARY EFFECTS IN QUANTUM ELECTRODYNAMICS

G. S. Agarwal

School of Physics, University of Hyderabad
Hyderabad -- 500134, INDIA

I. INTRODUCTION

Zero point fluctuations of the electromagnetic field are known to determine the characteristics of a radiating system.[1] Such characteristics are usually calculated by assuming that the radiating system is located in free space. However, any change in the environmental conditions is likely to affect the zero point fluctuations and hence the properties of the radiating system. Recently very interesting experiments[2,3] have been done which demonstrate the effect of environment on the characteristics of the atomic/molecular systems. For example Vaidyanathan et al.[3] have shown that the absorption of blackbody radiation by the highly excited states of the Na atom (29d → 30p) located between the plates of a parallel pate conductor is critically dependent on the distance between the plates and is very different from the free space value. The fluorescence spectra of dye molecules adsorbed on the metal and dielectric surface have been shown to exhibit a number of resonances which depend on the characteristics and shape of the metallic medium.[2,4-6]

From the foregoing it is clear that we need to know how the zero point fluctuations change if the medium is finite.[7-10] Obviously one would expect that the shape and the dielectric properties of the medium would be quite critical in such a calculation. Once the new vacuum field is known, the interaction of the atomic/molecular system with the modified vacuum can be studied by using perturbation theory. The following analysis is divided in two parts -- in Sec. II, we will calculate the correlation functions of the electromagnetic field of arbitrary order. Such correlation functions will be used in Sec. III to obtain the radiation characteristics of the atomic/molecular systems. It should be borne in mind that the formulation of this work is equally applicable to the case of free space.

II. QUANTUM CORRELATION FUNCTIONS OF THE ELECTROMAGNETIC FIELD IN PRESENCE OF DIELECTRIC BODIES

It will be seen in Sec. III that the scattering and absorption of radiation by an atomic system in presence of a dielectric body is determined from certain average characteristics of the electromagnetic field. Therefore instead of presenting the modal expansions for the fields and then quantizing these, we will present a method[7] that will enable us to calculate directly the quantum correlation functions of the electromagnetic field. For this purpose certain results from the linear response theory will be used.

A. Results from Linear Response Theory

Consider a quantum mechanical system in thermal equilibrium with Hamiltonian H_o. The density matrix of the system is

$$\rho = e^{-\beta H_o}/\mathrm{Tr}\, e^{-\beta H_o} \quad , \quad \beta = 1/K_B T \quad . \tag{2.1}$$

Let us perturb this system by external forces f_j; so that the perturbation Hamiltonian has the form

$$H_{ext} = -\int d^3r \sum_j A_j(\vec{r},t) f_j(\vec{r},t) \quad . \tag{2.2}$$

It is then known that the linear response of the system is given by

$$\delta\langle A_i(\vec{r},t)\rangle = \sum_j \int d^3r' \int dt' \, \chi_{ij}(\vec{r},\vec{r}',t-t') f_j(\vec{r}',t') \quad , \tag{2.3}$$

where

$$\chi_{ij}(\vec{r},\vec{r}',t-t') = 2i\eta(t-t')\chi''_{ij}(\vec{r},\vec{r}',t-t') \quad , \quad \eta(\tau) = 1 \text{ if } \tau > 0 \; ; \; = 0 \text{ if } \tau < 0$$

$$\chi''_{ij}(\vec{r},\vec{r}',t-t') = \frac{1}{2\hbar} \langle [A_i(\vec{r},t) \, , \, A_j(\vec{r}',t')]\rangle \quad . \tag{2.4}$$

The expectation values in (2.4) are with respect to (2.1) and all the operators are in Heisenberg picture (evolving with Hamiltonian H_o). In the frequency domain (2.3) becomes

$$\frac{\delta\langle A_i(\vec{r},\omega)\rangle}{\delta f_j(\vec{r}',\omega)} = \chi_{ij}(\vec{r},\vec{r}',\omega) \quad , \quad \psi(\omega) = \int_{-\infty}^{+\infty} dt \, \psi(t) \, e^{i\omega t} \quad . \tag{2.5}$$

On defining the symmetrized correlation function by

$$S_{ij}(\vec{r},\vec{r}',t-t') = \frac{1}{2} \langle\{A_i(\vec{r},t) - \langle A_i(\vec{r},t)\rangle, A_j(\vec{r}',t') - \langle A_j(\vec{r}',t')\rangle\}\rangle \;. \tag{2.6}$$

We have the relationship between S and χ

$$S_{ij}(\vec{r},\vec{r}',\omega) = \hbar \cot h \frac{\beta\omega\hbar}{2} \chi''_{ij}(\vec{r},\vec{r}',\omega) \quad . \tag{2.7}$$

It is to be noted that χ'' is the imaginary (-i Real) part of χ if the operators A_i and A_j have same (different) parity under time reversal.

B. Second-Order Electromagnetic Field Correlations

We will now apply the above results to the electromagnetic field fluctuations.[7] The basic dynamical variables of the field are taken to be $\vec{E}$ and $\vec{H}$. The external perturbations are taken to be external polarization $\vec{\mathcal{P}}$ and magnetization $\vec{\mathcal{M}}$, so that in place of (2.2) we will have

$$H_{ext} = -\int \{\vec{\mathcal{P}}(\vec{r},t) \cdot \vec{E}(\vec{r},t) + \vec{\mathcal{M}}(\vec{r},t) \cdot \vec{H}(\vec{r},t)\}\, d^3r \quad . \tag{2.8}$$

Because of a large number of dynamic variables one can introduce different kinds of response functions. For example, one can have the response of E to an applied polarization or to an applied magnetization. In order to keep the analysis simple, we concentrate on the response of $\vec{E}$ to an applied polarization. The details of other response functions can be found in Ref. 7. We now have from (2.4)-(2.7)

$$\mathcal{E}^{(S)}_{ij}(\vec{r},\vec{r}',t-t') = \frac{1}{2} \langle\{E_i(\vec{r},t) \;,\; E_j(\vec{r}',t')\}\rangle \quad ,$$

$$\delta\langle E_i(\vec{r},\omega)\rangle/\delta\, \mathcal{P}_j(\vec{r}',\omega) = \chi_{ijEE}(\vec{r},\vec{r}',\omega) \quad ,$$

$$\chi''_{ijEE}(\vec{r},\vec{r}',t-t') = \frac{1}{2\hbar} \langle[E_i(\vec{r},t) \;,\; E_j(\vec{r}',t')]\rangle \quad . \tag{2.9}$$

$$\mathcal{E}^{(S)}_{ij}(\vec{r},\vec{r}',\omega) = \hbar \cot h \frac{\beta\omega\hbar}{2} \,\mathrm{Im}\, \chi_{ijEE}(\vec{r},\vec{r}',\omega) \quad . \tag{2.10}$$

The response function χ will, shortly, be seen to depend on the nature and shape of the medium.

The relation (2.10) yields only the symmetrized correlation function of the electromagnetic field. However, for studying absorption and emission of photons we need the normally and antinormally ordered correlation functions of the field

$$\mathcal{E}_{ij}^{(N)}(\vec{r},\vec{r}',t-t') = \langle E_i^{(-)}(\vec{r},t)\, E_j^{(+)}(\vec{r}',t')\rangle$$

$$\mathcal{E}_{ij}^{(A)}(\vec{r},\vec{r}',t-t') = \langle E_i^{(+)}(\vec{r},t)\, E_j^{(-)}(\vec{r}',t')\rangle \quad , \tag{2.11}$$

where $E^{(\pm)}$ are the positive and negative frequency parts of the electric field operator. The correlations like (2.11) can be calculated by using the analytical signal concept, i.e. by writing

$$E_j^{(\pm)} = \frac{1}{2}(E_j \pm i\,\tilde{E}_j) \ , \ \tilde{E}_j(\vec{r},t) = \frac{1}{\pi} P \int_{-\infty}^{+\infty} \frac{dt'}{(t'-t)} E_j(\vec{r},t') \ . \tag{2.12}$$

Using (2.12) and (2.10), one can prove the following results

$$\mathcal{E}_{ij}^{\left\{\substack{N\\A}\right\}}(\vec{r}_1,\vec{r}_2,\omega) = \hbar\eta(\mp\omega)[1 + \cot h\,\frac{\beta\omega\hbar}{2}]\,\mathrm{Im}\,\chi_{ijEE}(\vec{r}_1,\vec{r}_2,\omega) \quad . \tag{2.13}$$

Thus antinormally ordered correlations have only positive frequency components.

Zero point fluctuations are determined from the antinormally ordered correlation functions. Taking the limit of zero temperature $\beta \to \infty$, we get

$$\mathcal{E}_{ij}^{(A)}(\vec{r}_1,\vec{r}_2,\omega) = 2\hbar\ \eta(\omega)\ \mathrm{Im}\ \chi_{ijEE}(\vec{r}_1,\vec{r}_2,\omega) \quad . \tag{2.14}$$

C. Linear Response Functions

We next show how the response function χ can be calculated. For this purpose we use Maxwell's equations -- which may be regarded as equations for averaged fields

$$\vec{\nabla} \times \langle\vec{E}\rangle = -\frac{1}{c}\frac{\partial}{\partial t}(\langle\vec{B}\rangle + 4\pi\vec{\mathcal{M}}) \quad , \quad \vec{\nabla}\cdot(\langle\vec{B}\rangle + 4\pi\vec{\mathcal{M}}) = 0 \quad ,$$

$$\vec{\nabla} \times \langle\vec{H}\rangle = \frac{1}{c}\frac{\partial}{\partial t}(\langle\vec{D}\rangle + 4\pi\vec{\mathcal{P}}) \quad , \quad \vec{\nabla}\cdot(\langle\vec{D}\rangle + 4\pi\vec{\mathcal{P}}) = 0 \quad , \tag{2.15}$$

Here $\langle\vec{B}\rangle$ and $\langle\vec{D}\rangle$ represent the induced values of the induction. Given the constitutive relations between $\langle\vec{D}\rangle$ and $\langle\vec{E}\rangle$, $\langle\vec{B}\rangle$ and $\langle\vec{H}\rangle$, such equations can be solved at least in cases when the medium is linear and when the geometry of the medium is relatively simple. In what follows we consider only the linear media, so that the relation between $\langle\vec{E}\rangle$ and $\vec{\mathcal{P}}$ is linear. Once the solution is known, the response can be obtained by using (2.9). χ has been calculated in the literature for various geometries such as slabs bounded by dielectric material on each side, spheres, cylinders, gratings, cubes, etc. The resulting expressions are extremely lengthy and so we will not give them here. The free space result for χ is relatively simple

$$\chi_{ijEE}(\vec{r},\vec{r}',\omega) = \left(\frac{\omega^2}{c^2}\delta_{ij} + \frac{\partial^2}{\partial x_i \partial x_j}\right) \frac{e^{(i\omega/c)|\vec{r}-\vec{r}'|}}{|\vec{r}-\vec{r}'|} \quad . \tag{2.16}$$

D. <u>Higher-Order Correlations of the Electromagnetic Field</u>

Problems involving multiphoton processes[11] require the higher-order correlations of the electromagnetic field. For example, the two-photon decay is determined by the fourth-order antinormally ordered correlation function of the field. In order to obtain such higher-order correlations, we again use statistical mechanics and the structure (2.15). Equation (2.15) shows that $\langle E \rangle$ is linear in the applied probes $\mathcal{P}$ and $\mathcal{M}$ and hence the nonlinear response of the present system is zero. If we combine this fact with the general result on the nonlinear response functions, we find that for a linear medium $\langle [E(t_1),[E(t_2)\ldots[E(t_{n-1}),E(t_n)]]\ldots]\rangle = 0$ for all $n > 2$. Hence we can conclude that

$$[E(t_1),E(t_2)] = \text{c-number} \quad . \tag{2.17}$$

Using (2.17) and (2.1) one can prove[10] a very important moment theorem for the electromagnetic field fluctuations in a medium. In the special case of fourth-order correlations one has

$$\langle A_i(t_1)A_j(t_2)A_k(t_3)A_\ell(0)\rangle = \langle A_i(t_1)A_j(t_2)\rangle\langle A_k(t_3)A_\ell(0)\rangle$$
$$+ \langle A_i(t_1)A_\ell(0)\rangle\langle A_j(t_2)A_k(t_3)\rangle + \langle A_i(t_1)A_k(t_3)\rangle\langle A_j(t_2)A_\ell(0)\rangle \quad . \tag{2.18}$$

Thus if we further specialize, then

$$\langle E_i^{(+)}(t_1)E_j^{(+)}(t_2)E_k^{(-)}(t_3)E_\ell^{(-)}(t_4)\rangle = \langle E_i^{(+)}(t_1)E_k^{(-)}(t_3)\rangle$$
$$\times \langle E_j^{+}(t_2)E_\ell^{(-)}(t_4)\rangle + \langle E_i^{(+)}(t_1)E_\ell^{(-)}(t_4)\rangle\langle E_j^{(+)}(t_2)E_k^{(-)}(t_3)\rangle \quad . \tag{2.19}$$

Each of the second-order correlation appearing in (2.19) can be calculated from (2.13).

III. INTERACTION OF ATOMIC/MOLECULAR SYSTEM WITH ELECTROMAGNETIC FIELDS IN PRESENCE OF DIELECTRICS

We will now consider a number of problems involving the interaction of atoms with the electromagnetic fields in finite geometries.

A. Perturbative Results on the Absorption and Emission of Radiation

Consider a quantum mechanical system in the state $|\psi_i\rangle$ at time t = 0 and let $|\psi_f\rangle$ be the final state. The interaction with the field can be written as

$$H = -\int \vec{P}_A(\vec{r}) \cdot \vec{E}(\vec{r})\, d^3r \quad , \tag{3.1}$$

where $\vec{P}_A$ is the polarization operator for the atomic system. In first-order perturbation, the rate of transition can be shown[8] to be

$$\gamma_{fi} = \frac{1}{\hbar^2} \sum_{m,n} \iint d^3r_1 d^3r_2 \langle\psi_i|P_m(\vec{r}_1)|\psi_f\rangle\langle\psi_f|P_n(\vec{r}_2)|\psi_i\rangle \times \begin{cases} \mathcal{E}^{(N)}_{mn}(\vec{r}_1,\vec{r}_2,\omega_{if}) & \text{if} \quad \omega_{if} < 0 \\ \mathcal{E}^{(A)}_{mn}(\vec{r}_1,\vec{r}_2,\omega_{if}) & \text{if} \quad \omega_{if} > 0 \end{cases} \quad . \tag{3.2}$$

For a single atom located at $\vec{b}$, the rate of spontaneous emission can be obtained by combining (3.2) with (2.14) with the result

$$\gamma^{spon}_{fi} = \frac{2}{\hbar} \sum_{mn} (d_m)_{if} (d_n)_{fi} \operatorname{Im} \chi_{mnEE}(\vec{b},\vec{b},\omega_{if}) \quad . \tag{3.3}$$

The corresponding result for the absorption of blackbody photons in presence of a medium will be

$$\gamma^{abs}_{fi} = \frac{2}{\hbar} \sum_{mn} (d_m)_{if} (d_n)_{fi} \left(e^{\beta|\omega_{if}|\hbar} - 1\right)^{-1} \operatorname{Im} \chi_{mnEE}(\vec{b},\vec{b},|\omega_{if}|) \; . \tag{3.4}$$

The multipolar transitions can be considered in a similar manner. For example, the quadrupole contribution to spontaneous emission can be shown to be

$$\gamma^{spon} = \frac{1}{2\hbar} \sum_{mn\alpha\beta} (Q_{m\alpha})_{if}(Q_{n\beta})_{fi} \frac{\partial^2}{\partial r_{1\alpha}\partial r_{2\beta}} \operatorname{Im} \chi_{mnEE}(\vec{r}_1,\vec{r}_2,\omega_{if})\Big|_{\vec{\gamma}_1=\vec{\gamma}_2=\vec{b}} . \tag{3.5}$$

The two-photon fluorescence rate, for example, the decay of an S-state to S-state can also be expressed in terms of χ by using (2.19). The derivation of the result is long and we quote the result

$$\gamma_{fi} = \frac{2}{\pi}\int_0^{\omega_{if}} d\omega [\mathrm{Im}\ \chi_{i\alpha EE}(\vec{r}_o,\vec{r}_o,\omega)][\mathrm{Im}\ \chi_{i\beta EE}(\vec{b},\vec{b},\omega_{if}-\omega)]$$

$$\Phi^*_{ij}(\omega-\omega_{if})(\Phi_{\alpha\beta}(\omega-\omega_{if}) + \Phi_{\beta\alpha}(-\omega)) \qquad (3.6)$$

where Φ is the two-photon matrix element

$$\Phi_{\alpha\beta}(-\omega) = \lim_{\varepsilon\to 0} \frac{i}{2\hbar^2} \sum_\mu \frac{(d_\alpha)_{f\mu}(d_\beta)_{\mu i}}{(i\varepsilon - \omega_{\mu i} - \omega)} \quad . \qquad (3.7)$$

An extra term in (3.6) arises as the moment theorem for the fourth-order correlation contains two terms.

We have thus expressed all the fundamental quantities associated with the decay of the atoms in terms of χ. We next examine some of the consequences of the above results. All the free space results are recovered by using (2.16) and its limiting terms

$$\lim_{\vec{r}\to\vec{r}'} \mathrm{Im}\ \chi_{ijEE}(\vec{r},\vec{r}',\omega) = \frac{2}{3}\frac{\omega^3}{c^3}\delta_{ij} \quad , \qquad (3.8)$$

$$\lim_{\vec{r}\to\vec{r}'} \frac{\partial^2}{\partial r_\alpha \partial r'_\beta} \mathrm{Im}\ \chi_{mnEE}(\vec{r},\vec{r}',\omega)$$

$$= \frac{\omega^5}{15c^5}[4\delta_{mn}\delta_{\alpha\beta} - (\delta_{m\beta}\delta_{n\alpha} + \delta_{n\beta}\delta_{m\alpha})] \quad . \qquad (3.9)$$

For the case of a parallel plate (with plates located at $z = 0$ and $z = -d$) conductor, the response functions are extremely complex and we quote the results[8] for the density of states $\rho(k)dk$ and the life time of an atom located at $z = -b$ whose dipole moment is randomly oriented:

$$\rho(k)dk = \frac{Vk^2}{\pi^2} dk[1 - \frac{\pi}{kd}(\frac{kd}{\pi} - \eta - \frac{1}{2})] \quad , \qquad (3.10)$$

$$\gamma/\gamma^{(0)} = \frac{\pi}{2kd} + \frac{\pi}{kd}\sum_1^\eta (1 - \frac{\pi^2 n^2}{d^2k^2}\cos(\frac{2b\pi n}{d})) \qquad (3.11)$$

where η is the largest integer less than kd/π. The cut-off frequency is defined by $\omega_c = \pi c/d$. Thus if $\omega < \omega_c$, then

$$\gamma/\gamma^{(0)} = \frac{\pi}{2kd} \quad , \quad \rho(k)dk = \frac{Vk^2}{\pi^2} dk\ (\frac{\pi}{2kd}) \qquad (3.12)$$

whereas for $\omega > \omega_c$, the other terms in the series (3.11) start contributing making $\gamma/\gamma^{(0)} > \pi/2kd$. These results are in qualitative agreement with the experimental results of Vaidyanathan et al.,[3] where the frequency of the transition was changed from below the cut-off frequency to a frequency higher than ω_c by applying a dc electric field. Note that $\gamma < \gamma^{(0)}$ if $\pi/2kd < 1$ leading to the inhibited spontaneous emission.

We now discuss another important case of the dye molecules coated on metallic spheres. We characterize the metal by a dielectric function

$$\varepsilon(\omega) = 1 - \frac{\omega_p^2}{\omega^2 + i\omega\Gamma} \tag{3.13}$$

and assume that the medium outside the metallic sphere has dielectric function ε_o. Detailed calculations using (3.3) and the appropriate response functions show[4] that the width (lifetime) is given by

$$\gamma/\gamma^{(0)} = 1 + \mathrm{Re}\, Q \quad , \tag{3.14}$$

$$Q = \frac{1}{2}\sum_{n=1}^{\infty} (2n+1)\Big\{A_n\big(h_n^{(1)}(k_o b)\big)^2 + \frac{B_n}{k_o^2 b^2}\big[(k_o b\, h_n^{(1)}(k_o b))'\big]^2$$

$$+ \frac{n(n+1)}{k_o^2 b^2} B_n\big(h_n^{(1)}(k_o b)\big)^2\Big\} = K_s + i\Omega_s \quad . \tag{3.15}$$

where

$$A_n = \frac{j_n(k_t a)[k_o a\, j_n(k_o a)]' - j_n(k_o a)[k_t a\, j_n(k_t a)]'}{h_n^{(1)}(k_o a)[k_t a\, j_n(k_t a)]' - j_n(k_t a)[k_o a\, h_n^{(1)}(k_o a)]'} \quad ,$$

$$B_n = \frac{\varepsilon\, j_n(k_t a)[k_o a\, j_n(k_o a)]' - \varepsilon_o j_n(k_o a)[k_t a\, j_n(k_t a)]'}{\varepsilon_o h_n^{(1)}(k_o a)[k_t a\, j_n(k_t a)]' - \varepsilon_t j_n(k_t a)[k_o a\, h_n^{(1)}(k_o a)]'} \quad ,$$

$$k_t = \frac{\omega}{c}\sqrt{\varepsilon(\omega)} \quad , \quad k_o = \frac{\omega}{c}\sqrt{\varepsilon_o} \quad . \tag{3.16}$$

In Fig. 1, we present the behavior of K_s for metallic spheres of different radius. Note the very interesting and predominant resonances in K_s. These resonances for such small sphere radii occur at

$$\varepsilon n + \varepsilon_o(n+1) = 0 \quad . \tag{3.17}$$

Figure 1 obviously shows the very dramatic effect of the metal on the lifetime of the excited states of the molecules adsorbed on metal surfaces. The frequencies given by (3.17) are just the surface plasmon frequencies. Figure 1 also gives the behavior of

Ω_s = Im Q, which also exhibits the resonance character (dispersive type) when (3.17) is satisfied. The quantity Ω_s (related to the real part of χ) can be shown[9] to give the frequency shift of an oscillator (i.e. molecule treated as an oscillator) placed in the vicinity of a metal surface. It may be added that a result simi-

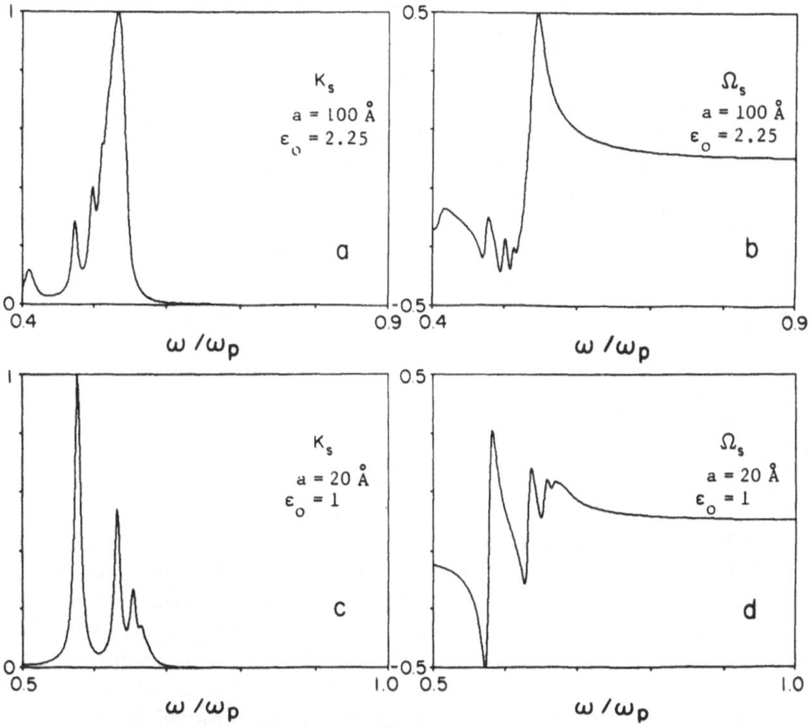

Fig. 1. The variation of K_s [a,c] and Ω_s [b,d] with $x = \omega/\omega_p$ when the dipole is located 20 Å from the sphere. The damping $\Gamma \sim 10^{-2}\ \omega_p$ (after Ref. 4). The ordinates in different figures are separately normalized. The actual extreme values of the ordinates are (a): 25530, (b) 17499 (c) 36186, (d) -22047. The sphere radius and the dielectric function of the medium are shown on each figure.

lar to (3.15) can be derived for spontaneous emission in a spherical cavity.

Finally note that the energy shifts of the levels of the atomic system can also be related to the response functions[9]

$$\delta E_j = -\frac{1}{\pi} \sum_{\ell\alpha\beta} P \int_0^{\infty} d\omega_o (\omega_o + \omega_{\ell j})^{-1} d_\alpha^{j\ell} d_\beta^{\ell j}$$

$$\mathrm{Im}\, \chi_{\alpha\beta EE}(\vec{b},\vec{b},\omega_o) \quad , \tag{3.18}$$

the explicit results for δE_j in the case of an atom near a metallic half space can be found in Ref. 5.

B. Dynamical Effects in the Interaction of Atomic System in Radiation Fields in Finite Domains

Certain dynamical effects in the atomic system near a metal surface can also be studied. For one can examine the structure of fluorescence spectra produced by an atom which is strongly driven by a coherent field. The dynamical behavior of the system is usually described in terms of the master equation.[1] To be specific consider an atomic system in presence of dielectric medium and interacting with a laser field. The atoms emit radiation by spontaneous emission. In such a case one can show that

$$\frac{\partial \rho}{\partial t} = -i[H_{coh},\rho] - \sum_{\alpha\beta} \iint d^3r_1 d^3r_2 \int_0^{\infty} d\tau \, \{\mathcal{E}_{\alpha\beta}^{(s)}(\vec{r}_1,\vec{r}_2,\tau)$$

$$\times [P_\alpha(\vec{r}_1,t),[P_\beta(\vec{r}_2,t-\tau),\rho(t)]] + \chi''_{\alpha\beta EE}(\vec{r}_1,\vec{r}_2,\tau)$$

$$\times [P_\alpha(\vec{r}_1,t),\{P_\beta(\vec{r}_2,t-\tau),\rho(t)\}]\} \quad , \tag{3.19}$$

where $\vec{P}$ is the atomic polarization operator in the interaction picture and H_{coh} represents the interaction with the external driving field. The parameters $\mathcal{E}^{(s)}$, χ depend on the nature of the medium. Similarly the medium affects the effective field that is acting on the atom. Note that (3.19) includes all cooperative as well as temperature effects. Thus (3.19) can be used to discuss the radiation either from a single atom or an ensemble of atoms. In deriving (3.19) we have also made the Markov approximation. If due to the presence of the medium, the coherence time[12] of the field is considerably increased, then we may have to work with nonMarkovian master equations.

For a single two-level atom (3.19) reduces to the usual form but with new γ and new α (Rabi frequency)

$$\frac{\partial \rho}{\partial t} = -i[\Delta s^z,\rho] - i[\alpha s^+ + \alpha^* s^-,\rho]$$

$$- \gamma(s^+s^-\rho - 2\, s^-\rho s^+ + \rho s^+s^-) \quad , \tag{3.20}$$

where γ, α and the detuning $\Delta = (\omega_o - \omega_\ell - \Omega)$ depend on the nature of the medium. These new parameters will have resonances associated with the nature of surface modes ω_s of the material medium --

(a) γ: Resonantly enhanced, $\omega_o \approx \omega_s$;

(b) α: Resonantly enhanced, $\omega_\ell \approx \omega_s$.

Spectrum of the scattered radiation is obtained by an analysis of the far zone behavior of the field. The far zone behavior[9] can again be related to the asymptotic form $\tilde{\chi}$ of the response function

$$E_\beta^{(+)}(\vec{r},t) \sim E_{o\beta}^{(+)}(\vec{r},t) + \sum_\alpha \tilde{\chi}_{\beta\alpha EE}(\vec{r},\vec{b},\omega)\, \vec{d}_\alpha\, S^{(-)}(t) \tag{3.21}$$

leading to

$$\lim_{t\to\infty} \langle E^{(-)}(t+\tau)E^{(+)}(t)\rangle = \sum \tilde{\chi}_{\beta\alpha EE}(\vec{r},\vec{b},\omega)\, \tilde{\chi}^*_{\beta\alpha EE}(\vec{r},\vec{b},\omega)$$

$$d_\alpha d^*_{\alpha'} \lim_{t\to\infty} \langle S^+(t+\tau)S^-(t)\rangle \quad . \tag{3.22}$$

Note that because of the presence of the metal surface or any medium, $\tilde{\chi}$ itself may have a resonant character. $\tilde{\chi}$ is known to have resonances for spheres and gratings though not for metallic half space. Figure 2 shows the resonant character of $\tilde{\chi}$ for the case of a grating.[13] The quantity $S^{(2)} \propto |\tilde{\chi}|^2$ and θ gives the direction of observation. From Eqs. (3.22) and (3.20), we would recover Mollow spectrum[14] say triplet structure provided that $\alpha \gg (\Delta^2 + \gamma^2)^{1/2}$. Of course, it might happen that $\alpha \ll (\Delta^2 + \gamma^2)^{1/2}$ even if $\alpha_o \gg (\Delta_o^2 + \gamma_o^2)^{1/2}$ depending on the closeness of ω_o and ω_ℓ to ω_s, in which case the spectrum of the scattered radiation in presence of a surface will have a simple peak.

In closing this section, we also mention that it is possible to generalize the concepts of radiation reaction to the case when the atom is emitting radiation in presence of a medium or in a cavity. Such a result for the radiation reaction, i.e. for the field acting back on the atom can be written[9] in terms of the new response function χ as

$$E_{RR\beta}^{(+)}(\vec{b},t) = \frac{i}{\pi}\int_0^t d\tau \sum_\alpha p_\alpha(t-\tau) \int_0^\infty d\omega\, e^{-i\omega\tau}$$

$$\mathrm{Im}\, \chi_{\beta\alpha EE}(\vec{b},\vec{b},\omega) \quad , \tag{3.23}$$

where p_α is the dipole moment operator whose dynamics is determined from the solution of the master equation.

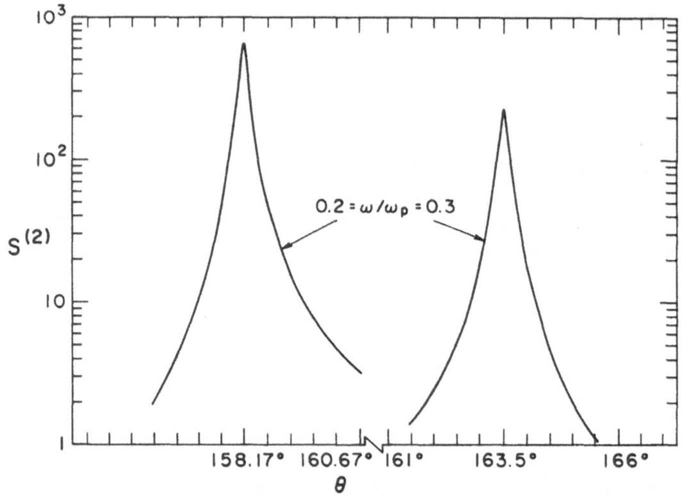

Fig. 2. Resonant behavior of $|\tilde{\chi}|^2$ as a function of the angle of observation for an atom with frequency ω located at (0,0,b) near a metallic grating $z = -h \sin gy$ with $\varepsilon(\omega) = 1 - \omega_p^2/(\omega^2 + i\omega\Gamma)$, $\Gamma = 10^{-2}\ \omega_p$ and $\omega_p b/c = 0.1$, $gc/\omega_p = 0.25$. The metallic medium occupies the domain $z + h \sin gy > 0$.

IV. CONCLUSIONS

We have shown how the characteristics of the quantized electromagnetic field fluctuations in a finite medium or in a finite cavity can be calculated in terms of certain response functions. Our calculations yield correlations of arbitrary order. Both normal and antinormal ordered correlations can be obtained. We have further shown how such correlations can be used in studying various characteristics of the radiating system such as lifetimes, shifts of the energy levels, transition rates in general and the spectrum of scattered light. The modes of the medium lead to very interesting resonances in such characteristics. It is clear that other types of nonlinear interactions between atoms/molecules and the medium can be considered similarly. For example, the spontaneous and stimulated Raman scattering by adsorbed atoms on the metal surface can be studied within the framework of a three-level model.

We would like to conclude this by mentioning that there is another area -- quantum electrodynamics in cavities with phase conjugate mirrors, where the vacuum fluctuations are considerably reduced. For example, we have shown[15] in a recent paper that it is possible to have complete inhibition of spontaneous emission if the phase conjugation is done without losses or gains. We hope to see considerable activity in this area in the future.

The author is grateful to the National Science Foundation for an international travel grant and to Jinks Cooper for his hospitality at the Joint Institute for Laboratory Astrophysics, University of Colorado, Boulder, where this work was prepared for publication.

REFERENCES

1. cf. G. S. Agarwal, "Quantum Optics," Springer Tracts in Modern Physics, Vol. 70, G. Höhler, ed., Springer, Berlin (1974).
2. R. E. Benner, P. W. Barber, J. F. Owen and R. K. Chang, Phys. Rev. Lett. 44:475 (1980); W. H. Weber and C. F. Eagen, Opt. Lett. 4:236 (1979).
3. A. G. Vaidyanathan, W. P. Spencer and D. Kleppner, Phys. Rev. Lett. 47:1592 (1982).
4. G. S. Agarwal and S. V. ONeil, Phys. Rev. B 28:xxx (1983).
5. G. S. Agarwal and H. D. Vollmer, Phys. Stat. Sol. 79b:249 (1977); ibid. 85b:301 (1978).
6. R. Ruppin, J. Chem. Phys. 76:1681 (1982); J. L. Gersten and A. Nitzan, J. Chem. Phys. 73:3023 (1980); ibid., 75:1139 (1981).
7. G. S. Agarwal, Phys. Rev. A 11:230 (1975).
8. G. S. Agarwal, Phys. Rev. A 11:253 (1975).
9. G. S. Agarwal, Phys. Rev. A 12:1475 (1975).
10. G. S. Agarwal, Phys. Rev. A 12:1974 (1975).
11. G. S. Agarwal, Phys. Rev. A 1:1445 (1970); B. R. Mollow, Phys. Rev. 168:1418 (1968).
12. Coherence times are calculated, for example, in W. Eckhardt, Phys. Rev. A 17:1093 (1978).
13. G. S. Agarwal and C. V. Kunasz, Phys. Rev. B 26:5832 (1982).
14. B. R. Mollow, Phys. Rev. 188:1969 (1969).
15. G. S. Agarwal, Opt. Commun. 42:205 (1982).

NONPERTURBATIVE QUANTUM ELECTRODYNAMICS OF BOUND STATES AND RADIATIVE PROCESSES

A. O. Barut

Department of Physics
University of Colorado
Boulder, CO. 80309

Some recent nonperturbative approaches to radiative processes (Lamb shift, spontaneous emission, anomalous magnetic moment,) and to two-body dynamics are reformulated in a simplified form in terms of an action principle without using the equations of motion.

INTRODUCTION

Quantum electrodynamics (QED) postulates the presence of a quantized electromagnetic field which exists separately from the charged particles. It is external to the charged particles, has an independent existence, but always coupled to the charged particles. The degrees of freedom of the system are those of the particles plus those of the field.

It is further assumed, at least in the practical calculations, that every QED-phenomenon proceeds via combinations of emission and absorption of real or vitual photons and e^+e^- -pairs.

While this picture is successful in the calculation of high energy scattering processes and some radiative corrections in connection with certain renormalization rules, it does not unfortunately apply to the calculation of bound states and resonances from first principles. These have to be put in by hand from our knowledge of unperturbed Schrödinger or Dirac bound state equations. Or, the perturbative Bethe-Salpeter equations have to be first approximated into a bound-state equation with a potential to which radiative corrections are applied as above. The perturbative approach has also the practical limitations when one has to evaluate large number of higher order graphs.[1]

There is another point of view to quantum electrodynamics. This, in the spirit of classical electrodynamics, describes the radiative effects solely in terms of the self-fields produced by accelerated charges. A uniformly moving charge has only its stationary field around it — thus in its rest frame a static field -- which is renormalized to its mass and charge, and does not radiate. In the presence of other charges, the particle accelerates, radiates and the self-field is modified which then leads to observable effects over and above the first quantized Schrödinger or Dirac equations describing two or more particles. In this approach, the idea of wave mechanics of de Broglie, Schrödinger and Dirac is continued to its logical end. In fact here, we only need to inclue into the successful Schrödinger and Dirac wave equations the self-field of the particles which has been left out to begin with.

The first purpose of these lectures is thus about the inclusion of the self-fields into the non-relativistic and relativistic wave mechanics. The electromagnetic field has no independent existence, it is created by the charged particles.

The potentials A_μ can be eliminated from the coupled Maxwell and Dirac equations so that we can work entirely with the wave functions of the particles, satisfying non-linear integro-differential equations. In order to do this we must start with at least two particles and their fields. The elimination of the electromagnetic potential leads to coupled equations for the two-particles from which relativistic potentials and relativistic two - or many-body equations can be derived.

The second purpose of these lectures is to give a new foundation to relativistic dynamics of two- or more particles in terms of the many body wave functions in configuration space with relativistic potentials. That this is the proper framework for bound state problems in nonrelativistic dynamics is clear. But that it can also be done in the relativistic case is perhaps not generally recognized. A fully covariant formulation of 2 or more-body dynamics is essential for the spectroscopy of moving atoms and nuclei, their form factors, and further for the relativistic treatment of the interactions of composite systems.

I am summarizing here the results of research done in the last several years on this nonperturbative approach to quantumelectrodynamics without the quantization of the electromagnetic field.[2-6] This approach also differs from models like stochastic electrodynamics where instead of the quantized radiation field, again an external random field (of unknown origin) is postulated. In the self-energy approach there are no separate external fields, quantized, stochastic or otherwise.

The main conceptually new aspect of the self-energy formulation of quantumelectrodynamics is the way renormalization is carried out. In QED, (and in stochastic electrodynamics), the self-field of the electron is first completely dropped and then added afterwards, photon by photon, together with counter-terms to cancel the infinities. In contrast we keep the complete self-field from the beginning, and extract only the observable part of this self-energy in the presence of the external fields.

BASIC EQUATIONS

We shall use an action principle which garanties consistency of the equations of motions with the conservation laws. the action is directly related to the invariant S-matrix for scattering problems, and to the total energy for bound sate problems.

For electrodynamics the action is ($c = 1$, $h = 1$)

$$A = \int dx[-\frac{1}{4} F_{\mu\nu}F^{\mu\nu} + \sum_j \{\overline{\Psi}_j(x)\ (i\gamma^\mu\partial_\mu - m_j)\ \psi_j(x)$$

$$- e_j\ \overline{\Psi}_j(x)\ \gamma^\mu\psi_j(x)A_\mu(x) - a_j\ \overline{\Psi}_j(x)\ \sigma_{\mu\nu}\psi_j(\chi)\ F^{\mu\nu}(x)\}] \quad (1)$$

Here we have introduced several distinct fermion fields $\psi_1(x)$, $\psi_2(x)$, . . . interacting with a single electromagnetic field $A_\mu(x)$, and both minimal and Pauli couplings with the coupling constants e_j(charge), and a_j (anomalous magnetic moment), $\sigma_{\mu\nu} = i/2\ (\gamma_\mu\gamma_\nu - \gamma_\nu\gamma_\mu)$. While for the usual local U(1) - gauge invariance, $\psi \to \psi\ e^{ie\theta(x)}$, $A_\mu \to A_\mu + \theta,\mu$ the minimal coupling is sufficient and almost exclusively used in QED, it has been shown recently,[7] that the more general spinorial realization of the U(1) - gauge transformation

$$\psi \to \exp\left[ie\theta(x) + \frac{1}{2}\ ia(1 \pm \gamma_5)\ \gamma^\mu\ \theta,\mu(x)\right]\ ,\ \theta(x) = 0 \quad (2)$$

necessitates both couplings.

The experimental limit on the value of the anomalous magnetic moment of the electron from (g-2) - measuremnts is $a_e < 10^{-7}\ \mu_o$, and for neutrino $a_\nu < 10^{-9}\ \mu_o$. Nevertheless, since even such a small anomalous magnetic moment can have remarkable effects at very short distances,[8] it seems important to carry such a coupling in the basic action.

Since we shall make use extensively of the Green's function of the wave operator, there is in this sense a preferred covariant gauge

$$A^\mu{}_{,\mu}(x) = 0 \quad , \quad (3)$$

because otherwise the Green's function of the operator ($\;\;A_\mu - \partial_\mu\partial^\nu A_\nu$) does not exist. In this gauge, the field equations $F_\mu{}^\nu{}_{,\nu} = - j_\mu$ become

$$A_\mu = j_\mu = \sum_j [e_j\ \overline{\Psi}_j\gamma_\mu\psi_j + 2a_j\ (\overline{\Psi}_j\sigma_{\mu\nu}\ \psi_j)^{,\nu}] \quad (4)$$

and

$$(i\gamma^{\mu}\partial_{\mu} - m_j)\,\psi_j - e_j\gamma^{\mu}A_{\mu}\,\psi_j - a_j\,\sigma^{\mu\nu}\,F_{\mu\nu}\,\psi_j = 0\ ,\ j = 1,2,3,\ldots \quad (5)$$

The general solution of eq. (4) is

$$A_{\mu} = A_{\mu}^{in} + \int dy\; D(x-y)\; j_{\mu}(y) \tag{6}$$

We shall specify later the boundary conditions on the choice of the Green's function D(x-y).

The effect of the anomalous magnetic term has been discussed in detail elsewhere.[3-4] For simplicity of writing we shall omit this term and mention the results at the end.

It is a general result that the contribution of the field Lagrangian, $-1/4\ F_{\mu\nu}\ F^{\mu\nu}$, to the action when expressed in terms of ψ-field, equals $-1/2$ times the interaction action $j_{\mu}A^{\mu}$. This is proved as follows:

$$A_F = -\frac{1}{4}\int F_{\mu\nu}\ F^{\mu\nu}\ dx = \frac{1}{2}\int dx\ F^{\mu\nu}\ A_{\mu,\nu} =$$

$$= \frac{1}{2}\int dx\ [F^{\mu\nu}\ A_{\mu}],\nu - \frac{1}{2}\int dx\ F^{\mu\nu},\nu\ A_{\mu} = \frac{1}{2}\int dx\ j^{\mu}\ A_{\mu}. \tag{7}$$

The only assumption is the vanishing of surface terms at infinity.

We wish now to express the action A, eq. (1), in terms of the particle fields ψ_i alone. Inserting (6) and (7) into (1), with $a_j = 0$, we get

$$A = \int dx \sum_j [\overline{\psi}_j\ (i\gamma^{\mu}\partial_{\mu} - m_j)\ \psi_j - \sum_k \frac{1}{2} e_j\ e_k \int dy\ D(x-y)$$

$$\times\ \overline{\psi}_j(x)\ \gamma^{\mu}\ \psi_j(x)\ \overline{\psi}_{\mu}(y)\ \gamma_{\mu}\psi_k(y)$$

$$-\ e_j\ \overline{\psi}_j(x)\ \gamma^{\mu}\ \psi_j(x)\ A_{\mu}^{in}(x)] \tag{8}$$

It is remarkable that eq. (8) has a direct classical limit: If ψ or $\overline{\psi}(x)\ \psi(x)$ is localized in space as a δ-function the first derivative term vanishes and we can replace

$$\int dx\ \overline{\psi}\ (x)\ \gamma^{\mu}\ \psi(x) \rightarrow \int ds\ \dot{z}^{\mu}(s)$$

$$\int dx\ \overline{\psi}\ i\ \gamma^{\mu}\partial_{\mu}\ \psi \rightarrow 0,\quad m\int dx\ \overline{\psi}(x)\ \psi(x) \rightarrow m\int ds \tag{9}$$

and obtain the Fokker-Schwartzschild-Tetrode action,[9] augmented by the self-interaction term[10]

$$A = -\sum_j \int m_j \, ds - \frac{1}{2} \sum_{\substack{k \\ j\neq k}} e_j \, e_k \int ds \, ds' \, \dot{z}_j^{\mu} (s) \, \dot{z}_{\mu k}(s') \, D(z_j - z_k)$$

$$- \frac{1}{2} e_j^2 \int ds \, ds' \, \dot{z}_{\mu}^{i}(s) \, \dot{z}_j^{\mu} \, (s') \, D(z_j(s) - z_j(s')) - e_j \dot{z}_j^{\mu} \, A_{\mu}^{in} \, (z_j). \quad (10)$$

If we choose $D = D^{ret}$, and because $D^{ret}(x) = D^{adv}(-x)$, we can also write

$$A = -\sum_j \int m_j \, ds - \frac{1}{2} \sum_{k\neq j} e_j \, e_k \int ds \, ds' \, \dot{z}_j^{\mu}(s) \, \dot{z}_{\mu k}(s') \, \overline{D}(z_j^{(s)}$$

$$- z_{\mu}^{(s')}) \; - \frac{1}{2} e_j^2 \int ds \, ds' \, \dot{z}_{\mu}^{j}(s) \, \dot{z}_j^{\mu}(s') \, D^{ret} \, (z_j(s) - z_j(s'))$$

$$- e_j \, \dot{z}_j^{\mu} \, A_{\mu}^{in} \qquad , \qquad (10')$$

where the symmetric function $\overline{D}$ is

$$\overline{D} = \frac{1}{2} (D^{ret} + D^{adv}) \quad .$$

In classical electrodynamics the third, self energy term provides an infinite mass renormalization term, $- \delta m_j \int ds$, and a finite radiation reaction term (which by the way is nonperturbative).[10]

Equations (9) can be taken directly to be the quantization procedure to transform the classical action (1) to the quantum action (8). Equation (8) has also a direct spin zero (and further a nonrelativistic limit), which are obtained first by the simple substitution

$$\gamma^{\mu} \rightarrow \frac{1}{m} (i \frac{\overrightarrow{\partial}_{\mu}}{2} - i \frac{\overleftarrow{\partial}_{\mu}}{2} - e \, A_{\mu})$$

The action (1) or (8) transforms then, with $\overline{\psi} \rightarrow \phi^*$, $\psi \rightarrow \phi$, into

$$A = \int dx \, \{ -\frac{1}{4} F_{\mu\nu} F^{\mu\nu} + \sum_j [\phi_j^* \frac{1}{m_j} (\frac{i}{2} \overrightarrow{\partial}_\mu - \frac{i}{2} \overleftarrow{\partial}_\mu - e A_\mu)$$

$$\times \; (i \, \partial^\mu - e A^\mu) \, \phi_j - m_j \, \phi_j^* \, \phi_j]\}$$

which gives correctly the equations of scalar electrodynamics, and the current is transformed into

$$m \, \overline{\psi} \, \gamma^\mu \, \psi \rightarrow \frac{i}{2} (\phi^* \, \phi^{,\mu} - \phi^{*,\mu} \, \phi) - e \, \phi^* \, \phi \, A^\mu \quad ,$$

the Klein-Gordon current. We further find that the magnetic coupling cancels in this limit, as expected: $\overline{\psi} \, \sigma_{\mu\nu} \, \psi \rightarrow 0$.

For nonrelativistic limit, we set directly,

$$\gamma^o \rightarrow 1, \quad \gamma^i = \frac{i}{2m} (\overrightarrow{\nabla} - \overleftarrow{\nabla}) \quad .$$

All these results and those in the following sections shows that it is most economical to work directly with the action integral itself.

SELF ENERGY

We shall show now that the action (8) can be used directly to evaluate radiative effects (Lamb shift, etc.), as well as the dynamics of interaction of several particles without going to the equations of motion, or Hamiltonian and perturbation theory. As far as I know this is a new elegant and economic theory of electromagnetic interaction in all its generality.

Consider a single self energy term ($j = k$) in (8):

$$A^{self} = -\frac{1}{2} e^2 \int dx \, dy \, \overline{\psi}(x) \, \gamma_\mu \, \psi(x) \, D(x-y) \, \overline{\psi}(y) \, \gamma^\mu \, \psi(y). \quad (11)$$

We perform a Fourier transform of the ψ-function in energy variable

$$\psi = \sum_n \psi_n(\overline{r}) \, e^{-i \, E_n \, t} \quad (12)$$

and use

$$D(x) = -\frac{1}{(2\pi)4} \int dq \, \frac{e^{+iqx}}{q_o^2 - \vec{q}^{\,2}} \quad (13)$$

and obtain with $e^2/4\pi = \alpha$.

$$A^{self} = -\frac{1}{2}\alpha\, 4\pi \int dx\, dy\, dq\, \frac{-1}{(2\pi)^4}\, \frac{e^{-i\vec{q}\cdot(\vec{r}-\vec{r}')}}{q_o^2 - \vec{q}^2}$$

$$\times \sum \overline{\psi}_n(\vec{r})\, \gamma_\mu\, \psi_m(\vec{r})\, \overline{\psi}_r(\vec{r}')\, \gamma^\mu\, \psi_s(\vec{r}')$$

$$\times\, e^{i(E_n - E_m + q_o)t}\, e^{i(E_r - E_s - q_o)t'}$$

We next integrate over t and t'

$$A^{self} = +\frac{\alpha}{2}\frac{4\pi}{(2\pi)^2}\int d\vec{r}\, d\vec{r}'\, d\vec{q}\, dq_o\, \delta(E_n - E_m + q_o)$$

$$\times \delta(E_r - E_s - q_o)\, \frac{1}{q_o^2 - \vec{q}^2} \sum_{n,m,r,s} (\overline{\psi}_n\, \gamma_\mu\, e^{-i\vec{q}\cdot\vec{r}}\, \psi_m)(\overline{\psi}_r\, \gamma^\mu\, e^{i\vec{q}\cdot\vec{r}'}\, \psi_s)$$

(14)

Now integrate one of the δ-functions with dq_o

$$A^{self} = \frac{\alpha}{2}\cdot\frac{2}{2\pi}\,\delta(E_n - E_m + E_r - E_s)\int \frac{d\vec{q}}{(E_r - E_s)^2 - \vec{q}^2}$$

$$\times \sum_{nmrs} T_{nm_\mu}(\vec{q})\, T_{rs}{}^{\mu}(-\vec{q}) \qquad (15)$$

where we have defined the vector form factors

$$T_{\mu,nm}(\vec{q}) = \int d\vec{r}\, \overline{\psi}_n(\vec{r})\gamma_\mu\, \psi_m(\vec{r})\, e^{-i\vec{q}\cdot\vec{r}} \qquad (16)$$

The action is dimensionless. For scattering problems we can write

$$A = (2\pi)^4\, \delta^{(4)}(P_{final} - P_{init})\, G \qquad (17)$$

and G has the physical meaning of invariant transition probability over all space and time. For bound state problems we write

$$A = (2\pi)\,\delta\,(E_{final} - E_{init})\,B \tag{18}$$

where B is the invariant total energy of the system, consisting, in order of the terms in eq. (8), of kinetic energy, interaction energies with other particles, self energies and interaction energies with fixed external fields, respectively.

In the case of self energies we write

$$A^{self} = [\sum_n (2\pi)\,\delta\,(E_n - E_m + E_r - E_s) \sum_{mrs} \Delta E_{nmrs}] \tag{19}$$

so that the self energy of the nth Fourier component may be defined to be

$$\Delta E_n = \frac{\alpha}{2\pi^2} \sum_{\substack{mrs \\ \delta(E_n - E_m + E_r - E_s)}} \int \frac{d\vec{q}}{(E_r - E_s)^2 - \vec{q}^2} T_{\mu,nm}(\vec{q})\, T^{\mu}{}_{rs}(-\vec{q}). \tag{20}$$

The theory up to now is exact.

As in classical electrodynamics, the major part of the self energy must be in the mass of the electron, because the free electron has also the same self energy terms. This is the process of renormalization.

We shall apply now an iteration method to interpret and calculate the energy shifts.

Let us take the Fourier expansion (12) to be over the energy levels of an electron in an external field, or more precisely, over the levels of a 2-body problem (i.e. the first two terms of the action (8) or (10)). There is, by the way, an exact two-body relativistic equation, thus including all recoil terms, which will be discussed in the next section.

The denominator of the $\vec{q}$-integration can be written as

$$\frac{1}{(E_r - E_s)^2 - \vec{q}^2} = \frac{1}{2q}\left(\frac{1}{(E_r - E_s - q} - \frac{1}{E_r - E_s + q}\right)$$

$$= -\frac{1}{q^2} + \frac{E_r - E_s}{2q^2}\left(\frac{1}{E_r - E_s - q} + \frac{1}{E_r - E_s + q}\right) \tag{21}$$

Hence

$$\Delta E_n = \frac{\alpha}{2\pi^2} \sum_{\substack{mrs \\ E_n - E_m = E_s - E_r}} \int dq d\Omega \left[- 1 + \frac{E_r - E_s}{2}\right.$$

$$\left.\times \left(\frac{1}{E_r - E_s - q} + \frac{1}{E_r - E_s + q} \right)\right] T_{nm,\mu}(\vec{q}) \; T^{\mu}_{rs}(-\vec{q}) \quad . \qquad (22)$$

There are two sets of terms in the above sum.

(a) $E_s = E_r$, hence $E_n = E_m$. Then the second term in the square bracket vanishes, and we get

$$- \sum_s \int dq \; d\Omega \; T_{nn,r}(\vec{q}) \; T_{ss}{}^{\mu}(-\vec{q})$$

(b) $E_n = E_s$, hence $E_m = E_r$. Then

$$\sum_m \int dq \; d\Omega \left[-1 + \frac{E_m - E_n}{2} \left(\frac{1}{E_m - E_n - q} + \frac{1}{E_m - E_n + q} \right)\right] T^{\mu}_{nm}(\vec{q}) \; T_{mn}(-\vec{q})_{\mu}$$

Here in the first term in the square bracket, with m = n, there is one term which is identical with the term s = n in (a). This term is the static energy of the level n itself (independent of all other levels) and should be renormalized into the kinetic energy terms

$$\Delta E_{renor.} = - \frac{\alpha}{2\pi^2} \int ds \; d\Omega \; T^{\mu}_{nn}(\vec{q}) \; T_{nn}(-\vec{q})_{\mu} \qquad (23)$$

Hence we have for the remaining terms

$$\Delta E^{(a)}_{(n)} = - \frac{\alpha}{2\pi^2} \sum_{s \neq n} \int dq \; d\Omega \; T_{nn}{}^{\mu}(\vec{q}) \; T_{ss}(-\vec{q})_{\mu} \qquad (24)$$

$$\Delta E^{(b)}_{n} = \frac{\alpha}{2\pi^2} \sum_{m \neq n} \int dq \; d\Omega \left[-1 + \frac{E_m - E_n}{2} \left(\frac{1}{E_m - E_n - q} + \frac{1}{E_m - E_n + q} \right)\right.$$

$$\left.\times \; T^{\mu}_{nm}(\vec{q}) \; T_{mn}(-\vec{q})_{\mu}\right. \qquad (25)$$

The renormalization is not just the term (23), however.

In the vacuum polarization term, (24), if the current is properly antisymmetrized only the 0-component of T^{μ} contribute[11]

$$\Delta E^{(a)}_{n} = - \frac{\alpha}{2\pi^2} \sum_{s \neq n} \int dq \; d\Omega \; T^{o}_{nn}(\vec{q}) \; T_{ss}(-\vec{q})_{o} \qquad (24')$$

We see this also from the nonrelativistic approximation of current,

$$\bar{\psi}\gamma^{\mu}\psi \rightarrow (\psi^{*}\psi, \quad \psi^{*}\vec{\alpha}\psi \rightarrow \psi^{*}\frac{i\overleftrightarrow{\nabla}}{m}\psi)$$

so that because of the antisymmetric derivatives $\overleftrightarrow{\nabla}$, the integrals T^{k}_{ss} cancel.

All integrals in (23) - (25) should be studied, including renormalization terms. The socalled Bethe-part of the energy-shift and the spontaneous emission terms are easily seen in eq. (25). If we use the decomposition

$$\frac{1}{E_n - E_m - q + i\varepsilon} = P\frac{1}{E_n - E_m - q} - i\pi\,\delta\,(E_n - E_m - q) \quad ,$$

we have, after angular integrations,

$$\Delta E_n^{(b)\ \mathrm{Bethe}} = \frac{4}{3}\frac{\alpha}{\pi}\cdot\frac{1}{m^2}\sum_{m\neq n}\frac{(E_n - E_m)}{2}\,P\int dq\,\frac{1}{E_m - E_n - q}$$

$$\times\ T^{k}_{nm}(\vec{q})\,T_{mn}(-\vec{q})_k \tag{25}$$

and

$$\Delta E_n^{(b)\ \mathrm{Spont.}} = i\,\frac{2\alpha}{3\pi m^2}\sum_{m\neq n}(E_n - E_m)\int dq\;\delta(E_m - E_n - q)\,T^{k}_{nm}(\vec{q})$$

$$\times\ T_{mn}(-\vec{q})_k \tag{26}$$

Although we have eliminated A_μ from the coupled Maxwell-Dirac eqs. in the particular gauge, $A^{\mu}{}_{,\mu} = 0$, both the self-action (11) and the results ΔE in Eq. (23) - (25) are invariant under the local transformation $\psi(r) \rightarrow \psi(r)e^{if(\vec{r})}$, f arbitary. In fact, the vacuum polarization term is even invariant under $\psi_n(\vec{r}) \rightarrow \psi_n(\vec{r})e^{if_n(\vec{r})}$ for each n separately. Thus only the real magnitudes of ψ_n contribute, not the phase.

Since we have analyzed the self-action (11) quite generally, our results (23) - (25) are valid not ony for any external field problem, but also for a free particle. In the latter case <u>all</u> ΔE is to be counted as a renormalization term. The evaluation of (23) - (25) depends on the choice of $\psi_n(r)$ in the expansion (12) via the form factors (16)).

For a localized free particle we would take a wave packet. In the presence of an external field this wave packet will be deformed. Total self interaction energy content of the deformed packet will be a bit different — this is the vacuum polarization.

TWO-BODY INTERACTIONS AND RELATIVISTIC 2-BODY EQUATIONS

Consider in the fundamental action (8) terms corresponding to two fields ψ_1 and ψ_2

$$\begin{aligned} A = \int dx\, [&\bar{\psi}_1(i\gamma^\mu \partial_\mu - m_1)\, \psi_1) + \bar{\psi}_2(i\, \gamma^\mu \partial_\mu - m_2)\, \psi_2 \\ &- \frac{1}{2} e_1 e_2 \int dy\, \bar{\psi}_1(x)\, \gamma^\mu\, \psi_1(x)\, D^{ret}(x-y)\, \bar{\psi}_2(y)\, \gamma_\mu\, \psi_2(y) \\ &- \frac{1}{2} e_1 e_2 \int dy\, \bar{\psi}_2(x)\, \gamma^\mu\, \psi_2(x)\, D^{ret}(x-y)\, \bar{\psi}_1(y)\, \gamma_\mu\, \psi_1(y) \\ &- \frac{1}{2} e_1{}^2 \int dy\, \bar{\psi}_1(x)\, \gamma^\mu\, \psi_1(x)\, D(x-y)\, \bar{\psi}_1(y)\, \gamma_\mu\, \psi_1(\gamma_\mu) \\ &- \frac{1}{2} e_2{}^2 \int dy\, \bar{\psi}_2(x)\, \gamma^\mu\, \psi_2(x)\, D(x-y)\, \bar{\psi}_2(y)\, \gamma_\mu\, \psi_2(y) \end{aligned} \tag{27}$$

In the last two terms we see the self-energies of the two-particles (Lamb shift, etc). These are self-consistent expressions, i.e. they depend on the type of interaction. Thus these corrections can be quite different in strong interactions than in weak interactions. The full treatment of both self-interactions and mutual interactions of two particles in one formalism does not yet exist. We shall here consider the solution of the 2-body interactions (the first two interaction terms) first and than solve the last two terms by an iterative method as in the previous section.

For this purpose, we define a new two-body one-time-wave function by

$$\Phi_{\pm}(\vec{x}, \vec{y}, t) \equiv \psi_1(\vec{x},t)\, \psi_2(\vec{y},t \pm \|\vec{x} - \vec{y}\|) \tag{28}$$

It is a product of fields at retarded (or advanced) times, and not the usual product of fields at two arbitrary times! The retardation is thus built in into the definition of Φ from the beginning. The interaction terms can be easily written in terms of Φ and $\bar{\Phi}$, because the two particles contribute to the action A only at retarded points. In order to write also the kinetic energy terms in terms of the two-body wave functions, we multiply them with the normalization integrals, like

$$\int_{t=cont} d\vec{y}\ \ \psi_2^\dagger(y)\ \psi_2(y) \ = \ 1 \tag{29}$$

Thus

$$A \ = \ \int dx\ \overline{\psi}_1(x)\ (\gamma^{(1)\mu}\ i\partial_\mu\ - m_1)\ \psi_1(x) \int_{y_o=x_o-r} d\vec{y}\ \psi_2^+(y)\ \psi_2(y)$$

$$+ \ \int dx\ \overline{\psi}_2(x)\ (\gamma^{(2)\mu}\ i\ \partial_\mu\ - m_2)\ \psi_2(x) \int_{y_o=x_o+r} d\vec{y}\ \psi_1^+(y)\ \psi_1(y)$$

In the second integral we change the integration variables

$$\int dy\ dy^o\ (\overline{\psi}_2(\vec{y},\ y^o)\ (\gamma^\mu\ i\partial_\mu\ - m_2)\ \psi_2(\vec{y},\ y^o)$$

$$\times \int_{x_o=y_o+r} d\vec{x}\ \psi_1^+(x)\ \psi_1(x)$$

$$A^{kin} = \int dx\ d\vec{y}\ \overline{\psi}_1(x)\ (\gamma^{(1)\mu}\ i\partial_\mu - m_1)\ \psi_1(x)\ \overline{\psi}_2(\vec{y},x_o-r)\ \gamma_o^{(2)}\psi_2(\vec{y},x_o-r)$$

$$+ \int dx\ d\vec{y}\ \overline{\psi}_1(\vec{x},x^o)\ \gamma_o^{(1)}\ \psi_1(\vec{x},x^o)\ \overline{\psi}_2(\vec{y},x_o-r)(\gamma^{(2)\mu} i\partial_\mu - m_2)\psi_2(\vec{y},x_o-r)$$

$$= \int dx\ d\vec{y}\ \overline{\Phi}_+\ [(\gamma^{(1)\mu} i\partial_\mu - m_1)\ \gamma_o^{(2)} + \gamma_o^{(1)}\ \ (\gamma^{(2)\mu}\ i\partial_\mu - m_2)]\ \Phi_+(\overline{x},\overline{y},x_o)$$

The same type of equation holds for Φ in which $y = x_o + r$. In the two interaction terms in (27) with $D^{ret}(x-y)$, we interchange in the second term the integration variables x and y and use the fact that $D_{ret}(y-x) = D_{adv}(x-y)$, and $\overline{D} = \frac{1}{2}(D_{ret} + D_{adv})$, and obtain

$$A^{int} = -\frac{1}{2}\ (e_1 e_2)\ \int dx\ dy\ \overline{\psi}_1(x)\ \gamma^\mu\ \psi_1(x)(D^{ret} + D^{adv})\ \overline{\psi}_2(y)\gamma_\mu\ \psi_2(y)$$

$$= \ -\ (e_1 e_2)\ \int dx\ dy\ \overline{\Phi}(\vec{x},\vec{y},t)\ \gamma^\mu\ \ \gamma_\mu\ \overline{D}(x-y)\ \Phi(\vec{x},\vec{y},t)$$

$$= -\ (e_1 e_2)\ \int dx\ dy\ \overline{\Phi}_\pm\ [\frac{\gamma_\mu\ \ \gamma^\mu}{4\pi r}\ (\frac{1}{2}\ \delta(x^o-y^o-r)\ + \frac{1}{2}\ (x^o-y^o+r))]\ \Phi_\pm \tag{30}$$

We write also A^{kin} with $1/2\ \delta(x^o - y^o - r) + 1/2\ \delta(x^o - y^o + r)$ Varying now the total action with respect to $\Phi_\pm$ gives us an equation for $\phi_\pm$:

$$[(\gamma^{(1)\mu}\, i\partial_\mu - m_1)\ \gamma_o^{(2)} + \gamma_o^{(1)}\ (\gamma^{(2)\mu}\, i\partial_\mu - m_2)$$

$$- \frac{e_1 e_2}{4\pi r}\ \gamma_\mu^{(1)}\ \gamma^{(2)\mu}\ ]\ \Phi\ (\vec{x}, \vec{y}, t) = 0 \quad , \tag{31}$$

or, in terms of the momenta of the two particles

$$[(\gamma^{(1)\mu}\, p_{1_\mu} - m_1)\ \gamma_o^{(2)} + \gamma_o^{(1)}\ (\gamma^{(2)\mu}\, p_{2_\mu} - m_2)$$

$$- \frac{e_1 e_2}{4\pi r}\ \gamma_\mu^{(1)}\ \gamma^{(2)\mu}\ ]\ \Phi = 0 \tag{32}$$

The Hamiltonian form of this equation is obtained by multiplying it with $\gamma_o^{(1)}\ \gamma_o^{(2)}$,

$$[p_{10} + p_{20} - \vec{\alpha}_1 \cdot \vec{p}_1 - m_1\beta_1 - \vec{\alpha}_2 \cdot \vec{p}_2 - m_2\beta_2$$

$$- \frac{e_1 e_2}{4\pi r}\ (1 - \vec{\alpha}_1 \cdot \vec{\alpha}_2\ )]\ \Phi = 0$$

or,

$$H = p_{10} + p_{20} = \vec{\alpha}_1 \cdot \vec{p}_1 + \vec{\alpha}_2 \cdot \vec{p}_2 + \beta_1 m_1 + \beta_2 m_2$$

$$+ \frac{e_1 e_2}{4\pi r}\ (1 - \vec{\alpha}_1 \cdot \vec{\alpha}_2) \quad . \tag{33}$$

It is remarkable that we are able in this formulation to make an exact covariant separation of center of mass and relative coordinates. Defining

$$P^\mu = p_1^\mu + p_2^\mu \quad , \quad p^\mu = (1-a)p_1^\mu - a\ p_2^\mu \tag{34}$$

with the conjugate variables $R_\mu = ax_\mu + (1-a)y_\mu$ and $r_\mu = x_\mu - y_\mu$ with

$$r^2 = r_\mu r = 0 \tag{35}$$

we obtain from (33) the following linear wave equation in the center of mass coordinates[12,4]

$$(\Gamma^{\mu} P_{\mu} - K) \ \Phi (R, \vec{r}) = 0 \quad , \tag{36}$$

where

$$\Gamma^{\mu} = a \, \gamma^{\mu} \quad \gamma^{o} + (1 - a) \, \gamma^{o} \quad \gamma^{\mu}$$

$$K = (\vec{\gamma} \quad \gamma^{o} - \gamma^{o} \quad \vec{\gamma}) \cdot \vec{p} + (I \quad \gamma^{o} m_{1} + \gamma^{o} \quad I \, m_{2}) + V$$

The operator K acts on the relative coordinate $\vec{r}$. Thus we have shown how a relativistic 2-body system can be treated as though it was a single relativistic particle with internal degrees of freedom. This feature can be seen more clearly, if we use a discrete basis, instead of the continuous variable r. Then

$$(\Gamma^{\mu} P_{\mu} + K_{nn'}) \ \Phi_{n'} (R) = 0 \quad . \tag{37}$$

Equations of this type are called infinite component wave equations; they describe composite objects.[13] They had been largely postulated phenomenologically. Here we have shown how they can be actually derived from a field theory. For further separation of the relativistic 2-body equation into center of mass and relative coordinates we refer to literature.[5]

The anomalous magnetic Pauli coupling terms can also be included into the 2-body action (27).[3-4] They give an additional potential in eqs. (32), (33) or (36). [Note that the normal magnetic moment terms are included in the term $\vec{\alpha}_{1} \cdot \vec{\alpha}_{2}$.] Although their effects are very small in low energy QED, we believe that they become very important at higher energies or short distances, and may become crucial fo the regularization of QED, as well as for high energy particle physics.[14]

REFERENCES

1. For a recent collection of different views and approaches to electrodynamics see "Foundations of Radiation Theory and Quantumelectrodynamics," (edit. A. O. Barut), Plenum, N.Y. 1980.
2. A. O. Barut and J. Kraus, Phys. Rev. D16, 161 (1977), Found. of Physics, 13 189 (1983).
3. A. O. Barut and B. W. Xu, Physica Scripta 26, 129 (1982).
4. A. O. Barut and S. Komy, Fortschritte der Physik (in press).
5. A. O. Barut and N. Unal, Fortschritte der Physik (in press).
6. A. O. Barut and J. V. van Huele (to be published).
7. A. O. Barut and J. McEwan, Physics Letters, 135B, 177 (1984).
8. A. O. Barut, J. Math. Phys. 21, 568 (1980); A. O. Barut and G. Strobel, KNAM, Revista de Fisica 4, 151 (1982).
9. H. Tetrode, Z. Phys. 10, 317 (1922), J. Frenkel, Z. Phys. 32 518 (1925); G. N. Lewis, Proc. Nat. Acad. Sci. 12 22 (1926).
10. A. O. Barut, Electrodynamics and Classical Theory of Fields and Particles, (Dover 1980), p. 217.
11. E. H. Wichmann and N. M. Kroll, Phys. Rev. 101, 843 (1956).
12. A. O. Barut, Proceedings of XIth Intern. Colloquium on Group Theoretical Methods in Physics, Lecture Notes in Physics, Vol. 180 (Springer 1983) p. 332.
13. For a review see A. O. Barut, in Groups, Systems and Many-Body Physics, Ch. VI. (Vieweg Verlag, 1980), edit. P. Kramer, p. 285-317.
14. For a recent review see A. O. Barut, in Quantum Electrodynamics of Strong Fields (Plenum Press, 1983), edit. W. Greiner, p. 755-781.

MASS AND CHARGE RENORMALIZATION IN QUANTUM ELECTRODYNAMICS

M. Berrondo

Instituto de Física, UNAM
Aptdo. Postal 20-364
01000 México D.F., MEXICO

INTRODUCTION

It is well known that the field theoretical approach to quantum electrodynamics is plagued with inconsistencies when we go from the formal properties found in the general theory to practical calculations in perturbation theory, for instance. The spectacular successes of the theoretical predictions of radiative corrections notwithstanding, the daring manipulations of infinities amid the calculations (or the need of redefining the basic parameters of the theory at the end of the calculation) are highly unsatisfactory in a theory which otherwise can be considered as one of the greatest intellectual acheivements of our times.

In this introduction we would like to concentrate on two particular inconsistencies, deeming that they pinpoint the path to be followed in order to have a consistent theory, without modifying however the basic principles of the field theoretical approach. In no way do we claim to present a rigorous mathematical theory, since we feel that the solution to these problems lies within the "physicist approach" to quantum field theory. The mathematical formalization of these concepts may come at a later stage.

The first example we would like to mention is the non-relativistic limit of the Coulomb interaction between two electrons[1]. Quantum electrodynamics yields an expression for the instantaneous Coulomb interaction[2,3] which reduces, in the non-relativistic approximation, to

$$H_{int} \xrightarrow[n.r.]{} \frac{1}{2} \int d\underline{r} \int d\underline{r}' \frac{\rho(\underline{r})\rho(\underline{r}')}{|\underline{r}-\underline{r}'|} \quad , \tag{1}$$

where $\rho(\underline{r}) = e\psi^{\dagger}(\underline{r})\psi(\underline{r})$ is the electric charge density operator in the non-relativistic approximation. This expression does not account for the Pauli principle in the interaction Hamiltonian. The difference between Eq. (1) and the correct expression:

$$H^{nr}_{int} = \frac{e^2}{2}\int d\underline{r}\int d\underline{r}' \frac{\psi^{\dagger}(\underline{r})\psi^{\dagger}(\underline{r}')\psi(\underline{r}')\psi(\underline{r})}{|\underline{r} - \underline{r}'|} \qquad (2)$$

is precisely the self-energy term, which must be included "by hand" in the usual mass renormalization procedure.

The second, and more relevant example, appears in the interaction representation[4]: Dyson's reordering of the iterated solution of the Schroedinger equation for the evolution operator is, in fact, not a solution![5] This can be seen as follows. Let $U(t,t_o)$ be the evolution operator from time t_o to time t in the interaction picture, fulfilling the usual integral equation[4]:

$$U(t,t_o) = I + i\int_{t_o}^{t} dt_1 \int d^3x_1 \mathcal{L}(x_1) U(t_1,t_o) \quad , \qquad (3)$$

where $\mathcal{L}$ is the interaction Lagrangian density. The usual iteration plus Dyson's time symmetrization of Eq. (3) yields the expansion:

$$U(t,t_o) = I + \sum_{n=1} \frac{1}{n!} \int_{t_o}^{t} d^4x_1 \ldots \int_{t_o}^{t} d^4x_n \; S_n(x_1,\ldots,x_n), \qquad (4)$$

where we have defined

$$S_n(x_1,\ldots,x_n) = i^n \; T(\mathcal{L}(x_1)\ldots\mathcal{L}(x_n)) \quad , \qquad (5)$$

T denoting Wick's time ordering operation[6]. It is well known[7] however, that already the second order term $S_2(x_1,x_2)$ contains delta functions (with divergent coefficients). The vacuum-vacuum contribution, in fact, gives a non-vanishing term in Eq. (4) in the limit when t approaches t_o:

$$\lim_{\tau\to 0} U2(t_o+\tau,t_o) =$$

$$\lim_{\tau\to 0} \int_{t_o}^{t_o+\tau} d^4x_1 \int_{t_o}^{t_o+\tau} d^4x_2 \; T(\mathcal{L}(x_1)\mathcal{L}(x_2)) \neq 0 \quad . \qquad (6)$$

Hence the initial condition $U(t_o,t_o)=I$ is <u>not</u> fulfilled in Eq. (4). In a similar fashion, the electron's self-energy second order term yields a non-vanishing derivative:

$$\left.\frac{\partial U^2(t,t_o)}{\partial t}\right|_{t_o} = \lim_{\tau\to 0} \frac{1}{\tau} \int_{t_o}^{t_o+\tau} d^4x_1 \int_{t_o}^{t_o+\tau} d^4x_2 \; T(\mathcal{L}(x_1)\mathcal{L}(x_2)) \neq 0, \qquad (7)$$

so that Eq.(4) is <u>not</u> a solution to the Schroedinger differential equation.

In what follows, we shall give two alternative solutions to these problems. The first one[5], within the interaction picture, recognizes the need to define Wick's ordering operation appropriately at coalescing times. In the second one[8], we choose the Heisenberg picture as the starting point and work with quantities which do <u>not</u> depend on the precise definition of T for coinciding times[9]. Both approaches maintain the original parameters (the empirical values of the electron's mass m and charge e) throughout the calculations, till the bitter end. Otherwise stated, they yield "already renormalized" results.

THE INTERACTION PICTURE

We have noticed in the introduction that the inclusion of hypersurfaces with coinciding times in Dyson's solution introduces additional (divergent) terms, which were not there in the original equation. Let us then insist on the vanishing of Eqs.(6) and (7), in order to ensure a correct solution of the original equation for the evolution operator. This can be acheived by extending[5] the original definition of the T operation for the (free) electron field ϕ:

$$T\bar{\phi}(x_1)\phi(x_2) = \begin{cases} \bar{\phi}(x_1)\phi(x_2) & t_1>t_2 \\ :\bar{\phi}(x_1)\phi(x_2): & x_1=x_2 \\ \phi(x_2)\bar{\phi}(x_1) & t_2>t_1 \end{cases} , \qquad (8)$$

which fixes the value of the T product at equal times*.

With this definition, we can take the free electron current density as

$$J^\mu(x) = T\;\bar{\phi}(x)\gamma^\mu\phi(x) \qquad , \qquad (9)$$

and the interaction Lagrangian density as:

$$\mathcal{L}(x) = T\;\bar{\phi}(x)\gamma^\mu A^o_\mu(x)\phi(x) \qquad . \qquad (10)$$

* The case $t_1=t_2$, $\underline{x}_1\neq\underline{x}_2$ is irrelevant.

The most important consequence of Eq.(8) is that it yields a vanishing contraction:

$$\overbrace{\phi(x_1)\bar{\phi}}(x_2)\Big|_{x_1=x_2} = 0 \quad . \tag{11}$$

We see immediately, with the aid of Wick's theorem[6], that the only surviving term in a T product with coinciding arguments is the normal product, since all contractions vanish in this case:

$$T\ \bar{\phi}(x_1)\phi(x_1)\ldots\bar{\phi}(x_1)\phi(x_1) = :\bar{\phi}(x_1)\phi(x_1)\ldots\bar{\phi}(x_1)\phi(x_1): \tag{12}$$

The vacuum-vacuum expectation value of Eq.(12) vanishes, so there are no vacuum-vacuum divergences. For n>1, the one-particle expectation value is also zero, and hence there are no divergent self-energies.

Let us illustrate this with the case of the electron self-energy to second order, given by:

$$S_2^{se}(x_1,x_2) = e^2:\bar{\phi}(x_1)\gamma^\mu\overbrace{\phi(x_1)\bar{\phi}}(x_2)\gamma^\nu\phi(x_2):D^c_{\mu\nu}(x_1-x_2) + \\ + (x_1 \longleftrightarrow x_2) \quad , \tag{13}$$

where D^c is the photon propagator.

For $x_1 \neq x_2$, we obtain the same result as in the usual approach. The mass shift δm however, can be read off as the coefficient of the four dimensional delta function $\delta(x_1-x_2)$ in Eq.(13). Taking into account Eq.(11), we see that there is no delta term in Eq.(13), and hence $\delta m=0$. Because of Eq.(12), this is valid to all orders.

A similar result obtains for the self-charge[5]. In general, we can write the S-matrix expansion as:

$$S = T\exp i\int\mathcal{L}(x)\,d^4x \quad , \tag{14}$$

where $\mathcal{L}(x)$ is given by Eq.(10), and provided we separate the contractions with coinciding arguments, and we use Eq.(11) in all Feynman diagrams.

VACUUM POLARIZATION IN THE HEISENBERG PICTURE

The need to specify the value of the T operation at the jump discontinuity, as was done in the preceding section, might seem awkward if we think in terms of distributions. It would be most gratifying to deal , instead, with quantities which yield definite answers, and are not sensitive to any particular choice.

In this section we proceed to show that this is indeed possible, when working in the Heisenberg picture, and using the correct expressions for the Green functions. In this case no special definition is needed for the T operation at equal times. This will be illustrated with the exact photon propagator and the radiative corrections due to the vacuum polarization. It will be self-evident at the end that an entirely similar result holds for the exact electron propagator[8].

The exact photon propagator can be defined as[10],

$$\mathcal{D}^{c}_{\mu\nu}(x_1-x_2) = i <T\, A_\mu(x_1) A_\nu(x_2)> \quad , \tag{15}$$

where A is the photon field operator in the Heisenberg picture, fulfilling the equation:

$$\Box^2 A_\mu(x) \equiv \partial_\nu \partial^\nu A_\mu(x) = e\, j_\mu(x) \tag{16}$$

in the Lorentz gauge, and where j is the current density in Heisenberg's picture also.

In order to have a simple covariant definition[11,12] of the T operation in Eq.(15), we shall rather work with the scalar quantity:

$$\mathcal{D}^{c}(x_1-x_2) = \frac{i}{3} <T\, A_\mu(x_1) A^\mu(x_2)> = \tag{17}$$

$$= \frac{i}{3} \left\{ \theta(t_1-t_2) < A_\mu(x_1) A^\mu(x_2)> + \theta(t_2-t_1) < A_\mu(x_2) A^\mu(x_1)> \right\}$$

where θ is the step function. We can thence reconstruct Eq.(15), using the fact that the difference $\mathcal{D}^{c}_{\mu\nu} - D^{c}_{\mu\nu}$ is purely transverse[10]. In terms of its Fourier transform, we have:

$$\mathcal{D}^{c}_{\mu\nu}(k) - D^{c}_{\mu\nu}(k) = \left(g_{\mu\nu} - \frac{k_\mu k_\nu}{k^2}\right) \left(\mathcal{D}^{c}(k) - D^{c}(k)\right) \tag{18}$$

independently of the fact that we have used the Lorentz gauge in Eq.(16).

The formal solution to Maxwell's equation (16) is given by[13]

$$A^\mu(x) = A^\mu_{in}(x) - e\int D_{ret}(x_1-y_1)\, j^\mu(y_1) dy_1 \tag{19}$$

in terms of the retarded free propagator D_{ret}:

$$\Box^2 D_{ret}(x_1-y_1) = -\delta(x_1-y_1) \quad . \tag{20}$$

Direct substitution of Eq.(19) into Eq.(17) yields:

$$\mathcal{D}^c(x_1-x_2) - D^c(x_1-x_2) = \frac{ie^2}{3}\{\theta(t_1-t_2)\int d^4y_1\int d^4y_2 D_{ret}(x_1-y_1)\, x$$

$$x D_{ret}(x_2-y_2) < j_\mu(y_1) j^\mu(y_2) > + (x_1 \longleftrightarrow x_2)\} \tag{21}$$

Taking the Fourier transform:

$$f(k) \equiv \mathcal{F} f(x) = \int f(x) e^{ikx}\, d^4x \quad , \tag{22}$$

the product appearing in Eq.(21) transforms into a convolution. The Fourier transform of the current-current Wightman function has the general form[14]:

$$\mathcal{F} < j_\mu(y_1) j^\mu(y_2) > = \mathcal{F} < j_\mu(y_1-y_2) j^\mu(0) > = 3J(k^2)\theta(k^o) \quad , \tag{23}$$

using translation invariance. Furthermore, recalling the Fourier representation of the step function:

$$\theta(t) = \frac{i}{2\pi}\int_{-\infty}^{\infty} \frac{e^{-i\omega t}}{\omega + i\,\varepsilon}\, d\omega \quad , \tag{24}$$

we finally get the Fourier transform of Eq.(21) as the convolution:

$$\mathcal{D}^c(k) - D^c(k) = -\frac{e^2}{2\pi}\int_{-\infty}^{\infty} d\omega\, \frac{J(k_\omega^2)}{k_\omega^4}\left\{\frac{\theta(\omega)}{k^o-\omega+i\varepsilon} + \frac{\theta(-\omega)}{-k^o+\omega+i\varepsilon}\right\} \quad , \tag{25}$$

in terms of the four vector:

$$k_\omega = (\omega, \vec{\underline{k}}) \tag{26}$$

Eq.(25) is trivially transformed into a positive ω integral. Changing the integration variable to an invariant square mass λ:

$$\lambda = \omega^2 - \vec{\underline{k}}\cdot\vec{\underline{k}} = k_\omega^{\ 2} \tag{27}$$

we finally get

$$\mathcal{D}^c(k) - D^c(k) = \frac{e^2}{2\pi}\int_0^{\infty} d\lambda\, \frac{J(\lambda)}{\lambda^2}\, \frac{1}{\lambda - k^2 - i\varepsilon} \quad . \tag{28}$$

---

* For our purposes the boundary condition imposed on Eq.(19) is irrelevant.

Expression (28) can be immediately recognized as the spectral representation[10] for the photon propagator "already renormalized". The calculational procedure in this scheme thus resembles the one used in the analytic S-matrix[15], rather than the usual Feynman diagrams: we first evaluate the spectral function $J(\lambda)$, and then substitute it into Eq.(28). The enormous difference however, is that we have worked entirely within the field theoretical framework, with no additional assumptions regarding the analytical properties of the propagators, or any hint about subtractions. The correct "convergence factor" in Eq.(28) follows directly from the convolution form of the Yang-Feldman equation (19), and the detailed form of the definition (17).

CONCLUSIONS

The conclusion we can draw from above is straightforward: it is indeed possible to obtain a finite, consistent field theory of quantum electrodynamics, as long as we are ready to deal with the T operation with due care. In the interaction representation this implies avoiding the introduction of singularities, by singling out the vertex of the light cone. In the Heisenberg picture, things look simpler, so we only have to keep track of the correct variables affected by the time ordering operation defining the causal propagators.

REFERENCES

1. L. I. Schiff, "Quantum Mechanics," McGraw-Hill, New York (1968), sect. 57.
2. J. D. Bjorken and S. Drell, "Relativistic Quantum Fields," McGraw-Hill, New York (1965), sect. 15.2.
3. I. Bialyncki-Birula, in this volume.
4. See e.g. D. Lurié, "Particles and Fields," Interscience, New York (1968), chap. 6.
5. M. Berrondo and R. Jáuregui, Kinam 5:163 (1983).
6. G. Wick, Phys. Rev. 80:268 (1950).
7. N. N. Bogoliubov and D. V. Shirkov, "Introduction to the Theory of Quantized Fields," Interscience, New York (1959), sect. 18.
8. M. Berrondo and R. Jáuregui, to be published.
9. J. Hilgevoord, Nucl. Phys. 15:657 (1960).
10. See e.g. E. M. Lifshitz and L. P. Pitaevskii, "Relativistic Quantum Theory," Pergamon Press, Oxford (1973), chap. XI.
11. See e.g. C. Itzykson and J. B. Zuber, "Quantum Field Theory," McGraw-Hill, New York (1980), sect. 5.1.7.
12. J. Schwinger, Phys. Rev. Lett. 3:296 (1959).
13. C. N. Yang and D. Feldman, Phys. Rev. 79:972 (1950).
14. G. Kallen, "Quantum Electrodynamics," Springer-Verlag, Berlin (1972).
15. T. Chou and M. Dresden, Rev. Mod. Phys. 39:142 (1967).

SIMPLE PHYSICAL PICTURES FOR RADIATIVE PROCESSES

VACUUM FLUCTUATIONS VERSUS RADIATION REACTION

Claude Cohen-Tannoudji

Ecole Normale Supérieure and Collège de France
24 rue Lhomond 75231 Paris Cedex 05 France

ABSTRACT

Radiative processes, such as spontaneous emission or radiative corrections (Lamb-shift, g-2), are usually described as resulting from the interaction of the electron with the quantum vacuum field (vacuum fluctuations), or with the self field of the electron (radiation reaction).

The first lecture, which is essentially a presentation of the results derived in reference [1], is devoted to the identification and to the discussion of the respective contributions of vacuum fluctuations and radiation reaction. The calculations are based on non relativistic Heisenberg equations (lowest order in 1/c). Spin and magnetic effects are neglected. Connections are established with quantum theory of damping and linear response theory.

The second lecture, which is essentially a presentation of the results derived in reference [2], extends the physical discussion of lecture 1 by including spin and magnetic affects and relativistic corrections. An effective hamiltonian, describing radiative corrections and valid up to order $(1/c)^2$, is derived for an electron moving slowly in external static fields. The positive sign of the spin anomaly g-2 is physically interpreted. Finally, a sketch of a completely relativistic calculation of g-2 is presented [3], allowing a discussion of the contribution to g-2 of the relativistic modes of the radiation field.

[1] J. DALIBARD, J. DUPONT-ROC and C. COHEN-TANNOUDJI
J. Physique 43, 1617 (1982)
[2] J. DUPONT-ROC, C. FABRE and C. COHEN-TANNOUDJI
J. Phys. B 11, 563 (1978)
see also same authors + P. AVAN
J. Physique 37, 993 (1976)
[3] J. DUPONT-ROC and C. COHEN-TANNOUDJI
To appear in "New Trends in Atomic Physics",
Les Houches 1982 Session XXXVIII, STORA R. and GRYNBERG G. ed. (North Holland, Amsterdam)

THE HAMILTONIAN OF QUANTUM ELECTRODYNAMICS

Iwo Bialynicki-Birula

Institute for Theoretical Physics
Polish Academy of Sciences
Lotnikow 32/46 02-668 Warsaw, Poland

1. INTRODUCTION

The purpose of these lectures is to establish a bridge between the full, relativistic QED and its nonrelativistic approximation, used in quantum optics and atomic physics. Usually, one introduces the nonrelativistic QED by minimally coupling the quantized electromagnetic field to the quantum-mechanical particle (or particles). Such a procedure, certainly, leads most directly to the final result, but it has two shortcomings. First, it does not directly generalize to an improved theory, in which the lowest order relativistic corrections are included. Second, it does not explain what is the connection between the fundamental theory of relativistic QED and its less sophisticated counterpart--the nonrelativistic QED.

These two shortcomings will be overcome here with the help of a clear mathematical procedure, based on a series of unitary transformations. I will introduce a systematic perturbation theory in the parameter v/c. The lowest order approximation will yield the nonrelativistic theory. Higher-order approximations give the relativistic corrections.

Various pieces of my approach, probably even all of them, may be found scattered in the literature. In these notes they are all joined into one logical reasoning for the benefit of those who prefer to learn it from one source.

2. THE ORIGIN OF THE QED HAMILTONIAN

Undoubtedly, the most important and universal physical quantity is the energy. The total energy of the system expressed in terms of canonical variables--the Hamiltonian--generates the time evolution in classical and in quantum theories. In quantum theories, the solution of the eigenvalue problem for the energy operator gives the energy spectrum and a very important family of quantum states--the stationary states.

The role of the Hamiltonian as the generator of time translations is embodied in the equation:

$$\dot{F} = \frac{1}{i\hbar} [F, H] \quad , \tag{1}$$

where F is any operator representing a physical variable, which does not depend explicitly on time. This fundamental equation is obtained directly from its classical counterpart by the standard prescription of canonical quantization--the replacement of the Poisson bracket by the commutator divided by $i\hbar$.

The Hamiltonian of a given physical system is not unique. Its form depends on the choice of canonical variables. In classical theories these variables may always be subject to a canonical transformation. In quantum theories the role of canonical transformations is played by unitary transformations. Under a general unitary transformation, which may also depend explicitly on time, Eq.(2) transforms as follows ($\hbar = 1$):

$$\dot{\tilde{F}} = \frac{1}{i} [\tilde{F}, \tilde{H}] \quad , \tag{2}$$

where

$$\tilde{F} = U^{-1}(t) \; F \; U(t) \quad , \tag{3}$$

and

$$\tilde{H} = U^{-1}(t) \; (H + i\partial_t) \; U(t) \quad . \tag{4}$$

Time-dependent unitary transformations result in a change of picture; the Schrödinger, Heisenberg and Dirac pictures being the most important special cases. Time-independent unitary transformations result merely in a change of representation within the same picture. The unitary transformations fo the Hamiltonian of the form:

$$H \to \tilde{H} = U^{-1} \; H \; U \quad , \tag{5}$$

do not change the energy spectrum, but do change the form of the Hamiltonian. The eigenvalue problem with the new Hamiltonian may be easier to handle. Of course, the form of $\tilde{H}$, as a function of the new canonical variables, is the same as the form of H expressed in terms of the old variables. Time-independent unitary transformations will serve as an important and convenient tool in our study of the Hamiltonian in QED.

The starting point of my derivation of the QED Hamiltonian is the full relativistic field theory, in which the electrons and the photons are described by the electron field ψ and the electromagnetic field, respectively. These fields obey the set of coupled Dirac and Maxwell equations ($\hbar = 1 = c$)

$$(i\gamma^\mu \partial_\mu - m)\psi = e\gamma^\mu A_\mu \psi \quad , \tag{6a}$$

$$\partial_\mu f^{\mu\nu} = j^\nu \quad , \tag{6b}$$

where

$$f_{\mu\nu} = \partial_\mu A_\nu - \partial_\nu A_\mu \quad , \tag{7}$$

and

$$j^\mu = \tfrac{1}{2}e[\bar{\psi}, \gamma^\mu\psi] \quad , \tag{8}$$

The commutator in the definition of j^μ is necessary to ensure the invariance under charge conjugation. More about charge conjugation will be said later. My notation is summarized in the Appendix A.

Since we will be studying the Hamiltonian, not the S matrix, an explicitly relativistic formalism will be of lesser importance. The following form of the field equations, in which the time variable is singled out, will be more suited for our considerations,

$$i\partial_t \psi = [m\beta - i\vec{\alpha}\cdot(\nabla - ie\vec{A})]\psi + eA_0\psi \quad , \tag{9a}$$

$$\partial_t \vec{E} = \nabla \times \vec{B} - \vec{j} \quad , \tag{9b}$$

$$\nabla\cdot\vec{E} = \rho \quad , \tag{9c}$$

$$\partial_t \vec{B} = -\nabla \times \vec{E} \quad , \tag{9d}$$

$$\nabla\cdot\vec{B} = 0 \quad . \tag{9e}$$

I have replaced here the formula (7) by its equivalent--the homogeneous Maxwell equations (9d) and (9e)--in order to stress the time-evolutionary character of the field equations.

When writing the field equations, I have disregarded all intricate problems connected with infinities and renormalization. They will be discussed later.

In theories with gauge fields and quantum electrodynamics is, of course, the most distinguished example of such a theory, the notion of time evolution requires an elaboration. Since the basic mathematical objects in QED, namely the electron field ψ, its hermitian conjugate $\psi^\dagger$ and the electromagnetic potentials A_μ are all affected by gauge transformations, the field equations (9) do not give a unique solution of the initial value problem. This complication is already present at the classical level of the theory. Given the set of variables ψ, $\psi^\dagger$, $\vec{A}$, A_0 and $\vec{E}$ at time t, we may evaluate with the help of Eqs. (9a), (9b) and (9d) the values of ψ, $\vec{B}$ and $\vec{E}$ at $t+\nabla t$, but we can not determine uniquely the values of the potentials A and A_0. The potentials are defined by $\vec{B}$ and $\vec{E}$ only up to gauge transformations:

$$\vec{A} \rightarrow \vec{A} + \nabla \Lambda \quad , \qquad A_0 \rightarrow A_0 - \partial_t \Lambda \quad . \tag{10}$$

Therefore, we must fix the potentials, by imposing a gauge condition, in order to obtain the solution of the initial value problem for Eqs.(9).

Following Heisenberg and Pauli, I shall choose the following, very simple gauge condition

$$A_0 = 0 \quad , \tag{11}$$

which is best suited for the canonical formulation. The potentials satisfying this condition are nowadays referred to as the potentials in the temporal gauge. The condition (11) does not fix the vector potential uniquely. We are still free to perform gauge transformations with an arbitrary, time-independent function Λ. However, the freedom of such time-independent gauge transformations does not affect the solution of the initial value problem, which now has a unique solution. In order to see explicitly how it works, let me write down in the temporal gauge Eqs.(9a), (9b) and the relation between the vector potential and the electric field,

$$\partial_t \psi = \frac{1}{i} (m \beta - i \vec{\alpha} \cdot \vec{D}) \psi \quad , \tag{12a}$$

$$\partial_t \vec{E} = \nabla \times (\nabla \times \vec{A}) - \vec{j} \tag{12b}$$

$$\partial_t \vec{A} = - \vec{E} \quad , \tag{12c}$$

where

$$\vec{D} = \nabla - i\,e\,\vec{A} \quad . \tag{13}$$

These are two basic time-evolution equations of our system. What remains to be done is to secure the validity of Gauss's law (9c). This will be done in the next Section.

The evolution equations (12) may be written in the canonical form (1). To this end ($\psi,\psi^{\dagger}$) and ($\vec{A},\vec{E}$) are chosen as the pairs of canonical variables obeying the following anticommutation and commutation relations:

$$\{ \psi_{\alpha}(\vec{r}) \; , \; \psi^{\dagger}_{\beta}(\vec{r}\,') \} = \delta_{\alpha\beta}\; \delta(\vec{r} - \vec{r}\,') \quad , \tag{14a}$$

$$[E_{i}(\vec{r}) \; , A_{j}(\vec{r}\,')] = i\, \delta_{ij}\; \delta(\vec{r} - \vec{r}\,') \quad . \tag{14b}$$

All the remaining commutators (anticommutators) vanish. The Hamiltonian is chosen as the energy operator in the temporal gauge, expressed in terms of canonical variables. It is simply equal to the sum of the energy of the electron field and the energy of the electromagnetic field,

$$H = \tfrac{1}{2}\int d^3r \; [\, \psi^{\dagger}, \, (\, m\, \beta - i\, \vec{\alpha}\cdot\vec{D}\,)\, \psi \,] + \tfrac{1}{2}\int d^3r \; (\, \vec{E}^2 + \vec{B}^2 \,) \quad . \tag{15}$$

Again, the commutator of the electron operators is needed for the charge-conjugation invariance.

It is now a matter of straightforward calculations to check that the commutators of ψ, $\vec{E}$ and $\vec{A}$ with the Hamiltonian (15) are equal to the r.h.s. of Eqs.(12) multiplied by i.

3. GAUSS'S LAW AND THE COULOMB INTERACTION

The time-evolution equations (12) do not guarantee the validity of Gauss's law, but it does follow from these equations that

$$\partial_t \, G(\vec{r}) = 0 \quad , \tag{16}$$

where

$$G(\vec{r}) = \nabla\cdot E(\vec{r}) - \rho(\vec{r}) \quad . \tag{17}$$

Thus, in classical physics the equation $G(r) = 0$ will be preserved in time, once it is assumed to hold for the initial conditions. In quantum physics such a simple implementation of Gauss's law is not possible; it would be inconsistent with the commutation relations. We can not just set $\nabla\cdot E$ equal to ρ, even for one value of t. What can be done instead is to select a subspace of states, which is annihilated by the operators $G(\vec{r})$,

$$G(\vec{r})\ \Psi = 0 \quad . \tag{18}$$

There is no clash between this condition and the eigenvalue problem for the Hamiltonian, because Eq.(16) tells us that all the operators $G(\vec{r})$ commute with the Hamiltonian, so that we may diagonalize H in the subspace of "good" states, on which Gauss's law is valid.

The commutativity of $G(\vec{r})$ and H has an interesting interpretation in the context of gauge transformations. The operator G,

$$G = \int d^3r\ \Lambda(\vec{r})\ G(\vec{r}) \quad , \tag{19}$$

is the generator of gauge transformations of the operators ψ and $\vec{A}$,

$$e^{iG}\ \psi(\vec{r})\ e^{-iG} = e^{\,ie\Lambda(\vec{r})}\ \psi(\vec{r}) \quad , \tag{20a}$$

$$e^{iG}\ A(\vec{r})\ e^{-iG} = \vec{A}(\vec{r}) + \nabla\Lambda(\vec{r}) \quad . \tag{20b}$$

Therefore, the fact that $G(\vec{r})$ are constants of motion means also that the Hamiltonian is invariant under the simultaneous gauge transformations of ψ and $\vec{A}$. In turn, the subsidiary condition (18) may now be interpreted as the gauge-invariance condition of the state vectors.

The subspace of gauge invariant state vectors can be explicitly found. To this end, let us introduce the field analog of the Schrödinger representation for the electromagnetic operators. In this representation $\vec{A}(\vec{r})$ acts as a multiplication and $\vec{E}(\vec{r})$ acts as a functional differentiation,

$$\vec{E}(\vec{r}) = i\delta/\delta\vec{A}(\vec{r}) \quad . \tag{21}$$

Next, let us decompose the vector potential $\vec{A}(\vec{r})$ into its longitudinal $\vec{A}_L$ and transverse $\vec{A}_T$ parts,

$$\vec{A} = \vec{A}_L + \vec{A}_T \ , \qquad \nabla\times\vec{A}_L = 0 \ , \qquad \nabla\cdot\vec{A}_T = 0 \quad . \tag{22}$$

This decomposition carries over to $\vec{E}$,

$$\vec{E}_L = i\delta/\delta\vec{A}_L \quad , \qquad \vec{E}_T = i\delta/\delta\vec{A}_T \quad . \tag{23}$$

Gauss's law involves only the longitudinal part of $\vec{E}$. In the Schrödinger representation the subsidiary condition (18) reads:

$$(\nabla\cdot i\delta/\delta\vec{A}_L - \rho)\ \Psi[\vec{A}_L,\vec{A}_T] = 0 \tag{24}$$

The general solution of this equation is

$$\Psi[\vec{A}_L,\vec{A}_T] = e^{-i\Phi}\ \Psi[\vec{A}_T] \quad , \tag{25}$$

where

$$\Phi = -\int d^3r\ \nabla\cdot\vec{A}_L\Delta^{-1}\ \rho \tag{26}$$

In my shorthand notation Δ^{-1} stands for the standard Green's function of the Laplacian.

Having established the dependence of the state vectors on $\vec{A}_L$, we may easily find the effective Hamiltonian H_C acting in the subspace spanned by the vectors $\Psi[\vec{A}_T]$. The subscript C stands here for the Coulomb, because, as we shall see below, in H_C the Coulomb interaction between the charges appears explicitly. In order to obtain H_C, we must pull through H the operator $\exp(-i\Phi)$. This is accomplished with the help of the following formulas:

$$e^{i\Phi}\ \psi\ e^{-i\Phi} = \exp(\ ie\Delta^{-1}\nabla\cdot\vec{A})\ \psi \quad , \tag{27a}$$

$$e^{i\Phi}\ \psi^{\dagger} e^{-i\Phi} = \psi^{\dagger}\ \exp(-ie\Delta^{-1}\nabla\cdot\vec{A}) \quad , \tag{27b}$$

$$e^{i\Phi}\ \vec{E}\ e^{-i\Phi} = \vec{E} + \Delta^{-1}\ \nabla\rho \quad , \tag{27c}$$

which follow directly from the canonical commutation relations (14). Eqs.(27) lead to the following expression for the Coulomb Hamiltonian:

$$H_C = e^{i\Phi}\ H\ e^{-i\Phi} = \tfrac{1}{2}\int d^3r[\psi^{\dagger},(\ m\ \beta - i\ \vec{\alpha}\cdot(\ \nabla - i\ e\ \vec{A}_T\ ))\ \psi]$$

$$+\frac{1}{8\pi}\int d^3r\ \int d^3r'\ \rho(\vec{r})\ \frac{1}{|\vec{r} - \vec{r}'|}\ \rho(\vec{r}') + \tfrac{1}{2}\int d^3r\ (\ \vec{E}_T^2 + \vec{B}^2\ ) \quad . \tag{28}$$

In this formula I have dropped the operator $\vec{E}_L^2$ in the electromagnetic field energy, because $\vec{E}_L$ gives zero when acting on the state vectors $\Psi[\vec{A}_T]$.

The operator H_C is often referred to as the Hamiltonian in the Coulomb gauge. However, it is clear from our derivation, that the operator H_C has a gauge invariant meaning. It is the Hamiltonian reduced to the subspace of gauge invariant state vectors.

The electromagnetic interactions, as described by the Hamiltonian H_C, consists of the interaction of charges with the electromagnetic field (interaction via the exchange of photons) and of the instantaneous interaction of charges via the Coulomb forces.

4. PARTICLES, ANTIPARTICLES, AND CHARGE CONJUGATION

Having established the form of the Hamiltonian, we may, in principle, proceed with the solution of the eigenvalue problem. However, this problem is so difficult that no progress can be made without a thorough understanding of the physical content of the theory. The difficulties are greatly enhanced by the fact that the theory is ill-defined and only a judicious use of the physical insight enables one to arrive at meaningful conclusions. Therefore, in addition to the definition of the Hamiltonian in terms of the canonical field operators, we also need an interpretation of the basic objects appearing in the Hamiltonian--the field operators ψ and $\psi^\dagger$ describing the charged particles.

I will approach this problem in cautious manner, making one step at a time, because a complete solution of QED is not known and aiming at ultimate answers may lead us astray.

The use of the field operators (or creation and annihilation operators) to describe the electron degrees of freedom is necessary in a relativistic theory. Otherwise, we immediately run into the problems caused by the negative-energy solutions of the Dirac equation. After all, this equation does not describe a single particle, but a many-particle system. Interactions are accompanied by the appearance of pairs of particles and antiparticles from the vacuum and their disappearance into the vacuum. Only in the nonrelativistic approximation we can separate the motion of particles from that of antiparticles. This nonrelativistic dynamics is deeply buried in the formalism of relativistic QED and unearthing it will require some effort. These complications arise, because the nonrelativistic particles are rather complex, collective excitations of the vacuum. Their electromagnetic interactions are much more involved than those of the "original particles", whose creation and annihilation operators enter the initial Hamiltonian.

These remarks bring us to the first important question: What are the main characteristics of those "original", or "bare", particles? In order to answer this question, I will tkae a closer look at the various terms in the Hamiltonian H_C. Let me start with the first term--the mass term.

In the Dirac representation (see the Appendix A), which I am using here, the β matrix is diagonal,

$$\beta = \text{diag}\ (\ 1,\ 1, -1, -1\ )\ . \tag{29}$$

We expect both particles and antiparticles to give positive contributions to the mass term. This can be achieved by interchanging the creation and the annihilation operators for the lower components of the Dirac bispinors ψ and $\psi\dagger$, which leads to the following decomposition of these operators,

$$\psi = \begin{bmatrix} \Phi \\ \tilde{\chi}\dagger \end{bmatrix}\ , \qquad \psi\dagger = (\ \Phi\dagger\ ,\ \tilde{\chi}\ )\ . \tag{30}$$

The conjugation, denoted by $\dagger$, means here both the hermitian conjugation of each component and the transition between the column- and row-arrangement of the components. The sign $\tilde{}$ denotes the transition between columns and rows (transposition) and an additional multiplication by the antisymmetric matrix $i\sigma_2$,

$$\tilde{\chi} = (\ \chi_1\ ,\ \chi_2\ )i\sigma_2 = (\ -\chi_2\ ,\ \chi_1\ )\ , \tag{31a}$$

$$\tilde{\chi}\dagger = -i\sigma_2 \begin{bmatrix} \chi_1^\dagger \\ \chi_2^\dagger \end{bmatrix} = \begin{bmatrix} -\chi_2^\dagger \\ \chi_1^\dagger \end{bmatrix} \tag{31b}$$

The multiplication by $i\sigma_2$ is necessary in order to have the same transformation properties for ϕ and χ under rotations.

In this new notation the mass term has the form

$$H_{mass} = m \int d^3r\ (\ \phi\dagger\phi + \chi\dagger\chi\ ) + \text{const}\ . \tag{32}$$

The (infinite) constant appearing in this formula results from the rearrangement of the creation and annihilation operators, in order to obtain their normal ordering. It is an analog of the zero-point energy of the electromagnetic field.

Now, let me show that our interpretation of ψ in terms of the annihilation operators of particles and the creation operators of antiparticles is also consistent with our understanding of the last term in H_C --the energy of the electrostatic interaction between the charge densities. With the use of the formulas (30), we obtain for the charge density,

$$\rho = \tfrac{1}{2} e [\psi^\dagger, \psi] = e (\phi^\dagger\phi - \chi^\dagger\chi) \quad . \tag{33}$$

Notice that there is no additional constant in this expression, because the two infinite contributions from the rearrangements of the field operators cancel out. This cancellation is a direct result of the charge-conjugation invariance of the theory. I have made sure from the very beginning that the theory be invariant under charge conjugation, consisting in the interchange of particles and antiparticles and the sign-reversal of the electromagnetic field:

$$\phi \leftrightarrow \chi \quad , \qquad \phi^\dagger \leftrightarrow \chi^\dagger \quad , \qquad \vec{E} \to -\vec{E} \quad , \qquad \vec{A} \to -\vec{A} \quad . \tag{34}$$

The charge density must be, clearly, odd under charge conjugation and, therefore, it can not contain a constant term.

The remaining term of the electron Hamiltonian has no simple interpretation. It contains the pair-creation and the pair-annihilation terms, which shows that the particles created by $\phi^\dagger$ and $\chi^\dagger$ are quite different from the electrons and positrons observed in experiments. However, since we have gotten the quantum numbers correctly, our tentative interpretation may serve as a sound and consistent starting point. The rest will, anyway, be determined by the detailed study of the Hamiltonian.

The full Hamiltonian expressed in terms of the operators ϕ and χ reads

$$H = m \int (\phi^\dagger\phi + \chi^\dagger\chi) + \tfrac{1}{2} \int (\vec{E}^2 + \vec{B}^2) - i \int (\phi^\dagger \vec{\sigma}\cdot\vec{D}\, \tilde{\chi}^\dagger + \tilde{\chi}\, \vec{\sigma}\cdot\vec{D}\, \phi) \quad , \tag{35}$$

where I have dropped the additive constant. Unmarked integrals will, from now on, denote the d^3r integration over the whole space.

The unitary transformations, which will be introduced in the next Section, are easier to apply when the Hamiltonian is not restricted to the subspace of gauge-invariant states, but at the end I will always impose Gauss's law and obtain the physical Hamiltonian H_C.

5. THE MAKING OF THE NONRELATIVISTIC HAMILTONIAN

Let me begin with making a somewhat unorthodox decomposition of the Hamiltonian (35) into H_0 and H_1,

$$H_0 = H_{mass} + H_{field} \quad , \qquad H_1 = H_{pair} \quad . \tag{36}$$

As I have already said before, the full Hamiltonian of QED will be subject to certain unitary transformations. The role of these unitary transformations is to bring the Hamiltonian to the diagonal form, which will contain only the terms conserving separately the number of particles and the number of antiparticles,

$$[\ H_{diag}\ ,\ \int \phi^\dagger\phi\ ] = [\ H_{diag}\ ,\ \int \chi^\dagger\chi\ ] = 0 \quad . \tag{37}$$

The terms *diagonal* and *off-diagonal* refer here to the decomposition of the full Hilbert space into the direct sum of subspaces labelled by the total number of particles and the total number of antiparticles. Each subspace is an invariant subspace of every diagonal operator. The H_0 part of H is already diagonal. The role of the unitary transformations will be to eliminate all the off-diagonal terms in the Hamiltonian.

Aiming at the description of low-energy phenomena, we may treat H_1 when compared to the mass term, as a small perturbation, because $-i\vec{D}/m$ is of the order of $\vec{v}/c$. We are unable to find the desired unitary transformation in one step, but the existence of this small parameter will enable us to effectively perform the elimination of the off-diagonal terms of the Hamiltonian in successive orders of perturbation theory in $1/m$.

The best method to explain this elimination procedure is to simply show how it works in practice. To this end, I will expand the unitary transformation of the QED Hamiltonian (35) into the power series in the (antihermitian) generator F of U,

$$e^F\ (\ H_0 + H_1\ )e^{-F} = H_0 + H_1 + [\ F,\ H_0\ ] + [\ F,\ H_1\ ]$$

$$+\tfrac{1}{2}[\ F,\ [\ F,\ H_0]]+\ldots \tag{38}$$

In perturbation theory this infinite series will be reduced to a finite sum, if F is a polynomial in the expansion parameter.

The off-diagonal small term H_1 will be eliminated in the lowest order of perturbation theory, if we choose F in such a way that

$$[\ F,\ H_0\ ] + H_1 = 0 \quad . \tag{39}$$

The effective method of solving this equation will be based on the lemma: Lemma. The solution of the commutator equation:

$$[\,X\,,\,A\,] = B \tag{40}$$

can be written in the following form:

$$X = \frac{i}{2} \int dt\ \mathrm{sgn}(t)\ e^{-\varepsilon|t|}\ e^{iAt}\ B\ e^{-iAt} \quad , \tag{41}$$

where the integral is extended over the whole real axis and the limit $\varepsilon \to 0$ is tacitly assumed.

The (formal) proof of this lemma consists in taking the commutator of the operator appearing under the integral with the operator A, noting that the result may be written as a derivative with respect to t, and finally integrating by parts.

The application of the lemma to Eq.(39) gives

$$F = -\frac{i}{2} \int dt\ \mathrm{sgn}(t)\ e^{-\varepsilon|t|}\ H_1(t) \quad , \tag{42}$$

where the dependence of $H_1(t)$ on t is generated by the Hamiltonian H_0. This time evolution may be viewed as a version of the Dirac (interaction) picture. In this picture the electron and positron operators depend on time as follows:

$$\phi(t) = e^{-imt}\ \phi \quad , \qquad \phi^\dagger(t) = e^{imt}\ \phi^\dagger \quad , \tag{43a}$$

$$\chi(\tau) = e^{-imt}\ \chi \qquad \chi^\dagger(t) = e^{imt}\ \chi^\dagger \quad . \tag{43b}$$

The explicit form of the time evolution of the vector potential will not be needed. It suffices to know its first derivative at $t = 0$,

$$\vec{A}(t) = \vec{A} + t\ \partial_t \vec{A} + \ldots = \vec{A} - t\ \vec{E} + \ldots \quad , \tag{44}$$

since the higher derivatives in the Taylor expansion contribute higher order terms in $1/m$. This is seen from the following formula:

$$-\frac{i}{2} \int dt\ \mathrm{sgn}(t)\ e^{-\varepsilon|t|}\ e^{\pm 2imt}\ t^n = \pm(\pm i)^n\ n!\ (2m)^{-n-1} \quad . \tag{45}$$

The first two terms in the expansion of F in inverse powers of m, calculated from the formula (42), are

$$F_1 = \frac{-i}{2m} \int (\phi\dagger \, \vec{\chi}\cdot\vec{D} \, \tilde{\chi}\dagger - \tilde{\chi} \, \vec{\sigma}\cdot\vec{D} \, \phi) \quad , \tag{46a}$$

$$F_2 = \frac{ie}{4m^2} \int (\phi\dagger \, \vec{\sigma}\cdot\vec{E} \, \tilde{\chi}\dagger + \tilde{\chi} \, \vec{\sigma}\cdot\vec{E} \, \phi) \quad . \tag{46b}$$

In order to obtain the transformed Hamiltonian,

$$\tilde{H} = H_0 + \tfrac{1}{2} [F_1 , H_1] + O(\tfrac{1}{m^2}) \tag{47}$$

we need only to calculate the commutator:

$$[F_1, H_1] = - \frac{1}{m} \iint [\phi\dagger \, \vec{\sigma}\cdot\vec{D} \, \tilde{\chi}\dagger , \; \tilde{\chi} \, \vec{\sigma}\cdot\vec{D} \, \phi]$$

$$= - \frac{1}{m} \int (\phi\dagger \, (\vec{\sigma}\cdot\vec{D})^2 \, \phi - \tilde{\chi} \, (\vec{\sigma}\cdot\vec{D})^2 \, \chi\dagger)$$

$$= - \frac{1}{m} \int (\phi\dagger \, (\vec{\sigma}\cdot\vec{D})^2 \, \phi + \chi\dagger \, (\vec{\sigma}\cdot\vec{D}^*)^2 \, \chi) + \frac{1}{m} \mathrm{Tr}(\vec{\sigma}\cdot\vec{D})^2 \quad , \tag{48}$$

where the trace operation refers to both the space variables and the spin indices. The positron term has been reordered with the help of the identity:

$$\sigma_2 \, \vec{\sigma} \, \sigma_2 = - \vec{\sigma}^T \tag{49}$$

and by integration by parts.

I will write the Hamiltonian (47) in the final, Coulomb form, which is obtained by performing the same transformation as in the formula (28). Note that this procedure is justified, because the subsidiary condition (18) is not modified by the unitary transformation (38) of the Hamiltonian (F is gauge invariant). In the final form of the Hamiltonian I will also reinstate c and $\hbar$.

$$\tilde{H}_C = mc^2 \int (\phi\dagger\phi + \chi\dagger\chi) - \frac{\hbar^2}{2m} \int (\phi\dagger \, \vec{D}_T^2 \, \phi + \chi\dagger \vec{D}_T^{*2} \, \chi)$$

$$- \frac{e\hbar}{2mc} \int (\phi\dagger\vec{\sigma}\phi - \chi\dagger\vec{\sigma}\chi)\cdot\vec{B} + \frac{1}{8\pi} \iint \rho(\vec{r}) \, |\vec{r} - \vec{r}\,'|^{-1} \, \rho(\vec{r}\,')$$

$$+ \tfrac{1}{2} \int (\vec{E}_T^2 + \vec{B}^2) + \text{vac} \quad , \tag{50}$$

where

$$\vec{D}_T = \nabla - \frac{ie}{\hbar c}\vec{A}_T \quad , \tag{51}$$

and vac stands for the divergent vacuum energy (the trace term in (48)), which does not depend on the electron and positron operators.

The interpretation of this Hamiltonian will be given in the next Section.

6. THE MEANING OF THE NONRELATIVISTIC HAMILTONIAN

I will begin this Section with the discussion of two distinct interpretations of the connection between the original Hamiltonian (25) and the transformed Hamiltonian (50).

According to the first interpretation the Hamiltonian is actually changed by the unitary transformation, only its spectrum remains the same. This point of view may be called the active interpretation of the unitary transformation of the Hamiltonian. Those who would prefer to deal with just one Hamiltonian, may choose the passive interpretation of the unitary transformation.

The distinction between the active and the passive interpretations is well known from geometry. There, we may view the transformation of the components, say, of a vector, as being due either to the rotation of the vector or to the (inverse) rotation of the coordinate system.

According to the passive interpretation of the unitary transformation of the Hamiltonian, the operator H does not change, but its form changes when one expresses H in terms of a different set of canonical variables. In other words, the canonical variables, not the Hamiltonian, are subject to the unitary transformation.

In order to see this more explicitly, let us go back to the transformation formula (5) and rewrite it in the form

$$H = U\tilde{H}U^{-1} \tag{52}$$

As I have noted before, the new Hamiltonian $\tilde{H}$ and the old Hamiltonian H have the same form when they are expressed in terms of new and old canonical variables, respectively. Let us regard the unitary transformation U in Eq.(52) as a function of the new canonical variables. Then, the right hand side of Eq.(52) looks very much the same as the unitary change of the Hamiltonian, which we have studied in the last Section, except that U is replaced by U^{-1}. Therefore, by just assuming that all the field operators are the new canonical variables, connected by a unitary transformation to the old ones, and by changing F and $-F$, we arrive at the same form of the Hamiltonian as in Eq.(50). However, according to this new, passive interpretation, this Hamiltonian is viewed as a new expression (in terms of new canonical variables) of the same old operator.

The two interpretations are not only completely equivalent, but even all the specific calculations are exactly the same in both approaches. I have chosen the active interpretation mostly for its typographical simplicity; one does not have to distinguish in print between the old and new canonical variables.

Our starting point required the use of field operators to describe the charged particles, but the final form of the Hamiltonian may easily be translated into the quantum-mechanical language. Since the new Hamiltonian (50) is diagonal, we may consider its action separately in each subspace of a given number of particles and antiparticles. In this subspace we may replace each one-particle operator by its quantum-mechanical counterpart according to the rule:

$$\int \phi^\dagger(\vec{r})\, O(\vec{r})\, \phi(\vec{r}) \rightarrow \sum_i O(\vec{r}_i) \quad , \tag{53a}$$

and each two-particle operator by the corresponding expression

$$\tfrac{1}{2} \int\int \phi^\dagger(\vec{r}_1)\, \phi^\dagger(\vec{r}_2)\, O(\vec{r}_1,\vec{r}_2)\, \phi(\vec{r}_2)\, \phi(\vec{r}_1) \rightarrow \sum_{i>j} O(\vec{r}_i,\vec{r}_j) \quad , \tag{53b}$$

where $O(\vec{r})$ and $O(\vec{r}_1,\vec{r}_2)$ may contain the derivatives, functions of $\vec{r}$, and spin operators. This procedure leads to the familiar Hamiltonian of the nonrelativistic quantum electrodynamics. For simplicity, I will write it down here only in the case when the positions are not present.

$$H = \frac{1}{2m} \sum_i (\vec{p}_i - \frac{e}{c} \vec{A}_T)^2 - \frac{e\hbar}{2mc} \sum_i \vec{\sigma}\cdot\vec{B}(\vec{r}_i)$$

$$+ \frac{e^2}{4\pi} \sum_{i>j} |\vec{r}_i - \vec{r}_j|^{-1} + \tfrac{1}{2} \int (\vec{E}_T^2 + \vec{B}^2). \tag{54}$$

In order to understand fully the connection between this Hamiltonian and its field-theoretic counterpart (50), we must clarify the problem of the electromagnetic mass of the electron. In the lowest order of perturbation theory in $1/m$ the electromagnetic mass is generated only by the Coulomb interaction.

The translation formula (53) for the two-particle operators assumes the normal ordering of the creation and annihilation operators. The Coulomb energy in the expression (50) is not normally ordered and the reordering introduces an additional term--the Coulomb self-energy:

$$\delta mc^2 \int (\phi^\dagger\phi + \chi^\dagger\chi) \quad , \tag{55}$$

where

$$\delta mc^2 = \frac{e^2}{2r}\Big|_{r=0} \cdot (4\pi)^{-1} \tag{56}$$

is the electrostatic-field energy of a point charge.

This is the first and the simplest example of the electromagnetic contribution to the electron's rest-mass (or rest-energy). Other terms of this type appear in higher order of perturbation theory in $1/m$. All these contributions may be absorbed into the mass term in the Hamiltonian by a suitable redefinition of the mass parameter (mass renormalization). The observed mass of the electron m_{obs}, therefore, must be identified not with the parameter m introduced in the original Hamiltonian, but with the final value of the coefficient of the mass term, which includes all electromagnetic corrections,

$$m_{obs} = m + \delta m \quad . \tag{57}$$

In higher orders of perturbation theory there appear the electromagnetic corrections to the mass of a more complicated character as compared to those given by the formulas (55) and (56). They result from the renormalization of the mass parameter occurring in other parts of the Hamiltonian. For example, in the second order of perturbation theory in $1/m$ there will appear the nonrelativistic kinetic-energy term multiplied by $-\delta m/m$, which can be absorbed into the first-order term by the replacement of $1/m$ by $1/m_{obs}$.

A similar process of renormalization takes place for the electron's charge. Due to vacuum polarization, the parameter e appearing in the initial Hamiltonian cannot be identified with the observed value of the charge e_{obs} . However, the correction terms appear for the first time in the third-order of perturbation in e and are of no concern to us here.

7. EXTERNAL ELECTROMAGNETIC FIELDS

So far I have treated QED as a theory of a closed system, made only of the electrons, the positrons, and the electromagnetic field. However, in most situations there are also other parts of the world, which can not be ignored. In many cases, these additional parts may be represented by some given configurations of the electromagnetic field, often called the external electromagnetic field. A very important example of such a larger system, which can be treated by the external field method, is the atom.

Let us first consider the case of a time-independent external field. An arbitrary configuration of the electrostatic and magnetostatic fields may be generated by an appropriate distribution of charges and magnetic moments (or current loops). Let me denote by ρ_{ext} and $\vec{m}_{ext}$ the corresponding charge density and the magnetism density. The total Hamiltonian will now have the form

$$H_{tot} = H_{ep} + H_{field} - \int \vec{B}\cdot\vec{m}_{ext} + H_{ext} \quad , \tag{58}$$

where H_{ext} describes the energy of the external sources of the field; it does not contain the electron and field degrees of freedom. Of course, the additional charges will also modify Gauss's,

$$\nabla\cdot\vec{E} = \rho + \rho_{ext} \quad . \tag{59}$$

Next, let me introduce the scalar potential V_{ext} and the vector potential $\vec{A}_{ext}$ generated by the external sources of the field,

$$- \Delta V_{ext} = \rho_{ext} \quad , \tag{60a}$$

$$\nabla \times \nabla \times \vec{A}_{ext} = \nabla\times\vec{m}_{ext} \quad . \tag{60b}$$

The Hamiltonian (58) will be subject to the following unitary transformation:

$$U = \exp(-i \int \vec{E}\cdot\vec{A}_{ext}) \quad , \tag{61}$$

whose purpose is to shift the vector potential operator by $\vec{A}_{ext}$,

$$\tilde{H} = U H_{tot} U^{-1} = H_{ep}[\vec{A} + \vec{A}_{ext}] + H_{field} - \tfrac{1}{2} \int \vec{m}^2_{ext} + H_{ext} \quad . \tag{62}$$

I have written here explicitly the dependence of the electron-positron Hamiltonian H_{ep} on the vector potential. The field Hamiltonian contains only the quantized $\vec{E}$ and $\vec{B}$ fields. For the static distribution of the external sources the last two terms may be dropped, because they only contribute a constant.

Finally, we have to restrict the Hamiltonian (62) to the gauge-invariant subspace, in order to obtain the Coulomb Hamiltonian. With the use of (59) and (60a), we obtain the well known expression:

$$\tilde{H}_C = H_{ep}[\ \vec{A} + \vec{A}_{ext}\ ] + H_{field} + \int \rho\, V_{ext} \ , \tag{63}$$

where I have omitted another constant--the energy of the electrostatic interaction of the external charges.

Time-dependent external fields may be introduced by assuming that the external Hamiltonian generates a nontrivial time evolution of the external charges and magnetic moments, which leads to a certain time variation of V_{ext} and $\vec{A}_{ext}$. This time variation may be treated as given a priori, if the back reaction on the charges and magnetic moments, generating the external field, may be neglected.

8. RELATIVISTIC CORRECTIONS

The method of unitary transformations is easily extended to give the relativistic corrections to the Hamiltonian (50). The lowest order relativistic corrections appear in the order $1/m^2$ of perturbation theory. The generator F of the unitary transformation has already been calculated to the desired order in Section 5. All we have to do now is to collect the terms of the order $1/m^2$ in the expansion (38). It turns out that such terms are produced only by the commutator $[\ F_2,\ H_1\ ]$. This commutator is slightly more complicated than (48), because it also contains a contribution from the commutation of $\vec{E}$ and $\vec{A}$,

$$[\ F_2,\ H_1\ ] = \frac{e}{4m^2} \int\int (\ [\ \phi^\dagger\ \vec{\sigma}\cdot\vec{E}\ \tilde{\chi}^\dagger\ ,\ \tilde{\chi}\ \vec{\sigma}\cdot\vec{D}\ \phi\ ] + \text{h.c.}\ )$$

$$+ \frac{e^2}{4m^2} \int (\ \phi^\dagger\vec{\sigma}\tilde{\chi}^\dagger \cdot \phi^\dagger\vec{\sigma}\tilde{\chi}^\dagger + \text{h.c.}\ ) \ , \tag{64}$$

where h.c. stands for the hermitian conjugate terms. The integral in the second line will be dropped, since it is off-diagonal and, therefore, it does not contribute to the diagonal part of the Hamiltonian in the order $1/m^2$. The same is also true about the double commutator $[\ F_1\ ,\ [F_1\ , H_1]]$. All such off-diagonal terms occurring in a given order of perturbation theory, in principle, should be eliminated by an appropriate unitary transformation, but after such a transformation we will obtain a contribution to the diagonal part at best in the next order. With the use of canonical commutation relations, the first integral in (64) may be rearranged as follows:

$$[\ F_2\ , H_1\ ] = -\frac{e}{4m^2} \int (\ \phi^\dagger\ [\vec{\sigma}\cdot\vec{D}\ , \vec{\sigma}\cdot\vec{E}\ ]\ \phi\ -\ \chi^\dagger\ [\vec{\sigma}\cdot D^*,\ \vec{\sigma}\cdot\vec{E}\ ]\ \chi\ )$$

$$+ \frac{e^2}{4m^2} \int \phi^\dagger\vec{\sigma}\tilde{\chi}^\dagger \cdot \tilde{\chi}\vec{\sigma}\phi - \frac{e^2}{4m^2}\, \text{Tr}([\vec{\sigma}\cdot\vec{D}\ ,\ \vec{\sigma}\cdot\vec{E}]) \ . \tag{65}$$

The first commutator on the right hand side gives:

$$[\ \vec{\sigma}\cdot\vec{D}\ ,\vec{\sigma}\cdot\vec{E}\ ] = \nabla\cdot\vec{E} + i(\vec{D}\times\vec{E})\cdot\vec{\sigma} - i(\vec{E}\times\vec{D})\cdot\vec{\sigma} - 3e\ \delta(0) \quad . \tag{66}$$

The analogous result for the second commutator is obtained by replacing e by $-e$.

Skipping some intermediate steps, I will write down the final expression for the Hamiltonian in the Coulomb form. To simplify the notation, I will be rather sloppy with the divergent expression--disregarding them all--and, consistently, I shall also identify m with the observed mass.

$$\begin{aligned} H &= mc^2 \int (\ \phi^\dagger\phi + \chi^\dagger\chi\ ) - \frac{\hbar^2}{2m} \int (\ \phi^\dagger\ (\vec{\sigma}\cdot\vec{D}_T)^2\phi + \chi^\dagger\ (\vec{\sigma}\cdot D_T^*)^2\chi\ ) \\ &+ \frac{1}{8\pi} \int\int \rho(\vec{r})\ |\vec{r} - \vec{r}\,'|^{-1}\ \rho(\vec{r}\,') + \tfrac{1}{2} \int (\ \vec{E}_T^2 + \vec{B}^2\ ) \\ &- \frac{ie\hbar^2}{8m^2c^2} \int (\ \phi^\dagger\ (\vec{D}_T\times\vec{E} - \vec{E}\times\vec{D}_T\ )\cdot\vec{\sigma}\ \phi - \chi^\dagger\ (\ \vec{D}_T^*\times\vec{E} - \vec{E}\times D_T^*\ )\cdot\vec{\sigma}\ \chi\ ) \\ &+ \frac{e^2\hbar^2}{4m^2c^2} \int \phi^\dagger\vec{\sigma}\tilde{\chi}^\dagger\ \cdot\ \tilde{\chi}\vec{\sigma}\phi - \frac{\hbar^2}{8m^2c^2} \int \rho(\vec{r})\ \rho(\vec{r}) \quad , \end{aligned} \tag{67}$$

where $\vec{E}$ contains both the transverse part and the longitudinal part determined by the application of Gauss's law,

$$\vec{E} = \vec{E}_T - \nabla \frac{1}{4\pi} \int |\vec{r} - \vec{r}\,'|^{-1}\ \rho(\vec{r}\,') \quad . \tag{68}$$

Following the procedure described in the previous Section, we may easily introduce the external field into the Hamiltonian (67).

Out of the three terms, which desciribe the relativistic corrections, the first term is the well-known spin-orbit coupling. In the quantum-mechanical notation it has the form:

$$\frac{e}{m^2c^2}\ ((\ \vec{p} - \frac{e}{c}\vec{A}\ ) \times \vec{E}\ )\cdot\vec{s} \quad , \tag{69}$$

from which we may see that it has a classical origin. In the two remaining terms Planck's constant cannot be absorbed into $\vec{p}$ and $\vec{s}$. They are of a quantum origin and cannot be obtained by the canonical quantization of the classical theory of charged particles.

Our Hamiltonian is still a QED Hamiltonian in the sense that the interaction with photons is fully included. We may perform another unitary transformation, which will eliminate all the terms linear in the photon creation and annihilation operators. This elimination will lead to new effective interaction terms describing in the instantaneous approximation the result of one-photon exchange.

9. HISTORICAL NOTES

There is a vast literature on many aspects of QED covered in my lectures and I will only mention here those contributions, which had the greatest influence on the subject, or at least on my view of the subject.

The version of the relativistic QED presented here is patterned after two classic papers of the founding fathers of quantum field theory--Heisenberg and Pauli.[1] In the second paper they introduced the concept of the gauge-invariant subspace, on which Gauss's law is valid.

My unitary transformations, leading to the nonrelativistic theory, are close relatives of the Foldy-Wouthuysen transformations in the Dirac theory of the relativistic electron.[2] There are two differences. The F-W transformations apply to the one-particle theory and they are studied in the presence of the external electromagnetic field only. The field-theoretic form of these transformations resembles the Bogolubov-Holstein-Primakoff transformations of the creation and annihilation operators, known from many-body theory.[3]

The problem of relativistic corrections to the nonrelativistic Hamiltonian has a long history, but early derivations were based either on classical theory or on the one-particle Dirac theory. The first derivation within the framework of QED was given by Itoh.[4] I believe that my derivation is more systematic and that the problems of gauge invariance are handled correctly.

---

[1]W. Heisenberg und W. Pauli, Zeitschrift f. Phys. 56, 1 (1929), 59, 168 (1930).

[2]J.D. Bjorken and S.D. Drell, Relativistic Quantum Mechanics, McGraw-Hill, 1964.

[3]A.L. Fetter and J.D. Walecka, Quantum Theory of Many-Particle Systems, McGraw-Hill, 1971.

[4]T. Itoh, Reviews of Mod. Phys. 37, 159 (1965).

Appendix A

My relativistic metric tensor is:

$$g_{\mu\nu} = \text{diag}\ (1,-1,-1,-1) \quad . \tag{A1}$$

The Dirac matrices are all used in the Dirac representation,

$$\beta = \rho_3 \times I \quad , \qquad \vec{\alpha} = \rho_1 \times \vec{\sigma} \quad , \tag{A2}$$

$$\gamma^0 = \beta \quad , \qquad \vec{\gamma} = \beta\,\vec{\alpha} \quad . \tag{A3}$$

I use the Heaviside-Lorentz system of electromagnetic units, in which the factor $1/4\pi$ appears in Coulomb's law, but not in Maxwell's equations.

The basic tool in the method of unitary transformations used in these lectures is the following operator identity:

$$e^{\lambda A}\, B\, e^{-\lambda A} = B + \lambda\, [A , B] + \frac{\lambda^2}{2!}\, [A, [A , B]] + \ldots, \tag{A4}$$

easily proven by differentiating both sides with respect to λ.

In the evaluation of several commutators, the following operator identities were found useful:

$$[AB , C] = A \{B , C\} - \{ C , A \} B \quad , \tag{A5}$$

$$[AB , CD] = A \{ B , C \} D - C \{ D , A \} B - ACBD + CADB \quad . \tag{A6}$$

NONPERTURBATIVE METHODS IN BOUND STATE QUANTUM ELECTRODYNAMICS

Peter J. Mohr

J.W. Gibbs Laboratory
Physics Department
Yale University
New Haven, CT 06520

INTRODUCTION

Quantum electrodynamics has been successful in all areas where it has been confronted with experiment. The most precise tests have been made in the regime where particles interact weakly either with each other as in hydrogen or muonium, or with an external magnetic field as in measurements of the electron or muon magnetic moment. For the calculation of the electron magnetic moment, the theory appears to be completely perturbative in the sense that Feynman diagrams with n virtual photons are of order $(\alpha/\pi)^n$ where α is the fine structure constant.[1] For bound state calculations, quantum electrodynamics is perturbative in α, but no longer strictly perturbative in the binding strength $Z\alpha$ as shown by the logarithmic terms in that parameter. However it is still possible to obtain accurate predictions for bound state energy levels by making an asymptotic expansion in powers of $Z\alpha$. A less precisely explored area is the more strongly bound regime in which Z is large and the asymptotic series in $Z\alpha$ is poorly convergent numerically. In this case, it is necessary to have expressions for quantum electrodynamics corrections which do not rely on expansion in $Z\alpha$, as provided by the Furry bound state interaction picture formalism. Methods for obtaining numerical predictions for radiative level shifts from the formal expressions in this formulation are described in this review, and a comparison of theory and experiment is made.

Another nonperturbative application of quantum electrodynamics is in the study of perturbed decaying states. Here, level widths and shifts proportional to the fine structure constant appear in energy denominators, so that infinite sums of self energy diagrams are needed. A convenient formulation of this problem based on Low's analysis of line shape is discussed in this review.

BASIC THEORY

Bound state quantum electrodynamics is formulated in the Furry bound interaction picture, where the potential binding the electron is taken into account completely by employing solutions to the Dirac equation for that potential.[2] In the Furry picture, the electron-positron field operator is expanded in terms of creation and destruction operators for electrons and positrons in bound and continuum states of the external potential (units in which $\hbar = c = m = 1$ are employed here)

$$\psi(x) = \sum_{n+} b_n \phi_n(\vec{x}) e^{-iE_n t} + \sum_{n-} d_n^* \phi_n(\vec{x}) e^{-iE_n t} \tag{1}$$

where $\phi_n(\vec{x})$ is a solution of the Dirac equation

$$[-i\vec{\alpha} \cdot \vec{\nabla} + V(\vec{x}) + \beta - E_n]\phi_n(\vec{x}) = 0. \tag{2}$$

In (1), b_n is a destruction operator for an electron in a positive energy state (n+), and d_n^* is a creation operator for a positron in a negative energy state (n-) in the external potential. The creation and destruction operators satisfy the usual Fermi anticommutation relations. Electron bound states are produced by creation operators acting on the vacuum

$$|n\rangle = \sum_{i_1 \cdots i_k} c_{i_1 \cdots i_k} b_{i_1}^* \cdots b_{i_k}^* |0\rangle. \tag{3}$$

For multi-electron states in a spherically symmetric potential, appropriate linear combinations of creation operators can be formed to produce states of well defined angular momentum.

Energy levels in the Furry picture are given by a perturbation expansion in α. The zero-order Hamiltonian for positrons and electrons interacting only with the binding field is

$$H_o = \int d\vec{x} : \psi^\dagger(x) H \psi(x) : = \sum_{n+} E_n b_n^* b_n - \sum_{n-} E_n d_n^* d_n . \tag{4}$$

The interaction Hamiltonian coupling the electron-positron current to the radiation field is

$$H_I(t) = \int d\vec{x} \mathcal{H}_I(x) \tag{5}$$

where

$$\mathcal{H}_I(x) = -\tfrac{1}{2}e[\bar{\psi}(x)\gamma^u, \psi(x)]A_\mu(x) - \tfrac{1}{2}\delta m[\bar{\psi}(x), \psi(x)]. \tag{6}$$

The zero-order energy for a bound state is

$$H_o|n_1,n_2,\ldots> = (E_{n_1} + E_{n_2} + \ldots)|n_1,n_2,\ldots>. \tag{7}$$

If, following Gell-Mann and Low, we define the energy level shift due to the interaction term in the Hamiltonian to be[3]

$$\Delta E_n = \lim_{\substack{\varepsilon\to 0\\ \lambda\to 1}} \frac{<n|U_\varepsilon(\infty,0)[H_o + \lambda H_I(0) - E_n]U_\varepsilon(0,-\infty)|n>}{<n|U_\varepsilon(\infty,-\infty)|n>} \tag{8}$$

where U_ε is the adiabatic time development operator

$$i\frac{\partial}{\partial t_2} U_\varepsilon(t_2,t_1) = \lambda e^{-\varepsilon|t_2|} H_I(t_2)U_\varepsilon(t_2,t_1) \tag{9}$$

that has the perturbation expansion

$$U_\varepsilon(t_2,t_1) = \sum_{j=0}^{\infty} U_\varepsilon^{(j)}(t_2,t_1) \tag{10}$$

where

$$U_\varepsilon^{(j)}(t_2,t_1) = \frac{(-i\lambda)^j}{j!}\int_{t_1}^{t_2} d^4x_j \cdots \int_{t_1}^{t_2} d^4x_1 e^{-\varepsilon|t_j|} \cdots e^{-\varepsilon|t_1|} \times T[\mathcal{H}_I(x_j)\cdots\mathcal{H}_I(x_1)], \tag{11}$$

then, according to the theorem of Gell-Mann and Low,[3] in the symmetric form of Sucher,[4] the level shift is given by

$$\Delta E_n = \lim_{\substack{\varepsilon\to 0\\ \lambda\to 1}} \frac{i\varepsilon}{2} \frac{\frac{\partial}{\partial\lambda}<n|S_\varepsilon|n>_c}{<n|S_\varepsilon|n>_c} + \text{constant} \tag{12}$$

where

$$S_\varepsilon = U_\varepsilon(\infty,-\infty). \tag{13}$$

In (12), the subscript c denotes the fact that only connected diagrams are included. This development of the energy level shifts leads to a Feynman diagrammatic representation of the perturbation expansion as in ordinary free particle electrodynamics except that here the electron-positron lines represent bound particles with the appropriate bound particle propagation function

$$S_F(x_2,x_1) = \begin{cases} \sum_{n+} \phi_n(\vec{x}_2)\overline{\phi}_n(\vec{x}_1)e^{-iE_n(t_2-t_1)} & t_2 > t_1 \\ -\sum_{n-} \phi_n(\vec{x}_2)\overline{\phi}_n(\vec{x}_1)e^{-iE_n(t_2-t_1)} & t_2 < t_1 \end{cases} \tag{14}$$

The main difficulty in the calculation of level shifts for bound particles is in the evaluation of the electron bound propagation function.

This evaluation is facilitated by applying the method of Wichmann and Kroll to obtain a convenient expression for the propagation function.[5] The propagation function can be written as

$$S_F(x_2,x_1) = \frac{1}{2\pi i}\int_{C_F} dz \sum_n \frac{\phi_n(\vec{x}_2)\overline{\phi}_n(\vec{x}_1)}{E_n - z} e^{-iz(t_2-t_1)} \tag{15}$$

where C_F is the Feynman contour, and the sum over states in the integrand is the Green's function

$$G(\vec{x}_2,\vec{x}_1,z) = \sum_m \frac{\phi_m(\vec{x}_2)\phi_m^\dagger(\vec{x}_1)}{E_m - z} \tag{16}$$

which satisfies the equation

$$[-i\vec{\alpha}\cdot\vec{\nabla}_2 + V(\vec{x}_2) + \beta - z]G(\vec{x}_2,\vec{x}_1,z) = I\delta(\vec{x}_2 - \vec{x}_1) \tag{17}$$

For a spherically symmetric binding potential, the eigenfunctions are eigenfunctions of the Dirac angular momentum operator

$$K = \beta(\vec{\sigma}\cdot\vec{L} + I) \tag{18}$$

with eigenvalue $-\kappa$. The wave functions factorize into radial and spin-angular parts as

$$\phi_n(\vec{x}) = \begin{bmatrix} f_1(x)\chi_\kappa^\mu(\hat{x}) \\ if_2(x)\chi_{-\kappa}^\mu(\hat{x}) \end{bmatrix} \tag{19}$$

In terms of these factors, the Green's function is

$$G(\vec{x}_2,\vec{x}_1,z) = \sum_{n\kappa\mu} \frac{1}{E_n - z}$$

$$\times \begin{bmatrix} f_1(x_2)\chi_\kappa^\mu(\hat{x}_2)f_1(x_1)\chi_\kappa^{\mu\dagger}(\hat{x}_1) & -if_1(x_2)\chi_\kappa^\mu(\hat{x}_2)f_2(x_1)\chi_{-\kappa}^{\mu\dagger}(\hat{x}_1) \\ if_2(x_2)\chi_{-\kappa}^\mu(\hat{x}_2)f_1(x_1)\chi_\kappa^{\mu\dagger}(\hat{x}_1) & f_2(x_2)\chi_{-\kappa}^\mu(\hat{x}_2)f_2(x_1)\chi_{-\kappa}^{\mu\dagger}(x_1) \end{bmatrix} \tag{20}$$

The sum over radial energy eigenfunctions for a given value of κ is denoted by

$$G_\kappa^{ij}(r_2,r_1,z) = \sum_n \frac{1}{E_n - z} f_i(r_2)f_j(r_1) \tag{21}$$

Then from Eq. (17), the G's satisfy the radial equation

$$\begin{bmatrix} V(r_2) + 1 - z & -\frac{1}{r_2}\frac{d}{dr_2} r_2 + \frac{\kappa}{r_2} \\ \frac{1}{r_2}\frac{1}{dr_2} r_2 + \frac{\kappa}{r_2} & V(r_2) - 1 - z \end{bmatrix} \begin{bmatrix} G_\kappa^{11} & G_\kappa^{12} \\ G_\kappa^{21} & G_\kappa^{22} \end{bmatrix} = I \frac{1}{r_2 r_1} \delta(r_2 - r_1) \tag{22}$$

For the case of the Coulomb potential, $V(x)=-\alpha/x$, the two-component solutions to the homogeneous version of (22) are

$$F_<(r) = \begin{bmatrix} \frac{\sqrt{1+z}}{2ar^{3/2}} [(\lambda-\nu)M_{\nu-\frac{1}{2},\lambda}(2ar) - (\kappa - \frac{\gamma}{a})M_{\nu+\frac{1}{2},\lambda}(2ar)] \\ \frac{\sqrt{1-z}}{2ar^{3/2}} [(\lambda-\nu)M_{\nu-\frac{1}{2},\lambda}(2ar) + (\kappa - \frac{\gamma}{a})M_{\nu+\frac{1}{2},\lambda}(2ar)] \end{bmatrix} \tag{23}$$

which is regular at $x=0$, and

$$F_>(r) = \frac{\Gamma(\lambda-\nu)}{\Gamma(1+2\lambda)} \begin{bmatrix} \frac{\sqrt{1+z}}{2ar^{3/2}} [(\kappa+\frac{\gamma}{a})W_{\nu-\frac{1}{2},\lambda}(2ar) + W_{\nu+\frac{1}{2},\lambda}(2ar)] \\ \frac{\sqrt{1-z}}{2ar^{3/2}} [(\kappa+\frac{\gamma}{a})W_{\nu-\frac{1}{2},\lambda}(2ar) - W_{\nu+\frac{1}{2},\lambda}(2ar)] \end{bmatrix} \tag{24}$$

which is regular at $x=\infty$, where

$$a = (1-z^2)^{1/2}; \ \gamma = Z\alpha; \ \nu = \gamma z/a; \ \lambda = (\kappa^2 - \gamma^2)^{1/2} \tag{25}$$

The Green's function is then constructed by writing

$$G_\kappa(r_2,r_1,z) = \theta(r_2-r_1)F_>(r_2)F_<^T(r_1) + \theta(r_1-r_2)F_<(r_2)F_>^T(r_1) \tag{26}$$

which is valid for appropriately normalized homogeneous solutions. The problem of evaluation of the electron propagation function is thereby reduced to one of numerically evaluating the functions M and W. This can be done by i. numerical integration of the differential equation in (22), ii. evaluating power series and asymptotic expansions for the M and W functions, or iii. by numerical integration of integral representations for the M and W functions.

Thus for the one-photon self-energy level shift

$$E_{SE}^{(2)} = -4\pi i\alpha \int d(t_2-t_1)\int d^3x_2 \int d^3x_1 \; D_F(x_2-x_1)$$
$$\times \bar{\phi}_n(\vec{x}_2)e^{iE_nt_2} \gamma_\mu S_F(x_2,x_1)\gamma^\mu \phi_n(\vec{x}_1)e^{-iE_nt_1} - \delta m\int d^3x \bar{\phi}_n(\vec{x})\phi_n(\vec{x}) \tag{27}$$

the calculation reduces to evaluation of expressions of the form

$$-\frac{i\alpha}{2\pi}\int_C dz \int_0^\infty dx_2 x_2^2 \int_0^\infty dx_1 x_1^2 \sum_\kappa \sum_{i,j=1}^{2} f_i(x_2) G_\kappa^{ij}(x_2,x_1,z) f_j(x_1) A_\kappa(x_2,x_1)$$
$$+ \; . \; . \; . \tag{28}$$

where for S states

$$A_\kappa(x_2,x_1) = ikbj_\kappa(bx_<)h_\kappa^{(1)}(bx_>) \tag{29}$$

for $\kappa > 0$, and $b = \sqrt{(E_n-z)^2+i\varepsilon}$

To carry out mass renormalization, the level shift is regulated by employing the photon propagator

$$D_F(x_2-x_1) = \frac{i}{(2\pi)^4}\int d^4k \; e^{-ik\cdot(x_2-x_1)}\left(\frac{1}{k^2+i\varepsilon} - \frac{1}{k^2-\Lambda^2+i\varepsilon}\right) \tag{30}$$

in Eq. (27). To isolate the divergent terms, the Green's function may be expanded as

$$G(z) = \frac{1}{H_o+V-z} \simeq \frac{1}{H_o-z} - \frac{1}{H_o-z}V\frac{1}{H_o-z} + \; . \; . \; .$$
$$= \frac{1}{\vec{\alpha}\cdot\vec{p}+\beta-z} - V\frac{1}{p^2+1-z^2} - z(\beta+z)V\frac{2}{[p^2+1-z^2]^2} + \; . \; . \; . \tag{31}$$

Here $H_o = \vec{\alpha}\cdot\vec{p}+\beta$. The leading terms are evaluated and renormalized in momentum-space, and the finite remainders are evaluated numerically as described above. Calculations of the ground state self energy in a Coulomb potential or a finite nucleus potential have been done with this general approach by several groups.[6-10] The results are in good overall agreement. In the following two sections, the self-energy results for n=2 states are compared to experiments in one- and two-electron atoms.

ONE-ELECTRON ATOMS

Theoretical energy levels in one-electron atoms are determined mainly by the Dirac eigenvalue, reduced-mass corrections, and Lamb-shift effects, i.e., the effects contributing to the $2P_{1/2}-2S_{1/2}$ splitting. The most precise present day test of quantum electrodynamics in high-Z one-electron atoms is measurement of the Lamb shift, either by laser spectroscopy or by electric field quenching of the metastable $2S_{1/2}$ state.[11] In this section, the theoretical contributions to the Lamb shift in high-Z one-electron atoms are compared to experiment.[12-15]

The largest radiative correction to electron energy levels is the self-energy correction of order α corresponding to the Feynman diagram in Fig. 1(a). The level shift is expressed as

$$E_{SE}^{(2)} = \frac{\alpha}{\pi} \frac{(Z\alpha)^4}{n^3} F(Z\alpha) m_e c^2 \tag{32}$$

where $F(Z\alpha)$ is a slowly varying function of Z. This function has been calculated with methods described in the preceding section The results for the states with n=2 are plotted in Fig. 2. Values for intermediate Z are obtained by interpolation.

The vacuum-polarization level shift of order α, corresponding

Fig. 1. Feynman diagrams for the (a) self energy and (b) vacuum polarization corrections in one-electron atoms.

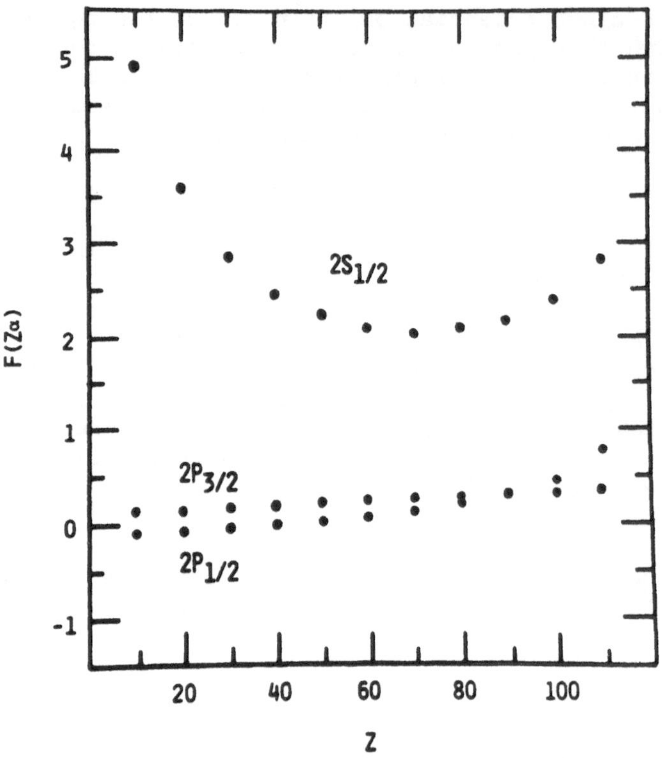

Fig. 2. Self energy level shift in one-electron atoms.

to the Feynman diagram in Fig. 1(b), is expressed as

$$E_{VP}^{(2)} = \frac{\alpha}{\pi}\frac{(Z\alpha)^4}{n^3}H(Z\alpha)m_e c^2 \tag{33}$$

The function $H(Z\alpha)$ can be expanded in the series

$$H(Z\alpha) = H_1(Z\alpha) + H_3(Z\alpha) + H_5(Z\alpha) + \dots \tag{34}$$

where $H_1(Z\alpha)$ arises from the expectation value of the Uehling potential $V_1(r)$,[16-17] and the higher-order terms $H_i(Z\alpha)$ arise from expectation values of vacuum-polarization operators $V_i(r)$

that are of order i in the external Coulomb potential.[5] Precise expectation values of the Uehling potential evaluated with Dirac-Coulomb wave functions have been tabulated for Z=10,20,..., 110.[13] Intermediate Z results are obtained by interpolation. Wichmann and Kroll have examined the vacuum-polarization effects that are of third and higher order in the external Coulomb potential. An evaluation of $<V_3>$ in powers of $Z\alpha$, based on results of their calculation yields accurate values for $H_3(Z\alpha)$ for the range of Z under consideration.[5,12] For the contribution to the Lamb shift one has

$$\Delta H_3(Z\alpha) = (Z\alpha)^2[0.0567-0.1374(Z\alpha) + 0.0580(Z\alpha)^2 \ln(Z\alpha)^{-2} + \ldots] \tag{35}$$

Radiative corrections with two virtual photons, of order α^2, have been calculated to lowest order in $Z\alpha$. They are characterized as vacuum polarization,[18,19] magnetic-moment,[20-22] and self-energy corrections.[23-30] The lowest-order total correction to the Lamb shift is

$$E_{HO} = \left(\frac{\alpha}{\pi}\right)^2 \frac{(Z\alpha)^4}{6} mc^2 \left[\pi^2 \ln 2 - \frac{37\pi^2}{144} - \frac{3767}{1728} - \frac{3}{2}\zeta(3)\right] \tag{36}$$

The field of the nucleus deviates from Coulomb behavior at short distances due to the finite charge radius of the nucleus. The corresponding shift in electron binding energy primarily affects states with j=1/2. A nonperturbative analysis for a uniformly charged nucleus model leads to a compact formula for the level shift that neglects terms of relative order $(Z\alpha)^4$ and $Z\alpha R/\lambda\!\!\!^-$

$$E_{NS} = [1+1.19(Z\alpha)^2] \frac{1}{12}(Z\alpha)^2 \left(\frac{Z\alpha R}{\lambda\!\!\!^-}\right)^{2s} mc^2 \tag{37}$$

where $s = [1-(Z\alpha)^2]^{1/2}$ and R is the rms charge radius of the nucleus.[12,31] It is of interest to note that the relativistic corrections are not completely given by first-order perturbation theory.

Relativistic recoil corrections are of order Zm/M relative to the lowest-order radiative corrections. To lowest order in $Z\alpha$, the correction is[32-35]

$$E_{RR} = \frac{m}{M} \frac{(Z\alpha)^5}{8\pi} mc^2 \left[\frac{1}{3}\ln(Z\alpha)^{-2} + \frac{8}{3}B(2,0) - \frac{8}{3}B(2,1) + \frac{97}{9}\right] \tag{38}$$

where $B(n,\ell)$ is the Bethe sum[36]

$$B(2,0) = -2.8118 \qquad B(2,1) = 0.0300 \tag{39}$$

Table I: Comparison of Theory and Experiment for the Lamb Shift for Z = 15,17,18, in Units of THz

Z	15	17	18
Self energy	21.393(5)	33.194(6)	40.546(7)
Vacuum polarization	-1.262(0)	-2.072(0)	-2.597(1)
Higher order	0.005(7)	0.008(11)	0.011(14)
Nuclear size	0.112(1)	0.206(3)	0.276(2)
Relativistic recoil	0.007(6)	0.010(9)	0.012(11)
Total	20.254(10)	31.347(16)	38.246(19)
Experiment	20.18(25)[37]	31.19(22)[38]	38.0(6)[39]

The various contributions to the Lamb shift in hydrogenlike atoms with Z = 15, 17, and 18 are summarized in Table I, where the results are compared to experiment. The estimated theoretical uncertainties are discussed in detail in Ref. 12. There is satisfactory agreement between theory and experiment.

TWO-ELECTRON ATOMS

Strong external field methods can be extended straightforwardly to atoms with more than one electron. In few-electron atoms with large Z, the electron-nucleus interaction dominates over the electron-electron interaction, so the natural zero-order approximation for the strong external field approach is one in which the bound electrons are noninteracting. Interaction with the quantized electromagnetic field generates the electron-electron interactions and the radiative corrections as perturbations. This approach produces a power series in 1/Z for energy levels, which is rapidly convergent for high-Z systems. Since the zero-order solutions are solutions to the Dirac equation in the external field, they are completely relativistic with no approximation in the parameter $Z\alpha$. For the two-electron case the leading terms in this perturbation expansion are known and give good agreement with experiment. In this section the theoretical contributions to the $2^3S_1-2^3P_0$ splitting, the two-electron analog of the Lamb shift, are compiled and compared to experiment.

The Feynman diagrams in the perturbation expansion fall into two categories: one-electron terms where photons interact with only one of the two external electron lines and two-electron terms where photons interact with both electron lines. The one-electron terms give the same level splitting as the one-electron corrections compiled in the preceding section. The leading Feynman diagrams are shown in Fig. 3. The two-electron one exchanged photon graph in that figure is accurately given by[40,41]

$$\Delta E_{pe}^{(2)}(2^3S_1-2^3P_0)=\alpha(Z\alpha)m_e c^2\left[\frac{248}{6561}+0.1428(Z\alpha)^2+0.1(Z\alpha)^4+\ldots\right] \quad (40)$$

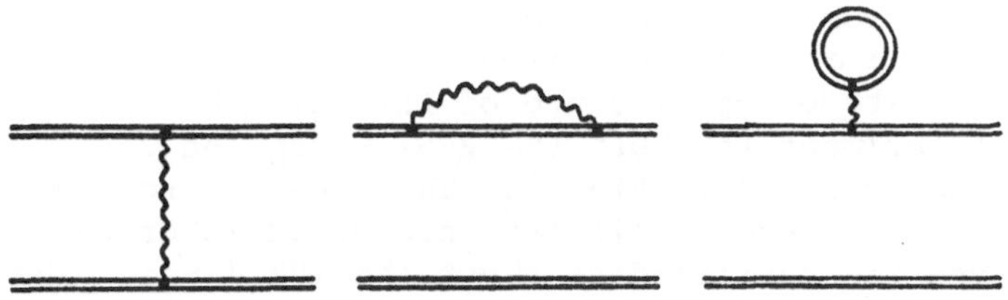

Fig. 3. Second-order Feynman diagrams for high-Z two-electron atoms.

for the $2^3S_1-2^3P_0$ splitting for the range of interest here ($10 \leq Z \leq 30$). The two exchanged photon diagrams in Fig. 4 have been approximately evaluated by variational calculations[42-47]

$$\Delta E_{pe}^{(4)}(2^3S_1 - 2^3P_0) = \alpha^2 m_e c^2[-0.0256 - 0.27(Z\alpha)^2 + \ldots] \quad (41)$$

The nonrelativistic limit of the three exchanged photon diagrams is also known from variational calculations[42-44]

$$\Delta E_{pe}^{(6)}(2^3S_1 - 2^3P_0) = \alpha^3(Z\alpha)^{-1} m_e c^2[-0.0117 + \ldots] \quad (42)$$

The largest uncalculated correction is the screening correction to the one-electron self energy or equivalently the vertex correction to the one exchanged photon diagram. The total splitting is

$$\Delta E(2^3S_1 - 2^3P_0) = -S + \Delta E_{pe}^{(2)} + \Delta E_{pe}^{(4)} + \Delta E_{pe}^{(6)} + \ldots \quad (43)$$

where S is the one-electron Lamb shift discussed in the preceding

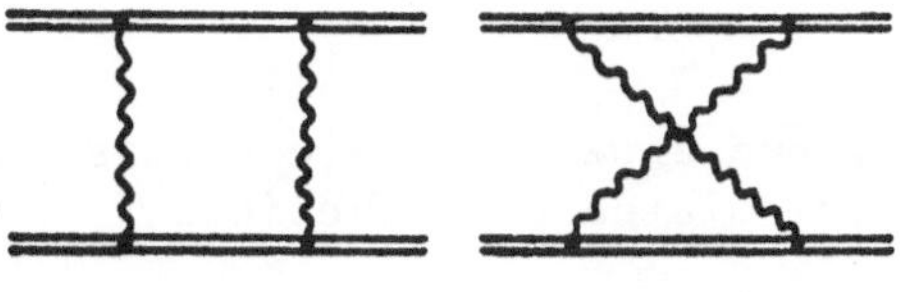

Fig. 4. Fourth-order Feynman diagrams with two exchanged photons.

section and the remaining terms are the dominant two-electron corrections. The orders of magnitude of Eqs. (40), (41), and (42) indicate the rate of convergence of the perturbation expansion. Theory and experiment for both the $2^3S_1 - 2^3P_0$ and $2^3S_1 - 2^3P_2$ splittings are compared in Fig. 5. The one-electron radiative corrections to the level splittings are about 1% of the total splittings and are confirmed at about the 10% level. The differ-

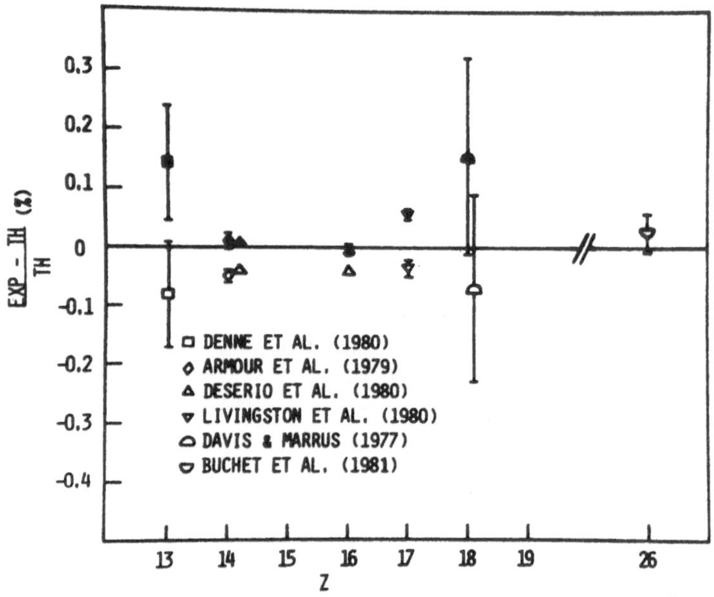

Fig. 5. Comparison of experiment and theory for the 2^3S_1-2^3P_0 and 2^3S_1-2^3P_2 energy level separations in high-Z two-electron atoms. The filled (open) shapes correspond to the 2^3S_1-2^3P_0 (2^3S_1-2^3P_2) transitions. The experimental values are from Ref. 48-53.

ences between theory and experiment are consistent with the magnitude of the uncalculated contributions.

UNSTABLE STATE PERTURBATION THEORY

One of the important topics in quantum electrodynamics and

quantum optics is the spectral line shape of quantum systems due to their interaction with virtual or real electromagnetic radiation. The natural width of an atomic level, due to interaction with virtual radiation, is proportional to the decay rate of the level, as shown in the classic paper of Weisskopf and Wigner.[54] The decay rate Γ appears in the imaginary part of a complex shift in the energy level $E \to E + \Delta E - i\Gamma/2$ which gives rise to the characteristic spectral line shape

$$\frac{1}{(k - E - \Delta E)^2 + \Gamma^2/4} \tag{44}$$

as a function of photon energy k. Weisskopf and Wigner also pointed out that there is a divergent integral in the real part of the level shift ΔE, which is known to be associated with the radiative correction, or Lamb shift, of the level. Low has formulated the question of natural line shape in the framework of modern renormalized quantum electrodynamics, in a way which gives an unambiguous finite result for both the real and imaginary parts of the level shift.[55] If the unstable level is weakly coupled to other unstable levels by a perturbation, then not only will the line shape be changed, but other properties of the level, such as the angular distribution of emitted radiation, will be modified. An example is the metastable $2S_{1/2}$ state of a hydrogenlike atom that is mixed with the nearby 2P states by a weak electric field. Hillery and Mohr have generalized Low's approach to describe this mixing within the framework of bound-state quantum electrodynamics.[56] In the following, Low's formulation of line shape is reviewed, and the generalization to weakly coupled resonant levels is described.

In order to formulate the line shape question in the framework of the S matrix of quantum electrodynamics, Low considers the unstable state as a resonance in a scattering process. An atom initially in the 1S state is excited by an effective potential and subsequently emits a photon in a transition back to the 1S state. If the excitation mechanism has a flat frequency spectrum near a particular resonance, then the emitted photon spectrum exhibits the line shape of the resonance. A Feynman diagram for this process is shown in Fig. 6. In that figure, the double line represents an electron bound to the nucleus in the ground state. The triple line represents the propagation function S_F' for an electron bound to the nucleus and interacting with the virtual radiation field. This propagation function is given in terms of the Coulomb propagation function S_F by

$$S_F' = S_F - iS_F \Sigma S_F' \tag{45}$$

The operator Σ is the renormalized sum over proper self energy and vacuum polarization corrections. Equation (45) sums the contribu-

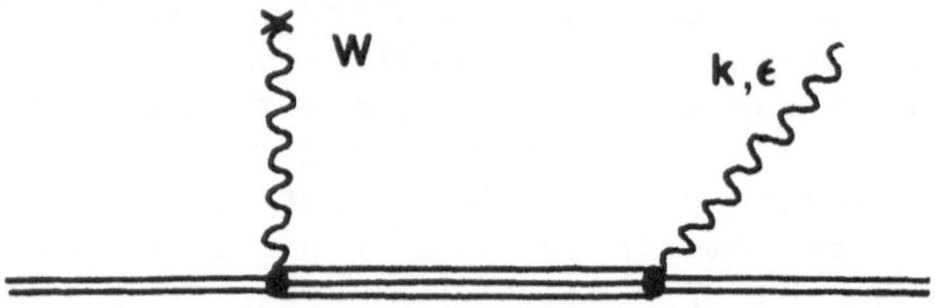

Fig. 6. Feynman diagram for excitation of an atom by an effective potential W, followed by emission of a photon.

tion of Σ to all orders. The complete propagator S_F' determines the line shape of the emitted photon. To obtain a convenient expression for S_F' we write Eq. (15) in the form

$$S_F(t_2-t_1) = \frac{1}{2\pi i}\int_{-\infty}^{\infty} d\omega \sum_n \frac{|n><n|\gamma^o}{E_n-\omega(1+i\varepsilon)}\, e^{-i\omega(t_2-t_1)} \tag{46}$$

and write a general expansion of S_F' as

$$S_F'(t_2-t_1) = \frac{1}{2\pi i}\int_{-\infty}^{\infty} d\omega \sum_{nm} f_{nm}(\omega)\,|n><m|\gamma^o\, e^{-i\omega(t_2-t_1)} \tag{47}$$

Then Eq. (45) yields

$$\sum_\ell \{[E_n - \omega(1+i\varepsilon)]\delta_{n\ell} + \Sigma_{n\ell}(\omega)\} f_{\ell m}(\omega) = \delta_{nm} \tag{48}$$

where

$$\Sigma_{nm}(\omega) = \int_{-\infty}^{\infty} dt\, e^{i\omega t} <n|\gamma^o\Sigma(t)|m> \tag{49}$$

The scattering in Fig. 6 corresponds to the matrix element

$$M=-2\pi ie\left(\frac{2\pi}{k}\right)^{\frac{1}{2}} <1S|\vec{\alpha}\cdot\hat{\varepsilon}_\lambda e^{-i\vec{k}\cdot\vec{x}} \sum_{nm} f_{nm}(E_{1S}+k)\,|n> \quad <m|\mathcal{W}(\vec{x},k)|1S> \tag{50}$$

where $\vec{k}$ and $\hat{\varepsilon}_\lambda$ are the momentum and polarization vectors for the emitted photon and

$$W(\vec{x},t) = \int_{-\infty}^{\infty} d\omega\, e^{-i\omega t}\, \mathcal{W}(\vec{x},\omega) \tag{51}$$

is the effective excitation potential. A first approximation for the functions $f_{nm}(\omega)$ is obtained by neglecting off-diagonal terms in $\Sigma_{nm}(\omega)$. Low justifies this approximation and solves for higher-order corrections in Ref. 55. With this approximation

$$f_{nm}(\omega) \approx \frac{\delta_{nm}}{E_n + \Sigma_{nn}(\omega) - \omega(1+i\varepsilon)} . \tag{52}$$

The resonance center is near the real value of k fixed by

$$E_n + \mathrm{Re}\,\Sigma_{nn}(E_{1S}+k) - (E_{1S}+k) = 0 \tag{53}$$

or approximately by

$$E_n + \mathrm{Re}\,\Sigma_{nn}(E_n) - (E_{1S}+k) = 0 . \tag{54}$$

Thus near the resonance

$$|M|^2 \propto \frac{1}{(E_n + \Delta E_n - E_{1S} - k)^2 + \Gamma_n^2/4} \tag{55}$$

where

$$\begin{aligned} \Delta E_n &= \mathrm{Re}\,\Sigma_{nn}(E_n) \\ \Gamma_n &= -2\,\mathrm{Im}\,\Sigma_{nn}(E_n) \end{aligned} \tag{56}$$

are the radiative level shift and width.

To study weakly coupled unstable levels, we consider the example of a metastable hydrogenlike atom in a weak constant electric field. Other approaches are discussed in Refs. 57-63. The mixing with the 2P states caused by the electric field affects the decays of the 2S state in a way that depends crucially on the radiative shifts of the levels. To account for these radiative shifts in a consistent description, we apply a generalization of Low's method. Here, the electric field is treated along with the radiative corrections in the expression for $S_F^{\frac{1}{2}}$ i.e., the operator Σ is replaced by $U = \Sigma + V$ where $V(x_2,x_1) = \delta^4(x_2-x_1)\gamma^0 e\vec{E}\cdot\vec{x}_2$, with $-e$ the electron charge and $\vec{E}$ the applied electric field. Radiative corrections to V are neglected. Let

$$V_{nm} = \langle n|e\vec{E}\cdot\vec{x}|m\rangle . \tag{57}$$

Then we obtain

$$\sum_{\ell}\{[E_n-\omega(1+i\varepsilon)]\delta_{n\ell}+\Sigma_{n\ell}(\omega)+V_{n\ell}\}f_{\ell m}(\omega) = \delta_{nm} \tag{58}$$

for the expansion coefficients in Eq. (47). To solve (58), we make the approximation that the matrix elements $\Sigma_{n\ell}(\omega)+V_{n\ell}$ are neglected unless n and ℓ are both in the set of states $S=\{2S_{1/2}, 2P_{1/2}, 2P_{3/2}\}$. This three-level approximation is justified in Ref. (56). The solutions $f_{nm}(\omega)$ are obtained by inverting the infinite dimensional matrix in curly brackets in Eq, (58). The matrix is diagonal except for matrix elements between states in the subset S. Hence for n or m not in the set S

$$f_{nm}(\omega) = \frac{\delta_{nm}}{E_n-\omega(1+i\varepsilon)} . \tag{59}$$

For n and m in the set S, the $f_{nm}(\omega)$'s are obtained by inverting the 3 × 3 matrix equation in the basis where the azimuthal quantization axis is chosen to be in the direction of the electric field $\vec{E}$. The result is an expression that factorizes in the neighborhood of the $2S_{1/2}$ resonance (denoting $2S_{1/2}$, $2P_{1/2}$, $2P_{3/2}$, by s,p,q, respectively)

$$\sum_{n,m\in S} f_{nm}(\omega)|n\rangle\langle m|$$

$$\propto \frac{1}{E''_s-\omega}\sum_{\mu=\pm\frac{1}{2}}\left(|s,\mu\rangle+\frac{V^{\mu}_{ps}}{E''_s-E'_p}|p,\mu\rangle+\frac{V^{\mu}_{qs}}{E''_s-E'_q}|q,\mu\rangle\right) \tag{60}$$

$$\times\left(\langle s,\mu|+\frac{V^{\mu}_{sp}}{E''_s-E'_p}\langle p,\mu|+\frac{V^{\mu}_{sq}}{E''_s-E'_q}\langle q,\mu|\right)$$

where $E'_n = E_n + \Sigma_{nn}(\omega)$ and the E''_n are the roots of

$$(E'_s-x)(E'_p-x)(E'_q-x)-(E'_q-x)|V^{(\frac{1}{2})}_{ps}|^2-(E'_p-x)|V^{(\frac{1}{2})}_{qs}|^2 = 0 \tag{61}$$

with $E''_n \to E'_n$ as $V \to 0$; $V^{(\mu)}_{nm} = \langle n,\mu|e\vec{E}\cdot\vec{x}|m,\mu\rangle$, and μ is the azimuthal quantum number.

The photon emission amplitude is obtained by substituting this expression into Eq. (50), and the emission line shape is determined mainly by the factor

$$\left|\frac{1}{E''_s-E_{1S}-k}\right|^2 = \frac{1}{(\mathrm{Re}\ E''_s-E_{1S}-k)^2+(\mathrm{Im}\ E''_s)^2} \tag{62}$$

in the square of the amplitude. The decay rate R for the resonance is related to the imaginary part of the pole in k by

$$R = -2 \text{ Im } E_s'' \tag{63}$$

evaluated at $k = E_s'' - E_{1S}$. Solving Eq. (61) for E_s'' to second order in V and making the approximation

$$E_n' \approx E_n + \Sigma_{nn}(E_n) = E_n + \Delta E_n - i\Gamma_n/2 \tag{64}$$

where ΔE_n and $-\Gamma_n/2$ are the real and imaginary parts of $\Sigma_{nn}(E_n)$, we have

$$R = \Gamma_s + \frac{|V_{ps}^{(\frac{1}{2})}|^2}{S^2 + \Gamma_p^2/4}\,\Gamma_p + \frac{|V_{qs}^{(\frac{1}{2})}|^2}{(\Delta E - S)^2 + \Gamma_q^2/4}\,\Gamma_q \tag{65}$$

where $S = \Delta E_s - \Delta E_p$ and $\Delta E = E_q + \Delta E_q - E_p - \Delta E_p$, and Γ_s is neglected compared to Γ_p or Γ_q in the last terms. This result is in agreement with the rate obtained by Lamb's method of derivation,[57] which is based on the Weisskopf-Wigner approach.[54] This formula is relevant to the interpretation of electric field quenching measurements of the Lamb shift.[39] The detailed angular distribution of emitted radiation depends on the matrix element of the photon emission operator between the ground state wavefunction and the effective resonant-state wavefunction

$$|s,\mu\rangle_R = |s,\mu\rangle + \frac{V_{ps}^{\mu}}{S + i\Gamma_p/2}|p,\mu\rangle - \frac{V_{qs}^{\mu}}{\Delta E - S - i\Gamma_q/2}|q,\mu\rangle \tag{66}$$

that appears in Eq. (60). Calculation of the angular distribution reveals a rich structure:

$$I(\hat{k}) = a + b(\hat{k}\cdot\hat{z})^2 + c\Gamma\hat{k}\cdot\hat{z} + d\vec{P}\cdot\hat{k}\times\hat{z} + e\Gamma\hat{k}\cdot\hat{z}\,\vec{P}\cdot\hat{k}\times\hat{z} \tag{67}$$

where $\Gamma \approx \Gamma_p \approx \Gamma_q$, $\vec{P}$ is the polarization vector of the atom in the resonant state, and $\hat{z}$ is the direction of the electric field $\vec{E}$. The structure is due to interference between M1 radiation from the S state and E1 radiation from the two P states. The terms proportional to Γ in (67) are due entirely to the imaginary part of the energy denominators in (66).

All except one of the terms in (67) has been measured experimentally. The term proportional to $(\hat{k}\cdot\hat{z})^2$ has provided an indirect means of measuring of the Lamb shift because the coefficient b depends strongly on S through the first energy denominator in (66).[64] There has been no observation of the term proportional to $\Gamma\hat{k}\cdot\hat{z}$, although it may be prominent in high-Z one-electron atoms.[65] van Wijngaarden and Drake have observed the effect of the term proportional to $\vec{P} \cdot \hat{k} \times \hat{z}$ in He^+ atoms.[66] The first observation of the term proportional to $\Gamma \hat{k} \cdot \hat{z} P \cdot \hat{k} \times \hat{z}$ was recently made by van Wijngaarden et al. in He^+, from which they infer a value for Γ_p that agrees with theory to about 1%.[67] This observation is of particular interest, because it checks the imaginary part of the complex level shift in the energy denominators in Eq. (66). Damping-dependent asymmetry effects have been observed by Lévy and Williams.[68]

CONCLUSION

In one-electron atoms, there is agreement between a nonperturbative calculation of the theory and experiment for the Lamb shift for Z up to 18. For two-electron atoms at high Z, the first few terms in a relativistic perturbation expansion in 1/Z give accurate predictions for energy levels. Theory and experiment agree at about the level of 10% of the radiative corrections. Line width and decay asymmetries of the 2S state in hydrogenlike atoms in an electric field are accurately described in quantum electrodynamics by an extension of Low's treatment of line shape.

ACKNOWLEDGEMENT

This research was supported by the National Science Foundation Grant No. PHY80-26549.

REFERENCES

1. T. Kinoshita and W.B. Lindquist, Phys. Rev. Lett. 47, 1573 (1981).
2. W.H. Furry, Phys. Rev. 81, 115 (1951).
3. M. Gell-Mann and F. Low, Phys. Rev. 84, 350 (1951).
4. J. Sucher, Phys. Rev. 107, 1448 (1957).
5. E.H. Wichmann, N.M. Kroll, Phys. Rev. 96, 232 (1954); 101, 843 (1956).
6. G.E. Brown, J.S. Langer, and G.W. Schaefer, Proc. Roy. Soc. (London) A 251, 92 (1959).
7. A.M. Desiderio and W.R. Johnson, Phys. Rev. A 3, 1267 (1971).
8. P.J. Mohr, Ann. Phys. (N.Y.) 88, 26, 52 (1974).
9. K.T. Cheng and W.R. Johnson, Phys. Rev. A 14, 1943 (1976).
10. G. Soff, P. Schlüter, B. Müller, and W. Greiner, Phys. Rev. Lett. 48, 1465 (1982).
11. S.J. Brodsky and P.J. Mohr, in Structure and Collisions of Ions and Atoms, ed. by I.A. Sellin (Springer, Berlin, 1978), Topics in Current Physics, Vol. 5, p. 3.
12. P.J. Mohr, Atomic Data and Nuclear Data Tables, to be published.
13. P.J. Mohr, Phys. Rev. A 26, 2338 (1982).
14. G.W. Erickson, Phys. Rev. Lett. 27, 780 (1971); J. Phys. Chem. Ref. Data 6, 831 (1977).
15. J.D. Garcia and J.E. Mack, J. Opt. Soc. Am. 55, 654 (1965).

16. R. Serber, Phys. Rev. 48, 49 (1935).
17. E.A. Uehling, Phys. Rev. 48, 55 (1935).
18. M. Baranger, F.J. Dyson, and E.E. Salpeter, Phys. Rev. 88, 680 (1952).
19. G. Källén and A. Sabry, Dan. Mat. Fys. Medd. 29, no. 17 (1955).
20. R. Karplus and N.M. Kroll, Phys. Rev. 77, 536 (1950).
21. C.M. Sommerfield, Phys. Rev. 107, 328 (1957).
22. A. Petermann, Helv. Phys. Acta 30, 407 (1957), Nucl. Phys. 3, 689 (1957).
23. J. Weneser, R. Bersohn, and N.M. Kroll, Phys. Rev. 91, 1257 (1953).
24. R.L. Mills and N.M. Kroll, Phys. Rev. 98, 1489 (1955).
25. M.F. Soto, Jr., Phys. Rev. Lett. 17, 1153 (1966); Phys. Rev. A 2, 734 (1970).
26. T. Appelquist and S.J. Brodsky, Phys. Rev. Lett. 24, 562 (1970); Phys. Rev. A 2, 2293 (1970).
27. B.E. Lautrup, A. Peterman, and E. de Rafael, Phys. Lett. 31B, 577 (1970).
28. R. Barbieri, J.A. Mignaco, and E. Remiddi, Lett. al Nuovo Cim. 3, 588 (1970).
29. A. Peterman, Phys. Lett. 35B, 325 (1971).
30. J.A. Fox and D.R. Yennie, Ann. Phys. (N.Y.) 81, 438 (1973).
31. J.L. Friar, Ann. Phys. (N.Y.) 122, 151 (1979).
32. E.E. Salpeter, Phys. Rev. 87, 328 (1952).
33. T. Fulton and P.C. Martin, Phys. Rev. 95, 811 (1954).
34. H. Grotch and D.R. Yennie, Rev. Mod. Phys. 41, 350 (1969).
35. G.W. Erickson and D.R. Yennie, Ann. Phys. (N.Y.) 35, 271 (1965).
36. S. Klarsfeld and A. Maquet, Phys. Lett. 43B, 201 (1973), and references therein.
37. P. Pellegrin, Y. El Masri, L. Palffy, and R. Prieels, Phys. Rev. Lett. 49, 1762 (1982).
38. O.R. Wood, II, C.K.N. Patel, D.E. Murnick, E.T. Nelson, M. Leventhal, H.W. Kugel, and Y. Niv, Phys. Rev. Lett. 48, 398 (1982).
39. H. Gould and R. Marrus, Phys. Rev. Lett. 41, 1457 (1978).
40. H.T. Doyle, Advances in Atomic and Molecular Physics, (Academic, New York, 1969), Vol. 5, p. 337.
41. P.J. Mohr, unpublished (1975).
42. R.E. Knight and C.W. Scherr, Rev. Mod. Phys. 35, 431 (1963).
43. F.C. Sanders and C.W. Scherr, Phys. Rev. 181, 84 (1969).
44. K. Aashamar, G. Lyslo, and J. Midtdal, J. Chem. Phys. 52, 3324 (1970).
45. Y. Accad, C.L. Pekeris, and B. Schiff, Phys. Rev. A 4, 516 (1971).
46. P.J. Mohr, in Beam-Foil Spectroscopy, eds, I.A. Sellin and D.J. Pegg, (Plenum Press, New York, 1976) p. 97.
47. A.M. Ermolaev and M. Jones, J. Phys. B 7, 199 (1974).
48. B. Denne, S. Huldt, J. Pihl, and R. Hallin, Physica Scripta 22, 45 (1980).

49. I.A. Armour, E.G. Myers, J.D. Silver, and E. Träbert, Phys. Lett. 75A, 45 (1979).
50. R. DeSerio, H.G. Berry, and R.L. Brooks, in Proceedings of the Workshop on Foundations of the Relativistic Theory of Atomic Structure, Argonne National Laboratory, Dec., 1980, Argonne Report ANL-80-126, p. 240.
51. A.E. Livingston, S.J. Hinterlong, J.A. Poirier, R. DeSerio, and H.G. Berry, J. Phys. B 13, L139 (1980).
52. W.A. Davis and R. Marrus, Phys. Rev. A 15, 1963 (1977).
53. J.P. Buchet, M.C. Buchet-Poulizac, A. Denis, J. Désesquelles, M. Druetta, J.P. Grandin, and X. Husson, Phys. Rev. A 23, 3354 (1981).
54. V. Weisskopf and E. Wigner, Z. Phys. 63, 54 (1930).
55. F. Low, Phys. Rev. 88, 53 (1952).
56. M. Hillery and P.J. Mohr, Phys. Rev. A 21, 24 (1980).
57. W.E. Lamb, Jr., and R.C. Retherford, Phys. Rev. 79, 549 (1950); W.E. Lamb, Jr., Phys. Rev. 85, 259 (1952).
58. C.Y. Fan, M. Garcia-Munoz, and I.A. Sellin, Phys. Rev. 161, 6 (1967).
59. P.R. Fontana and D.J. Lynch, Phys. Rev. A 2, 347 (1970).
60. H.K. Holt and I.A. Sellin, Phys. Rev. A 6, 508 (1972).
61. M.T. Grisaru, H.N. Pendleton and R. Petrasso, Ann. Phys. (N.Y.) 79, 518 (1973).
62. G.W.F. Drake and R.B. Grimley, Phys. Rev. A 11, 1614 (1975); G.W.F. Drake, P.S. Farago, and A. van Wijngaarden, Phys. Rev. A 11, 1621 (1975).
63. E.J. Kelsey and J. Macek, Phys. Rev. A 16, 1322 (1977).
64. A. van Wijngaarden and G.W.F. Drake, Phys. Rev. A 17, 1366 (1978); G.W.F. Drake, S.P. Goldman, and A. van Wijngaarden, Phys. Rev. A 20, 1299 (1979).
65. P.J. Mohr, Phys. Rev. Lett. 40, 854 (1978).
66. G.W.F. Drake, Phys. Rev. Lett. 40, 1705 (1978); A. van Wijngaarden and G.W.F. Drake, Phys. Rev. A 25, 400 (1982).
67. A. van Wijngaarden, R. Helbing, J. Patel, and G.W.F. Drake, Phys. Rev. A 25, 862 (1982).
68. L.P. Lévy and W.L. Williams, Phys. Rev. Lett. 48, 1011 (1982).

RECENT ADVANCES IN MUONIUM HYPERFINE SPLITTING CALCULATIONS

J. R. Sapirstein

Newman Laboratory of Nuclear Studies
Cornell University
Ithaca, NY 14853

ABSTRACT

Recent advances in the theoretical calculation of ground state muonium hyperfine splitting are described. The role of the known nonrelativistic Coulomb Green's function as the basis of approximation schemes is emphasized.

I wish to discuss today the use of an old solution to a classic problem, the evaluation of the Green's function for a nonrelativistic (NR) particle in a static Coulomb field[1-3], in calculations testing Quantum Electrodynamics (QED) in atomic systems. While the system that will be discussed here is muonium, the bound state of an electron and a positive muon, the calculational techniques developed are just as applicable to the systems where the muon is replaced by a proton, hydrogen, or by a positron, positronium. In all these atoms the presence of binding emphasizes a non-perturbative aspect of QED, namely the necessity of including an infinite number of Coulomb exchanges. This is in contrast to the aspect of QED probed in the study of radiative corrections to lepton g-2 factors or high-energy scattering, where one works with a finite number of loops. There are several reasons why these bound state calculations are particularly important in modern field theory that I would like to emphasize.

Firstly, by going to increasingly higher order in α, the calculational techniques one uses are pushed to their limits. If an approximation scheme is barely adequate in a given order, it can be expected to be hopeless in the next[4], which forces the development of more sophisticated techniques. This fact is particularly valuable in providing a strong constraint on one's approach to the bound state problem. While there are an infinite number of ways to formulate bound state equations, most of which are adequate for low order calculations, the requirement that one can actually complete a higher order calculation provides a powerful criterion for selecting one's approach.

Secondly, while years of success have made it obvious that there is no gross breakdown of QED, it is always necessary to check the internal consistency of QED. A powerful way to do this, that will be discussed below, is to determine the fine structure constant from QED experiments, assuming the theory is correct. Already this procedure for the electron g-2 has generated a fine structure constant about two standard deviations away from the Josephson junction value.[5] To clarify this situation it is vital to infer α^{-1} from other QED systems, in particular muonium.

Thirdly, any development of QED can be expected to have some analog in the gauge theories that have been modeled on QED, in particular QCD. While an atomic picture of the bound states of light quarks may or may not be useful, the atomic picture of the bound states of the heavy c and b quarks with their antiparticles, analogous to e^+e^-, has been quite successful.[6] While the presence of a confining potential is bound to produce quantitative changes, qualitative features of positronium, such as the existence of hyperfine splittings and Lamb shifts should have counterparts in heavy quark bound states, and QED theory can provide insights into the behavior of these more exotic states.

Lastly, once one has accepted the proposition that QED is valid to a certain degree of accuracy, one can use hydrogen as a source of information for proton structure that is competitive with the scattering experiments usually used to probe the proton. The Lamb shift is quite sensitive to the size of the proton, and hydrogen hyperfine splitting is also sensitive to the polarizability of the proton. While the present status of theory and experiment does not yet allow a precise determination of these terms, in the next few years it is quite likely that hydrogen spectroscopy will become one of the most sensitive probes of proton structure.

To begin the discussion of muonium hyperfine splitting, we recall that if the muon is treated as a source of a dipole magnetic field producing the vector potential

$$\vec{A} = \frac{\vec{\mu}\times\vec{r}}{4\pi r^3} \tag{1}$$

then first order perturbation theory gives

$$E_F = \frac{16}{3} \cdot \alpha^2 \cdot c \cdot R_\infty \cdot \frac{\mu_\mu}{\mu_p} \cdot \frac{\mu_p}{\mu_B^e} \cdot \left(1 + \frac{m_e}{m_\mu}\right)^{-3}$$

.22 .004 .003 .30 .01 .007

Under each constant the error in parts per million (ppm) is displayed. The usual factor μ_μ/μ_B^e has been broken up into a product of directly measured ratios. Table I indicates the corrections to E_F that have been calculated to date. Note that Z = 1 for muonium: the theoretical contributions are given for general Z in order to differentiate α's coming from exchange between the electron and muon from α's arising from emission and reabsorption on the electron line. The underlined terms are new results.[7,8] The constant 15.1±0.3 had previously been estimated[9] as 18.36±5. The experiment and theory for ground state muonium hfs are presently

$$\Delta\nu_{exp} = 4\ 463\ 302.88(16)\ \text{kHz}^{10}$$

$$\Delta\nu_{th} = 4\ 463\ 304.5(1.6)\ \text{kHz}$$

and are consistent with one another. The Fermi splitting was evaluated with the value of the fine structure constant determined using the Josephson junction effect,

$$\alpha_J^{-1} = 137.035\ 963(15)^{11} \tag{2}$$

It is interesting to turn the muonium hfs experiment around, and to ask what range of α_{hfs}^{-1} produces agreement between theory and experiment. Then

$$\alpha_{hfs}^{-1} = 137.035\ 988(21) \tag{3}$$

with the error resulting primarily from the uncertainty of the muon mass. These values should be contrasted with the values of α^{-1} determined from the quantum Hall effect and from the electron g-2 calculation:

$$\alpha_H^{-1} = 137.035\ 968(23)^{12} \tag{4}$$

$$\alpha_{g-2}^{-1} = 137.035\ 993(10)^{5} \tag{5}$$

No inconsistency is really implied among these four determinations given the uncertainties in each. However, a reduction of error by a factor of two or three could easily reveal real discrepancies that would either signal the presence of as yet unnoticed solid state corrections, or else the first true signs of a breakdown of QED.

The calculation of muonium hfs can be divided into a static part, in which the muon is treated as a fixed source of charge and

Table I

Coefficient of E_F	Contribution in KHz
1	4 459 034.6(1.6)
$\frac{\alpha}{2\pi}$	5 178.8
$-.328\ (\frac{\alpha}{\pi})^2$	-7.9
$1.2\ (\frac{\alpha}{\pi})^3$	.1
$\frac{3}{2}\ (Z\alpha)^2$	356.2
$\alpha(Z\alpha)(\ln 2 - \frac{5}{2})$	-429.0
$\frac{\alpha}{\pi}\ (Z\alpha)^2[-\frac{2}{3}\ \ln^2\ (Z\alpha)^{-2}$	-35.6
$+\ (\frac{37}{72} + \frac{4}{15} - \frac{8}{3}\ \ln\ 2)\ \ln(Z\alpha)^{-2}$	-5.8
$+\ \underline{15.1\pm 0.3}]$	8.3(2)
$-\ \frac{3\alpha}{\pi}\ \frac{m_e m_\mu}{m_\mu^2 - m_e^2}\ \ln\ \frac{m_\mu}{m_e}$	-801.2
$(\frac{\alpha}{\pi})^2\ \frac{m_e}{m_\mu}\ [-2\ \ln^2\ \frac{m_\mu}{m_e}$	-6.6
$+\ \frac{31}{12}\ \ln\ \frac{m_\mu}{m_e} - \frac{28}{9} - \frac{\pi^2}{3} + 1.9$	1.1
$+\ \underline{22.68(58)}]$	2.6(1)
$\alpha^2\ \frac{m_e m_\mu}{m_\mu^2 - m_e^2}\ [\ln(Z\alpha)^{-2} - 8\ \ln\ 2 + 3\ \frac{11}{18}]$	9.1

magnetic moment, and a recoil part, where the non-vanishing of m_e/m_μ is taken into account. This can be discussed in terms of the expansion

$$\Delta E = \left(\sum_{n,m=0}^{\infty} \alpha^n (m_e/m_\mu)^m A_{n,m} \right) E_F \tag{6}$$

where $A_{n,m}$ can contain logarithms of α and m_e/m_μ. The presence of $\ln \alpha$ indicates that a direct expansion in powers of the Coulomb potential will fail. Thus, in the graphs that contribute to the Schwinger correction, at some level one must consider the effect of

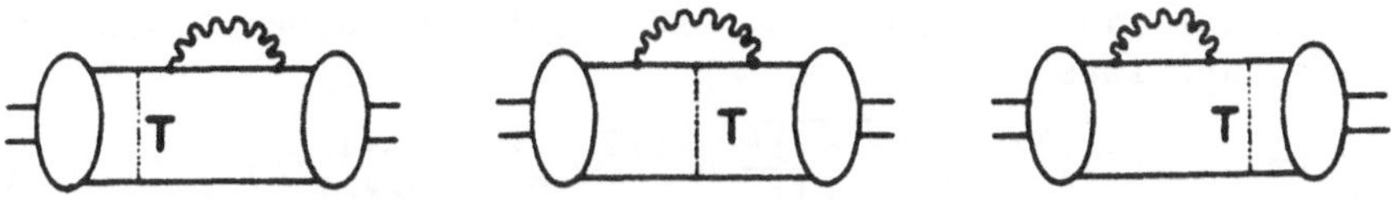

Fig. 1 T represents a transverse photon; the circles represent the bound state wave function

arbitrarily many Coulomb interactions. It turns out that this level is $\alpha^3 E_F$, the present order of interest, so a technique to handle this complication must be developed. The need to consider an arbitrarily large number of Coulomb interactions is equivalent to needing an understanding of the electron's motion in a Coulomb bound state, which information is contained in the electron Coulomb Green's function. This function is known exactly as a partial wave expansion[13], and is a quite complicated object. However, since the velocity of an electron bound to a nucleus of charge $+Z|e|$ is $v/c \sim Z\alpha$, it should be possible to exploit a nonrelativistic approach for atoms with small Z. That it is indeed possible to obtain a rather simple Green's function will now be shown.

Consider for purposes of illustration the Coulomb Green's function for a charged spin zero particle in a Coulomb field,

$$\left[\left(E + \frac{Z\alpha}{r}\right)^2 + \vec{\nabla}^2 - m^2\right] G(\vec{r},\vec{r}\,';E) = \delta^3(\vec{r}-\vec{r}\,') \tag{7}$$

Defining $k^2 = m^2 - E^2$, which will be positive or negative depending on whether the particle is bound or in a scattering state, we can rewrite this as

$$\left(\vec{\nabla}^2 + \frac{2EZ\alpha}{r} - k^2 + \left[\frac{Z^2\alpha^2}{r^2}\right]\right) G(\vec{r},\vec{r}\,';E) = \delta^3(\vec{r}-\vec{r}\,') \tag{8}$$

In the approximation that we ignore the $Z^2\alpha^2/r^2$ term and replace $2EZ\alpha$ with $2mZ\alpha$, this becomes the equation for the NR Coulomb Green's function. Now, it is well known that the NR Coulomb problem has special symmetries: it is less widely known that they can be applied to the solution of the NR Coulomb Green's function. We present here a coordinate space analog to the classic momentum space discussion by Schwinger.[1] The approach is as follows: suppose we can find a set of basic functions such that

$$\sum_{n\ell m} \psi_{n\ell m}(\vec{r})\psi^*_{n\ell m}(\vec{r}\,') = \delta(\vec{r}-\vec{r}\,') \tag{9a}$$

$$(\vec{\nabla}^2 + \frac{2mZ\alpha}{r} - k^2)\psi_{n\ell m}(\vec{r}) = (An - B)\psi_{n\ell m}(\vec{r}); \tag{9b}$$

then

$$G(\vec{r},\vec{r}\,') = \sum_{n\ell m} \frac{\psi_{n\ell m}(\vec{r})\psi^*_{n\ell m}(\vec{r}\,')}{An - B} \tag{9c}$$

These functions can be found, and are simply the usual hydrogen bound state wave functions

$$\psi_{n\ell m}(\vec{r}) = 2(2k)^{3/2} \sqrt{\frac{(n-\ell-1)!}{(n+\ell)!}}\, e^{-\rho/2} \cdot \rho^{\ell} \cdot L^{2\ell+1}_{n-\ell-1}(\rho) Y_{\ell m}(\theta,\phi) \tag{10}$$

with the vital exception that the radial dependence on r is $\rho = 2kr$, rather than $\rho = 2r/na_o$. It is the presence of 1/n that prohibits using the hydrogen bound state wavefunctions as a complete set and requires the inclusion of scattering states. We now prove (9a), which requires a piece of mathematical machinery, the Hille-Hardy formula, that will also prove useful in evaluating (9c).

$$\sum_{s=0}^{\infty} \frac{s!}{(s+2\ell+1)!} z^s L^{2\ell+1}_s(x) L^{2\ell+1}_s(y) = \frac{1}{1-z}\, e^{-\frac{z(x+y)}{1-z}} (xyz)^{-\frac{1}{2}(2\ell+1)} I_{2\ell+1}\left(2\frac{\sqrt{xyz}}{1-z}\right) \tag{11}$$

To use this formula in (9a), rearrange

$$\sum_{n=1}^{\infty} \sum_{\ell=0}^{n-1} = \sum_{\ell=0}^{\infty} \sum_{n=\ell+1}^{\infty} \tag{12}$$

and replace n with $s = n-\ell-1$. In the limit $z \to 1$, (11) becomes

$$\sum_{s=0}^{\infty} \frac{s!}{(s+2\ell+1)!} L^{2\ell+1}_s(2kr) L^{2\ell+1}_s(2kr') = \frac{e^{2kr}}{2k} (2kr)^{-2\ell-1} \delta(r-r'), \tag{13}$$

the statement of completeness of the Laguerre polynomials. Putting this into (9a) gives

$$\sum_{n\ell m} \psi_{n\ell m}(\vec{r})\psi^*_{n\ell m}(\vec{r}') = \frac{1}{r}\,\delta(r-r') \sum_{\ell m} Y_{\ell m}(\theta,\phi)Y^*_{\ell m}(\theta',\phi')$$
$$= r\delta^3(\vec{r}-\vec{r}') \ . \tag{14}$$

While we have not quite produced a delta function, this is made up for by $\psi_{n\ell m}$ satisfying

$$(\vec{\nabla}^2 + \frac{2mZ\alpha}{r} - k^2)\psi_{n\ell m}(r) = \frac{2k}{r}\,(\frac{mZ\alpha}{k} - n)\psi_{n\ell m}(\vec{r}) \tag{15}$$

which allows us to write down the Green's function

$$G(\vec{r},\vec{r}';k) = \frac{1}{2k} \sum_{n\ell m} \frac{\psi_{n\ell m}(\vec{r})\psi^*_{n\ell m}(\vec{r}')}{mZ\alpha/k - n} \tag{16}$$

This form exhibits poles at $k = mZ\alpha/n$ which, since $k = \sqrt{2m|E|}$, occur at the correct nonrelativistic energy levels, and at these poles G factorizes into the hydrogen atom bound state wave functions, with the correct r/na_0 radial dependence restored. To proceed further, we want to use the Hille-Hardy formula, which we can use with the following identity

$$\frac{1}{n - mZ\alpha/k} = \int_0^1 d\rho\ \rho^{\,n - \frac{mZ\alpha}{k} - 1} = \int_0^1 d\rho\ \rho^{\,\ell - \frac{mZ\alpha}{k}}\,[\rho^s] \, . \tag{17}$$

The s sum in (16) can now be performed, yielding a complicated expression that can be simplified using the Neumann sum

$$k \sum_{\ell=0}^{\infty} (2\ell+1)P_\ell(\cos\gamma)I_{(2\ell+1)}(z) = \frac{1}{2}\,kz\ I_0(kz) \tag{18}$$

and the following expression, first presented by Hostler[2] emerges:

$$G(\vec{r},\vec{r}';k) = -\frac{k}{2\pi}\int_0^1 \frac{d\rho\,\rho^{-\frac{mZ\alpha}{k}}}{(1-\rho)^2}\, e^{-k(r+r')\frac{1+\rho}{1-\rho}}\ I_0(\frac{2k\sqrt{2\rho}}{1-\rho}\sqrt{rr' + \vec{r}\cdot\vec{r}'}) \tag{19}$$

Were it not for the factor $\rho^{-\frac{mZ\alpha}{k}}$, this would be an unusually complicated representation of a free propagator in coordinate space. It is this factor that compactly takes into account all the addition physics of the Coulomb problem. The ρ integration can actually be performed directly, leading to a product of Whittaker functions, but this is not necessary for our purposes here.

An important feature of this Green's function is most evident in momentum space. The Fourier transform of the NR approximation to Eq. (8) is

$$(\vec{p}^2 + k^2)G(\vec{p},\vec{p}';k) - \frac{mZ\alpha}{\pi^2}\int \frac{d\vec{q}}{|\vec{p}-\vec{q}|^2} G(\vec{q},\vec{p}';k) = -\delta^3(\vec{p}-\vec{p}') \qquad (20)$$

and the solution is the Fourier transform of (19)[3],

$$G_{NR}(\vec{p},\vec{p}';k) = -\left[\frac{\delta^3(\vec{p}-\vec{p}')}{\vec{p}^2 + k^2} + \frac{mZ\alpha}{\pi^2}\,\frac{1}{(\vec{p}^2+k^2)\,|\vec{p}-\vec{p}'|^2\,(\vec{p}'^2+k^2)}\right.$$

$$\left. + \frac{4(mZ\alpha)^2 k}{\pi^2(\vec{p}^2+k^2)(\vec{p}'^2+k^2)}\int_0^1 d\rho\, \frac{\rho^{-\frac{mZ\alpha}{k}}}{4\rho|\vec{p}-\vec{p}'|^2k^2 + (1-\rho)^2(\vec{p}^2+k^2)(\vec{p}'^2+k^2)}\right] \qquad (21)$$

The reason that this form is particularly useful is the natural separation of the free propagator and the single Coulomb interaction term from the last term, which takes into account two or more Coulomb interactions. The divergences associated with self mass, wave function, and vertex renormalization can then be treated using standard perturbation theory techniques, while the non-perturbative aspect is ultraviolet finite. Returning to the relativistic Green's function defined by Eq. (8), we see it can be expanded as

$$G_{KG}(\vec{r},\vec{r}';E) = G_{NR}(\vec{r},\vec{r}';m\to E) - Z^2\alpha^2\int \frac{d\vec{x}}{|\vec{x}|^2} G_{NR}(\vec{r},\vec{x},E)G_{NR}(\vec{x},\vec{r}';E) + \ldots \qquad (22)$$

where only the first iteration is needed in practice. A very similar formalism can be developed for the electron Green's function, and is the main tool for the evaluation of the electron self-energy in a Coulomb field,

$$\Delta E = -ie^2\int \frac{d^4k}{(2\pi)^4}\frac{1}{k^2}\int d\vec{p}\,d\vec{p}'\,\overline{\psi}(p)\gamma_\mu S(\vec{p}-\vec{k},\vec{p}'-\vec{k};E-k_0)\gamma^\mu\psi(\vec{p}') \qquad (23)$$

If S is a pure Coulomb propagator, this self energy contributes to the Lamb shift[14]: for muonium hfs, S is expanded to first order in the dipole magnetic field of the muon.

While the calculation is too complex to be described in detail here, the heart of the calculation is where the second term of eq. (22) is used in eq. (23). One has the structure of an arbitrary number of Coulomb potentials acting, followed by a perturbation $Z^2\alpha^2/r^2$, followed by yet another string of Coulomb interactions. This structure can be treated in a totally NR approximation, and is quite similar to the integral that gives the Bethe logarithm. The availability of a form that takes into account these strings of Coulomb interactions at the cost of a single ρ integration, eq. (21), allows for the straightforward numerical evaluation of the related integrals, which in previous calculations had only been argued to be small.[9,15] While indeed small, they turn out to contribute at the order of a kilohertz, and had therefore to be evaluated.

The Green's function discussed above refers to the constant Coulomb field produced by a static source. The finite mass of the muon implies, however, that the source has dynamic properties, and a relativistic bound state formalism must be set up to take recoil into account. What we would like to show here is that G_{NR} can be used in the development of a particular abound state formalism.

Schematically, the Bethe-Salpeter equation can be written as

$$G = S + SKG$$

$$= S + SKS + SKSKS + \ldots \tag{24}$$

where G is a full four point amplitude, S the product of two free propagators, and K an irreducible kernel. In general there is no known solution to this equation. However, suppose that with an approximated propagator S_o and an approximated kernel K_o one can obtain the solution to

$$G_o = S_o + S_o K_o G_o \tag{25}$$

Now, by applying to eq. (4.1) ((4.2)) the operator S^{-1} (S_o^{-1}) on the left and G^{-1} (G_o^{-1}) on the right one finds

$$S^{-1} = G^{-1} + K$$

$$S_o^{-1} = G_o^{-1} + K_o \tag{26}$$

or $$G_o^{-1} - G^{-1} = S_o^{-1} - S^{-1} + K - K_o \equiv \delta K$$

which has the solution

$$G = G_o + G_o \delta K G \, . \tag{27}$$

To solve this equation for bound states we recall that at a bound state energy, a Green's function must have the behavior

$$G(E \approx E_n) = \frac{\psi_n \psi_n^*}{E-E_n} \tag{28}$$

which behavior we have already encountered in the discussion of the Coulomb Green's function. Therefore, we take for G_o

$$G_o = \frac{\phi\phi^*}{E-E_n^o} \equiv \frac{|0\rangle\ \langle 0|}{E-E_n^o} \tag{29}$$

in the neighborhood of E_n, which should be close to E_n^o if G_o has been well chosen. If we now expand eq. (27) in a power series in δK and take the expectation value between the states $|0\rangle$, further inserting a complete set of states $|n\rangle$ in between products of G_o and δK, we find

$$\begin{aligned} \langle 0|G|0\rangle &= \frac{1}{E-E_{n_o}} + \frac{\langle 0|\delta K|0\rangle}{(E-E_{n_o})^2} + \frac{|\langle 0|\delta K|0\rangle|^2}{(E-E_{n_o})^3} + \ldots \\ &\quad + \sum_n{}' \frac{\langle 0|\delta K|n\rangle\ \langle n|\delta K|0\rangle}{E-E_n} \frac{1}{(E-E_n^o)^2} + \ldots \\ &= \frac{1}{E-E_n^o - \langle 0|\delta K|0\rangle - \sum_{n'} \frac{|\langle 0|\delta K|n\rangle|^2}{E-E_n} + \ldots} \end{aligned} \tag{30}$$

Each successive power of δK in the expansion either is part of a previous arithmetic progression or is beginning a new one. G is thus seen to have a pole at energy E_n, where

$$E_n = E_n^o + \langle 0|\delta K|0\rangle + \langle 0|\delta K G_o' \delta K|0\rangle + \ldots \tag{31}$$

An unusual feature of this approach is the presence of the structure $S^{-1} - S_o^{-1}$ in δK. The net effect of this is to change S_o's to S's in the perturbation expansion, yielding the standard graphical representation of the BS perturbation expansion.[16]

It should be no surprise that the G_o we want to work with is chosen to be related to the nonrelativistic propagator previously discussed. To see how it emerges as a natural approximation to the relativistic theory, consider the graph in Fig. 2. We want G to describe the propagation of an electron and a positive muon near the ground state of muonium, so we choose the indicated routing of momentum with the expectation that p and p' will be small. Working in Feynman gauge, we find

$$\begin{aligned} \mathcal{M} \sim (e)^4 \int \frac{d^4k}{(2\pi)^4} \frac{1}{(k-p)^2(k-p')^2} &\left[\frac{1}{\not{t}+\not{p}'-m_e} \gamma_\mu \frac{1}{\not{t}+\not{k}-m_e} \gamma_\nu \frac{1}{\not{t}+\not{p}-m_e}\right]_e \\ &\left[\frac{1}{\not{K}-\not{p}'-m_\mu} \gamma^\mu \frac{1}{\not{K}-\not{k}-m_\mu} \gamma^\nu \frac{1}{\not{K}-\not{p}-m_\mu}\right]_\mu \end{aligned} \tag{32}$$

As it stands, this is a very complicated integral. However, consider the muon denominator after rationalization, and define $m_\mu^2 - E''^2 = \gamma^2$. Near the ground state pole, where p^2 and γ^2 are both small, we make

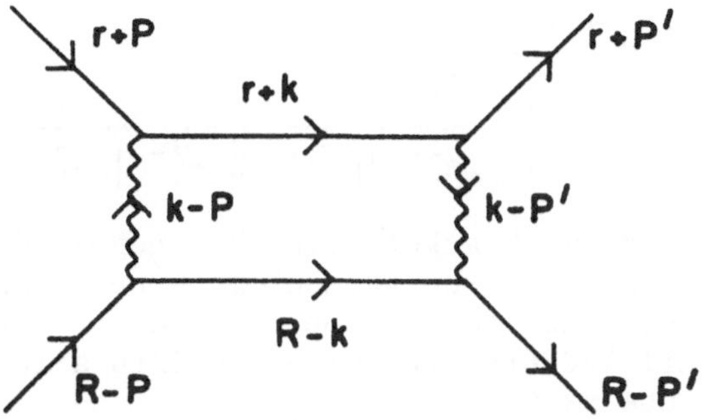

Fig. 2 One loop ladder contribution to electron-muon scattering with $r \equiv (E',\vec{0})$ and $R \equiv (E'',\vec{0})$ for the electron and muon line respectively.

the following approximation, due to Yennie[17]:

$$\frac{1}{p^2 - 2p_o E'' - \gamma^2 + i\varepsilon} \equiv - \frac{1}{2m_\mu}\left(\frac{1}{p_o - i\varepsilon} - \frac{1}{p_o + i\varepsilon}\right) + R$$
$$\approx - \frac{2\pi i\delta(p_o)}{2(m_e+m_\mu)} \tag{33}$$

If one further approximates

$$\begin{aligned} \not{R} - \not{p} + m_\mu &\to m_\mu(1+\gamma_o) \\ \not{r} + \not{p} + m_e &\to m_e(1+\gamma_o) \\ \gamma_\mu\cdot\gamma^\mu &\to \gamma_o\cdot\gamma^o \ , \end{aligned} \tag{34}$$

then

$$[\gamma_\mu \frac{1}{\not{r}+\not{p}-m_e}\gamma^\nu]_e[\gamma^\mu \frac{1}{\not{R}-\not{p}-m_\mu}\gamma_\nu]_\mu \approx \frac{2\pi i\delta(p_o)m_e m_\mu}{2(m_e+m_\mu)}\frac{1}{\vec{p}^{\,2}+\gamma^2}$$
$$\cdot[1+\gamma_o]_e\cdot[1+\gamma_o]_\mu \ . \tag{35}$$

The net result of these approximations is that the four-dimensional k integration collapses to a three-dimensional integral, and

$$\mathcal{M} \sim (m_R Z\alpha)^2 \delta(p_o)(1+\gamma_o)_e (1+\gamma_o)_\mu \delta(p_o')$$

$$\left[\frac{1}{\vec{p}^2+\gamma^2}\,\frac{1}{\vec{p}'^2+\gamma^2}\int \frac{d^3k}{(2\pi)^3}\,\frac{1}{|\vec{p}-\vec{k}|^2}\,\frac{1}{\vec{k}^2+\vec{\gamma}^2}\,\frac{1}{|\vec{k}-\vec{p}'|^2}\right] \tag{36}$$

However, the term in brackets can be recognized as the two Coulomb interaction part of the NR Coulomb Green's function in momentum space. The other terms simply serve to "dress up" what becomes G_{NR} when all ladders are summed up into a four-dimensional form suitable for use as G_o,

$$G_o \to \sum_{n\ell m} \frac{\psi_{n\ell m}(\vec{p})\overline{\psi}_{n\ell m}(\vec{p}')}{E - E_n}$$

$$\psi_{n\ell m} = \delta(p_o)\begin{pmatrix}1\\0\end{pmatrix}\phi_{NR}^{n\ell m}(\vec{p}) \tag{37}$$

The particular form presented here is a little too drastic an approximation, since it eliminates the lower component of the approximate wave function, which makes the discussion of hfs awkward. It is not difficult to remedy this, and a modification of this approach is presently being used in a positronium hfs calculation.

REFERENCES

1. J. Schwinger, J. Math. Phys. (N.Y.) 5, 1606 (1964).
2. L. Hostler, J. Math. Phys. (N.Y.) 5, 591 (1964).
3. L. Hostler, J. Math. Phys. (N.Y.) 5, 1235 (1964).
4. Quoted from G. Erickson and D. R. Yennie, Ann. Phys. (N.Y.) 35, 271 (1965).
5. T. Kinoshita and W. B. Lindquist, Phys. Rev. Lett. 42, 1575 (1979).
6. See, for example, W. Buchmuller and S-H. H. Tye, Phys. Rev. D24, 132 (1981).
7. J. R. Sapirstein, E. A. Terray and D. R. Yennie, Phys. Rev. Lett. 51, 982 (1983).
8. J. R. Sapirstein, Phys. Rev. Lett. 51, 985 (1983).
9. S. J. Brodsky and G. W. Erickson, Phys. Rev. 148, 26 (1966).
10. F. G. Mariam, W. Beer, P. R. Bolton, P. O. Egan, C. J. Gardner, V. W. Hughes, D. C. Lu, P. A. Souder, H. Orth, J. Vetter, U. Moser, and G. zu Putlitz, Phys. Rev. Lett. 49, 993 (1982).
11. E. R. Williams and P. T. Olsen, Phys. Rev. Lett. 42, 1575 (1979).
12. D. C. Tsui, A. C. Gossard, B. F. Field, M. E. Cage and R. F. Dziuba, Phys. Rev. Lett. 48, 3 (1982).
13. E. H. Wichmann and N. M. Kroll, Phys. Rev. 101, 843 (1956).
14. J. R. Sapirstein, Phys. Rev. Lett. 47, 1773 (1981).
15. Many potential terms were explicitly evaluated in a related problem by D. Zwanziger, Phys. Rev. 121, 1128 (1960).

16. G. P. Lepage, Phys. Rev. A16, 863 (1977).
17. Private communication from D. R. Yennie.

CHAOS IN RADIATIVE INTERACTIONS

J. R. Ackerhalt and H. W. Galbraith

Theoretical Division (T-12)
Los Alamos National Laboratory
Los Alamos, New Mexico 87545

P. W. Milonni

Department of Physics
University of Arkansas
Fayetteville, AR 72701

INTRODUCTION

Even a casual perusal of the literature reveals a great interest, in many different fields, in chaotic dynamics. Our purpose here is to indicate why chaotic dynamics may be of interest in the study of radiative interactions in quantum optics.

There are at least two excellent general reviews of chaotic dynamics for physicists. [1,2] In the first part of these notes we will cover some well-traveled ground, but with a slightly different emphasis. Specifically, we will emphasize how to systematically identify chaos in various dynamical systems.

The first important point is that the "chaotic behavior" we are talking about is, in fact, deterministic. We will always be talking about a dynamical system, in which, loosely speaking, the future is determined uniquely from the past (initial conditions). What does "chaos" mean in such a deterministic system? We will emphasize what is perhaps the hallmark of chaotic behavior: Very sensitive dependence on initial conditions.

We will begin by discussing two celebrated examples of discrete mappings, the logistic equation and the Hénon mapping. Since these have been studied and reviewed many times in the literature, we will skip with impunity over some very interesting points, and focus mainly on what "chaos" means in such systems.

These preliminary examples will serve to introduce some useful words and concepts. There is actually a close similarity between these discrete mappings and the dynamical systems of differential equations of primary interest to us here. In a sense this goes back to Poincaré, who introduced the "Poincaré map" as a way of studying the breakdown of integrability in classical Hamiltonian systems. The area-preserving nature of the Poincaré map is related to the preservation of phase-space volumes in such systems. Similarly the area-contracting nature of the Hénon map, and the appearance of a strange attractor, is analogous to the contraction of phase-space volume and the appearance of a strange (chaotic) attractor in dissipative systems of differential equations.

Chaos in both discrete mappings and systems of differential equations is characterized by aperiodicity and very sensitive dependence on initial conditions. Orderly, or regular dynamics, on the other hand, is characterized by periodicity (or quasi-periodicity) and "not very" sensitive dependence on initial conditions.

Chaos as such is not as new a concept as the current flurry of interest might suggest. The so-called C-systems of classical Hamiltonian dynamics, for instance, have been known for some time to have the quality of very sensitive dependence on initial conditions and "random" behavior. [3] One of the things that is new is the identification of certain characteristic "routes" to chaos from an orderly regime of behavior, as exemplified by Feigenbaum's period doubling route to chaos. [4,5] This period doubling route is found in the example of the logistic mapping, which we now briefly review.

THE LOGISTIC MAPPING

This dynamical system is the noninvertible mapping

$$x_{n+1} = 4\lambda x_n(1-x_n) = f(x_n), \quad 0 \le \lambda, x_o \le 1 \qquad (2.1)$$

of the line segment [0, 1] into itself. Feigenbaum [4,5] has used this example in deriving some quantitatively universal characteristics of a period doubling route to chaos in maps with quadratic maxima. It is most instructive to study (2.1) using a programmable pocket calculator, which is in fact how some of its properties were initially discovered.

The <u>fixed points</u> of the mapping are defined by $x^* = f(x^*)$. These are the points that are mapped into themselves. For the logistic mapping we have the fixed points

$$x_0^* = 0 \quad \text{and} \quad x_1^* = 1 - 1/4\lambda \tag{2.2}$$

It is of interest to know whether a fixed point is stable. Suppose $x^* = f(x^*)$, and consider a small perturbation ϵ_0 on x^*:

$$f(x^* + \epsilon_0) \cong f(x^*) + \epsilon_0 f'(x^*) = x^* + \epsilon_1 \tag{2.3}$$

Thus

$$\left| \frac{\epsilon_1}{\epsilon_0} \right| = |f'(x^*)| \tag{2.4}$$

If the perturbation grows with successive applications of the mapping, the fixed point x^* is unstable. If it shrinks, the fixed point is stable. From (2.4) it follows that a fixed point x^* is stable if $|f'(x^*)| < 1$, unstable if $|f'(x^*)| > 1$, and marginally stable if $|f'(x^*)| = 1$.

For the logistic mapping, we deduce from this stability condition that:
(a) For $0 < \lambda < 1/4$ the only fixed point on the interval [0, 1] is $x_0^* = 0$, and it is stable. Regardless of the initial "seed" value x_0, we eventually settle on the fixed point at the origin on successive applications of the logistic mapping. The fixed point x_0^* is therefore called an attractor of period one, or a one-cycle. If $\lambda = 1/4$, x_0^* is only marginally stable.
(b) For $1/4 < \lambda < 3/4$ the fixed point x_0^* is unstable, but now there is a second fixed point $x_1^* = 1 - 1/4\lambda$, and it is stable. Regardless of the initial seed value x_0, we settle after a few iterations of the mapping onto this attractor. If $\lambda = 3/4$, x_1^* is marginally stable.
(c) For $3/4 < \lambda \leq 1$ both fixed points are unstable. We remark again that it is instructive (and entertaining) to check these properties of the logistic mapping, and those discussed below, in numerical "experiments" with a programmable pocket calculator. For $\lambda < 3/4$, successive iterations of the mapping converge onto a stable fixed point.

Just beyond $\lambda = \Lambda_1 = 3/4$, a period doubling occurs; successive iterates x_n bounce alternately between two points after initial "transients" have died out. Let $x_1 = f(x_0)$. Then $x_2 = f(x_1) = f(f(x_0)) = f^2(x_0)$, where f^2 is called the second iterate of f. The period doubling that occurs just beyond $\lambda = 3/4$ is associated with the appearance of two stable fixed points, x_2^* and x_3^*, of f^2. The points

x_2^* and x_3^* form an attractor of period two, or two-cycle, of f:

$$f(x_2^*) = x_3^* \ , \ f(x_3^*) = x_2^* \tag{2.5}$$

f^2 is a fourth-order polynomial in x. The stability of its fixed points is determined by its derivative $f^{2\prime}$, exactly as for f. From the chain rule for derivatives it follows that $f^{2\prime}(x_0) = f'(x_0)f'(x_1)$. Similarly we can define the <u>nth</u> iterate f^n of f, and the chain rule implies that

$$f^{n\prime}(x_0) = \prod_{i=0}^{n-1} f'(x_i) \tag{2.6}$$

where $f(x_0) = x_1$, $f(x_1) = x_2$, ...

It is found that the fixed points x_2^* and x_3^* of f^2 are stable until λ is increased to $\Lambda_2 = 1/4(1 + \sqrt{6}) = .862372...$ Then just beyond $\lambda = \Lambda_2$ another period doubling <u>bifurcation</u> occurs, and we now have a four-cycle of f. Further period doublings are found as λ is increased, and the "window" of λ's associated with a given n-cycle gets rapidly narrower as n increases. Feigenbaum has established that the sequence $\{\Lambda_n\}$ converges geometrically:

$$\lim_{n\to\infty} \frac{\Lambda_n - \Lambda_{n-1}}{\Lambda_{n+1} - \Lambda_n} = \delta = 4.6692016091... \tag{2.7}$$

δ is "universal" in that it applies to all maps with <u>quadratic maxima</u>. The set $\{\Lambda_n\}$ has a limit point $\lambda^* = .891593...$ Beyond this value of the "knob" λ, the generated sequence $\{x_n\}$ appears to be a <u>chaotic</u> sequence without a periodic attractor. For $\lambda = \lambda^*$ the period has evidently doubled <u>ad infinitum</u>, and the sequence $\{x_n\}$ no longer has any perceivable periodicity, or "regularity." (Even for $\lambda > \lambda^*$, however, there are periodic "windows," as discussed below.)

So by "chaos" we mean the absence of any regularities (periodicities) in the sequence $\{x_n\}$ generated by the logistic mapping. But true chaos requires something stronger: <u>Very sensitive dependence on initial conditions</u>. This aspect of chaos may be given a quantitative measure in terms of the <u>Lyapunov characteristic exponent</u>. Suppose we perturb the initial seed value x_0 slightly: $x_0 \to x_0 + \epsilon_0$. After n iterations of the mapping the perturbation will be ϵ_n:

$$f^n(x_0 + \epsilon_0) = f^n(x_0) + \epsilon_0 f^{n\prime}(x_0) = f^n(x_0) + \epsilon_n \tag{2.8}$$

where

$$|\epsilon_n| = |\epsilon_0|\cdot|f^{n\prime}(x_0)| = |\epsilon_0|\cdot\prod_{i=0}^{n-1}|f'(x_i)|$$
$$= |\epsilon_0|\cdot e^{\sum_{i=0}^{n-1}\log|f'(x_i)|} = |\epsilon_0|\, e^{n\chi_n} \qquad (2.9)$$

The limit

$$\chi = \lim_{n\to\infty}\chi_n = \lim_{n\to\infty}\frac{1}{n}\sum_{i=0}^{n-1}\log|f'(x_i)| \qquad (2.10)$$

is called the Lyapunov characteristic exponent (LCE) of the mapping. If $\chi > 0$ we have exponential separation of initially close "trajectories" as $n \to \infty$, i.e., very sensitive dependence on initial conditions; this behavior we call chaotic. If $\chi < 0$ on the other hand, the separation shrinks exponentially to zero.

Chaotic behavior is associated with a positive LCE. For the logistic mapping, we compute $\chi > 0$ for $\lambda > \lambda^*$, except for narrow periodic windows where $\chi \leq 0$. The orderly case $\chi \leq 0$ is associated with periodicity: The sequence $\{x_n\}$ recurs with some period n, i.e., we have an n-cycle (ignoring initial transients). This occurs for all $\lambda < \lambda^*$, for instance. It is obvious, in particular, that $\chi < 0$ for $\lambda < 3/4$, for in this case we have a one-cycle, corresponding to a stable fixed point x^* of f; if x^* is a stable fixed point then $|f'(x^*)| < 1$, $\log|f'(x^*)| < 0$, and hence $\chi < 0$. It is just as easy to prove that any periodic attractor (n-cycle) has $\chi < 0$, and therefore that $\chi > 0$ implies nonperiodicity. (Actually the converse is not necessarily true; aperiodicity need not imply a positive LCE.)

Figure 2.1 shows the result of the computation (2.10) for $\lambda = 0.95$ and three different seeds x_0. It is found empirically that the limit (2.10) is independent of x_0. Huberman and Rudnick [6] have plotted computed values of χ vs. λ for the logistic mapping.

The extreme sensitivity to initial conditions in the chaotic regime can be derived explicitly in the particular case $\lambda = 1$. [7] Letting $x_n = \sin\theta_n$, we obtain $\theta_{n+1} = 2\theta_n \pmod{\pi}$ and thus $\theta_n = 2^n\theta_0$, obviously having the property of very sensitive dependence on initial conditions for n large.

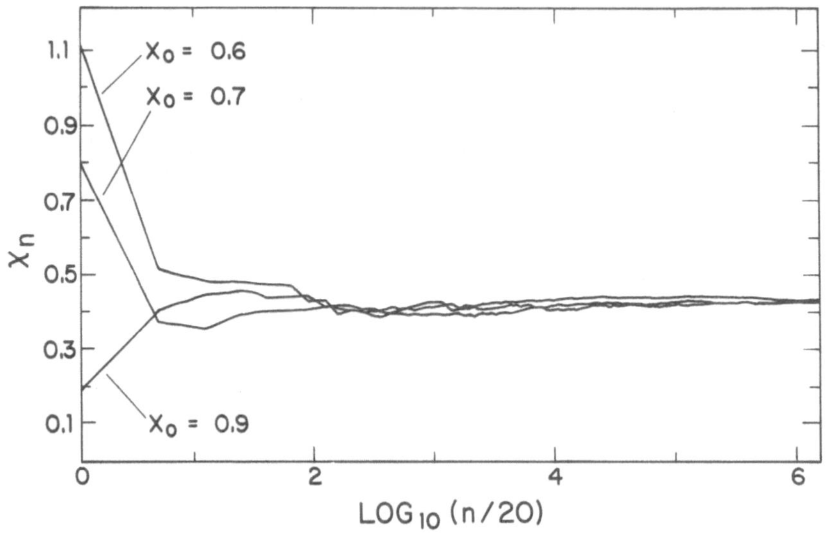

Fig. 2.1. Lyapunov exponent for the logistic map.

We can use this relation for $\lambda = 1$ to derive a probability distribution P(x) such that P(x)dx is the relative number of terms in the sequence $\{x_n\}$ appearing in the interval [x, x + dx]. For large n the θ interval from 0 to $\pi/2$ must evidently be uniformly populated by the sequence $\{\theta_n\}$. Thence $P(x)dx = (2/\pi)d\theta$, implying the probability distribution

$$P(x) = \frac{1/\pi}{\sqrt{x - x^2}} \tag{2.11}$$

generated by the logistic mapping with $\lambda = 1$. In general it is not known how to determine probability distributions on chaotic attractors without resorting to numerical experiments. [8]

We have noted that periodic windows with $\chi < 0$ appear in the "chaotic regime" $\lambda > \lambda^*$. Actually, given an n-cycle, we can infer the existence of other cycles from Sarkovskii's theorem. Specifically, let A be the ordered set $2^0 \cdot 3$, $2^0 \cdot 5$, $2^0 \cdot 7$, ... $2^1 \cdot 3$, $2^1 \cdot 5$, $2^1 \cdot 7$, ... $2^2 \cdot 3$, $2^2 \cdot 5$, $2^2 \cdot 7$, ... i.e., the set of integers $2^n \ell$, $n \geq 0$ and $\ell \geq 3$, ℓ odd, ordered as above. Let B be the set of integers 2^m, $m \geq 0$, and write the following ordered sequence with A written out to the left of B:

$$3 \Leftarrow 5 \Leftarrow 7 \ldots 6 \Leftarrow 10 \Leftarrow 14 \Leftarrow \ldots 12 \Leftarrow 20 \Leftarrow 28 \ldots$$
$$\ldots 32 \Leftarrow 16 \Leftarrow 8 \Leftarrow 4 \Leftarrow 2 \Leftarrow 1$$

Then Sarkovskii's theorem [9-11] says that, if a continuous mapping f of an interval into itself has an n-cycle, then it has an m-cycle for every m such that $n \Leftarrow m$ above. The existence of a 32-cycle for instance, implies the existence of a 16-, 8-, 4-, 2-, and 1-cycle, as in the example of the logistic mapping. In particular, a 3-cycle implies the existence of all m-cycles as well as chaos associated with period-doubling ad infinitum . This explains the title of a paper by Li and Yorke [11]: "Period Three Implies Chaos."

We have by no means said everything that can be said about the logistic mapping! Our introduction to its properties, however, has brought out certain general themes that will recur throughout these lectures: (1) The sequence $\{x_n\}$ generated by the logistic mapping is either periodic or "chaotic". (2) Chaotic behavior is characterized by a positive Lyapunov characteristic exponent, implying "very sensitive dependence on initial conditions." (3) There is a period-doubling route to chaos.

THE HÉNON MAPPING

Consider now the two-dimensional, invertible Hénon mapping [12]

$$x_{n+1} = y_n + 1 - ax_n^2 = f(x_n, y_n) \tag{3.1a}$$

$$y_{n+1} = bx_n = g(x_n, y_n) \tag{3.1b}$$

where a, b are constants. This example of a mapping with chaotic behavior is of interest to us because it illustrates nicely the notion of a strange attractor with fractional dimension.

We will fix our attention on the choice b = 0.3. $x_0 = y_0 = 0$. For a = 1.3, for example, we settle after several tens of iterations on a 7-cycle:

x_n	y_n
1.227	-0.0745
-1.031	0.368
-0.0143	-0.309
0.690	-0.0043
0.0376	0.207
1.023	0.113
-0.248	0.307
1.227	-0.0745
-1.031	0.368
-0.143	-0.309

The set of seven points is called an attractor of period 7 (or 7-cycle) because it is reached for a wide range of initial points (x_0, y_0).

For a = 1.4, however, the corresponding attractor seems to be nonperiodic. Successive points appear to wander erratically on the attractor, which is shown in Figure 3.1. That is, we appear to have a chaotic attactor.

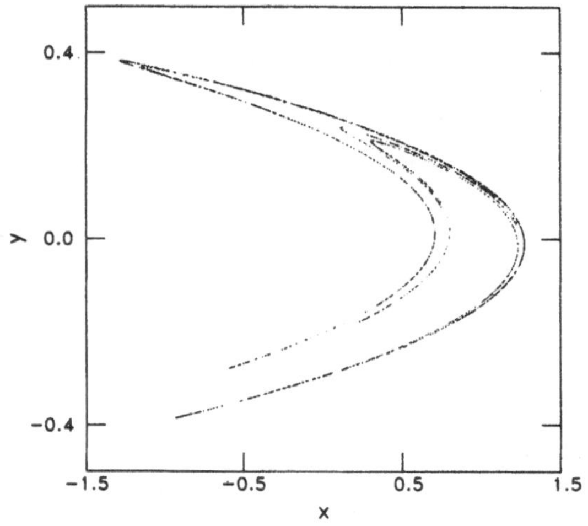

Figure 3.1 The Hénon attractor.

Let us consider this in more detail. The Jacobian of the Hénon mapping is

$$\left|\frac{\partial(x_{n+1}, y_{n+1})}{\partial(x_n, y_n)}\right| = \begin{vmatrix} -2ax_n & 1 \\ b & 0 \end{vmatrix} = -b \qquad (3.2)$$

For $b = 0.3$, therefore, we have an area-contracting map (modulus of Jacobian < 1). That is, an infinitesimal region of the xy plane is contracted in area by the mapping. After many iterations we have a zero-area attractor.

Now this is not hard to understand in the case of the 7-cycle when $a = 1.3$ and the attractor is just a set of 7 points. But what about the attractor shown in Figure 3.1 for $a = 1.4$? Evidently it must have zero area, but it is not an "obvious" zero-area set like a set of n points or a line, corresponding to dimension 0 or 1. It turns out that the zero-area attractor of Figure 3.1 has fractional (or "fractal") dimension.

Of course the concept of dimension here must be carefully defined. In this context the Hausdorff dimension is used. In an n-dimensional space the Hausdorff dimension d of a set is

$$d = -\lim_{\epsilon \to 0} \frac{\log N(\epsilon)}{\log \epsilon} \qquad (3.3)$$

where $N(\epsilon)$ is the number of (n-dimensional) cubes of side ϵ needed to cover the set. For instance, a point ($N(\epsilon) = 1$) has dimension 0, a line ($N(\epsilon) = \epsilon^{-1}$) dimension one, and a plane ($N(\epsilon) = \epsilon^{-2}$) dimension two, as one might expect. A famous example of a set with non-integral dimension is the Cantor set, which may be defined by the following construction: Divide the line segment $0 \le x \le 1$ into thirds and remove the middle third; do the same with the remaining two line segments and continue the process (Figure 3.2). For the set so constructed it is easy to see, using "cubes" (line segments) of side $\epsilon = (1/3)^m$, that $N(\epsilon) = 2^m$, and thus the Hausdorff dimension $d = \log 2/\log 3 \cong 0.631$.

The chaotic Hénon attractor of Figure 3.1 has such a fractional Hausdorff dimension. In fact Hénon showed that there is "structure within structure" as in the Cantor set. Following Hénon, we show this structure within structure in Figure 3.3. Russell et al. [13] have computed a Hausdorff dimension $d \cong 1.26$ for the Hénon chaotic attractor.

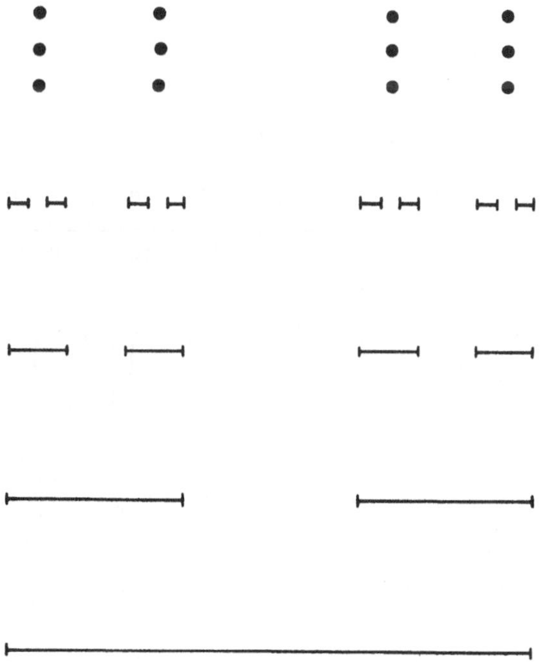

Figure 3.2 Cantor set construction.

Frequently the term strange attractor means an attractor of fractional (non-integral) Hausdorff dimension. Another definition calls an attractor strange if its Hausdorff dimension is greater than its "topological dimension." For our purposes a strange attractor may just as well be taken to be synonomous with a chaotic attractor.

The chaos of the Hénon mapping is associated with very sensitive dependence on initial conditions, just as in the logistic map. Suppose the initial seed values x_0, y_0 are perturbed by ϵ_0, δ_0. Then $x_1 \rightarrow x_1 + \epsilon_1$ and $y_1 \rightarrow y_1 + \delta_1$, where (for small ϵ_0, δ_0)

$$\begin{pmatrix} \epsilon_1 \\ \delta_1 \end{pmatrix} = J(x_0, y_0) \begin{pmatrix} \epsilon_0 \\ \delta_0 \end{pmatrix} \tag{3.4}$$

and the Jacobian matrix

$$J(x, y) = \begin{pmatrix} \frac{\partial f}{\partial x} & \frac{\partial f}{\partial y} \\ \frac{\partial g}{\partial x} & \frac{\partial g}{\partial y} \end{pmatrix} \tag{3.5}$$

The determinant $|J| \neq 0$ because the mapping is invertible. It is easy to show that the perturbations ϵ_n, δ_n after n iterations of the mapping are given by

$$\begin{pmatrix} \epsilon_n \\ \delta_n \end{pmatrix} = J(n) \begin{pmatrix} \epsilon_0 \\ \delta_0 \end{pmatrix} \tag{3.6}$$

where $J(n) = J(x_{n-1}, y_{n-1})J(x_{n-2}, y_{n-2})\ldots J(x_0, y_0)$.

The question is whether the $|\epsilon_n|$, $|\delta_n|$ grow or shrink with increasing n. By making a similarity transformation to diagonalize J(n), this question may be answered by looking at the eigenvalues $\lambda_1(n)$, $\lambda_2(n)$ of the matrix J(n):

$$\begin{pmatrix} \epsilon_n' \\ \delta_n' \end{pmatrix} = \begin{pmatrix} \lambda_1(n) & 0 \\ 0 & \lambda_2(n) \end{pmatrix} \begin{pmatrix} \epsilon_0' \\ \delta_0' \end{pmatrix} \tag{3.7}$$

or

$$|\epsilon_n'| = |\lambda_1(n)| \cdot |\epsilon_0'| = |\epsilon_0'| e^{n\chi_n^{(1)}}$$

$$|\delta_n'| = |\lambda_2(n)| \cdot |\delta_0'| = |\delta_0'| e^{n\chi_n^{(2)}} \tag{3.8}$$

The limits [13]

$$\lambda^{(1,2)} = \lim_{n \to \infty} [\text{magnitude of eigenvalues of } J(n)]^{1/n} \qquad (3.9)$$

are called the Lyapunov numbers of the mapping, and

$$\chi^{(1,2)} = \log \lambda^{(1,2)}$$

are its Lyapunov characteristic exponents. The condition for very sensitive dependence on initial conditions (exponential separation of "trajectories") is that $\chi^{(1)}$ or $\chi^{(2)} > 0$, i.e., the maximal LCE is positive. (For an area-contracting map like (3.1) the sum $\chi^{(1)} + \chi^{(2)}$ will be < 0.)

Feit [14] has computed a maximal LCE of $\cong 0.42$ for the Hénon mapping with a = 1.4, b = 0.3, confirming its chaotic behavior in the sense of "very sensitive dependence on initial conditions".

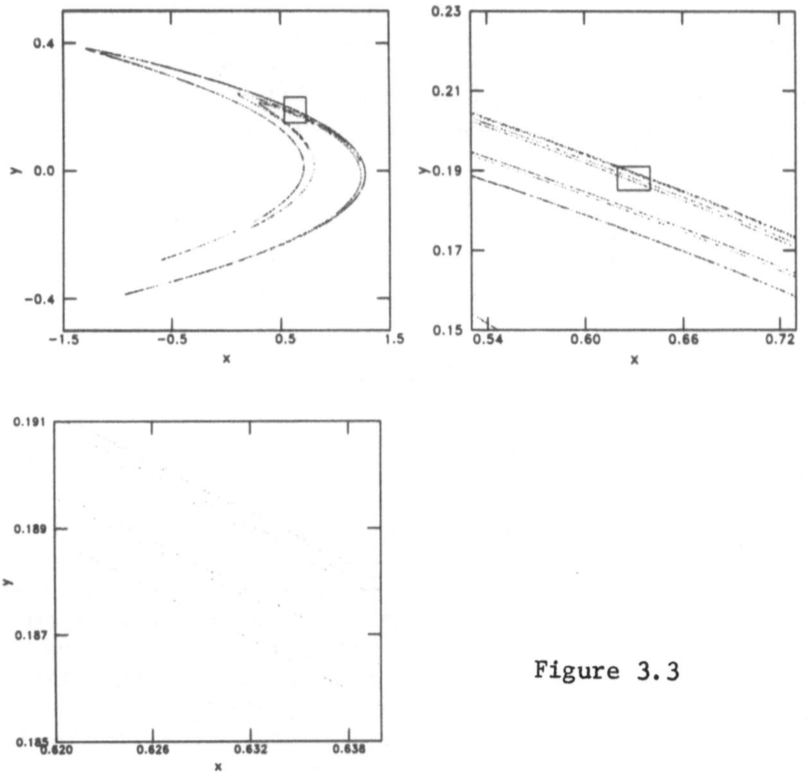

Figure 3.3

DYNAMICAL SYSTEMS OF DIFFERENTIAL EQUATIONS

With the logistic and Hénon mappings as backdrops, we now turn our attention to a different type of (deterministic) dynamical system, namely a set of ordinary differential equations. Once again we will have orderly (periodic) behavior or chaotic (aperiodic) behavior, the latter corresponding to a positive LCE.

It is convenient to consider autonomous systems, which do not contain the independent variable t ("time") explicitly. This is not much of a restriction, because any ordinary differential equation can (in principle) be written as an autonomous one of one higher order: Just solve for t in terms of the dependent variables and their derivatives, and then take the derivative with respect to t. Furthermore we can write an nth order autonomous system as a set of n first-order equations. So we will focus our attention on autonomous systems of the form

$$\dot{y}_n = F_n(y_1, y_2, \ldots y_N), \quad n = 1, 2, 3, \ldots N \tag{4.1}$$

The functions F_n are assumed to have continuous first derivatives with respect to each of their arguments, so that the existence and uniqueness theorem for differential equations is applicable.

Following Poincaré, it is convenient to regard the solution of (4.1) as a trajectory in the n-dimensional phase space $(y_1, y_2, \ldots, y_N)$. The uniqueness theorem tells us that different trajectories in phase space (corresponding to different initial conditions) do not cross. In the special case of Hamiltonian systems, where the system (4.1) takes the form

$$\dot{q}_i = \frac{\partial H}{\partial p_i}, \quad \dot{p}_i = -\frac{\partial H}{\partial q_i} \tag{4.2a}$$

$$H = H(q_1, q_2, \cdots q_m; p_1, p_2, \cdots p_m) \tag{4.2b}$$

the phase space is the usual 2m-dimensional phase space of classical Hamiltonian mechanics.

As in the case of discrete maps we define fixed points (or critical points or equilibrium points) of the dynamical system (4.1). These are points $(y_1^*, y_2^*, \ldots, y_N^*)$ of phase space for which all F $(y_1^*, y_2^*, \ldots, y_N^*) = 0$. Similarly it is again of interest to look into the stability of a fixed point, and this may (usually) be done by linearizing about the fixed point in the usual way. Mathematically, both stable and unstable fixed points are regarded as attractors, the one for $t \to \infty$ and the other for $t \to -\infty$. Frequently it is possible to infer something about the "global" properties of a trajectory by this "local" analysis about fixed points. [15]

A point in phase space is an ω-limit point of a trajectory if the points on the trajectory approach it as $t \to \infty$. It is an α-limit point if the points on the trajectory approach it as $t \to -\infty$. (These are not precise definitions.) The set of all ω-limit (α-limit) points of a trajectory is called its ω-limit (α-limit) set (or attractor).

A limit cycle is a closed loop in phase space, isolated in the sense that no nearby trajectory is also closed. More precisely, it is a closed, periodic trajectory that is the ω-limit (or α-limit) set of another trajectory. If it is the ω-limit (α-limit) set for trajectories both inside and outside the loop, it is a stable (unstable) limit cycle. Limit cycles arise only in nonlinear, nonconservative systems. For such systems they are, like fixed points, generic "steady-state" solutions. Within the "basin of attraction" the amplitude of a limit cycle is independent of initial conditions and is determined by the parameters in the equations. Example: The self-sustained oscillations of various electronic devices and clocks are independent (within bounds) of the starting impulse. [16]

The logistic and Hénon mappings are nonlinear mappings. Similarly we will be interested in nonlinear systems of differential equations, and "chaotic" behavior in such systems is characterized by aperiodicity and "very sensitive dependence on initial conditions." There are thus two important tools for the identification of chaos: (1) The frequency spectra of the $y_n(t)$. A series of sharp spikes in the Fourier spectrum indicates well-defined periodicities, so that the time evolution must be "orderly" or "regular" rather than chaotic. In numerical computations the spectrum is efficiently computed by taking the Fast Fourier Transform (FFT) of the time series generated by numerical integration. We will see examples of this below. (2) The maximal LCE. If this is positive, there is exponential separation of initially close trajectories i.e., chaotic time evolution. This quantity may be computed along the same lines as for the logistic and Hénon mappings. We have used the method described by Galgani et al., [17] as described below.

It is not known in general how to tell whether a system is chaotic without doing numerical computations. However, it is obvious that a system (4.1) with N = 1 cannot be chaotic; the non-crossing of trajectories means that for N = 1 the trajectory is a fixed point, approaches a fixed point as $t \to \infty$, or approaches $\pm\infty$ as $t \to \infty$. The latter case is of no interest to us because we are concerned only with bounded trajectories. Similarly there can be no chaos for N = 2, either. In this case a (bounded) trajectory is a fixed point, approaches a fixed point as $t \to \infty$, or is a closed curve, which may be a limit cycle; this is the main import of the Poincaré-Bendixson theorem. [18] All these cases represent regular motion. It should also be mentioned that a linear system (4.1) will have only regular motion.

For chaotic time evolution the system (4.1) must therefore be nonlinear with order $N \gtrsim 3$.

We have already mentioned the Poincaré map as a way of reducing the study of a dynamical system like (4.1) (a continuous "flow" in phase space) to the study of a discrete mapping. We will see an example below. From a mathematical viewpoint, one might just focus attention on discrete mappings. However, the step from a continuous flow to a discrete mapping, via a Poincaré map or otherwise, still requires the integration of differential equations, and it is the detailed solution of such equations that is of most interest to physicists.

With this introduction to chaotic dynamics behind us, we will now go on to consider some examples of chaos in the radiative interactions of quantum optics.

THE JAYNES-CUMMINGS MODEL

It is difficult to imagine a more basic model in quantum optics than the Jaynes-Cummings model. It consists of a two-level atom (TLA) interacting with a single mode of the electromagnetic field in the electric-dipole and rotating-wave approximations.

For a two-level atom the state vector at any time t may be written

$$|\psi(t)\rangle = c_1(t)\,|1\rangle + c_2(t)\,|2\rangle \qquad (5.1)$$

where $|1\rangle$ and $|2\rangle$ are the eigenstates of lower and higher energy, respectively. From the Schrödinger equation it follows that [19]

$$\dot{x}(t) = -\omega_0\, y(t) \qquad (5.2a)$$

$$\dot{y}(t) = \omega_0\, x(t) + \left(\frac{2d}{\hbar}\right) E(t)\, z(t) \qquad (5.2b)$$

$$\dot{z}(t) = -\left(\frac{2d}{\hbar}\right) E(t)\, y(t) \qquad (5.2c)$$

where $x(t) = c_1{}^*c_2 + c_1 c_2{}^*$, $y(t) = i(c_1{}^*c_2 - c_1{}^*c_2)$, and $z(t) = |c_2|^2 - |c_1|^2$. $\omega_0 = (E_2 - E_1)/\hbar$ is the transition frequency, and d is the electric-dipole matrix element, assumed for simplicity to be real. $E(t)$ is the applied electric field. These are the well-known "optical Bloch equations." They have the first integral $x^2 + y^2 + z^2 = 1$, which means that any trajectory of (5.2) must lie on the "Bloch sphere." Furthermore the system is linear, and therefore cannot be chaotic.

Instead of regarding $E(t)$ as a prescribed external field, let us suppose that it is the field generated by a set of N two-level atoms per unit volume. Then we can write the Maxwell equation

$$\ddot{E}(t) + \omega^2 E(t) = -4\pi N d \ddot{x}(t) \qquad (5.3)$$

for a single-mode field of frequency ω_0. The system (5.2) plus (5.3) couples the atoms and the field together "self-consistently." That is, E determines the motion on the Bloch sphere via (5.2), and the atoms in turn act as a field source via (5.3). Let $\tau = \omega_0 t$ and $e(t) = (2d/\hbar\omega_0)E(t)$. Then our dynamical system is

$$\dot{x} = -y \qquad (5.4a)$$

$$\dot{y} = x + ez \qquad (5.4b)$$

$$\dot{z} = -ey \qquad (5.4c)$$

$$\ddot{e} + \mu^2 e = \beta \dot{y} \qquad (5.4d)$$

which of course may be written in the generic form (4.1). We have defined

$$\mu = \frac{\omega}{\omega_0}, \quad \beta = \frac{8\pi N d^2}{\hbar\omega_0} \qquad (5.5)$$

and the derivatives in (5.4) are with respect to τ.

Let us restrict ourselves to the case of exact resonance ($\mu = 1$) between the TLAs and the field, and initially excited atoms ($z(0) = 1$). Figure 5.1 shows the results of numerical integration for $z(\tau)$, assuming $\beta = 1$, $e(0) = 10^{-6}$, and $\dot{e}(0) = 0$ ($z(0) = 1$ implies $x(0) = y(0) = 0$). The time evolution of $z(\tau)$ looks chaotic, but we cannot ascertain by just looking at Figure 5.1 whether it is chaotic or merely complicated. In Figure 5.2 we show the power spectrum of $z(\tau)$, obtained by applying a cosine bell window to the time series and then taking a 4096-point FFT. This represents clear evidence that the time evolution of z is indeed chaotic, for the spectrum is fairly broadband with no evident periodicities. Finally we have computed the maximal LCE. (The details of such a computation are given in Section 7). Our result, $\chi \cong .087$, confirms that the system is chaotic in the sense of "very sensitive dependence on initial conditions." [20]

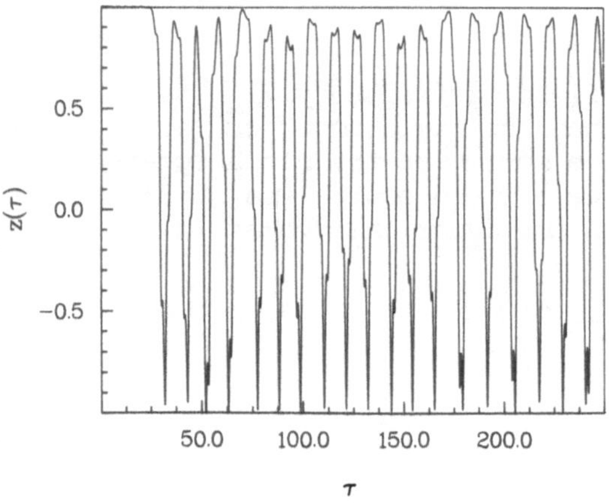

Figure 5.1

In Figure 5.3 we show x(t) vs. y(t) for the case $\beta = 1$. This plot shows clearly the irregular time evolution of the system; compare Figure 5.3 with Figure 5.4, which is for the manifestly non-chaotic case $\beta = 0$ (where the atoms and the field are uncoupled and we are looking at the second-order system consisting of the optical Bloch equations).

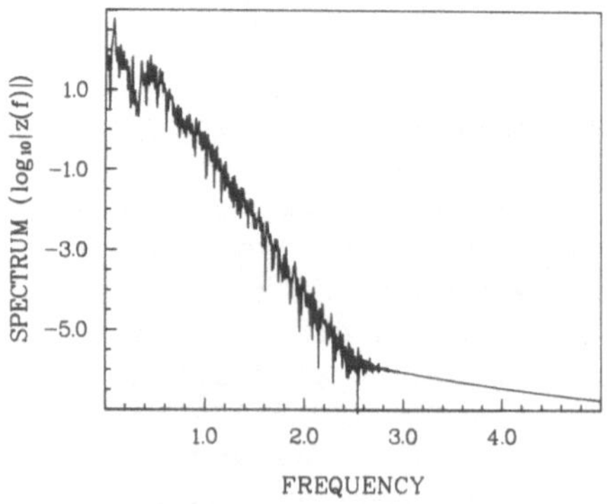

Figure 5.2

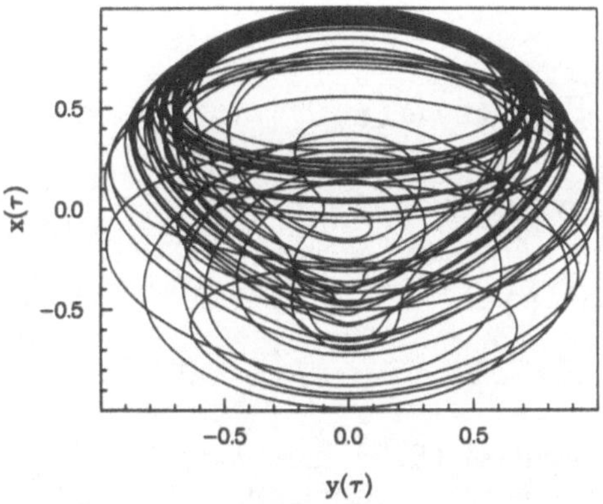

Figure 5.3

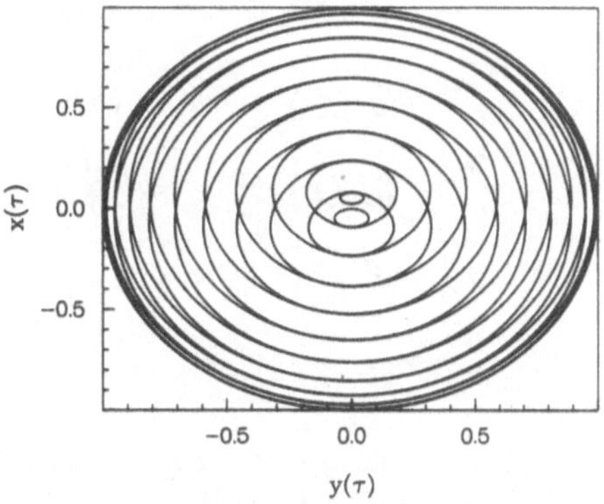

Figure 5.4

We have found that the chaos becomes more pronounced as β increases. For small values of β the computation for the maximal LCE shows that χ is close to 0, but it was difficult to obtain convergence. The power spectra for small values of β showed marked periodicities with some "noise" superimposed. We have not found a sharp boundary between order and chaos as β is varied.

In (5.2) and (5.3) let

$$x(t) = u(t)\cos\omega t - v(t)\sin\omega t \tag{5.6a}$$

$$y(t) = u(t)\sin\omega t + v(t)\cos\omega t \tag{5.6b}$$

$$z(t) = w(t) \tag{5.6c}$$

$$E(t) = \mathcal{E}(t)\cos\omega t \tag{5.6d}$$

and assume u, v, w, and $\mathcal{E}$ are slowly varying compared with $\sin\omega t$, $\cos\omega t$. We thus obtain a "rotating-wave approximation" (RWA) to (5.2) and (5.3) [19]:

$$\dot{u}(t) = -\Delta v(t) \tag{5.7a}$$

$$\dot{v}(t) = \Delta u(t) + \left(\frac{d}{\hbar}\right) \mathcal{E}(t) w(t) \tag{5.7b}$$

$$\dot{w}(t) = -\left(\frac{d}{\hbar}\right) \mathcal{E}(t) v(t) \tag{5.7c}$$

$$\dot{\mathcal{E}}(t) = (2\pi N d\omega) v(t) \tag{5.7d}$$

where the detuning $\Delta = \omega_0 - \omega$. The system (5.7) has two integrals: $u^2 + v^2 + w^2 = 1$ and $(N\hbar\omega)w + \mathcal{E}^2/4\pi$ = energy = constant. The RWA system (5.7) is therefore equivalent to an autonomous system of order two. The Poincaré-Bendixson theorem therefore precludes chaotic behavior. Indeed there are known analytic solutions of (5.7).

In other words, when an RWA is made there may be no prediction of chaos, although the original dynamical system without any approximations may be chaotic. Nevertheless the RWA may still be a good approximation! This can happen if there is little relative energy in the broadband, chaotic portion of the power spectrum. We will discuss this later in connection with the Duffing oscillator. In the present problem the strongly chaotic regime ($\beta \gg 1$) is just the region where N is large enough ($N \gtrsim 4.2 \times 10^{22}$ atoms/cm^3) for the RWA to be of limited validity. This is because the field generated by the atoms in this case is not so small that the Rabi frequency $2dE/\hbar$ is negligible compared with the transition frequency ω_0.

The Jaynes-Cummings problem can be solved exactly in the RWA, even when the field is quantized. The same applies for the Tavis-Cummings model [21] of N TLAs interacting with a single-mode field. It is therefore interesting to extend our non-RWA, <u>semiclassical</u> problem to treat a quantized field. This would be of considerable interest in the context of "quantum chaos": How (if at all) does classical chaos manifest itself quantum mechanically? Note that the semiclassical Jaynes (Tavis)-Cummings model described by (5.4) is not dissipative. The question of whether it is Hamiltonian raises some interesting points that will be discussed elsewhere.

LASERS

Consider the RWA equations (5.7) with $\Delta = 0$ and $w(0) = \pm 1$. Then $u(t) \equiv 0$ and, if we add phenomenological damping terms in the usual way, [19] we obtain the system

$$\dot{v}(t) = -\beta v(t) + \left(\frac{d}{\hbar}\right) \mathcal{E}(t) w(t) \tag{6.1a}$$

$$\dot{w}(t) = \gamma\,[w_{eq} - w(t)] - (\tfrac{d}{\hbar})\,\mathcal{E}(t)\,v(t) \tag{6.1b}$$

$$\dot{\mathcal{E}}(t) = (2\pi N d\omega)\,v(t) - \gamma_c\,\mathcal{E}(t) \tag{6.1c}$$

Here β is essentially the homogeneous linewidth of the transition, γ is the rate at which the population difference relaxes to its equilibrium value w_{eq} in the absence of any applied field, and γ_c is the field loss rate.

Equations (6.1) may be used to model a single-mode (on resonance), homogeneously broadened laser. In this case γ_c is the cavity loss rate, inversely proportional to the cavity Q.

The fixed points of the dynamical system (6.1), (v_s , w_s , $\mathcal{E}_s$), satisfy the equations

$$v_s = (\tfrac{d}{\hbar\beta})\,\mathcal{E}_s\,w_s \tag{6.2a}$$

$$w_s = w_{eq} - (\tfrac{d}{\hbar\gamma})\,\mathcal{E}_s\,v_s \tag{6.2b}$$

$$\mathcal{E}_s = \left(\frac{2\pi N d\omega}{\gamma_c}\right) v_s \tag{6.2c}$$

whose solutions are well known. [19] Introduce the new variables

$$\tilde{\mathcal{E}} = \mathcal{E}/\mathcal{E}_s \quad , \quad \tilde{v} = v/v_s \quad , \quad \tilde{w} = w/w_s \tag{6.3}$$

in terms of which (6.2) becomes:

$$\dot{\tilde{v}} = -\beta\tilde{v} + \beta\tilde{\mathcal{E}}\tilde{w} \tag{6.4a}$$

$$\dot{\tilde{w}} = -\gamma\tilde{w} + \gamma(\lambda+1) - \gamma\lambda\tilde{\mathcal{E}}\tilde{v} \tag{6.4b}$$

$$\dot{\tilde{\mathcal{E}}} = \gamma_c(\tilde{v} - \tilde{\mathcal{E}}) \tag{6.4c}$$

where

$$\lambda = w_{eq}/w_s - 1 \tag{6.5}$$

Finally let

$$r = \lambda + 1 \tag{6.6a}$$

$$\sigma = \gamma_c / \beta \tag{6.6b}$$

$$b = \gamma / \beta \tag{6.6c}$$

introduce the new independent variable $\tau = \beta t$, and define

$$x = (b\lambda)^{1/2} \tilde{\mathcal{E}} \tag{6.7a}$$

$$y = (b\lambda)^{1/2} \tilde{v} \tag{6.7b}$$

$$z = r - \tilde{w} \tag{6.7c}$$

In terms of these new quantities we may write the system (6.4) in the form

$$\dot{x} = -\sigma(x - y) \tag{6.8a}$$

$$\dot{y} = -y - xz + rx \tag{6.8b}$$

$$\dot{z} = xy - bz \tag{6.8c}$$

The system (6.8) is known as the Lorenz model. The isomorphism of the single-mode laser equations (6.1) to the Lorenz model equations was noted by Haken [22] in 1975.

Lorenz [23] in 1963 discussed the remarkable properties of the dynamical system (6.8). (He derived these equations as a rather severe approximation to a set of partial differential equations in fluid mechanics.) In the current parlance we can say that the system (6.8) has a "chaotic" regime, i.e., a regime of aperiodic time evolution with very sensitive dependence on initial conditions. The title of Lorenz's paper, "Deterministic Nonperiodic Flow," summarizes rather well these properties.

A trivial fixed point of (6.8) is (0, 0, 0). For $r > 1$ we have two additional fixed points:

$$x = y = \pm [b(r-1)]^{1/2}, \quad z = r - 1 \tag{6.9}$$

The three fixed points for $r > 1$ are all unstable if

$$\sigma > b + 1 \tag{6.10a}$$

and

$$r > \frac{\sigma(b + \sigma + 3)}{\sigma - b - 1} \tag{6.10b}$$

The system (6.8) contracts volumes in the phase space (x, y, z), since the divergence of the flow in phase space is negative (σ, $b > 0$):

$$\frac{\partial \dot{x}}{\partial x} + \frac{\partial \dot{y}}{\partial y} + \frac{\partial \dot{z}}{\partial z} = -(\sigma + b + 1) \tag{6.11}$$

Thus the phase-space volume of the Lorenz system must contract to zero, but simple fixed points in phase space are not stable attractors when (6.10) is satisfied. It is exactly in this case that the Lorenz model has such fascinating properties.

Figures 6.1 and 6.2 show the results of numerical integration of equations (6.8) for $\sigma = 3$, $b = 1$, $x(0) = y(0) = z(0) = 1.0$, and $r = 0.5$ and 22, respectively. For simplicity we show only y(t). For $r = 0.5$ there is only one fixed point, namely (0, 0, 0), and it is stable. Figure 6.1 shows how the solution converges onto this simple attractor. For $r > 21$ the fixed points are all unstable. Figure 6.2 shows y(t) in the "chaotic regime," specifically for $r = 22$.

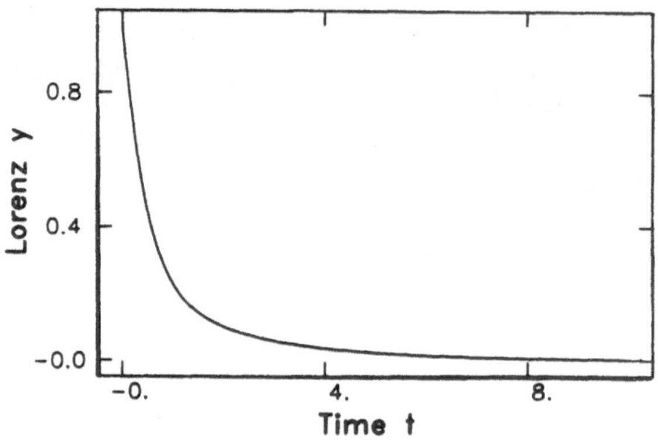

Figure 6.1

We have emphasized that sensitive dependence on initial conditions is one of the features of chaotic behavior. In Figure 6.3 we plot y(t) for the same parameters as Figure 6.2, except that the initial conditions are x(0) = y(0) = z(0) = 1.001. In Figure 6.4 we plot the quantity

$$\beta(t) = [(y(t) - y'(t))^2]^{1/2} \tag{6.12}$$

where y(t) and y'(t) are given in Figures 6.2 and 6.3, respectively. It is evident that there is a rather dramatic sensitivity to initial conditions. This may be formulated rigorously in terms of the Lyapunov characteristic exponents, as in Section 7 below.

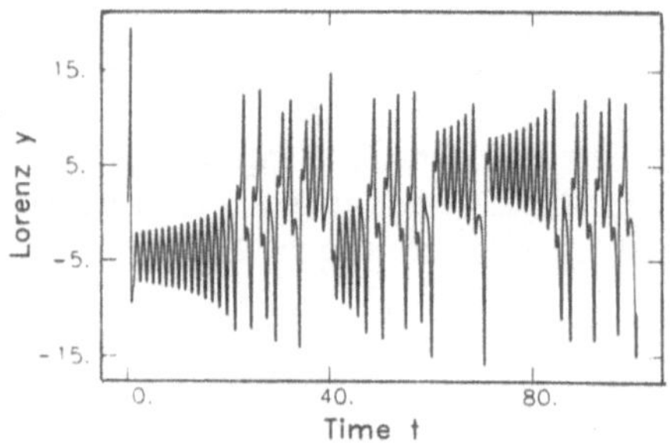

Figure 6.2

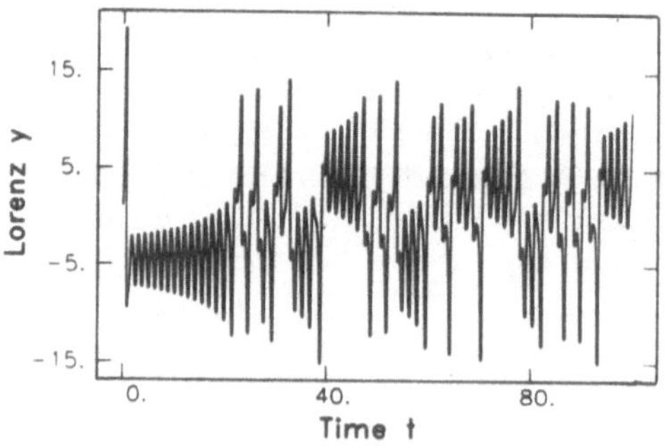

Figure 6.3

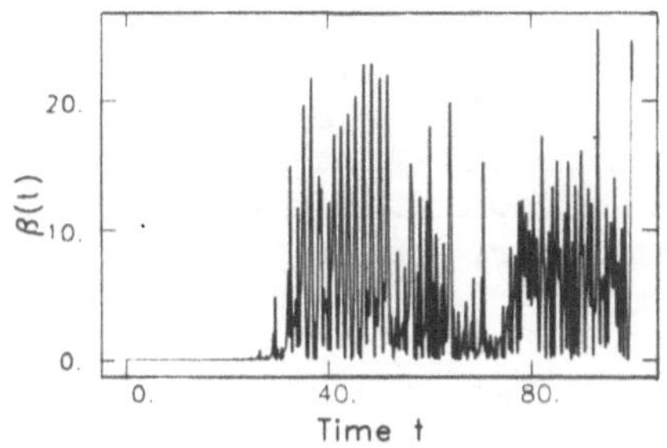

Figure 6.4

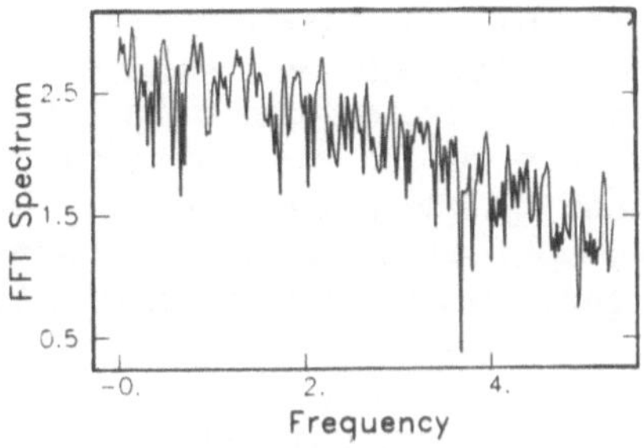

Figure 6.5

Another feature of chaos is nonperiodicity. In Figure 6.5 we show the power spectrum of the function in Figure 6.2. This was computed with a 2048-point Fast Fourier Transform algorithm. Before transforming, the average value of y over the data set was subtracted out and a cosine bell window was applied. The spectrum is a broadband wash rather than a series of sharp spikes that would characterize periodic (or almost-periodic) time evolution.

In the laser case the conditions (6.10) for chaos become

$$\gamma_c > \beta + \gamma \qquad (6.13a)$$

$$g_o > \gamma_c \left(\frac{\gamma_c}{\beta}\right) \frac{\gamma_c/\beta + \gamma/\beta + 3}{\gamma_c/\beta - \gamma/\beta - 1} \qquad (6.13b)$$

The first condition shows that this is "bad-cavity" chaos: The chaos arises only if the field damping rate exceeds the homogeneous linewidth plus the decay rate of excited-state population. The second condition shows that the small-signal gain coefficient g_o must exceed a value determined by the cavity loss, the homogeneous linewidth, and the excited-state decay rate. Loosely speaking, therefore, the Lorenz chaos in a laser should arise only in high-gain lasers with low-Q cavities.

For "typical" lasers like CO_2, CO, HF/DF, numerical estimates indicate that the conditions (6.13) for chaos are not normally realized. In fact, the condition (6.13a) requires the cavity bandwidth to exceed the gain bandwidth, a highly atypical situation. In any case we are not aware of <u>any</u> experimental observations of chaos described by the Lorenz model. But the message is clear: Laser and quantum-optical problems are usually nonlinear, and may be attractive for experimental studies of chaotic systems.

Frequently β is much larger than γ, and we replace (6.1a) by the "quasi-steady" or adiabatic approximation to v(t):

$$v(t) \cong \left(\frac{d}{\hbar}\right) \mathcal{E}(t)\, w(t)/\beta \tag{6.14}$$

Then (6.1b) and (6.1c) become

$$\dot{w}(t) = \gamma[w_{eq} - w(t)] - \left(\frac{8\pi d^2\omega}{\hbar\beta V}\right) n(t)\, w(t) \tag{6.15a}$$

$$\dot{n}(t) = \left(\frac{4\pi d^2\omega}{\hbar\beta}\right) N\, n(t)\, w(t) - 2\gamma_c\, n(t) \tag{6.15b}$$

where

$$I(t) = \frac{c}{8\pi}\mathcal{E}(t)^2 = \left(\frac{c\hbar\omega}{V}\right) n(t) \tag{6.16}$$

is the cavity intensity, n(t) the cavity photon number, and V the mode volume. Let

$$\Delta = NV\, w(t) \quad , \quad G = \frac{4\pi d^2\omega}{\hbar\beta V}$$

$$R = NV\gamma w_{eq} \quad , \quad K = 2\gamma_c \tag{6.17}$$

Then (6.15) has the form

$$\dot{\Delta} = R - 2Gn\Delta - \gamma\Delta \tag{6.18a}$$

$$\dot{n} = Gn\Delta - Kn \tag{6.18b}$$

Unlike the original system (6.1), the system (6.18) obtained by making the rate-equation approximation (REA) (6.14) has no unstable fixed points. This point was made by Shirley, [24] who also wrote the instability condition (6.10) (or (6.13)) for the "exact" system (6.1). Indeed the REA equations (6.18) cannot, according to the Poincaré-Bendixson theorem, have chaotic time evolution, because they form an autonomous system of order two. The situation is changed completely, however, if the cavity loss is time-dependent, for then the order of the equivalent autonomous system is > 2.

Arecchi, et al. [25] have modulated the cavity loss of a CO_2 laser in such a way that

$$K \rightarrow K_1 (1 + m \cos \Omega t) \qquad (6.19)$$

In their device $\beta/\gamma \approx 10^5$, and so the REA equations (6.18) are applicable, with K given by (6.19). Numerical simulation of (6.18) predicts a sequence of period doublings and a chaotic regime for a certain range of modulation frequencies Ω. Period doublings and chaotic behavior were indeed found experimentally, apparently providing "the first experimental evidence of these phenomena in a quantum-optical molecular system."

In the experiments the modulation index m is of the order of a few per cent. For $m \rightarrow 0$ the system (6.18) has the fixed point

$$\bar{\Delta} = K_1/G, \quad \bar{n} = R/2K_1 - \gamma/2G \qquad (6.20)$$

and a linearized stability analysis gives a natural oscillation frequency $\Omega_r/2\pi = (2G^2\bar{\Delta}\bar{n})^{1/2}/2\pi \approx 43$ kHz for $G \cong .24 \times 10^7\ \mathrm{sec}^{-1}$, $K \cong 3 \times 10^7$ sec , and $\bar{n} \approx 10^3$. [25] Based on this linearized frequency, Arecchi et al. used modulation frequencies $\Omega/2\pi$ in the range from 40 to 150 kHz, the modulation being done electro-optically. Numerical solution of equations (6.18) shows that a period doubling sequence should be found as Ω is increased, and that there is a chaotic regime beyond roughly $\Omega/2\pi = 64$ kHz. This is borne out quite well experimentally, suggesting that further laser and quantum-optical experiments might provide a useful testing ground for theoretical studies of chaos.

There are experimental data on Xe lasers that also suggest chaotic time evolution. [26] The simulation of these experimental results is considerably complicated, however, by the inhomogeneous broadening of the laser transition. We are not aware at present of any detailed theoretical work in this direction.

RAMAN SCATTERING

Chaos may also appear in the propagation of laser radiation in a nonlinear optical medium. As an example of this possibility, we will consider the coupled wave equations for plane-wave, steady-state propagation of pump, Stokes, and anti-Stokes modes in Raman scattering. Under certain conditions this system of equations has a strange (chaotic) attractor. [27]

Letting A_1, A_2 and A_3 be the electric field amplitudes for the anti-Stokes, pump, and Stokes modes, respectively, and assuming perfect phase matching, we write [28]

$$\dot{A}_1 = -\gamma_1 A_1 - \beta_1 \left(|A_2|^2 A_1 + A_2^2 A_3^* \right) \tag{7.1a}$$

$$\dot{A}_2 = -\gamma_2 A_2 + \beta_2 \left(|A_1|^2 - |A_3|^2 \right) A_2 \tag{7.1b}$$

$$\dot{A}_3 = -\gamma_3 A_3 + \beta_3 \left(|A_2|^2 A_3 + A_1^* A_2^2 \right) \tag{7.1c}$$

The γ_i are phenomenological loss or gain coefficients, and the β_i are mode-coupling coefficients involving Raman cross sections. [28] Introduce real variables a_i and θ_i:

$$A_i(z) = a_i(z)\, e^{i\theta_i(z)} \tag{7.2}$$

and assume for simplicity that $2\theta_2 - \theta_3 - \theta_1 = 0$ at the initial point $z = 0$. Then the system (7.1) can be written as a third-order autonomous system:

$$\dot{a}_1 = -\gamma_1 a_1 - \beta_1 (a_1 + a_3) a_2^2 \tag{7.3a}$$

$$\dot{a}_2 = -\gamma_2 a_2 + \beta_2 (a^2 - a_3^2) a_2 \tag{7.3b}$$

$$\dot{a}_3 = -\gamma_3 a_3 + \beta_3 (a_1 + a_3) a_2^2 \tag{7.3c}$$

We will focus our attention first on the specific case

$$\gamma_1 = -1, \quad \gamma_2 = \gamma_3 = 1, \quad \beta_1 = 9, \quad \beta_2 = 5, \quad \beta_3 = 1 \tag{7.4}$$

The system (7.3) plus (7.4) is dissipative in the sense that volumes in the phase space (a_1, a_2, a_3) contract to zero over a wide range of initial conditions. Furthermore there are five fixed points, and they are all unstable, as in the Lorenz model in the chaotic regime. In this case our dynamical system (7.3) plus (7.4) has chaotic time evolution, although "time" in this example is really the distance z of propagation in the Raman medium.

Figure 7.1 shows the pump amplitude computed numerically for the initial condition $a_1(0) = a_2(0) = a_3(0) = 1$. The evolution looks chaotic, and this is confirmed by taking the FFT of the time series. Figure 7.2, for instance, shows the low-frequency portion of the power spectrum of the anti-Stokes amplitude a_1; it is broadband with no evident periodicities.

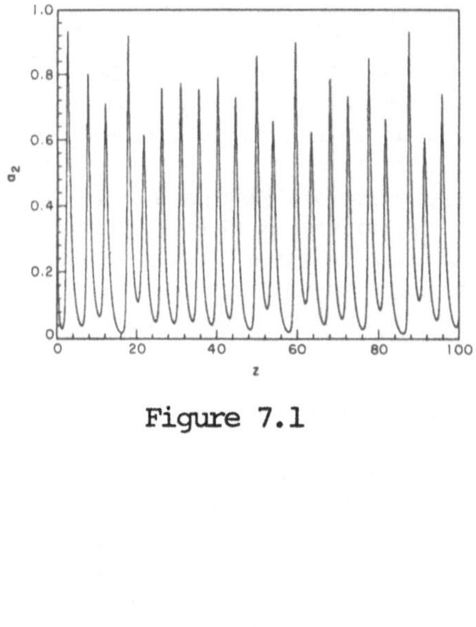

Figure 7.1

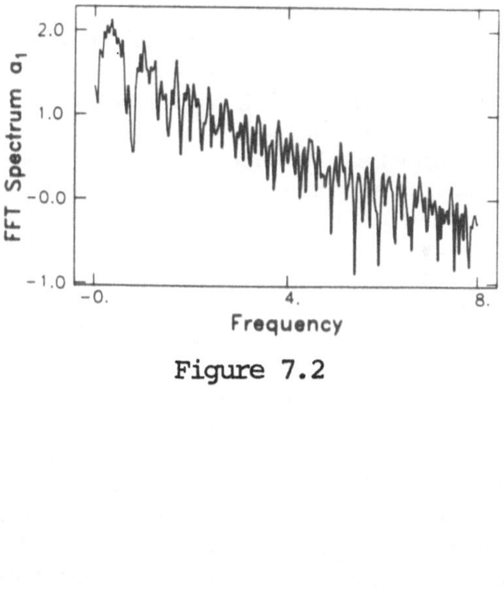

Figure 7.2

In Figure 7.3 we plot $a_2(z)$ versus $a_1(z)$. The orbit winds around one unstable fixed point and then another, and when one looks at its development in time it appears to be switching erratically between the two loops. A similar result is found when $a_3(z)$ is plotted versus $a_2(z)$. These results are somewhat like the corresponding results in the Lorenz model.

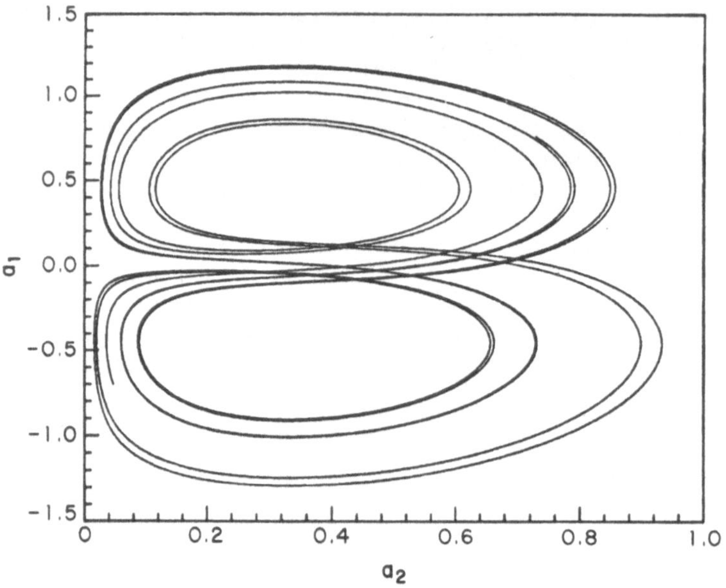

Figure 7.3

Let us now consider a Poincaré mapping obtained from the dynamical system (7.3) plus (7.4). As mentioned earlier, a Poincaré mapping is a way of developing a discrete two-dimensional mapping from the continuous flow in phase space. Figure 7.4 shows the Poincaré map obtained by plotting $a_3(z)$ versus $a_2(z)$ for points z at which $a_1 = 0$ and $\dot{a}_1 < 0$ (Figure 7.5). Successive points appear in the a_2a_3 plane in an apparently random fashion. We also show in Figure 7.6 a magnified section of the Poincaré plot. When the section inside the rectangle of Figure 7.6 is further magnified (Figure 7.7), we can again distinguish two lines, suggesting that the Poincaré map has a Cantor-like structure reminiscent of the "structure with structure" of the Hénon strange attractor.

Poincaré maps are less useful in higher-dimensional systems, and even in a low-dimensional system like (7.3) the generation of a Poincaré plot requires large integration times. In any event a Poincaré mapping is a two-dimensional, invertible mapping; the invertibility follows from the uniqueness theorem for a continuous flow in phase space. In Hamiltonian systems, furthermore, the Poincaré map is area preserving. Indeed area-preserving mappings have for some time been used to mimic the properties of Hamiltonian systems.

Finally let us look at the system (7.3) plus (7.4) with regard to sensitivity on initial conditions. As in our example with the Lorenz model we can do this very crudely by simply following the separation in time of two initially close trajectories. Figure 7.8 shows the

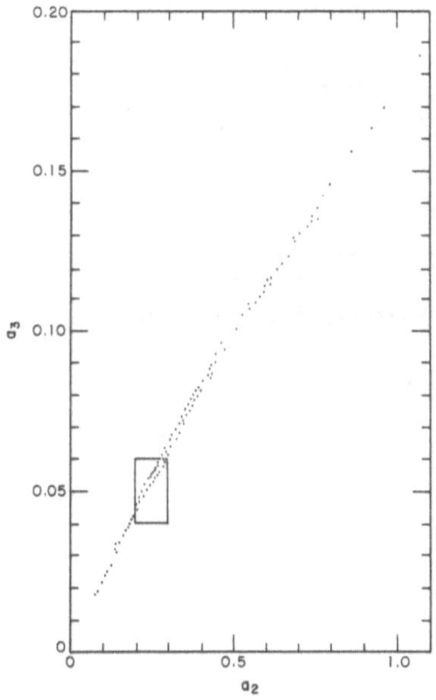

Figure 7.4

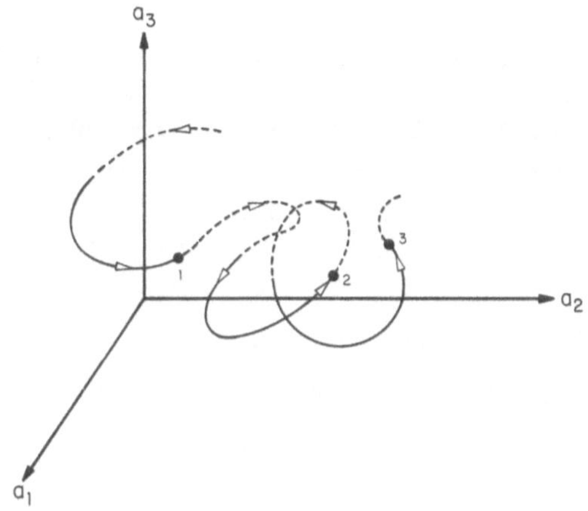

Figure 7.5

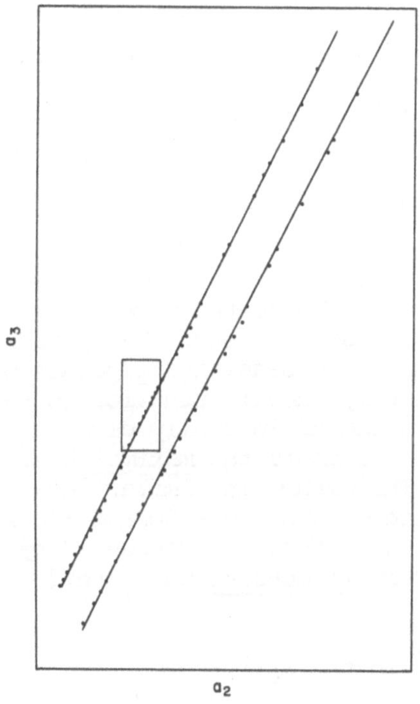

Figure 7.6

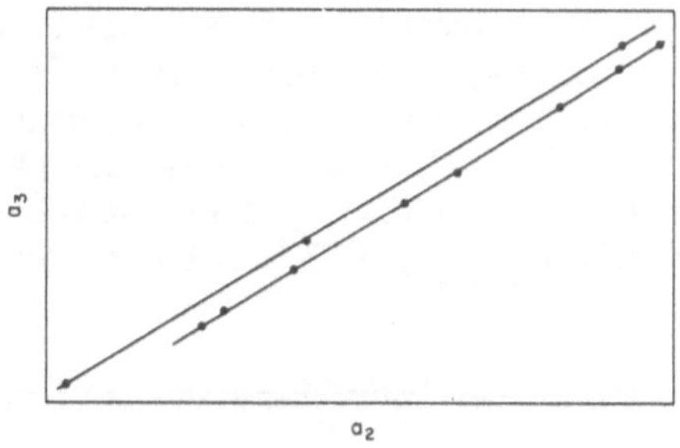

Figure 7.7

"distance"

$$\beta(z) = \left[\left(a_2(z) - a_2'(z) \right)^2 \right]^{1/2} \tag{7.5}$$

between two trajectories with initial conditions $(a_1, a_2, a_3) = (1.0, 1.0, 1.0)$ and $(1.01, 1.01, 1.01)$. The result for the relative separation is similar to that shown in Figure 6.4 for the Lorenz model.

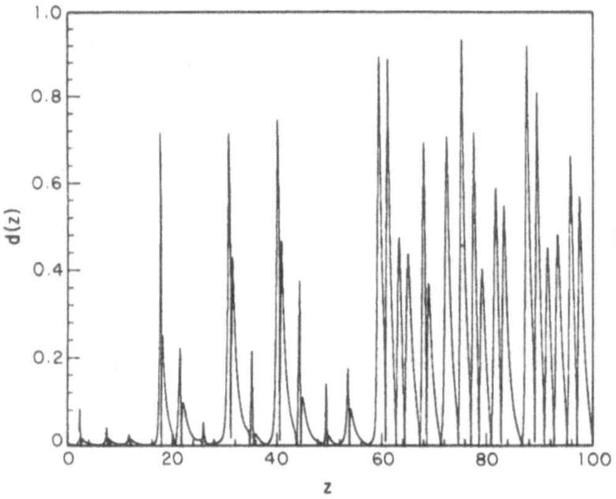

Figure 7.8

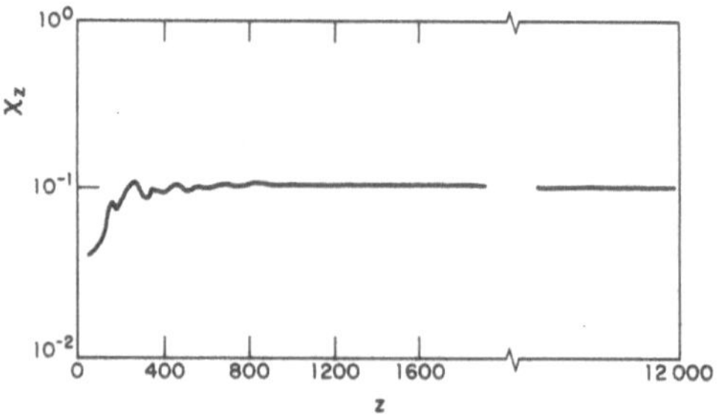

Figure 7.9

Sensitivity to initial conditions may be tested more rigorously by computing the Lyapunov characteristic exponents of the system. If any of the LCE's are positive, we have "very sensitive dependence on initial conditions." In order to identify chaotic behavior in the sense of an average exponential separation of initially close trajectories, therefore, it suffices to compute the maximal LCE, which is in fact the simplest to compute. The method is similar in principle to that described in the case of the logistic and Hénon mappings. The theory and details of the method are discussed by Galgani, et al., [17] and here we merely illustrate the method for the example (7.3).

First we introduce variations ξ_i in the a_i:

$$a_i(z) \rightarrow a_i(z) + \xi_i(z)$$

The $\xi_i(z)$ are taken to be small, and so from (7.3) we obtain the linearized equations

$$\dot{\xi}_1 = -\gamma_1 \xi_1 - \beta_1 a_2^2 (\xi_1 + \xi_3) - 2\beta_1 (a_1 + a_3) a_2 \xi_2 \qquad (7.7a)$$

$$\dot{\xi}_2 = -\gamma_2 \xi_2 + \beta_2 (a_1^2 - a_3^2) \xi_2 + 2\beta_2 a_2 (a_1 \xi_1 - a_3 \xi_3) \qquad (7.7b)$$

$$\dot{\xi}_3 = -\gamma_3 \xi_3 + \beta_3 a_2^2 (\xi_1 + \xi_3) + 2\beta_2 a_2 (a_1 + a_3) \xi_2 \qquad (7.7c)$$

for the variations. Equations (7.3) and (7.7) may be solved simultaneously for the ξ_i, assuming small initial values for the variations. We compute

$$\chi_z = \frac{1}{z} \log \| \xi(z) \| \qquad (7.8)$$

where $\| \xi \|$ denotes the Euclidean norm:

$$\| \xi \| = [\xi_1^2 + \xi_2^2 + \xi_3^2]^{1/2} \qquad (7.9)$$

The maximal LCE is given by

$$\chi = \lim_{z \to \infty} \chi_z \qquad (7.10)$$

If $\chi > 0$ we have "very sensitive dependence on initial conditions," and the motion in the space (a_1, a_2, a_3) is chaotic. For a dissipative system like (7.3) or the Lorenz model, the motion is chaotic over the attractor. Hamiltonian systems have no attractors as such, but χ may nevertheless be computed in the same way as illustrated in our example.

Figure 7.9 shows the result of the computation for the system (7.3) plus (7.4). χ converges to about 0.10, thus confirming the chaotic nature of the system. In fact results like those shown in Figures 7.1-7.4 suffice for most practical purposes to classify the system as chaotic. Computations of Poincaré maps and Lyapunov characteristic exponents, while perhaps more explicit and more rigorous, are far more expensive.

If we reduce γ_2 and γ_3 to .05, while keeping $\gamma_1 = -1$, we obtain for $a_2(z)$ the result shown in Figure 7.10. The system appears to reach a limit cycle, as confirmed both by spectral analysis and Poincaré maps. In Figure 7.11 we show $a_1(z)$ versus $a_2(z)$ for this case of orderly behavior. As time evolves the orbit in the a_1a_2 plane simply keeps retracing its path over the figure-eight.

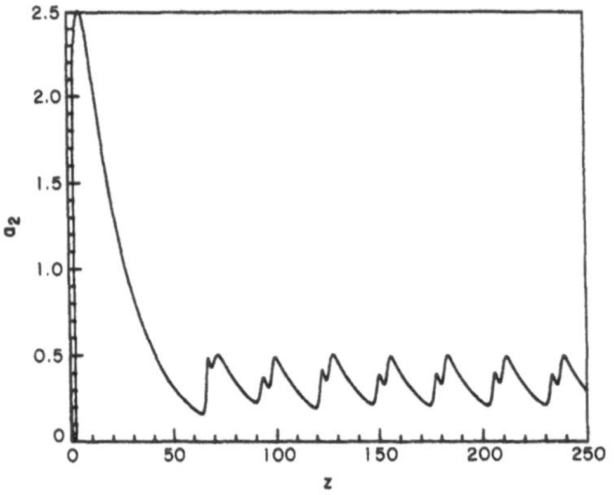

Figure 7.10

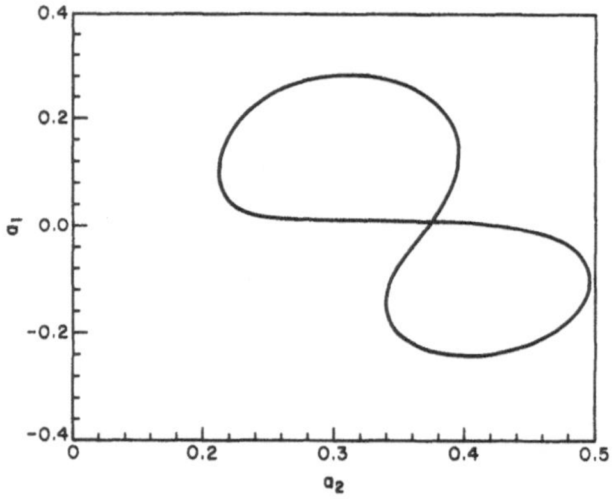

Figure 7.11

This example of order and chaos in a nonlinear optical problem will not be easy to study experimentally, as it assumes (nonsaturable) gain at the anti-Stokes frequency and loss at the pump and Stokes frequencies. Furthermore large propagation distances, or multipass configurations, would be required for typical Raman cross sections. In any event it certainly suggests that chaotic behavior may be found in other nonlinear optical phenomena. We might mention that this example was in fact the first (and only) example we have studied in the area of nonlinear wave propagation.

OPTICAL BISTABILITY

A system is optically bistable if it has two possible output states for the same input intensity. We refer the reader to the comprehensive review by Abraham and Smith. [29] Here we consider the possibility, first discussed by Ikeda, [30] of chaotic behavior (or "optical turbulence") in an optically bistable device. The device is basically of a type considered earlier by Szöke et al. [31] and McCall. [32]

The system of interest is shown in Figure 8.1. Mirrors 1 and 2 have reflectivity $R < 1$ ($R = 1 - T$), whereas mirrors 3 and 4 are assumed to be perfect reflectors. The cell inside this ring cavity contains N two-level atoms per unit volume. E_I is the complex amplitude of a monochromatic input field, and the transmitted and intracavity amplitudes are denoted E_T and E, respectively. The cell length is L, the total ring circuit length is $\mathcal{L}$, and we let $\ell = \mathcal{L} - L$.

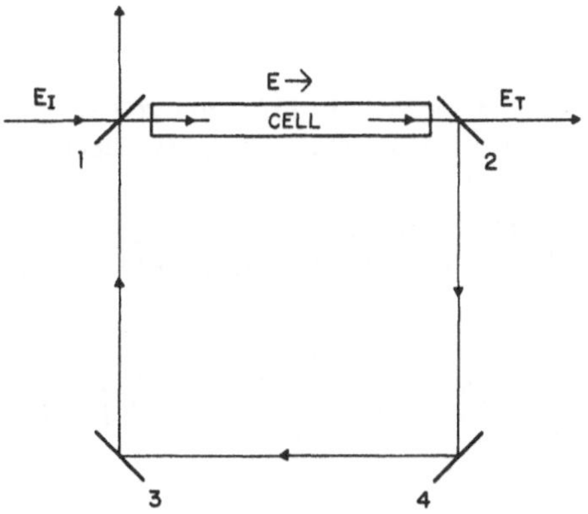

Figure 8.1

The system is described by the Maxwell-Bloch equations plus the cavity boundary conditions. In this sense the device is like a laser. The only major differences are that there is an injected field (E_I), and the medium is an absorber rather than an amplifier (i.e., $w_{eq} = -1$).

We allow for a z dependence of $\mathcal{E}$, so that

$$\dot{\mathcal{E}} = \frac{\partial \mathcal{E}}{\partial t} + c\frac{\partial \mathcal{E}}{\partial z} \tag{8.1}$$

in (5.7d). We also include a z-dependent phase ϕ in the electric field; this term is of interest for "dispersive" optical bistability. In terms of the independent variables $\tau = t - z/c$ and z, the appropriate Maxwell-Bloch equations are [33]

$$\frac{\partial u}{\partial \tau} = -\Delta v - \beta u \tag{8.2a}$$

$$\frac{\partial v}{\partial \tau} = \Delta u - \beta v + \left(\frac{d}{\hbar}\right)\mathcal{E} w \tag{8.2b}$$

$$\frac{\partial w}{\partial \tau} = -\gamma(w+1) - \left(\frac{d}{\hbar}\right)\mathcal{E} v \tag{8.2c}$$

$$\frac{\partial \mathcal{E}}{\partial z} = (2\pi N d k) v \tag{8.2d}$$

$$\mathcal{E}\frac{\partial \phi}{\partial z} = -(2\pi N d k) u \tag{8.2e}$$

We are assuming field loss only at the mirrors, so that $\gamma_c = 0$, and the absorber is assumed to be homogeneously broadened.

In the REA we assume β is large and replace u and v by their quasi-steady values. This leaves us with equations for w, $\mathcal{E}$, and ϕ:

$$\frac{\partial w}{\partial \tau} = -\gamma(w+1) - \left(\frac{\beta d^2/\hbar}{\Delta^2+\beta^2}\right)|E|^2 w \tag{8.3a}$$

$$\frac{\partial E}{\partial z} = \frac{2\pi N d^2 k/\hbar}{\Delta^2+\beta^2}(i\Delta+\beta) E w \tag{8.3b}$$

where $E = \mathcal{E}e^{i\phi}$. From these equations we easily derive the following:

$$E\left(\tau+\frac{z}{c}, z\right) = E(\tau,0)\exp\left[\frac{2\pi N d^2 k/\hbar}{\Delta^2+\beta^2}(i\Delta+\beta) W(\tau,z)\right] \tag{8.4a}$$

$$\frac{\partial}{\partial \tau} W(\tau,z) = -\gamma[W(\tau,z)+z] - \frac{1}{4\pi N\hbar k}|E(\tau,0)|^2 \cdot \left\{\exp\left[\frac{4\pi N d^2 k\beta/\hbar}{\Delta^2+\beta^2} W(\tau,z)\right] - 1\right\} \tag{8.4b}$$

where

$$W(\tau, z) = \int_0^z dz' \, w(\tau + \frac{z'}{c}, z') \tag{8.4c}$$

We must also satisfy the cavity boundary conditions. It is convenient to write these in terms of the fields at the ends of the absorption cell: [30]

$$E(t,0) = \sqrt{T}\, E_I(t) + R e^{ik\mathcal{L}} E(t - \frac{\ell}{c}, L) \tag{8.5a}$$

$$E_T(t) = \sqrt{T}\, E(t,L)\, e^{ikL} \tag{8.5b}$$

Equations (8.4) and (8.5) are the basis of Ikeda's analysis. In his paper the population difference is defined in such a way that w = 1/2(-1/2) for an atom in the upper (lower) level. To make the connection with Reference [30] easier, therefore, it is convenient to replace W in (8.4) by $2W$, so that W will be the same as Ikeda's. Defining $\theta = 2\pi N d^2 k/\hbar$, $\Delta\omega = \Delta$, $\gamma_T = \beta$, $\gamma_{\parallel} = \gamma$, and $\mu = d/\hbar$, we can write (8.4a) and (8.4b) as equations (4) and (5) of Reference [30]:

$$E(\tau + \frac{z}{c}, z) = E(\tau, 0) \exp\left[2\theta W(\tau, z) \frac{i\Delta\omega + \gamma_{\perp}}{\Delta\omega^2 + \gamma_{\perp}^2}\right] \tag{8.6a}$$

$$\frac{\partial}{\partial \tau} W(\tau, z) = -\gamma_{\parallel}(W + \frac{z}{2}) - \frac{\mu^2}{4\theta} |E(\tau, 0)|^2 \cdot \left\{ \exp\left[\frac{4\theta\gamma_{\perp} W}{\Delta\omega^2 + \gamma_{\perp}^2}\right] - 1 \right\} \tag{8.6b}$$

Finally, define the dimensionless variables [30]

$$x = \gamma_{\parallel} t \quad , \quad \kappa = \gamma_{\parallel}\mathcal{L}/c \tag{8.7a}$$

$$\phi(t) = \frac{1}{L} W(t - \frac{\mathcal{L}}{c}, L) \tag{8.7b}$$

$$\epsilon(t, z) = \frac{\mu}{2} E(t, z)\left[\gamma_{\perp}\gamma_{\parallel}(1 + \Delta^2)\right]^{-1/2} \tag{8.7c}$$

$$\epsilon_T(t) = \frac{\mu}{2} E_T(t-\frac{\ell}{c}) [\gamma_\perp \gamma_\parallel (1+\Delta^2)]^{-1/2} \quad (8.7d)$$

$$\epsilon_I(t) = \frac{\mu}{2} E_I(t) [\gamma_\perp \gamma_\parallel (1+\Delta^2)]^{-1/2} \quad (8.7e)$$

where now $\Delta = \Delta\omega/\gamma_\perp$. In terms of these variables, we can combine (8.5) and (8.6) to write the time-delay equations [30]

$$\epsilon(x,0) = \sqrt{T}\,\epsilon_I(x) + R\,\epsilon(x-\kappa,0)\, e^{aL\phi(x)} \cdot e^{i[aL\Delta(\phi(x)+\frac{1}{2}) - \delta_0]} \quad (8.8a)$$

$$\frac{d\phi}{dx} = -(\phi(x)+\tfrac{1}{2}) - 2|\epsilon(x-\kappa,0)|^2 \cdot \frac{1}{aL}[e^{2aL\phi(x)} - 1] \quad (8.8b)$$

$$\epsilon_T(x) = \sqrt{T}\,\epsilon(x-\kappa,0)\, e^{aL\phi(x)}\, e^{i[aL\Delta(\phi(x)+\frac{1}{2}) - (\delta_0+k\ell)]} \quad (8.8c)$$

where

$$a = \frac{2\Theta\gamma_\perp}{\Delta\omega^2+\gamma_\perp^2} \quad (8.9a)$$

$$\delta_0 = -k[(1-\frac{2\pi N\mu^2\Delta\omega}{\Delta\omega^2+\gamma_\perp^2})L + \ell] + 2\pi M \quad (8.9b)$$

and $2\pi M$ is introduced as the integral multiple of 2π closest to the first term on the right side of (8.9b).

Consider the steady-state limit of (8.8), obtained by setting $d\phi/dx = 0$ and $\epsilon(x,0) =$ constant. If $\delta_0 = \Delta = 0$ we obtain after some algebra the relation

$$\log[1+T(\frac{\epsilon_I}{\bar{\epsilon}_T}-1)] + \frac{1}{2T}\bar{\epsilon}_T^2\{[1+T(\frac{\epsilon_I}{\bar{\epsilon}_T}-1)]^2 - 1\} = \frac{1}{2}aL \quad (8.10)$$

where $\bar{\epsilon}_T$ denotes the steady-state value of ϵ_T. This is precisely the relation for "absorptive" bistability obtained by Bonifacio and Lugiato. [34] They show that, for R greater than some critical value depending on αL, there can be two values of $\bar{\epsilon}_T$ for the same value of ϵ_I, i.e., optical bistability. (This is called "absorptive" bistability because for $\Delta = 0$ there is no resonant dispersion.) If αL, $\alpha\Delta L$, $|\delta_0|$, and $|\bar{\epsilon}_T|^2/T$ are all small compared with unity, we have the case of "dispersive" bistability that has been observed experimentally by Gibbs et al. [35] For a review of optical bistability per se we again refer the reader to the article by Abraham and Smith, [29] which contains an exhaustive list of references. Henceforth we focus our attention on chaos in optical bistability.

If $\kappa \ll 1$, i.e., if the population difference has a relaxation time much shorter than the ring transit time $\mathcal{L}/c$, we may set the left side of (8.8b) equal to zero. In this limit the time-delay equations (8.8) reduce to the difference equations [30]

$$\epsilon_{on} = \sqrt{T}\,\epsilon_I + R\epsilon_{on-1} e^{\alpha L \phi_n} e^{i[\alpha L \Delta(\phi_n + 1/2) - \delta_0]} \tag{8.11a}$$

$$\epsilon_{Tn} = \sqrt{T}\,\epsilon_{on-1} e^{\alpha L \phi_n} e^{i[\alpha L \Delta(\phi_n + 1/2) - (\delta_0 + k\ell)]} \tag{8.11b}$$

$$\phi_n + \frac{1}{2} = \frac{2}{\alpha L} |\epsilon_{on-1}|^2 (1 - e^{\alpha L \phi_n}) \tag{8.11c}$$

where $\epsilon_{on} = \epsilon(x_0 + n\kappa, 0)$ and $\epsilon_{Tn} = \epsilon_T(x_0 + n\kappa)$. Thus the delay equations (8.8) have been reduced to a discrete mapping.

Ikeda [30] iterated the mapping (8.11) for the case $\alpha L\Delta = 6$, $\alpha L = 4$, $\delta_0 = 0$, and $R = .95$. Within a range of values of $|\epsilon_I|$, all the fixed points of (8.11) are unstable. Moreover, the sequence $\{\epsilon_{Tn}\}$ of transmitted field values is chaotic. This possibility of chaos in optical bistability has initiated a lively interest in "optical turbulence." We remark parenthetically that time-delay equations have been especially popular in the study of chaos more generally, because they may be reduced to discrete mappings as in the present example. For instance, Casati et al. [36] have studied a model of quantum chaos by casting a time-delay Schrödinger equation in the form of a discrete mapping.

Consider now a medium of two-level atoms far removed from

resonance with the input field. Suppose the absorption coefficient of the medium is nonsaturable and that the two-level atoms, being far off resonance, make no significant contribution to it. Then in (8.8) we replace $\alpha L\phi$ by $\alpha L(-1/2)$. Letting

$$\epsilon(x,0) = [2\Delta(1-e^{-\alpha L})]^{-1/2} e(x) \tag{8.12a}$$

$$\phi(x) + \frac{1}{2} = \left(\frac{1}{\alpha L\Delta}\right)\Theta(x) \tag{8.12b}$$

and using (8.7a), we may write (8.8a) and (8.8b) in the form

$$e(t) = A + Be(t-t_R)e^{i[\Theta(t)-\delta_0]} \tag{8.13a}$$

$$\frac{1}{\gamma_{\parallel}}\dot{\Theta}(t) = -\Theta(t) + |e(t-t_R)|^2 \tag{8.13b}$$

where $A = [2T\Delta(1-e^{-\alpha L})]^{1/2}\epsilon_I$, $B = Re^{-\alpha L/2}$, and $t_R = \mathcal{L}/c$.

Equations (8.13) are of exactly the same form as those of Ikeda et al. [37] for the case in which the cell contains a medium with a nonlinear refractive index. In their case the longitudinal relaxation rate $\gamma_{\parallel}$ in the Bloch equations is replaced by the relaxation rate used in the Debye theory of dielectric relaxation, and A is defined differently. They show that the equations (8.13) admit chaotic behavior, and Nakatsuka et al. [38] have observed this optical turbulence experimentally.

For $\gamma_{\parallel}t_R \gg 1$, i.e., when the cavity transit time is much larger than the material relaxation time, we may use an adiabatic approximation to $\Theta(t)$ by setting the left side of (8.13b) to zero. Then we have a discrete mapping rather than a continuous flow:

$$e(t) = A + Be(t-t_R)e^{i[|e(t-t_R)|^2-\delta_0]} \tag{8.14}$$

As A is increased with B, δ_0 fixed, there is a period-doubling route to chaos and the mapping (8.14) has a strange attractor.

Ikeda et al. [37] note that, with increasing t, equation (8.14) eventually becomes inapplicable, even if the condition $\gamma_{\parallel}t_R \gg 1$ is satisfied. This is a consequence of "very sensitive dependence on initial conditions." For if e(t) has any variations, small differences between $e(t_1)$ and $e(t_2)$ $(0 \le t_1 \ne t_2 < t_R)$ become magnified, and their images under the mapping (8.14) eventually have shorter and shorter "correlation time." When this time becomes as small as the material relaxation time, we can no longer make an adiabatic elimination of $\Theta(t)$.

If $B \ll 1$ but $A^2 B \sim O(1)$, we can write the following approximate version of (8.13): [37]

$$\frac{1}{\gamma_{\parallel}} \dot{\Theta}(t) = -\Theta(t) + A^2\{1 + 2B\cos[\Theta(t-t_R) - \delta_0]\} \tag{8.15}$$

Note that the nonlinear, dissipative dynamical system (8.15), being a time-delay differential equation, is equivalent to an autonomous system of effectively infinite order. The Poincaré-Bendixson theorem is therefore not applicable. Indeed, by numerically integrating (8.15) and Fourier transforming the time series so obtained, Ikeda et al. [37] find that the system (8.15) exhibits a transition to chaos as A is varied. That is, the power spectrum of $\Theta(t)$ changes from a series of sharp spikes to a broadband spectrum.

The experiments of Nakatsuka et al. [38] support this prediction of chaos in an "all-optical" bistable system. (Previous observations of chaos in optical bistability [39,40] employed electronic delay lines to obtain the delayed feedback.) Their experimental arrangement is indicated in Figure 8.2. A single-mode optical fiber, having a quadratic nonlinear index, is used as the nonlinear "cell" in the ring cavity. The high power level necessary for the observation of chaos was obtained from the second harmonic of a mode-locked YAG laser (pulse separation = 7.6 nsec). Since the bifurcations to chaos stem from the interference of the input and cavity fields, it was necessary to match the ring transit time to the period of the mode-locked pulses (7.6 nsec). The ring cavity was such that B = .4-.5, and the parameter A could be increased sufficiently (by raising the input power level) to realize the chaotic regime of the system (8.14), modified so that A is a (Gaussian) function of time. When the peak power of the input pulse train was raised from 50 to 160 W, a period doubling of the output pulse train was observed. At a power level of 300 W the output was chaotic. The results were in good agreement with the theoretical model. [38]

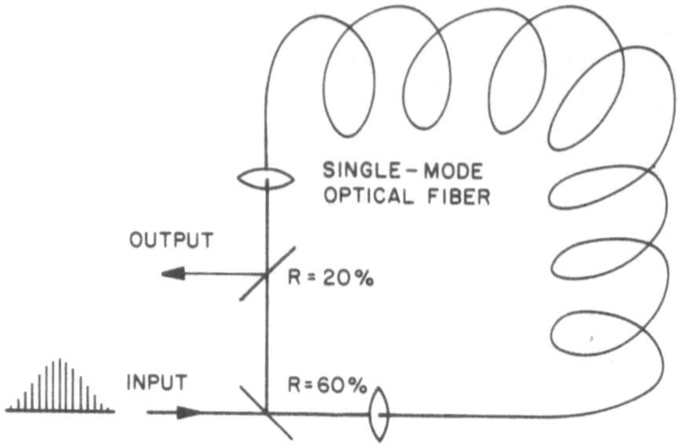

Figure 8.2

INFRARED LASER-MOLECULE INTERACTIONS

Molecular Structure

Let us consider the gas-phase energy-level diagram of a typical molecule as shown in Figure 9.1. We observe that the largest energy level spacings correspond to electronic transitions which are in the visible and uv spectral regions, 10^4 or 10^5 cm^{-1}. Superimposed on these states with the next largest energy level spacing are vibrational transitions in the infrared region, 10^3 cm^{-1}. The smallest energy level spacing corresponds to rotational transitions in the microwave region, 1 cm^{-1}. [41] The physical motions corresponding to these spectral regions are illustrated in Figure 9.2 for a diatomic molecule. [42] The rotational structure is due to the ability of this molecule to rotate about the x and y axes. The vibrational structure is due to the bonding between the atoms. If the binding force is simply proportional to the separation of the atoms from their equilibrium position, then the motion is always described by a simple harmonic oscillator. In a real molecule the harmonic region exists only for small displacements from equilibrium.

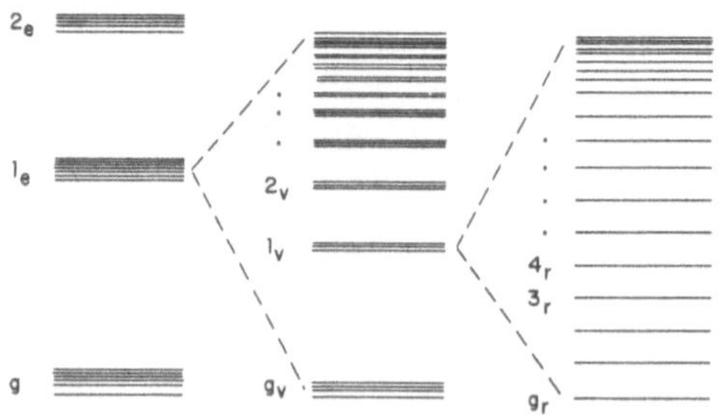

Figure 9.1. Molecular energy levels.

Since the energy level spacing for electronic, vibronic and rotational transitions differ by orders of magnitude, it is safe to assume the separability of these motions. The Born-Oppenheimer approximation is simply a statement of the separation of the electronic structure from the vibrational-rotational structure. [43]

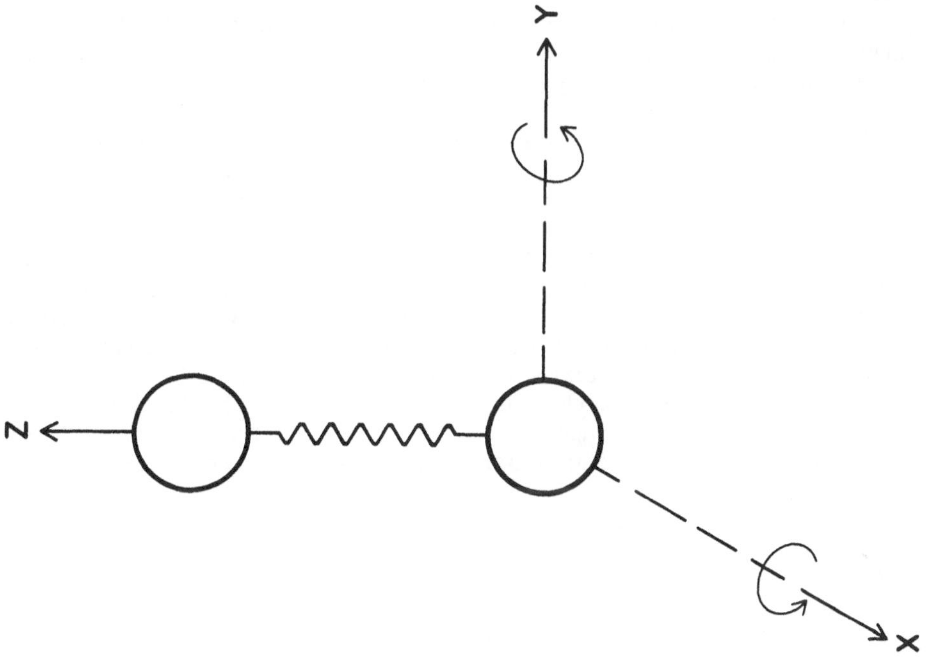

Figure 9.2

Since we are presently interested in infrared laser-molecule interactions, we will henceforth neglect the molecule's electronic structure.

The Hamiltonian for a diatomic molecule interacting with an infrared laser is taken to be

$$H = \Delta a_z^\dagger a_z + B_o(J_x^2 + J_y^2) + \Omega \hat{\epsilon}_Z \mathbb{C}_{Zz}(a_z + a_z^\dagger) \quad (9.1)$$

where the first (second) term represents the vibrational (rotational) motion of the molecule. The last term represents the dipole coupling of the molecule to the laser. The operators $a_z^\dagger$ and a_z are the creation and annihilation operators for the vibrational motion along the $\hat{z}$ axis of the molecule. The operators J_x and J_y are the operators representing the two rotational degrees of freedom of the molecule. The quantity C_{Zz} is the direction cosine diadic which takes the laser along the $\hat{Z}$ axis in the laboratory frame, into the molecular dipole moment along the $\hat{z}$ axis in the molecule frame. The unit vector $\hat{\epsilon}_Z$ describes the orientation of the laser. The parameters Δ and Ω are the laser-molecule frequency detuning and the laser Rabi frequency. B_o is proportional to the inverse of the molecule's moment of inertia. The Hamiltonian (9.1) describes the interaction of the laser with a non-degenerate mode of a polyatomic molecule.

Let us now consider SF_6 as an example of a polyatomic molecule. In Figure 9.3 we observe the 2 infrared active normal modes of vibration, ν_3 and ν_4. Since these modes are separated by $\sim 300\ cm^{-1}$, a laser can selectively interact with a specified mode. Since the mode is in the $10\,\mu m$ region, it has been studied extensively using a CO_2 laser. [44] As opposed to our diatomic molecule the ν_3 mode is triply degenerate, and this leads to some additional features in the Hamiltonian, as we will see later. In order to fully account for molecular multiple photon absorption we must include the non-laser-coupled normal modes of the molecule in the Hamiltonian because these modes are coupled to the laser-pumped ν_3 mode. The complete generic Hamiltonian describing infrared absorption in a polyatomic molecule is

$$H = H_{pm} + H_{bm} + H_{ic} + H_R + H_{pmf} \quad (9.2)$$

where

$$H_{pm} = \Delta \underset{\sim}{a}^\dagger \cdot \underset{\sim}{a} + \chi(\underset{\sim}{a}^\dagger \cdot \underset{\sim}{a})^2 + G(\underset{\sim}{a}^\dagger \times \underset{\sim}{a})^2 + T\sum_i (a_i^\dagger a_i)^2 \quad (9.3a)$$

$$H_{bm} = \sum_m (\Delta + \epsilon_m)\, \underset{\sim}{b}_m^\dagger \cdot \underset{\sim}{b}_m + \sum_m \chi_{mm} (\underset{\sim}{b}_m^\dagger \cdot \underset{\sim}{b}_m)^2 \quad (9.3b)$$

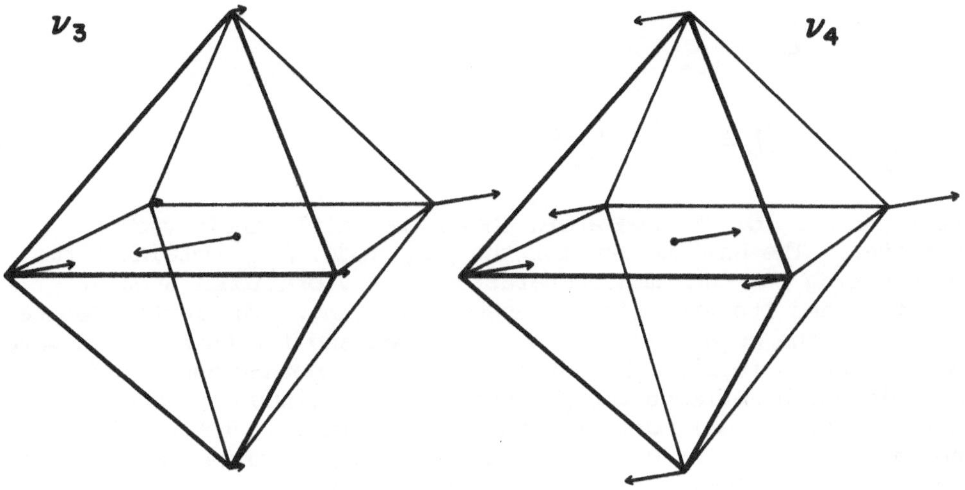

Figure 9.3

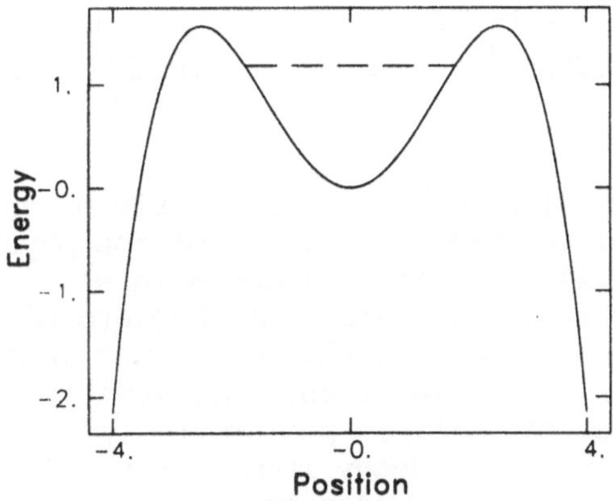

Figure 9.4

$$H_{ic} = \sum_m \beta_m (\underset{\sim}{a}^{\dagger} \cdot \underset{\sim}{b}_m + \underset{\sim}{b}_m^{\dagger} \cdot \underset{\sim}{a}) + \sum_n \sum_{m \neq n} \chi_{mn} (\underset{\sim}{b}_n \cdot \underset{\sim}{b}_m^{\dagger})(\underset{\sim}{b}_m \cdot \underset{\sim}{b}_n^{\dagger}) \tag{9.3c}$$

$$H_R = B_0 \underset{\sim}{J}^2 \tag{9.3d}$$

$$H_{pmf} = \Omega \hat{\varepsilon} \cdot \overleftrightarrow{C} \cdot (\underset{\sim}{a}^{\dagger} + a) \tag{9.3e}$$

The twiddle below the operators shows explicitly their vector character. The Hamiltonian for the pump mode, H_{pm}, contains terms representing its anharmonic character, its vibrational angular momentum, and its vibrational tensor splitting. The latter two terms are due to the degeneracy of the pump mode and for this reason were not present in (9.1). The anharmonicity is another aspect of a real molecule which reflects the presence of a continuum at some higher energy, i.e., a real molecule can be dissociated. The Hamiltonian for the background modes, H_{bm} is similar to the pump mode, but we have neglected to include the analogous terms to those with G and T in (9.3a) for simplicity. The background modes, $\underset{\sim}{b}_m$, are therefore a generic representation for the non-laser-coupled modes of the molecule. We believe this model for H_{bm} is adequate to understand general features of intramolecular coupling. Eq. (9.3c) represents the intramolecular coupling between all the modes where the first term is the coupling between the pump mode and the background modes, and the second term is a mixing amongst the background modes. Eqs. (9.3d) and (9.3e) are the rotational and the laser-pump mode Hamiltonians, respectively.

In the next sections we will consider model problems which relate to some of the physical features of multiple photon absorption. We believe the most significant aspect of our latter two examples is the existence of chaotic behavior for realistic values of the molecular parameters, i.e., chaos is a fundamental aspect of the actual physics.

The Duffing Oscillator

The Duffing Oscillator is a damped, driven anharmonic oscillator. [45,46] This model represents a subset of the physics contained in the Hamiltonian (9.2), i.e. it includes a laser, a non-degenerate pumped mode and irreversible decay into background modes. The coupling of the pumped mode and the background modes in (9.3c) is not irreversible. We will see later the relationship of this model to a more realistic model of multiple photon absorption. If for the moment we exclude the irreversible decay from this model, we can write a Hamiltonian for the Duffing system as

$$H = \frac{x^2}{2} + \frac{p^2}{2} - \frac{\beta x^4}{4} - x \cos \mu \tau \tag{9.4}$$

The potential energy is $V(x) = x^2/2 - \beta x^4/4$. In (9.4) we have not made the rotating-wave-approximation so that the $\cos \mu \tau$ oscillation of the monochromatic laser is shown explicitly. In these units $\hbar = 1$. We can recognize the relationship between this Hamiltonian and our previous Hamiltonians by writing

$$x \sim a + a^{+}$$

$$p \sim i(a - a^{+})$$

which are the standard relations for the momentum and position and the creation and annihilation operators for an oscillator. The parameter β represents the anharmonicity of the oscillator, and μ represents the laser frequency. Resonance for this oscillator occurs at $\mu = 1$.

The equations of motion for this system are obtained from the Hamiltonian (9.4) in the usual way,

$$\dot{x} = p \tag{9.5a}$$

$$\dot{p} = -x + \beta x^3 + \cos \mu \tau - \gamma \dot{x} \tag{9.5b}$$

where we have inserted a decay term with parameter γ by hand. Eqs. (9.5) are usually written as a single differential equation in x:

$$\ddot{x} + \gamma \dot{x} + x - \beta x^3 = \cos \mu \tau \tag{9.6}$$

A more physical picture of this system can be obtained by studying Figure 9.4 for this symmetric potential well, $\beta = .1587$. We are always going to assume that the oscillator is initially at rest at the bottom of the well and that the excitation is solely due to the laser pumping. Because the pumped mode is damped the system tends to reach a close balance between laser energy into the molecular pumped mode and intramolecular energy loss to the background modes of the molecule. A dashed line has been drawn in Figure 9.4 to show an example of a bound orbit with constant energy. This balance is exact if the rotating-wave approximation is made. It is also possible to dissociate the molecule if the energy loss is not too rapid. In this case the laser will drive the oscillator out of the well and the coordinate displacement will diverge.

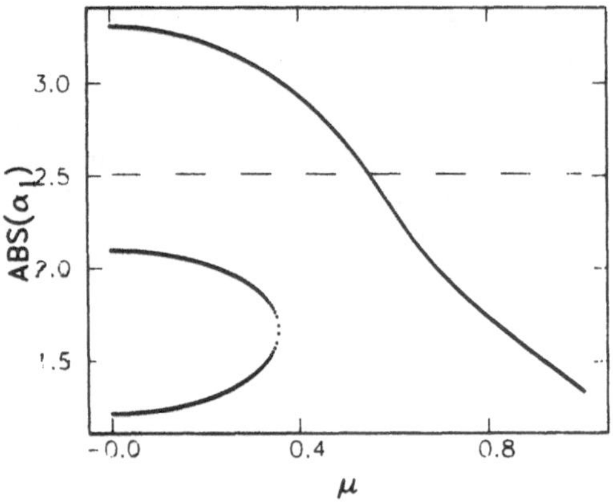

Figure 9.5

In order to get a better feel for the system's dynamics we will make the rotating-wave approximation (RWA). We assume a solution of the form

$$x(\tau) \cong \frac{1}{2}\{\alpha_1(\tau)e^{i\mu\tau} + c.c.\} \tag{9.7}$$

where $\alpha_1(\tau)$ is assumed slowly varying with respect to the oscillations at period μ. Upon substitution of (9.7) into (9.6) we find the RWA equation of motion to be

$$(\gamma + 2i\mu)\dot{\alpha}_1 + \{1-\mu^2 + i\mu\gamma - \frac{3\beta}{4}|\alpha_1|^2\}\alpha_1 = 1 \tag{9.8}$$

where we have neglected higher harmonics and $|\ddot{\alpha}_1| \ll \mu^2|\alpha_1|$. Since (9.8) is equivalent to two real first-order differential equations, we see from the Poincaré-Bendixson theorem that this system cannot exhibit chaos. Eqs. (9.8) relax to a steady state which is given by

$$\alpha_1\{1-\mu^2 + i\gamma\mu - \frac{3\beta}{4}|\alpha_1|^2\} = 1 \tag{9.9}$$

Eq. (9.9) is equivalent to a cubic equation for $|\alpha_1|^2$. Since $|\alpha_1|^2$ is positive and real, there will be either one solution or three solutions to the cubic equation. In Figure 9.5 we show a plot of $|\alpha_1|^2$ vs. μ, for $\beta = .1587$ and $\gamma = .72$. We have also indicated the dissociation threshold, i.e. the value of $|\alpha_1|^2$ which is at the turnover in the potential well. For these parameter values we observe a region where there exist three solutions. Both the RWA solutions to (9.8) and the full solutions to (9.6) can show hysteresis in this region. Therefore, any chaos observed in the solutions to (9.8) cannot be due solely to the bistable behavior of the dynamics. We should also point out that the middle solution shown in Figure 9.5 is always unstable. This is implied by turning up the strength of the laser and noticing that this curve decreases in amplitude, which is unphysical. [46] The upper and lower curves both increase.

In order to observe the onset of chaos in the solution to (9.6) we will plot graphs of x(t) vs. $\dot{x}(t)$ after the system has reached a steady state, i.e., look at the behavior of the trajectory after it has reached a limit cycle. In Figure 9.6 we show the limit cycle for $\beta = .1587$, $\gamma = .72$ and $\mu = 1$ (resonance). In Figure 9.7 we show the FFT of the dynamics of Figure 9.6. We see not only the fundamental at $\mu = 1$, but all higher odd harmonics, $\mu = 3,5,7\ldots.$ Since the spectral amplitudes are on a log plot, we observe the dominance of the fundamental, emphasizing the validity of the RWA. As we decrease and move along the upper branch of the steady-state curve in Figure 9.5, we slowly approach the turnover in the potential well. When we get near this point of becoming unbounded, we observe the onset of a dc shift of the limit cycle and the appearance of all the even harmonics as shown in Figures 9.8 and 9.9. The onset occurs near $\mu =$ .541 and it has been studied in some detail by Novak and Frehlich. [47] As we decrease μ further, we eventually observe the onset of the 1/2 subharmonic and all its harmonics, 3/2, 5/2, 7/2... as shown in Figures 9.10 and 9.11, $\mu = .5255$ (the onset for the 1/2 harmonic sequence occurs for $\mu \cong .5277$). This period doubling sequence continues until the dynamics becomes chaotic. [48] Once in the chaotic regime there will exist windows where stable limit cycles are observed. The existence of these windows is expected from the theorem of Sarkovskii. [9-11]

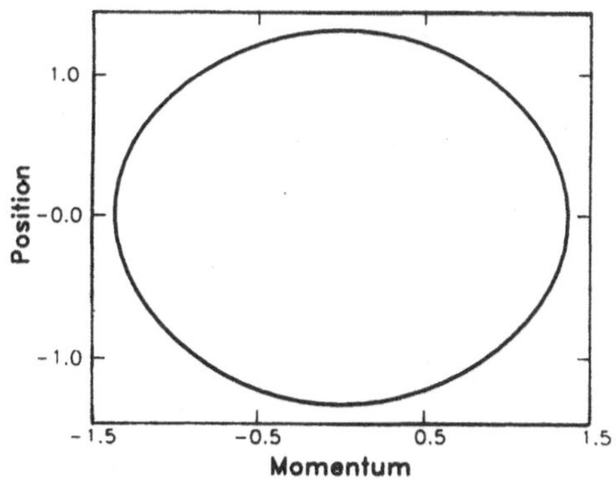

Figure 9.6

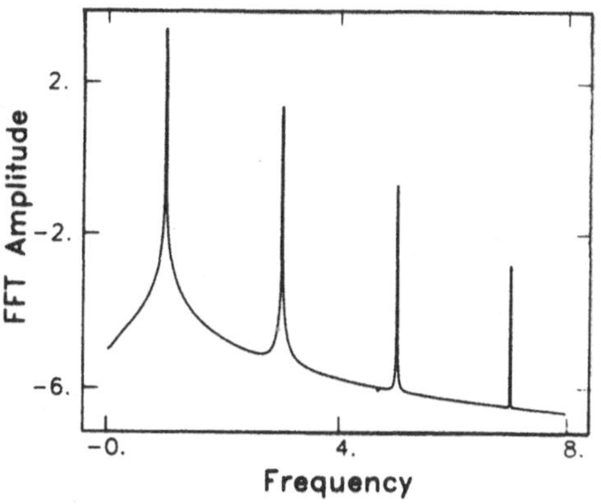

Figure 9.7

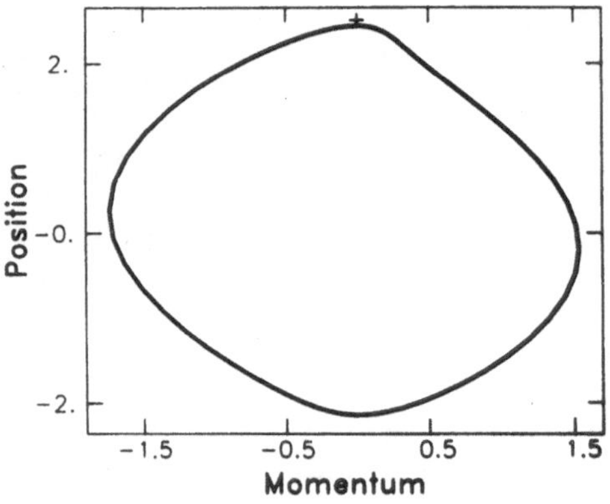

Figure 9.8

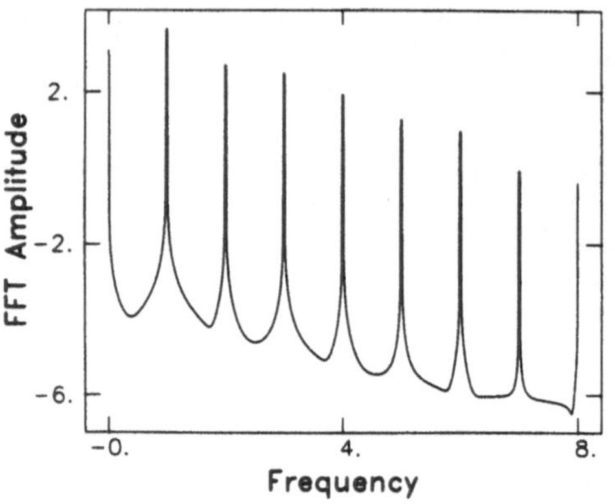

Figure 9.9

Now that we have seen the structure of the solutions via the FFT, it would appear useful to assume a solution of the form

$$x(\tau) = a_0 + \frac{1}{2}\sum_n (a_n e^{in\mu\tau} + c.c.) \qquad (9.10)$$

where n takes on whatever discrete values are necessary to satisfy the complete solution. While it is possible in practice to begin at $\mu = 1$ and assume the series only requires n = 1,3, one quickly finds that more and more terms are necessary to predict the successive period-doubling onset conditions accurately.

Let us, therefore, go into the mathematical structure of the system in more detail, assuming none of the harmonics exist. In order to predict the onset for the dc shift we assume a solution of the form

$$x(\tau) = a_0 + \frac{1}{2}(a_1 e^{i\mu\tau} + c.c.) \qquad (9.11)$$

Substituting (9.11) into (9.6) gives

$$\{A_1 - 3\beta a_0^2 - \frac{3\beta}{4}|a_1|^2\} a_1 = 1 \qquad (9.12a)$$

$$a_0\{1 - \beta[a_0^2 + \frac{3|a_1|^2}{2}]\} = 0 \qquad (9.12b)$$

where

$$A_n \equiv 1 - (n\mu)^2 + in\mu\gamma \tag{9.13}$$

Before and up until the onset, (9.12a) reduces to (9.9) and determines the value of α_1 for all μ in this region. Eq. (9.12b) determines the onset, since for all μ lower than the α_0 onset the { } must be identically zero. Therefore, α_0 onsets when

$$1 = \frac{3\beta}{2}|\alpha_1|^2 \tag{9.14}$$

Substituting (9.14) into the modulus square of (9.9) gives the condition on μ for the onset,

$$\mu^2 = \frac{1}{2}(1-\gamma^2) \pm \frac{1}{2}\sqrt{(1-\gamma^2)^2 - (1-6\beta)} \tag{9.15}$$

There are two solutions for μ from (9.15); however, only the + sign corresponds to the region we are studying. The other solution occurs at an intercept with the unstable branch and is not physical. For the parameter values β = .1587 and γ = .72 we find μ = .67480134 at the onset of α_0. This number is not in agreement with the solution of the full problem because of the influence of the higher harmonics, both even and odd. In the past-onset region where only α_0 and α_1 exist we can solve for these parameters using the curly bracket in (9.12b) and substituting into (9.12a) for α_0^2. This results in a new cubic equation for $|\alpha_1|^2$,

$$\left(\frac{15\beta}{4}\right)^2|\alpha_1|^6 - \frac{15\beta}{4}(2+\mu^2)|\alpha_1|^4 + \left[(2+\mu^2)^2 + (\mu\gamma)^2\right]|\alpha_1|^2 - 1 = 0 \tag{9.16}$$

whose solution gives both α_0 using (9.12b) and α_1 using (9.12a). We have found that when three roots exist the central root is the physically meaningful one. Since these expressions do not contain all the higher harmonics, they are not a valid description of the full system (9.6). Can we realize our approximate "analytic" results? The answer is yes if we modify the original system. Our approach to modifying (9.6) is to modify the laser source term in such a way that the higher harmonics are exactly zero. For example in the region μ = 1 let us generate the "RWA" equation for α_3 assuming only α_1 exists,

$$\left(A_3 - \frac{3\beta}{2}|\alpha_1|^2 - \frac{3\beta}{4}|\alpha_3|^2\right)\alpha_3 - \frac{\beta\alpha_1^3}{4} = 0 \tag{9.17}$$

If we add a source of the form

$$S_3 \equiv -\beta \frac{a_1^3}{8} e^{3i\mu\tau} + c.c. \qquad (9.18)$$

to (9.6) then this source will remove the generation of a_3 due to a_1. Therefore, all the higher odd harmonics are decoupled from a_1. Using the solution to (9.9) for $\mu = 1$ we can solve the modified equation (9.6). In Figure 9.12 we show the FFT of this limit cycle which exhibits only the fundamental. Using this same technique to remove a_2 once a_0 exists we can probe the a_0 onset regime. In Figures 9.13a,b we show two values of μ, one just below the onset and one just above. The circles on the FFT are the analytic solutions. The agreement is excellent, showing the validity of the method.

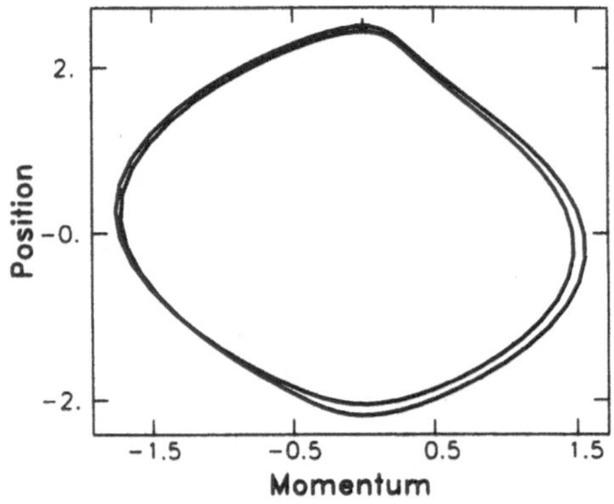

Figure 9.10

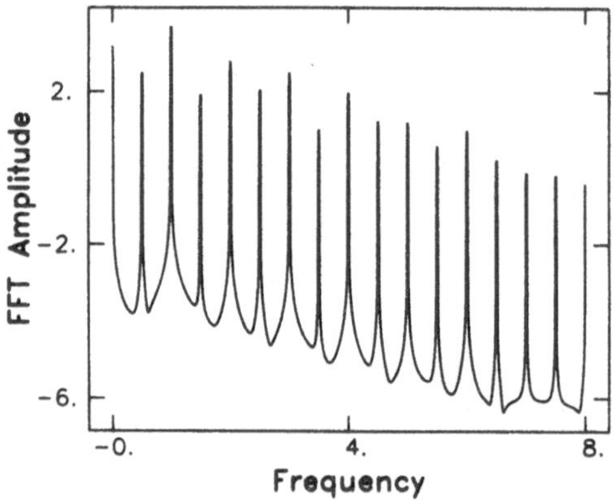

Figure 9.11

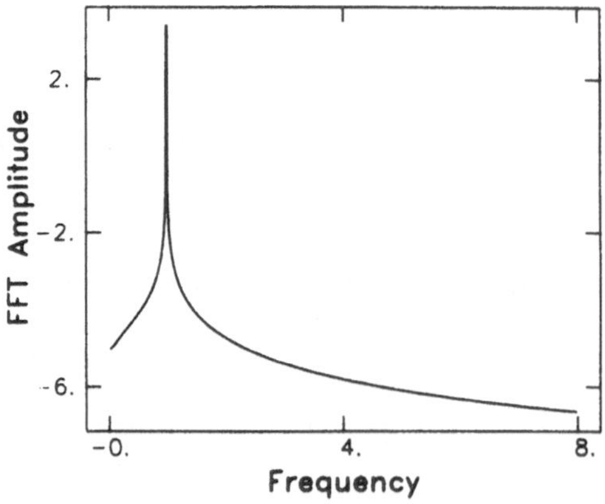

Figure 9.12

We have illustrated the period-doubling sequence to chaos in the Duffing oscillator and also showed a window of period three. The onset of the dc component of the solution and the period-doubling sequence arise as the limit cycles approach the top of the well and become unbounded. [47,49] We have demonstrated a method for realizing these onset conditions free of the complicating higher harmonics. Our last remark concerns the physical reality of this model to MPE. In a real molecule, it is not possible for a laser to excite the pumped mode very near to the top of the potential well because of the anharmonicity and the rapid intramolecular dynamics. The parameters have been extended beyond their physically realistic values. In addition, the irreversible decay of energy into the background is not physically correct. As we will see in our next example of MPE, energy is continually transferred between the laser pumped mode and the background modes.

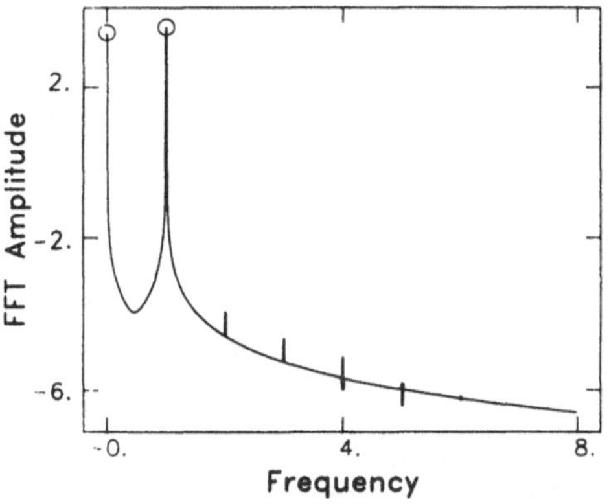

Figure 9.13a

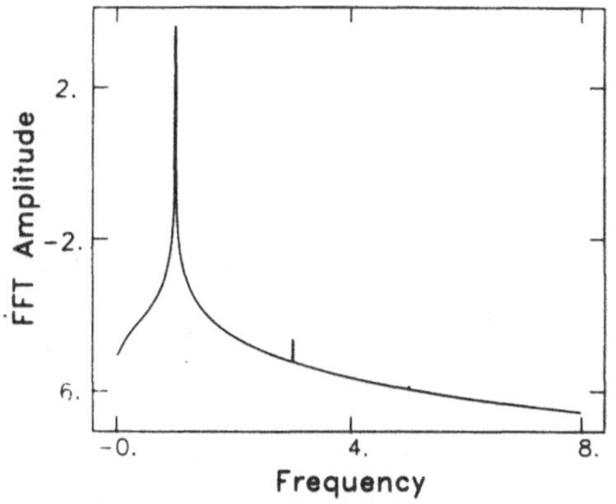

Figure 9.13b

A Basic Model of Multiple Photon Excitation

A general feature of multiple photon excitation is that the absorption of laser photons is essentially fluence dependent, i.e. it depends on the pulse energy and not the pulse intensity. The generality of this result has been shown by Lyman, Quigley and Judd for ~ 50 different polyatomic molecules. [50] Since averaging over homogeneous and inhomogeneous linewidths would lead to an obvious explanation of this effect, it has been assumed that hotband and rotational averaging are responsible for this effect. As theorists we made the obvious suggestion to the experimentalists: If you perform cold experiments, you will see both intensity dependence and aspects of each molecule's molecular structure. You should note that low temperature reduces the number of hotbands initially excited and also collapses the rotational distribution to low J values.

To describe these low temperature experiments we need a model which includes the basic physics of the process: First, an anharmonic mode which is coupled to the laser and second, a group of background models which are coupled to the laser-pumped mode. The essence of this model is contained in a subset of the full Hamiltonian (9.3). The Hamiltonian is

$$H = H_{pm} + H_{bm} + H_{ic} + H_{pmf} \tag{9.19}$$

where

$$H_{pm} = \Delta a^{\dagger} a - \chi (a^{\dagger} a)^2 \tag{9.20}$$

$$H_{bm} = \sum_m (\Delta + \epsilon_m) b_m^{\dagger} b_m \tag{9.20}$$

$$H_{ic} = \sum_m \beta_m (a^{\dagger} b_m + b_m^{\dagger} a) \tag{9.20}$$

$$H_{pmf} = \frac{\Omega}{\sqrt{n}} (a c^{\dagger} + c a^{\dagger}) \tag{9.20}$$

In (9.3) we have made the RWA and assumed a semiclassical laser field. In this Hamiltonian we have assumed a quantized laser represented by an harmonic oscillator with creation and annihilation operators $c^{\dagger}$ and c. We have done this so that a constant of the motion exists for this system, i.e., the total excitation number

$$a^{\dagger} a + \sum_m b_m^{\dagger} b_m + c^{\dagger} c = constant \tag{9.21}$$

By having (9.21) we need only calculate two of the quantities in (9.21) in order to obtain the third.

Let us now consider some approximations which are useful in dealing with the system (9.19). Without loss of generality we will assume the coupling parameters β_m are all equal to β. In radiationless transition theory it is also convenient to assume the background is uniformly distributed. Here we have

$$\epsilon_m = \bar{\Delta}_0 + m\rho^{-1} \qquad (9.22)$$

where $\bar{\Delta}_0$ is the separation between the band origins of the nearest background oscillator and the pumped mode. [51] The density of background states is ρ. For simplicity we have assumed the background oscillators are harmonic. In SF_6 the closest background mode to ν_3 is $\nu_2 + \nu_6$, and the anharmonicity of this mode is much smaller than the anharmonicity of ν_3. It is also known from the work of Bixon and Jortner that the strength of the coupling of the pumped mode with the background, when diagonalized, is distributed with a Lorenzian profile. [52] For small coupling constant β, which we have in a real molecule, we can safely extend the number of background oscillators to infinity. Since we have many more photons than molecules, it is reasonable to solve the dynamics for the molecule semiclassically. Our last approximation is to ignore quantum effects and solve the entire dynamics classically, i.e., $a^{\dagger} \rightarrow a^*$, $c^{\dagger} \rightarrow c^*$, etc. We believe that multiple photon excitation which involves the absorption in many molecules of $\sim$30 photons cannot be dependent on quantum effects, which correlate to one or two photons absorbed. In addition, the generality of the phenomenon implies the dominance of gross features of the absorption as being important.

The effect of all these approximations is to allow the equations of motion to be written in the form

$$\dot{a} = -i\Delta a + 2i\chi |a|^2 a - i\Omega - i\beta \sum_m b_m \qquad (9.23a)$$

$$\dot{b}_m = -i(\Delta + \Delta_0 + m\rho^{-1}) b_m - i\beta a \qquad (9.23b)$$

$$\dot{c} = -\frac{i\Omega}{\sqrt{n}} a \qquad (9.23c)$$

where at t = 0 we have n photons in the laser field. The semiclassical approximation allows (9.23a) to be solved independently of c, where c is computed from our knowledge of a. A formal solution of (9.23c) is

$$c(t) = \sqrt{n} - \frac{i\Omega}{\sqrt{n}} \int_0^t dt' \, a(t') \qquad (9.24)$$

which allows us to easily compute $|c|^2$:

$$|c(t)|^2 = n + 2\Omega \int_0^t dt' \, Im\,(a(t')) \tag{9.25}$$

Using (9.21) we obtain the total number of photons absorbed by the molecule,

$$n - |c(t)|^2 = -2\Omega \int_0^t dt' \, Im\,(a(t')) \tag{9.26}$$

where I_m refers to the imaginary part. The linear form of (9.23b) allows us to write the formal solution for $b_m(t)$ as

$$b_m(t) = -i\beta \int_0^t dt' \, e^{-i(\Delta + \bar{\Delta}_0 + m\rho^{-1})(t-t')} a(t') \tag{9.27}$$

where all the modes are initially in their ground states, eliminating the homogeneous solution in (9.27). Since in (9.23a) we need to evaluate the quantity $\beta \sum_m b_m$, let us look at this term now. Using (9.27) we obtain

$$\sum_m b_m = -i\beta \int_0^t dt' \, e^{-i(\Delta + \bar{\Delta}_0)(t-t')} \left\{ \sum_m e^{-im\rho^{-1}(t-t')} \right\} a(t') \tag{9.28}$$

where the only dependence on the background m level is contained in the curly bracket. The expression in the curly bracket using the Poisson sum rule can be rewritten as

$$\sum_{m=-\infty}^{\infty} e^{-im\rho^{-1}\tau} = 2\pi\rho \sum_{m=-\infty}^{\infty} \delta(\tau - 2\pi m\rho) \tag{9.29}$$

The equally spaced background forces the input of information from the background modes to the pumped mode to occur in regular intervals. While the integral in principle requires all past history to influence the future continuously, this model exhibits a strong rephasing at well-defined intervals. [55] Substituting (9.29) and (9.28) into (9.23a) gives a delay differential equation for a(t),

$$\dot{a}(t) = -i\Delta a(t) + 2i\chi |a(t)|^2 a(t) - i\Omega - \frac{\gamma}{2} a(t)$$
$$- \gamma \sum_{m=1}^{\infty} \bar{\beta}^m a(t - m\tau_R)\, \Theta(t - m\tau_R) \qquad (9.30)$$

where $\Theta(t)$ is the Heavyside step function, $\gamma = 2\pi\beta^2\rho$ is the Golden Rule rate, $\bar{\beta} = e^{-i(\Delta+\Delta_0)\tau_R}$ is a phase factor which determines whether the rephasing of the background modes leads to absorption or emission, and $\tau_R = 2\pi\rho$ is the time interval for background rephasing. [55]

If we set the anharmonicity χ to zero in (9.30), the equation becomes linear and has an exact analytic solution. This system cannot exhibit chaos because it is linear even though it is infinite dimensional. If we let $\tau_R \to \infty$, then we are always on the first interval and the summation over m vanishes. In this case (9.30) reduces essentially to the RWA Duffing equation (9.8). The limit $\tau_R \to \infty$ is equivalent to letting the spacing between background states $\to 0$ and $\beta \to 0$ such that γ is finite. As we learned before this equation does not exhibit chaos. Notice on the first interval we have exponential irreversible decay of energy from the pumped mode into the background modes. Only after τ_R do we begin to see energy transfer back into the pumped mode. The phase $\bar{\beta}$ determines in detail how the relative location of the band origins affects absorption and emission of photons in the molecule. While the model overall is very sensitive to this parameter, the integrated absorption for both harmonic and anharmonic systems is independent of $\bar{\beta}$. In the harmonic case the absorption is always linear.

Equation (9.30) exhibits some scaling relations which both illustrate some physics and also simplify the parameter space. If we define

$$\bar{a} \equiv \frac{\gamma}{\Omega} a \qquad (9.31a)$$

$$\alpha \equiv \frac{\Omega^2}{\gamma^3} \chi \qquad (9.31b)$$

$$T \equiv \gamma t \qquad (9.31c)$$

$$T_R \equiv \gamma \tau_R \qquad (9.31d)$$

then (9.30) becomes

$$\dot{\bar{a}}(T) = -i\frac{\Delta}{\gamma}\bar{a}(T) - \frac{\bar{a}(T)}{2} + 2ia|\bar{a}(T)|^2\bar{a}(T)$$

$$- i - \sum_{m=1}^{\infty}\bar{\beta}^m \bar{a}(T-mT_R)\,\Theta(T-mT_R) \tag{9.32}$$

where $\bar{a}$ is the scaled oscillator annihilation variable and α is the scaled anharmonicity. On resonance (9.32) depends only on two variables α and $\bar{\beta}$. For all further discussions we will only consider this case, $\Delta = 0$. The physics of this scaling is primarily contained in (9.31b). If we know the molecular anharmonicity χ, then the ratio of Ω and γ determine the scaled system. If the relaxation is very fast and the laser cannot pump the molecule very far up the ladder, then the scaled system will have a small anharmonicity. If the laser power is very intense, then the molecule can be pumped fairly far up the ladder and the scaled model will have a large anharmonicity. Both of these effects are physically intuitive and show us the equivalence of many different molecules with respect to their relaxation, laser power, and anharmonicity.

If we consider a case where the relaxation is very fast such that on each interval τ_R the dynamics quickly approaches a steady-state, then we can set $\dot{a}(t) = 0$ and reduce (9.32) to an equation for the Nth interval steady-state amplitude as a function of all previous steady-state amplitudes,

$$\left\{\frac{1}{2} - 2i\alpha|\bar{a}^N|^2\right\}\bar{a}^N = -i - \sum_{m=1}^{N}\bar{\beta}^m\,\bar{a}^{N-m}$$

$$= -i - \sum_{m=0}^{N-1}\bar{\beta}^{N-m}\,\bar{a}^m \tag{9.33}$$

The superscript on $\bar{a}$ references the appropriate interval, and a superscript of zero refers to the first interval. By writing the amplitudes on interval N and N-1 we can rewrite (9.33) as

$$\left\{\frac{1}{2} - 2i\alpha|\bar{a}^N|^2\right\}\bar{a}^N = -i(1-\bar{\beta})$$

$$- \bar{\beta}\left\{\frac{1}{2} + 2i\alpha|\bar{a}^{N-1}|^2\right\} \tag{9.34}$$

where

$$\left\{ \frac{1}{2} - 2i\alpha |\bar{a}^0|^2 \right\} \bar{a}^c = -i \tag{9.35}$$

This iteration scheme is uniquely defined as the implied cubic in (9.34) and (9.35) for $|\bar{a}^N|^2$ always has only one real root. If we define

$$c^N \equiv \left\{ \frac{1}{2} - 2i\alpha |\bar{a}^N|^2 \right\} \bar{a}^N \tag{9.36a}$$

$$d^N \equiv \left\{ \frac{1}{2} + 2i\alpha |\bar{a}^N|^2 \right\} \bar{a}^N \tag{9.36b}$$

then

$$|c_N| = |d_N| \tag{9.37}$$

We can define $|\bar{a}^N|^2$ as

$$|\bar{a}^N|^2 \equiv f(|c^N|^2) = f(|d^N|^2) \equiv f_N \tag{9.38}$$

where the function f is the unique cubic equation for $|\bar{a}^N|^2$. Using the definition of f_N we can rewrite (9.34) as

$$c^N = -i(1-\bar{\beta}) - \bar{\beta} \left(\frac{\frac{1}{2} + 2i\alpha f_{N-1}}{\frac{1}{2} - 2i\alpha f_{N-1}} \right) c^{N-1} \tag{9.39}$$

where we have used (9.36) and (9.38). The factor in brackets is simply a phase factor,

$$e^{i\phi_N} \equiv \frac{\frac{1}{2} + 2ia f_N}{\frac{1}{2} - 2ia f_N} \tag{9.40}$$

In this form the iteration scheme (9.34) and (9.35) can be rewritten as

$$c^N = -i(1-\bar{\beta}) - \bar{\beta} e^{i\phi_{N-1}} c^{N-1} \tag{9.41}$$

where

$$c^0 = -i \tag{9.42}$$

We should remember the solution for c^{N-1} is needed to obtain $e^{i\phi_{N-1}}$. If the system is harmonic, then $a = 0$ and $e^{i\phi_N} = 1$ giving us

$$c^N = -i(1-\bar{\beta}) - \bar{\beta} c^{N-1} \tag{9.43}$$

whose solution is

$$c^N = -i\left\{\frac{-2(-\bar{\beta})^{N+1} + 1 - \bar{\beta}}{1+\bar{\beta}}\right\} \tag{9.44}$$

As we did earlier for our other discrete mappings, let us find the condition on $\bar{\beta}$ for having an N cycle, i.e., $c^N = c^0$, which reduces to

$$(-\bar{\beta})^N = 1 \tag{9.45}$$

N-cycles occur at the Nth roots of unity. The solution (9.44) is apparently singular at $\bar{\beta} = -1$, but using L'Hospital's Rule we obtain at this point

$$c^N = -i(2(N+1) - 1) \tag{9.46}$$

which is an open diagram, not an N-cycle. We must, therefore, exclude $\bar{\beta} = -1$ from the allowed solutions of (9.45). While the steady-state diagrams have a rich structure, let us end here and simply comment that neither the harmonic nor anharmonic mappings are chaotic. [56]

Before we study some of the time dynamics let us consider some parameter values characteristic of real molecules. Laser intensities of 1-16 MW/cm^2 are common in multiple photon excitation experiments and this translates into $.1 \leq \Omega \leq .4$ cm^{-1}. We would expect densities of background states in the 1-4 cm^{-1} regime and possibly higher. Coupling parameters in the range .1-.4 cm^{-1} are also expected. These numerical values lead to ranges of .8-1 cm^{-1} for γ(~5ps) and $\tau_R \sim$ 30-130 ps. A reasonable value for χ would be around 2 cm^{-1} which gives $\alpha \lessapprox .5$ cm^{-1}. These numbers also imply an average absorption which is roughly a factor of five smaller than that obtained from the scaled system.

We will only consider resonant pumping, which means $\bar{\beta} = e^{-i\bar{\Delta}_0 \tau_R}$. Since the relative location of the pump mode band origin is multiplied by τ_R to give a phase, we can rewrite the exponent as

$$\bar{\Delta}_0 \tau_R = (\bar{\Delta}_0 \rho) 2\pi \equiv \Phi \tag{9.47}$$

where ρ is the inverse of the background level spacing. Therefore, if $\bar{\Delta}_0$ is halfway between the background levels, $\bar{\Delta}_0\rho = 1/2$ and $\Phi = \pi$. If $\bar{\Delta}_0$ is on a background level, $\bar{\Delta}_0\rho = 0$ and $\Phi = 0$. Because of the equal spacing the background Φ is periodic with period 2π.

In Figure 9.14 we see the average number of photons absorbed by the pumped mode plotted for $5\tau_R$. The density of states is 4 and $\Phi = \pi/2$. On the first interval we observe the smooth exponential decay to the steady-state value shown by a box plotted at the end of the interval. Using (9.45) we can see this choice of Φ corresponds to a 4-cycle. Notice, however, the approach to steady state is different on the fourth interval as opposed to the first interval. Since the steady states are a good representation of the average dynamics, in Figure 9.15 we observe that the total photons absorbed are approximately linear on each interval. The steady-state predicted number of photons absorbed is indicated by a box plotted at the end of each interval. The breakdown in the validity of the steady-states is due to the carryover in the full dynamical solution of one interval's dynamics into the next interval. The physics of the dynamics is that the pumped mode may reach steady state on each interval and stop absorbing photons, but the background modes in the molecule will continue to take up energy from the laser using the pumped mode as the intermediate transfer mechanism. Whether the system absorbs or emits photons depends on the choice of Φ and the interval. We can observe the importance of Φ by comparing the previous Figure 9.14 for $\Phi = \pi/2$ with Figures 9.16 and 9.17 for $\Phi = 0$ and $\Phi = \pi$.

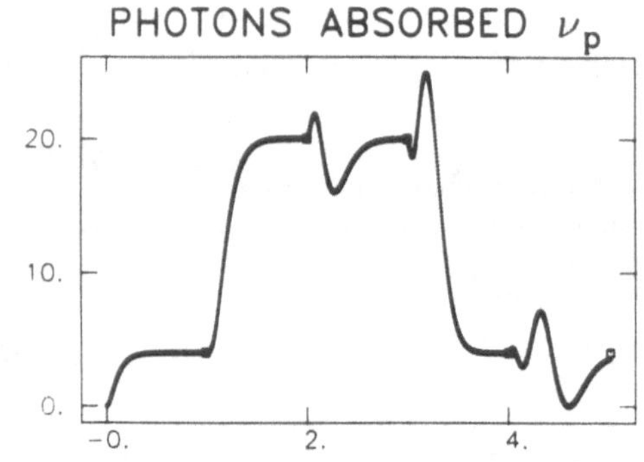

Figure 9.14

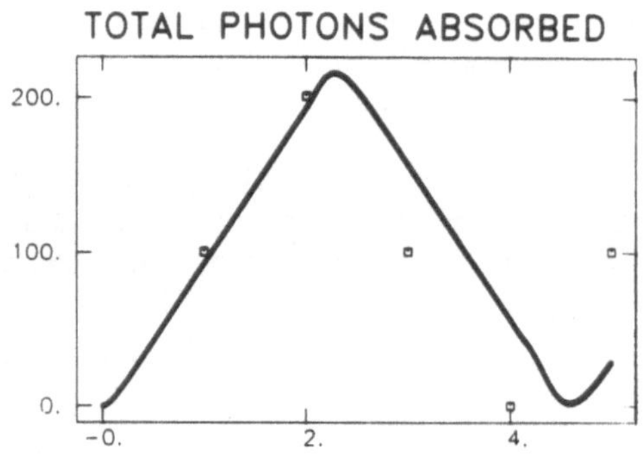

Figure 9.15

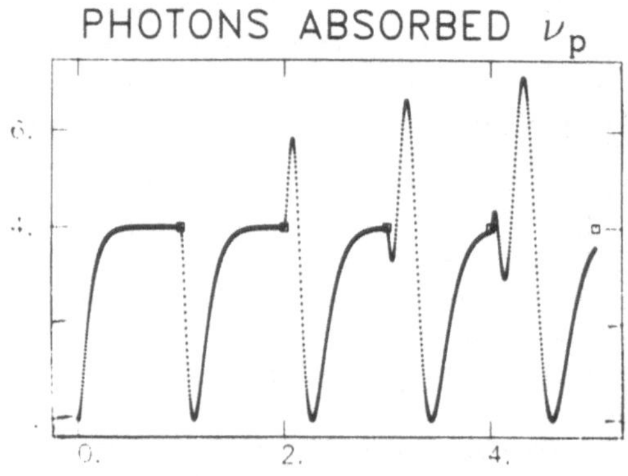

Figure 9.16

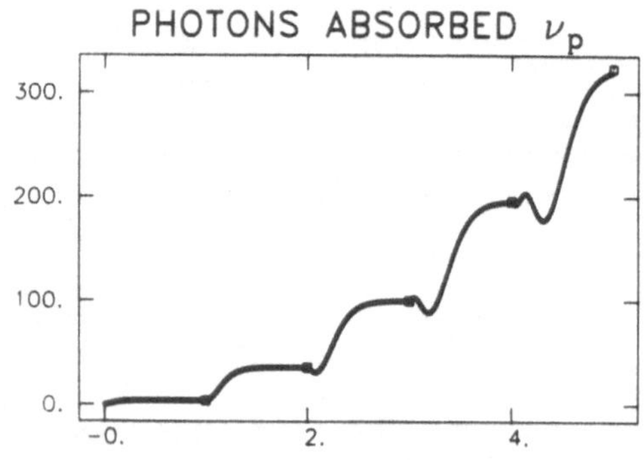

Figure 9.17

In Figure 9.18 all the parameter values are the same as in Figure 9.14 except we have introduced an anharmonicity, α = .5. We observe the anharmonicity apparently adds more ringing to the dynamics; however, the steady-state values still appear to be valid at the ends of the intervals. In Figure 9.19 we look at the total photons absorbed for the same dynamics as in Figure 9.18 but extended to 150 τ_R. Notice the dynamics goes apparently chaotic and shows a roughly

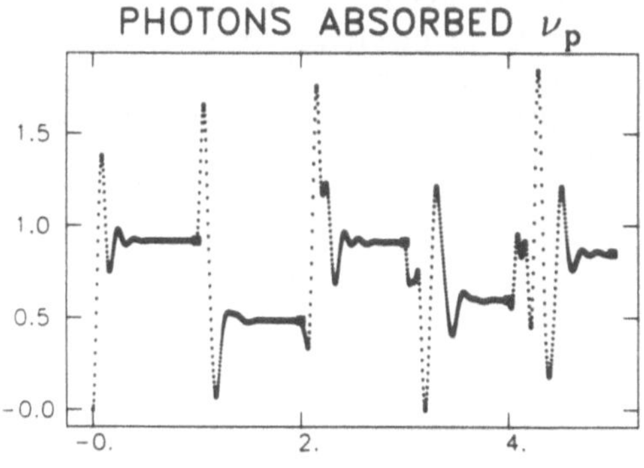

Figure 9.18

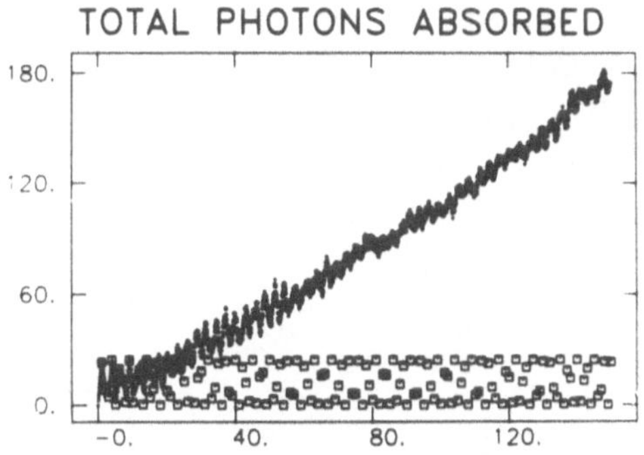

Figure 9.19

linear absorption with a "random" noise superimposed. In Figure 9.20 we show an FFT spectrum of the dynamics which indicate chaos. Notice also the steady-states no longer characterize the system. We have also computed the maximal Lyapunov exponents for this system confirming our observation of chaos. We should note that for different values of Φ the system still goes chaotic; however, the slope is sensitive to Φ. The fact that the average absorption becomes linear means that the absorption is fluence dependent.

Since we are in the chaotic regime, we would like to illustrate the sensitivity of the dynamics to a small change in one of the parameters or equivalently in the initial conditions. In Figure 9.21 we show the FFT of this same dynamics, but with α = .499. Notice that the detailed structure is substantially different in Figure 9.20 (α =.5 versus Figure 9.21.

In conclusion we have observed that the interplay between intramolecular energy transfer and the pumped mode's anharmonicity conspire to give chaotic absorption of photons. On average this absorption is linear giving a fluence-dependent absorption for a single molecule. We have found no threshold for the onset of this chaos. In addition, since we believe this chaotic behavior is a generic feature of multiple photon absorption, we can no longer expect cold experiments to exhibit intensity dependence and possibly any details of a particular molecule's structure.

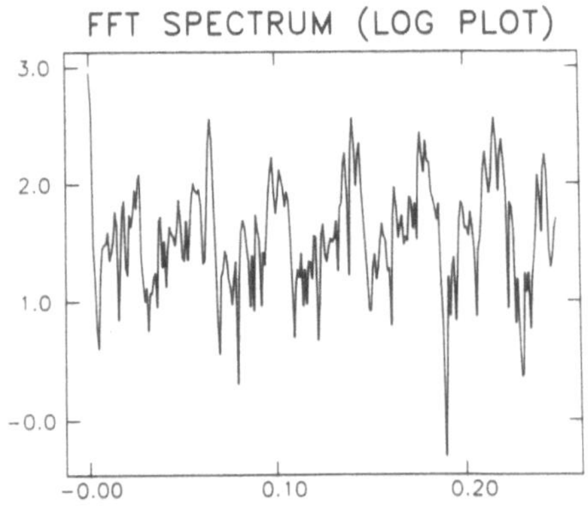

Figure 9.20

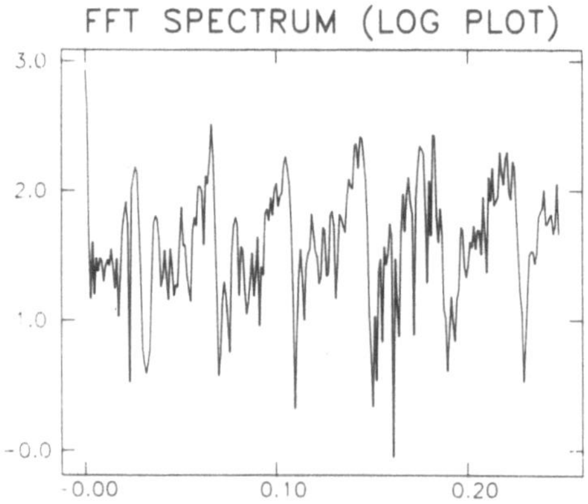

Figure 9.21

A Rotation-Vibration Model

In our previous example we considered a non-rotating model. Let us now consider a rotating model, but for convenience in isolating the physics of this process we will neglect intramolecular coupling. Our model will consist of an harmonic mode which is triply degenerate, coupled to the laser, and can rotate. From (9.2) we can identify the Hamiltonian for this system as

$$H = \Delta \underset{\sim}{a}^{+} \cdot \underset{\sim}{a} + B_0 \underset{\sim}{J}^2 + \Omega \hat{\epsilon} \cdot \overleftrightarrow{\mathbb{C}} \cdot (\underset{\sim}{a}^{+} + \underset{\sim}{a}) \tag{9.48}$$

The classical equations of motion obtained from this Hamiltonian are

$$i \dot{\underset{\sim}{a}} = \Delta \underset{\sim}{a} + \Omega \hat{\epsilon} \cdot \overleftrightarrow{\mathbb{C}} \tag{9.49a}$$

$$\hat{\epsilon} \cdot \dot{\overleftrightarrow{\mathbb{C}}} = -2B_0 \underset{\sim}{J} \times (\hat{\epsilon} \cdot \overleftrightarrow{\mathbb{C}}) \tag{9.49b}$$

$$\dot{\underset{\sim}{J}} = \Omega \hat{\epsilon} \cdot \overleftrightarrow{\mathbb{C}} \times (\underset{\sim}{a}^{*} + \underset{\sim}{a}) \tag{9.49c}$$

where we observe (9.49b) is physically a precession of the body-fixed components of the laser polarization vector, $\hat{\epsilon} \cdot \overleftrightarrow{\mathbb{C}}$, about the total angular momentum vector $\underset{\sim}{J}$.

If $\underset{\sim}{J}$ is a constant of the motion and chosen to be equal to J_z, then a solution to (9.49b) is

$$\hat{\epsilon} \cdot \overleftrightarrow{\mathbb{C}}(t) = \frac{1}{\sqrt{2}} \hat{x} \cos 2B_0 J_z t + \frac{1}{\sqrt{2}} \hat{y} \sin 2B_0 J_z t + \frac{1}{\sqrt{2}} \hat{z} \tag{9.50}$$

where at t=0 we have arbitrarily chosen $\hat{\epsilon} \cdot \overleftrightarrow{\mathbb{C}}$ to be in the x-z plane. Substitution of (9.50) into (9.49a) and recombining equations gives

$$\dot{a}_{+} = -i\Delta a_{+} - \frac{i\Omega}{\sqrt{2}} e^{2iB_0 J_z t} \tag{9.51a}$$

$$\dot{a}_{-} = -i\Delta a_{-} - \frac{i\Omega}{\sqrt{2}} e^{-2iB_0 J_z t} \tag{9.51b}$$

$$\dot{a}_{z} = -i\Delta a_{z} - \frac{i\Omega}{\sqrt{2}} \tag{9.51c}$$

where

$$a_{\pm} \equiv a_x \pm i a_y \tag{9.52}$$

Since we are interested in evaluating the absorption of photons, we want to calculate $\underset{\sim}{a}^{*} \cdot \underset{\sim}{a}$ which is equal to $(|a_+|^2 + |a_-|^2)/2 + |a_z|^2$. Solving (9.51) in the standard way we can substitute into the above expression for $\underset{\sim}{a}^{*} \cdot \underset{\sim}{a}$ and obtain

$$\underset{\sim}{a}^{*} \cdot \underset{\sim}{a} = \Omega^2 \left\{ \frac{1 - \cos(2B_0 J_z + \Delta)t}{2(\Delta + 2B_0 J_z)^2} + \frac{1 - \cos(\Delta - 2B_0 J_z)t}{2(\Delta - 2B_0 J_z)^2} + \frac{1 - \cos \Delta t}{\Delta^2} \right\} \quad (9.53)$$

This expression shows resonances at $\Delta = 0, \pm 2B_0 J$ which illustrates the well-known P, Q and R branch structure of a rotating molecule. [41] In Figures 9.22 a, b, c we see the energy in the pumped mode vs. time, an FFT of this dynamics, and a plot at $(\hat{\epsilon}\cdot\overleftrightarrow{C})_1$ vs. $(\hat{\epsilon}\cdot\overleftrightarrow{C})_2$. The peaks in Figure 9.22 are the off-resonant pumping of the P, Q and R branches of the molecule. We have chosen J=5 and an initial detuning $\Delta = .5$ cm^{-1}. With J held fixed in time we see that Eqs. (9.49a) and (9.49b) are a linear system so that only quasiperiodic motion is possible.

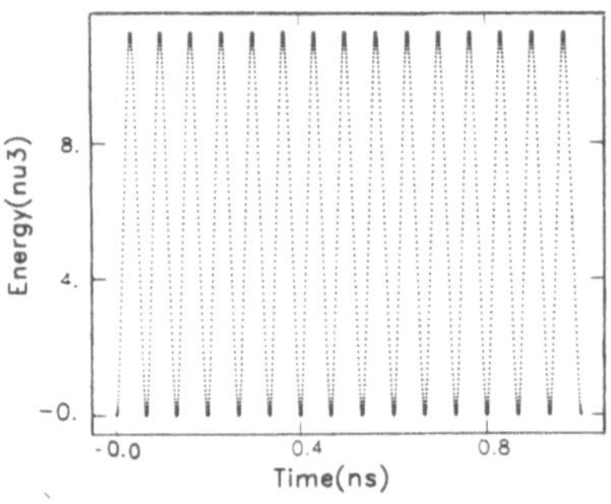

Figure 9.22a

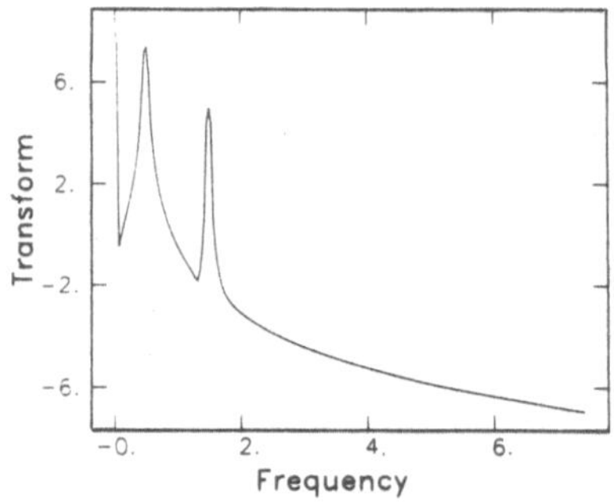

Figure 9.22b

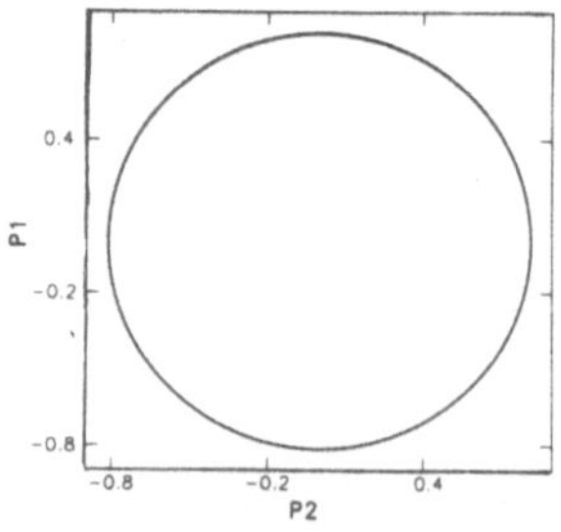

Figure 9.22c

Returning now to (9.49), we observe from (9.49c) that $\underset{\sim}{J}$ is not a constant of the motion, but is coupled by the laser to the vibrational excitation in the molecule. In Figure 9.23a,b,c we have chosen the same parameter values as in Figure 9.22a,b,c but J=5 is now only an initial condition, as J is no longer a constant of the motion. In Figure 9.23a we observe a greater (on average) absorption and a more erratic absorption than in Figure 9.22a where J is constant. The FFT, Figure 9.23b, implies the dynamics are chaotic as the spectrum has become very broadband. In Figure 9.23c the polarization vector no longer executes a simple motion. In Figures 9.24a,b we have the same dynamics as in Figures 9.23 but with $\Delta = .501$. A comparison of these two cases shows the dynamics is very sensitive to a very small change in detuning. We have calculated the maximal Lyapunov exponent verifying that we are observing chaotic dynamics. Furthermore a comparison of Figures 9.23a and 9.24a is indicative of chaos setting in for times $\gtrsim .7$ ns in the energy absorption vs. time. In the sequence of Figures 9.25a,b,c we have increased the initial J value to 20. Figure 9.25a shows the absorption dynamics are once again fairly regular, but with a small random variation in amplitude. The FFT, Figure 9.25b, shows the expected Coriolis splittings, but with a low-level broadband noise superimposed. We have verified that this small random variation in amplitude is not chaotic by again calculating the maximal Lyapunov exponent. The polarization vector is also well-behaved, but with a small variation in its circular orbit. Here the photon absorption is much less than the initial rotational energy so that $\underset{\sim}{J}$ changes by only a fraction of its initial value giving a gyroscopic stability to the system. Our conclusion is that rotational chaos in driven systems is significant when the photon absorption is comparable to the initial angular momentum value.

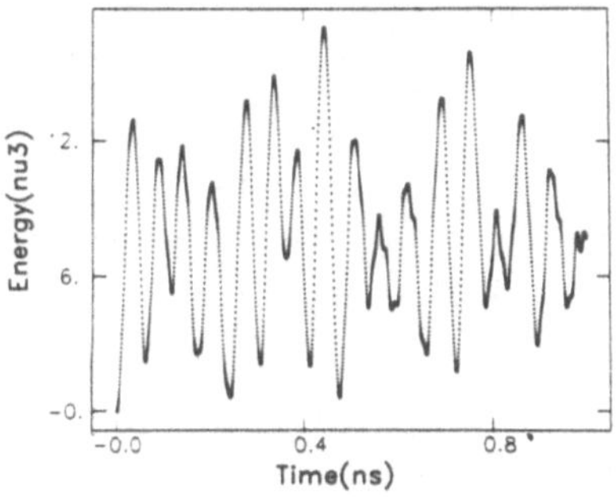

Figure 9.23a

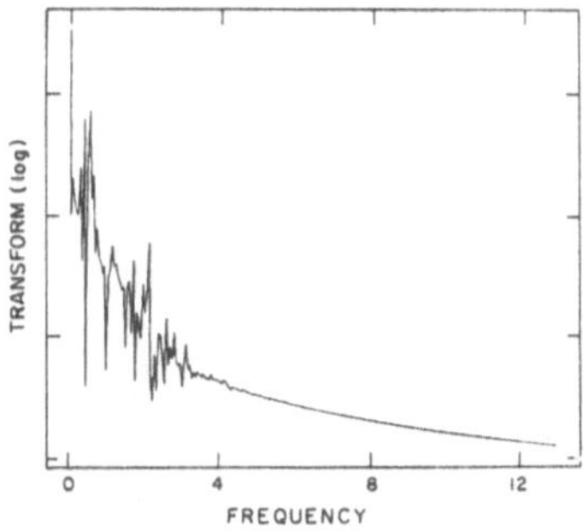

Figure 9.23b

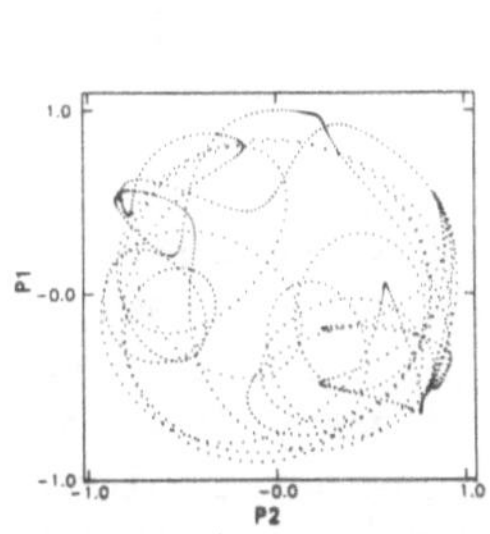

Figure 9.23c

We can conclude this section by remarking on the observation of strong rotational-vibrational chaos in multiple photon absorption for low J. This type of chaos would be important in cold experiments where the rotational distribution is compact in the region of low J. At high temperatures in heavy molecules it would not be important, except at very high laser powers.

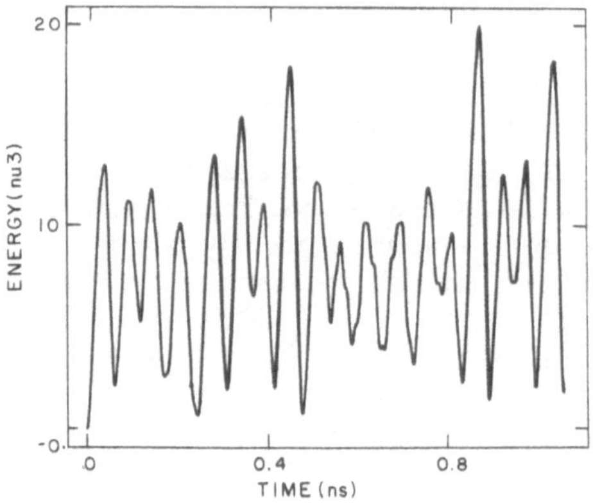

Figure 9.24a

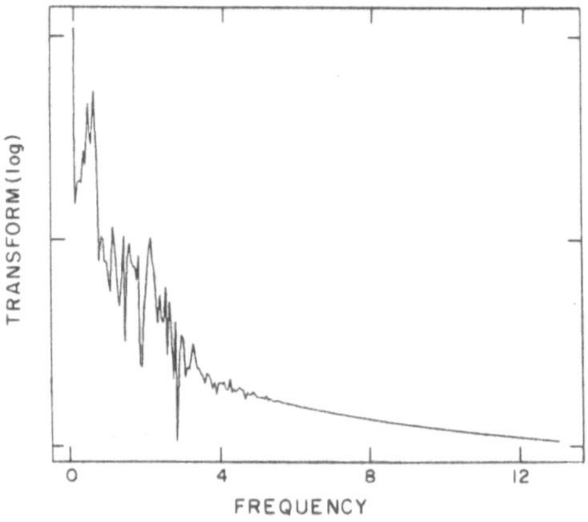

Figure 9.24b

In summary we believe there is chaos in the absorption of infrared photons by polyatomic molecules. The chaos is very likely the reason for the observed generic features of the process which will persist in experiments done at very low temperatures. In addition, we should reiterate that the parameter region, where chaos is observed in our model, is realistic for typical multiple photon absorption experiments.

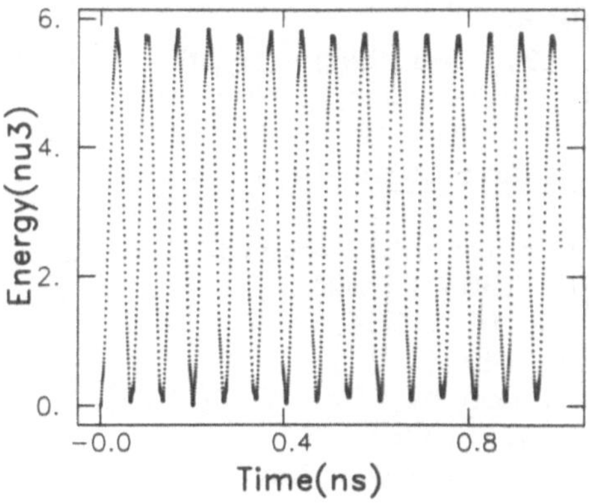

Figure 9.25a

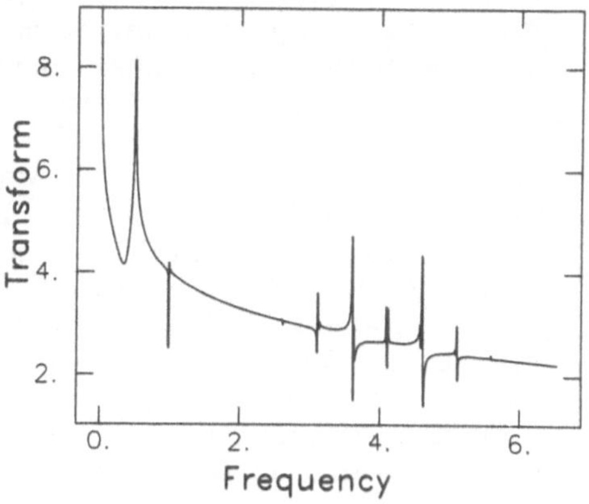

Figure 9.25b

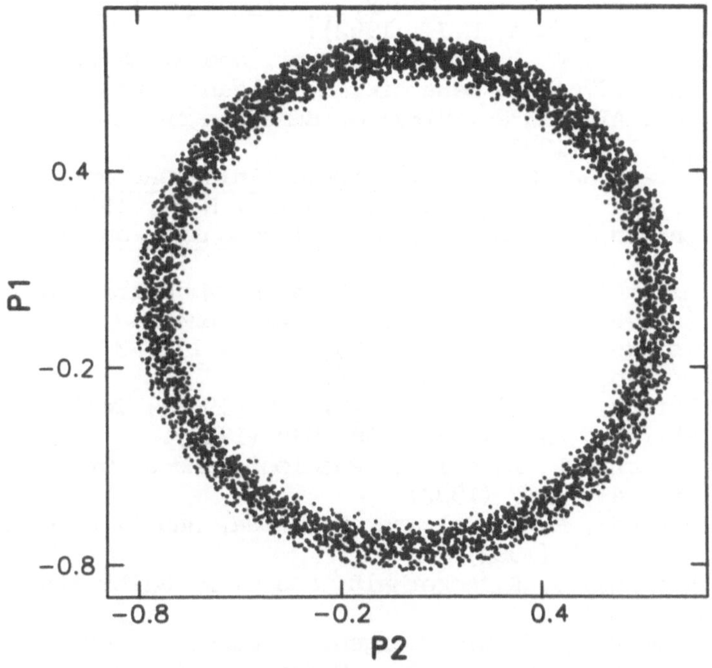

Figure 9.25c

REFERENCES

1. J.-P. Eckmann, Rev. Mod. Phys. 53, 643 (1981).
2. E. Ott, Rev. Mod. Phys. 53, 655 (1981).
3. V. I. Arnol'd and A. Avez, Ergodic Problems of Classical Mechanics (Benjamin, New York, 1968).
4. M. J. Feigenbaum, J. Stat. Phys. 19, 25 (1978); 21, 669 (1979).
5. M. J. Feigenbaum, Los Alamos Science (Summer, 1980), pp. 4-27.
6. B. A. Huberman and J. Rudnick, Phys. Rev. Lett. 45, 154 (1980).
7. M. Kac and S. M. Ulam, Mathematics and Logic (New American Library, New York, 1969), p. 123.
8. An approximate form of P(x) for the logistic equation is given in J. D. Farmer, Z. Naturforsch. 37a, 1304 (1982).
9. A. N. Sarkovskii, Ukr. Mat. Zh. 16, 61 (1964).
10. P. Stefan, Commun. Math. Phys. 54, 249 (1977).
11. T.-Y. Li and J. Yorke, Amer. Math. Monthly 82, 985 (1975).
12. M. Hénon, Commun. Math. Phys. 50, 69 (1976).
13. D. A. Russel, J. D. Hanson, and E. Ott, Phys. Rev. Lett. 45, 1175 (1980).
14. S. D. Feit, Commun. Math. Phys. 61, 249 (1978).
15. See, for example, C. M. Bender and S. A. Orszag, Advanced Mathematical Methods for Scientists and Engineers (McGraw-Hill, N.Y., 1978), pp. 171-197.
16. See, for example, N. Minorsky, Nonlinear Oscillations (Van Nostrand, Princeton, N.J., 1962).
17. M. Casartelli, E. Diana, L. Galgani, and A. Scotti, Phys. Rev. A13, 1921 (1976); G. Benettin, L. Galgani, and J. M. Strelcyn, Phys. Rev. A14, 2338 (1976); G. Benettin and L. Galgani, J. Stat. Phys. 27, 153 (1982).
18. M. W. Hirsch and S. Smale, Differential Equations, Dynamical Systems, and Linear Algebra (Academic, N.Y., 1974), Chapter 11.
19. L. Allen and J. H. Eberly, Optical Resonance and Two-Level Atoms (Wiley, N.Y., 1975).
20. P. W. Milonni, J. R. Ackerhalt, and H. W. Galbraith, Phys. Rev. Lett. 50, 966 (1983), and references therein.
21. M. Tavis and F. W. Cummings, Phys. Rev. 170, 379 (1968).
22. H. Haken, Phys. Lett. 53A, 77 (1975).
23. E. N. Lorenz, J. Atmos. Sci. 20, 130 (1963); Tellus 16, 1 (1964).
24. J. H. Shirley, Am. J. Phys. 36, 949 (1968).
25. F. T. Arecchi, R. Meucci, G. Puccioni, and J. Tredicce, Phys. Rev. Lett. 49, 1217 (1982).
26. N. B. Abraham, M. D. Coleman, M. Maeda, and J. C. Wesson, Appl. Phys. B 28, 169 (1982).
27. P. W. Milonni, J. R. Ackerhalt, and H. W. Galbraith, Phys. Rev. A28, 32 (1983).
28. J. A. Armstrong, N. Bloembergen, J. Ducuing, and P. S. Pershan, Phys. Rev. 127, 1918 (1962). Under certain circumstances analytical solutions are possible; see J. R. Ackerhalt, Phys. Rev. Lett. 46, 922 (1981).
29. E. Abraham and S. D. Smith, Rep. Prog. Phys. 45, 815 (1982).

30. K. Ikeda, Opt. Commun. 30, 257 (1979).
31. A. Szöke, V. Daneu, J. Goldhar, and N. A. Kurnit, Appl. Phys. Lett. 15, 376 (1969).
32. S. L. McCall, Phys. Rev. A9, 1515 (1974).
33. S. L. McCall and E. L. Hahn, Phys. Rev. 183, 457 (1969).
34. R. Bonifacio and L. A. Lugiato, Lett. Nuovo Cimento 21, 505 (1978).
35. H. M. Gibbs, S. L. McCall, and T. N. C. Venkatesan, Phys. Rev. Lett. 36, 1135 (1976).
36. G. Casati, B. V. Chirikov, F. M. Izraelev, and J. Ford, in Stochastic Behavior in Classical and Quantum Hamiltonian Systems, ed. by G. Casati and J. Ford (Springer, Berlin, 1979).
37. K. Ikeda, H. Daido, and O. Akimoto, Phys. Rev. Lett. 45, 709 (1980).
38. H. Nakatsuka, S. Asaka, H. Itoh, K. Ikeda, and M. Matsuoka, Phys. Rev. Lett. 50, 109 (1983).
39. H. M. Gibbs, F. A. Hopf, D. L. Kaplan, and R. L. Shoemaker, Phys. Rev. Lett. 46, 474 (1981).
40. F. A. Hopf, D. L. Kaplan, H. M. Gibbs, and R. L. Shoemaker, Phys. Rev. A25, 2172 (1982).
41. It is always useful to look at Molecular Spectra and Molecular Structure, volumes II and III, by Gerhard Herzberg (Van Nostrand Reinhold Co., N.Y., 1966).
42. For diatomic molecules an excellent discussion is given in volume 1 of Herzberg's Molecular Spectra and Molecular Structure.
43. M. Born and R. Oppenheimer, Ann. Physik 84, 457 (1927).
44. Laser-Induced Chemical Processes, ed. by J. I. Steinfeld (Plenum, N.Y., 1981). See Chapter 1 by H. W. Galbraith and J. R. Ackerhalt for an extensive list of references on multiple photon excitation in SF .
45. A recent study has been made by A. Huberman and J. P. Critchfield, Phys. Rev. Lett. 43, 1743 (1979).
46. For a traditional approach to these systems see W. J. Cunningham, Introduction to Nonlinear Analysis (McGraw-Hill, N.Y., 1958), p. 173; L. D. Landau and L. M. Lifshitz, Mechanics (Addison-Wesley, Reading, 1980), Chapter V; N. Minorsky, op. cit.
47. S. Novak and R. G. Frehlich, Phys. Rev. A26, 3660 (1982).
48. A study of period doubling including the Duffing oscillator cannot be complete without reference to M. J. Feigenbaum, Reference (5).
49. J. N. Elgin, D. Forster, and S. Sarkar, Phys. Lett. 94A, 195 (1983).
50. J. L. Lyman, G. P. Quigley, and O. P. Judd, in Multiple-Photon Excitation and Dissociation of Polyatomic Molecules, ed. by C. D. Cantrell (Springer-Verlag, Heidelberg, 1982).
51. The earliest reference we have found to a model of this type, with equal couplings and equal spacings, is in the context of spontaneous emission: V. F. Weisskopf and E. P. Wigner, Z. Phys. 63, 54 (1930). An English translation can be found in W. R. Hindmarsh, Atomic Spectra (Pergamon, Oxford, 1967).
52. M. Bixon and J. Jortner, J. Chem. Phys. 48, 715 (1968).

53. M. J. Lighthill, Introduction to Fourier Analysis and Generalized Functions (Cambridge University Press, Cambridge, 1970), p. 67.
54. Related work has been done by G. C. Stey and R. W. Gibbert, Physics (Utrecht) 60, 7 (1972); R. Lefebvre and J. Savolainen, J. Chem. Phys. 60, 2509 (1974); J. H. Eberly, J. J. Yeh and C. M. Bowden, Chem. Phys. Lett. 86, 76 (1982); and J. J. Yeh, C. M. Bowden and J. H. Eberly, Chem. Phys. 76, 5930 (1982).
55. Our own work references this method as applied to oscillator systems. See P. W. Milonni, J. R. Ackerhalt, H. W. Galbraith, and M. L. Shih, Phys. Rev. A26 (July, 1983).
56. We have made a maximal LCE computation on this iteration scheme, but have not found chaos.

ELECTRON SPECTRUM AND THE TIME-DEPENDENCE OF INITIAL STATE POPULATION IN A MODEL FOR LASER INDUCED AUTOIONIZATION

D. Agassi

Department of Physics and Astronomy
University of Rochester
Rochester, New York 14627

1. THE MODEL AND THE HEISENBERG EQUATIONS OF MOTION

1.1 Introduction

There has been recently a surge of interest in laser-induced autoionization, motivated by the hope to observe experimentally strong field effects.[1-7] Such features, in particular the existence of very narrow and substantially skewed line shapes in the electron spectrum are predicted by a simple, soluble, semi-classical model.[1-3] The same model also predicts Rabi-oscillations (with decaying amplitude) of the initial-state population.[2,3,6] The scope of these lectures is to extend the semi-classical model to include spontaneous decay back to the initial state (recycling) and to survey systematically the different qualitative features of the exact solution of the extended model as we scan over the parameters space. The analysis of the results suggests a reformulation of the model in a representation that incorporates the central physical ingredients from the outset. These are the "dressed resonances", the resonances resulting from embedding the "dressed states"[8-9] into the continuum of the problem.

The new representation, dubbed as the "Dressed Resonance Representation" (DRR)[10] offers some distinct advantages. For one, the equations of motion simplify considerably, and in particular, the solution of the semi-classical model is immediate. The structure of the expressions becomes transparent. An important bonus is the identification of an approximation scheme, attempted here only for the electron spectrum. The "small parameter" associated with the relevant expansion is "1/q" for the $q \gg 1$

regime, and "q" for the $q \ll 1$ regime where "q" is the Fano asymmetry parameter.[11] Establishing an approximation scheme may prove helpful for future applications, when an exact solution is unavailable.

The lectures are organized as follows. First the model is defined and the basic Heisenberg equations of motion are derived. We then outline the method used to obtain the exact solution. The second lecture is devoted to the interpretation of a systematic survey of the electron spectrum $S(\omega)$, and initial state population $P_o(t)$ over the various regions of the physical-parameter space. The analysis highlights the importance of the dressed states and provides the motivation to reformulate the model in a representation that incorporates these states from the outset (DRR). This is done in the third lecture, as well as the introduction of an approximation scheme for the electron spectrum associated with the (q, 1/q) expansion.

1.2 The Model

Consider the simplest possible model for laser induced autoionization:[1-3] One autoionizing state $|1\rangle$, one continuum , one bound state $|0\rangle$ where all the population is initially and one driving laser. The model and its couplings are schematically depicted in Fig.(1.a). The difference between $|0\rangle$ and $|1\rangle$ is that the latter is coupled by a static interaction to the continuum.[12] This is the very coupling that makes the $|1\rangle$ state "autoionizing". The "Semi-Dressed Representation" (SDR)[1-3,5-7] is based on a preferential treatment of this coupling: First the state $|1\rangle$ and continuum $|\omega\rangle$ are admixed by the static interaction (the "embedding"), to give a perturbed, non-flat continuum $|\tilde{\omega}\rangle$. The model is then formulated in terms of stimulated transitions and spontaneous decay between the $|\tilde{\omega}\rangle$ continuum and the ground state $|0\rangle$ (Fig. 1.b). The "embedding" step is recognized to be the Fano problem[11] involving one bound state and one continuum, hence the 'non flat' continuum has a dipole-strength distribution $\tilde{F}(\omega)$ of the Fano type (see below). Note that in the SDR the model takes the form of a two-level atom where the excited state is replaced a a continuum. This suggests that the two-level atom paradigm of incorporating spontaneous decays[13] can be followed in the present context.

Assume, for simplicity, the model to represent a single electron atom. The Hamiltonian is then

$$H = H_A + H_R + \vec{A}(r;t) \cdot \vec{p} \tag{1.1}$$

where H_A is the diagonalized atomic Hamiltonian

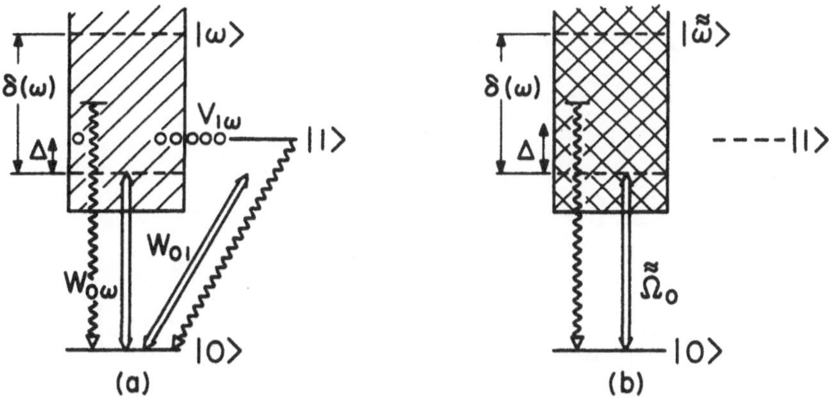

Fig. 1. Schematic outlay of the model-space and couplings in the unperturbed (a) and semi-dressed (SDR) representations (b). Stimulated transitions are marked by double-arrows, static-coupling by a string of circles and spontaneous decay transitions by a directed wiggly arrow.

$$\hat{H}_A = E_0|o><o| + \int d\omega \hbar\omega \, |\tilde{\tilde{\omega}}><\tilde{\tilde{\omega}}|, \tag{1.2}$$

$\hat{H}_R$ is the free radiation field Hamiltonian

$$H_R = \sum_\lambda \hbar\,\omega_\lambda \,\hat{a}^\dagger_\lambda \;\hat{a}_\lambda \tag{1.3}$$

where $\lambda = [\vec{k}\,, \vec{\epsilon}_s(\hat{k})]$, s = 1,2 denote the modes of a quantized electromagnetic field in a large box of volume V, characterized by a momentum $\vec{k}$, a polarization vector $\vec{\epsilon}$ and energy $\omega_\lambda = k\,c$. The field-atom interaction term $\vec{p}\cdot\vec{A}$ involves the electromagnetic potential $\vec{A}$ in the Coulomb gauge $\nabla\cdot\vec{A} = 0$. Inserting the well known mode expansion of[13,14] $\vec{A}(r,t)$ in (1.1), straightforward manipulations give

$$\hat{H}_I = \sum_\lambda \int d\omega \left[\hbar\tilde{\tilde{\Omega}}^*_\lambda(\omega)\;\hat{\tilde{\tilde{B}}}^\dagger(\omega)\hat{a}_\lambda\;(t) + \hbar\tilde{\tilde{\Omega}}_\lambda(\omega)\;\hat{a}^\dagger_\lambda(t)\hat{\tilde{\tilde{B}}}(\omega)\right] \tag{1.4}$$

where $\hat{\tilde{\tilde{B}}}(\omega) = |0><\tilde{\tilde{\omega}}|$ $\quad \hat{\tilde{\tilde{B}}}^\dagger(\omega) = |\tilde{\tilde{\omega}}><0|$

$$\tilde{\tilde{\Omega}}_\lambda(\omega) = -ig_\lambda \tilde{\tilde{F}}(\omega), \quad g_\lambda = \left[\frac{2\pi\hbar c^2}{\omega_\lambda V}\right]^{1/2} (\vec{\epsilon}_s(\vec{k}) \cdot \hat{z}) \tag{1.5}$$

and

$$\tilde{\tilde{F}}(\omega) = \frac{|e| \langle 0|r|\tilde{\tilde{\omega}}\rangle (\omega - \omega_0)}{\hbar c}$$

$$= \frac{\tilde{\tilde{\Omega}}_0}{\sqrt{4\pi\gamma_1}} \left[\frac{\gamma_1}{\omega - \omega_1 + i\gamma_1} - \frac{1}{1+iq} \frac{\sigma}{\omega - \omega_1 + i\sigma} \right] \tag{1.6}$$

The profile $\tilde{\tilde{F}}(\omega)$ Eq. (1.6), is equated to the Fano profile in the spirit of the SDR. The (γ_1, ω_1) parameters denote the width and energy of the autoionizing state $|1\rangle$, "q" is the Fano asymmetry parameter, σ is a cutoff parameter set to infinity at the end of the calculation, and $\tilde{\tilde{\Omega}}_0$ is the strength parameter. This parameterization of the Fano profile has been used in the semi-classical model[2,3] except for $\tilde{\tilde{\Omega}}_0 \to \Omega_0$, where Ω_0 is of dimentionality [frequency). The relation between these two parameterizations is

$$\tilde{\tilde{\Omega}}_0 = \frac{i\Omega_0}{\alpha^* g_L} \tag{1.7}$$

where $\hat{a}_\lambda|\alpha\rangle = \delta_{L,\lambda}\alpha|\alpha\rangle$ and $|\alpha\rangle$ is the initial (coherent) state of the radiation field in mode "L". The combination α^* g_L must appear,[14] to remove the V-dependence in g_L, Eq. (1.5).

The transition from the $\vec{A}\cdot\vec{p}$ form to (1.4 - 1.6) involves several standard approximations worth mentioning:

a) The dipole approximation which assumes that a 'typical' $1/k_\lambda$ is much larger than the dimensions of the atomic wave function. In the present context, where bound-continuum transitions are considered this is still valid since the integration range in a matrix-element of the form $\langle 0 | e^{ikr} | \tilde{\tilde{\omega}}\rangle$ is determined by the extent of the bound state wave function. For continuum-continuum transitions this approximation is invalid.

b) To simplify the algebra we assumed only $\Delta m = 0$ transitions (m here is the magnetic quantum number). This is the origin of the $(\vec{\epsilon}_s(k) \cdot \hat{z})$ factor in (1.5), where $\hat{z}$ is a unit vector in the z-direction.

c) We assumed the Rotating-Wave Approximation (RWA) thus retaining in (1.4) only the terms which can conserve energy. In the context of a two-level atom, a "small parameter" associated with the RWA is $n = \Delta/(\Delta + 2\omega_L)$ where $\Delta = \omega_{12} - \omega_L$ is the detuning and ω_L is the laser frequency. This implies that for bound-continuum transitions the RWA is valid for a band in the continuum of width $2\omega_L$. Whenever the bound-continuum transitions are confined e.g. by $F(\omega)$, to a band narrower than $2\omega_L$ the RWA is valid automatically. Otherwise it is necessary to introduce extraneously some cutoff procedure (or an equivalent to it) to be consistent with the RWA.

d) Note that the radiation field operators appear in normal order in (1.4). The particular order has no effect on the final results.[15] The present choice is motivated by convenience.

1.3 The General Heisenberg Equations of Motion

Our focus is on 'atomic' observables associated with the expectation value of an atomic operator $\hat{O}_A(t)$. To solve the dynamic problem the approach is to set the Heisenberg equation of motion and then try and solve them exactly.[13,15] For interactions of the type (1.4), which are linear in the operators of the two interacting subsystems (the atom and radiation field) this is a valid proposition.

The equation of motion of a general time-dependent (in the Schrodinger picture) atomic operator $\hat{O}_A(t)$ is

$$i\hbar \frac{d}{dt}(\hat{O}_A(t))_H = i\hbar\left(\frac{\partial}{\partial t}\hat{O}_A(t)\right)_H + \left([\hat{O}_A(t), \hat{H}(t)]\right)_H$$

$$= i\hbar\left(\frac{\partial}{\partial t}\hat{O}_A(t)\right)_H + \left([\hat{O}_A(t), \hat{H}_A + \hat{H}_I(t)]\right)_H \qquad (1.8)$$

where, obviously $[\hat{O}_A(t), \hat{H}_R] = 0$. The "H" subscript indicates the Heisenberg representation, i.e., multiply from the left and right by $\hat{U}^{+}(t)$, $\hat{U}(t)$ respectively, where $\hat{U}(t)$ is the evolution operator pertaining to $\hat{H}(t)$.

Consider now the interaction commutator $[\hat{O}_A(t), \hat{H}_I]$. Inserting (1.4) renders an expression that depends on the field operators $\hat{a}_\lambda$, $\hat{a}^{+}_\lambda$. These, in turn, must be eliminated if we to obtain a "closed" equation in terms of atomic variables only. The equation of motion for $\hat{a}_\lambda$ is

$$i\hbar \frac{d}{dt} (\hat{a}_\lambda)_H = ([\hat{a}_\lambda , \hat{H}_R])_H + ([\hat{a}_\lambda , \hat{H}_I])_H$$
$$= \hbar \omega_\lambda (\hat{a}_\lambda)_H + \int d\omega \hbar \tilde{\tilde{\Omega}}_\lambda(\omega) (\tilde{\tilde{\hat{B}}}(\omega))_H \quad (1.9)$$

can be formally solved:

$$(\hat{a}_\lambda)_H(t) = \hat{a}_\lambda e^{-i\omega_\lambda t} - i\int_0^t d\tau G_\lambda^R(t-\tau)\int d\omega \tilde{\tilde{\Omega}}_\lambda(\omega) (\tilde{\tilde{\hat{B}}}(\omega))_H(\tau)$$
$$= \hat{a}_\lambda^{(v)}(t) + \hat{a}_\lambda^{(s)}(t) \quad (1.10)$$

where $G^R(t-\tau) = \theta(t-\tau)\, e^{-i(\omega_\lambda - i\eta)(t-\tau)}$ is the Green function of the homogeneous equation. The two terms in (1.10) describe different physical processes: The first term, $\hat{a}_\lambda^{(v)}$, is the only one that survives in the semiclassical limit, while the "source-term", $\hat{a}_\lambda^{(s)}$, represents a feedback contribution: The field induces an atomic dipole, which in turn effects back the radiation field.

The exact expression (1.10) is seen to depend on atomic variables at previous times $\tau \leq t$. Hence would it be inserted back into (1.8) the resulting equation would be non-local in the time variable, which usually is a difficult problem to solve. Fortunately, for optical processes the time non-local contribution can be neglected to a very good approximation.[10,13,16] This step is known, quite appropriately, as the Markov-Born approximation and is equivalent to the replacement

$$(\tilde{\tilde{\hat{B}}}(\omega''))_H(\tau) = e^{-i\omega''(\tau - t)} (\tilde{\tilde{\hat{B}}}(\omega''))_H(t) \quad (1.11)$$

in the $\hat{a}_\lambda^{(s)}(t)$ term, Eq. (1.10), which gives

$$\hat{a}_\lambda^{(s)}(t) = -i\pi\int d\omega \tilde{\tilde{\Omega}}_\lambda(\omega)\, \delta(\omega_\lambda - \omega)\, (\tilde{\tilde{\hat{B}}}(\omega''))_H(t) \quad (1.12)$$

The validity of the Markov-Born approximation (1.11-1.12) hinges on the slowness of "Atomic Frequencies", like Rabi oscillations, etc. as compared to the fast optical frequency oscillations. In deriving (1.12) two more steps have been invoked: a) We discard a (divergent) Lamb-shift term, based on the argument that the Lamb-shift is very small, and the divergence here is an artifact of the dipole approximation. b) We assume that the "running time" - "t" is very long compared to "typical" inverse frequencies in the problem, so that oscillating factors complete many oscillations in the time-interval [o,t].

With the provision of the Markov-Born approximation, the general equation of motion (1.8) takes the form

$$i\hbar \frac{d}{dt} (\hat{O}_A(t))_H = i\hbar(\frac{\partial}{\partial t} \hat{O}_A(t))_H + [(\hat{O}_A(t))_H, \hat{H}_A(\text{eff.};t)] + \hat{S}_A(t) \tag{1.13}$$

where

$$[(\hat{O}_A(t))_H, \hat{H}_A(\text{eff.};t)] = ([\hat{O}_A(t), \hat{H}_A])_H$$

$$+\sum_\lambda \int d\omega \Big\{ \hbar \tilde{\tilde{\Omega}}_\lambda^*(\omega) \;\; ([\hat{O}_A(t), \hat{\tilde{\tilde{B}}}^+(\omega)])_H \, \hat{a}_\lambda^{(v)}(t)$$

$$+ \hbar \tilde{\tilde{\Omega}}_\lambda(\omega) \;\; \hat{a}_\lambda^{+(v)}(t) \; ([\hat{O}_A(t), \hat{\tilde{\tilde{B}}}(\omega)])_H \Big\} \tag{1.14a}$$

and

$$\hat{S}_A(t) = \sum_\lambda \int d\omega \Big\{ \hbar \tilde{\tilde{\Omega}}_\lambda^*(\omega) \; ([\hat{O}_A(t), \hat{\tilde{\tilde{B}}}^+(\omega)])_H \, \hat{a}_\lambda^{(s)}(t)$$

$$+ \hbar \tilde{\tilde{\Omega}}_\lambda(\omega) \;\; \hat{a}_\lambda^{+(s)}(t) \; ([\hat{O}_A(t), \hat{\tilde{\tilde{B}}}(\omega)])_H \Big\} \tag{1.14b}$$

The two terms in (1.13) describe different aspects of the model. The commutator, Eq. (1.14.a), represent the rate of change due to "coherent" processes; it is retained in the semi-classical limit when the radiation field is treated as a c-number. The second term, (1.14.b), represents the "statistical" aspects of the model because it describes the feedback processes, which bring in quantum fluctuations of the radiation field ("vacuum polarization") associated with the spontaneous decay channels. Other stochastic relaxations will add, incoherently, to the source-term $\hat{S}_A$. Equation (1.13) bears similarity to the Master-Equation approach.[4,17]

Equation (1.13) is still an operator equation. Its solution should be averaged with the t = 0 density matrix $\hat{\rho}$ (t=0). For the present problem we have

$$\hat{\rho}(t = 0) = \Big[\; |0\rangle_{A\;A}\langle 0| \; \Big] \otimes \Big[\; |0\rangle .. |\alpha\rangle_{\lambda=L} .. b\rangle_R \, {}_R\langle 0| \; .. \langle\alpha|_{\lambda=L} \langle 0| \Big] \tag{1.15}$$

i.e., the atom is in its ground state $|0\rangle$ and all modes of the radiation field are empty except one mode λ = L which is in a coherent state $|\alpha\rangle$. In the case that stochastic entries are added to the model ["phase-jitter" relaxation, see section (2.2)] an additional ensemble-average of the solution is required.[2,3]

1.4 The Equations of Motion and Exact Solution.

As mentioned above, we are intersted in evaluating the electron spectrum $S(\omega)$, and the time-dependent initial state ($|0\rangle$) population $P_o(t)$. For this purpose it is necesary to solve for the evolution of a complete set of atomic projection operators since

$$P_o(t) = \langle (|0\rangle\langle 0|)_H(t)\rangle$$
$$S(\omega) = \lim_{t\to\infty} \langle (|\tilde{\tilde{\omega}}\rangle\langle\tilde{\tilde{\omega}}|)_H(t)\rangle \quad (1.16)$$

where "$\langle\ldots\rangle$" stands for averaging with respect to $\hat{\rho}$ (t = 0) and ensemble-averaging. In the present context, it is particularly convenient to consider the following complete set of projection operators (in the Schrodinger picture):[3]

$$\hat{P}_o = |0\rangle\langle 0| \qquad \tilde{\tilde{C}}(\omega_o,\omega_o') = |\tilde{\tilde{\omega}}_o\rangle\langle\tilde{\tilde{\omega}}_o'|$$
$$\hat{\tilde{\tilde{B}}}(\omega_o) = |0\rangle\langle\tilde{\tilde{\omega}}_o| \qquad \hat{\tilde{\tilde{B}}}^+(\omega_o) = |\tilde{\tilde{\omega}}_o\rangle\langle 0| \quad (1.17)$$

When phase-jitter relaxation is included the definition of $\hat{\tilde{\tilde{B}}}$, $\hat{\tilde{\tilde{B}}}^+$ is slightly modified [see section (2.2)]. Transforming the set (1.17) to the rotating frame (which gives a time-dependent set of operators) and evaluating the corresponding equations of motion (1.13), yields[6,10]

$$\frac{d}{dt} P_o(t) = -\int d\omega [\tilde{\tilde{F}}^*(\omega)\,\tilde{\tilde{B}}^*(\omega,t) + \tilde{\tilde{F}}(\omega)\,\tilde{\tilde{B}}(\omega,t)] + \int d\omega\, d\omega'\, [\tilde{\tilde{F}}^*(\omega)\,\tilde{\tilde{R}}(\omega') + \tilde{\tilde{F}}(\omega')\,\tilde{\tilde{R}}^*(\omega)]\,\tilde{\tilde{C}}(\omega,\omega';t)$$
$$\frac{d}{dt}\tilde{\tilde{B}}(\omega,t) = [-i(\omega-\omega_L)-\gamma_T]\,\tilde{\tilde{B}}(\omega,t) + \tilde{\tilde{F}}^*(\omega)\,P_o(t) - \int d\omega'\tilde{\tilde{F}}^*(\omega')\,\tilde{\tilde{C}}(\omega',\omega;t) - \tilde{\tilde{F}}^*(\omega)\int d\omega'\,\tilde{\tilde{R}}(\omega')\,\tilde{\tilde{B}}(\omega',t)$$
$$\frac{d}{dt}\tilde{\tilde{C}}(\omega,\omega';t) = i(\omega-\omega')\,\tilde{\tilde{C}}(\omega,\omega';t) + \tilde{\tilde{F}}^*(\omega')\tilde{\tilde{B}}^*(\omega,t) + \tilde{\tilde{F}}(\omega)\,\tilde{\tilde{B}}(\omega',t) - \tilde{\tilde{F}}^*(\omega')\int d\omega''\,\tilde{\tilde{R}}(\omega'')\,\tilde{\tilde{C}}(\omega,\omega'';t) - \tilde{\tilde{F}}(\omega)\int d\omega''\,\tilde{\tilde{R}}^*(\omega'')\,\tilde{\tilde{C}}(\omega'',\omega';t) \quad (1.18)$$

In (1.18) we used the notation $O(t) = \langle[\hat{U}_o(t)\,\hat{O}\,\hat{U}_o^+(t)]_H(t)\rangle$ where $\hat{U}_o(t)$ is the transformation to the rotating frame. The dipole-strength profile $\tilde{\tilde{F}}$ is given by the Fano parameterization (1.6) and

$$\tilde{\tilde{R}}(\omega) = Q(\omega_1)\, \tilde{\tilde{F}}(\omega)$$

$$Q(\omega_1) = \pi \sum_\lambda g_\lambda^2 \delta(\omega_\lambda - \omega_1) = 2\hbar\omega/3c \qquad (1.19)$$

There are two types of terms in (1.18): Those that do not involve $R(\omega)$ and those that do. The former constitute the equations of motion of the semiclassical model, and are generated by the commutator term in the general Heisenberg equation (1.13). The terms involving $R(\omega)$ originate from the feedback term S_A, Eq. (1.13), and describe the "recycling" due to spontaneous decay back to the ground state $|0\rangle$. We have also included the phase-jitter relaxation contribution, (the "γ_T"-term) defined in section (2.2).

The initial conditions associated with (1.18), in keeping with the t=0 density matrix (1.15), are

$$P_o(t = 0) = 1, \quad \tilde{\tilde{B}}(\omega, t = 0) = \tilde{\tilde{C}}(\omega, \omega'; t = 0) = 0 \qquad (1.20)$$

Note also that (1.18) conserves unitarity, i.e.:

$$\frac{d}{dt}\left[P_o(t) + \int d\omega\, \tilde{\tilde{C}}(\omega, \omega; t)\right] = 0 \qquad (1.21)$$

The equations of motion (1.18) are recognized to be linear (as a result of the Markov approximation), of first order and with time-independent coefficients since the radiation field is assumed to be CW. To solve this type of coupled system, the Laplace-transform method is particularly suitable: First transform the operator to the z-domain $F(z) = \int_0^\infty dt\, e^{-zt} F(t)$ to obtain a set of "algebraic" equations, and then solve the equations, exactly if possible, in the z-domain. The strategy for obtaining the exact solution is quite tedious though straightforward: First we "solve" the third equation in (1.18) for $\tilde{C}(\omega, \omega'; z)$ in terms of the "$\tilde{\tilde{B}}(\omega, z)$", "$\tilde{\tilde{B}}^*(\omega, z)$". This is possible because the Fano profile $\tilde{\tilde{F}}(\omega)$ is a sum of poles, which in turn leads to a separable equation amenable to an exact solution. Thus it is possible to eliminate "$\tilde{\tilde{C}}(\omega, \omega'; z)$" from the other two equations. The next step is to solve for the "$\tilde{\tilde{B}}(\omega, z)$", "$\tilde{\tilde{B}}^*(\omega, z)$" in terms of $P_o(z)$. This, again, is possible due to the simple pole structure of $\tilde{\tilde{F}}(\omega)$. Finally we solve an algebraic equation for $P_o(z)$, subject to the initial condition $P_o(t=0) = 1$.

The exact solutions of (1.18) are very complex and hardly transparent.[10] Three comments are in order: a) It turns out that $\tilde{\tilde{B}}(\omega, z) \sim \tilde{\tilde{F}}^*(\omega)\, P_o(z)$, $\tilde{\tilde{C}}(\omega, \omega'; z) \sim \tilde{\tilde{F}}(\omega)\, \tilde{\tilde{F}}^*(\omega')\, P_o(z)$ which imply in particular the preservation of Fano zero [vanishing $\tilde{\tilde{F}}(\omega)$] at all times. This is expected recalling the origin of the Fano zero[11] from destructive interference of two transition amplitudes connecting the same pair of initial-final states. b) The strength parameter associated with the spontaneous decay is the following combination:

$$\gamma_s = \left| \frac{\tilde{\tilde{\Omega}}_o}{2} \right|^2 Q(\omega_1) \tag{1.22}$$

of dimentionality [frequency]. In the limit of a two-level system [$\gamma_1 \to 0$, $q \to \infty$ in $F(\omega)$] a simple calculation[13] gives $\gamma_s = A/2$ where "A" is Einstein coefficient. We can interpret γ_s as half the Einstein coefficient for continuum-bound transitions where the continuum has a Fano profile.

c) The structure of $P_o(z)$ is

$$P_o(z) = \frac{1}{z + \frac{L(z)}{R(z)}} \tag{1.23}$$

where L(z), R(z) are polynomials of order 3 when $\gamma_s = 0$ (no spontaneous decay) and of order 10 for $\gamma_s \neq 0$. "Trapping" can be defined to occur whenever $P_o(z)$ developes a pole at z = 0 (which contributes in the $t \to \infty$ limit), i.e., when $L(z=0) = 0$. This is impossible unless $\gamma_s = \gamma_T = 0$.

2. DISCUSSION OF RESULTS

Due to the complex structure of the exact solution we try now to gain insight into the salient features of the model by interpreting the numerical results. The discussion is heuristic in nature, yet indicative of the central role of the "Dressed Resonances". This observation in turn provides the motivation for a reformulation of the model in the DRR which is described in the next section.

2.1 Classification of the "Coherent Parameters"

The model Hamiltonian depends on six parameters

$$\Omega_o,\ q,\ \gamma_1,\ \Delta = \omega_1 - \omega_L;\ \gamma_s,\ \gamma_T \tag{2.1}$$

and the computed quantities are $P_o(t)$ and $S(\omega)$ eq. (1.16). In order to cover systematically the various regions in the parameter space (2.1) we classify the parameters into the first four "coherent" and the last two "statistical" ones. The "coherent" parameters specify the stimulated transitions and static couplings. Of particularly usefulness is "q", the Fano asymmetry parameter, defined as (see Fig. 1.a)

$$q = \left[W_{01} + \int d\omega\, V_{1\omega} \frac{P}{\bar{\omega}-\omega} W_{0\omega}\right] / \pi W_{0\bar{\omega}} V_{1\bar{\omega}}$$

$$\cong \frac{W_{01}}{\pi W_{0\omega} V_{1\omega}} = \left(\frac{W_{01}}{\gamma_1}\right) / \left(\sqrt{\frac{\pi}{\gamma_1}}\, W_{0\bar{\omega}}\right) \tag{2.2}$$

where

$$\gamma_1 = \pi |V_{1\omega}|^2 \tag{2.3}$$

and in keeping with the assumed weak ω-dependence of all coupling constants, all entries into (2.2-2.3) are considered constants. Consequently the (Ω_0, q, γ_1) plane can be classified into the $q \gg 1$ sector, where the $(|0\rangle - |1\rangle)$ transitions dominate and the $q \ll 1$ sector where the $(|0\rangle - |\omega\rangle)$ transitions dominate (Fig. 2). This is a useful classification since the underlying "physics" in both sectors is different.

Consider first the $q \gg 1$ regime, and to simplify the point take the $q = \infty$ case, i.e., when $W_{0\omega} = 0$, Fig. (3.a). In this instance the continuum is pumped up from $|0\rangle$ indirectly, via the autoionizing state $|1\rangle$ which decays to the continuum at a rate γ_1. Therefore two different situations may arise: When the decay rate γ_1 is much larger (faster) than the $|0\rangle - |1\rangle$ population-flipping frequency W_{01}, the $|0\rangle$-population is drained out to the continuum, since there is no chance for the electron to complete a "lap" $|0\rangle - |1\rangle - |0\rangle$. This is obviously a "weak field" situation[2,3] (Fig. 2). The other extreme occurs when the $|0\rangle - |1\rangle$ population flipping rate (W_{01}) is large compared to γ_1. In this case the electron completes a few "laps" between $|0\rangle$ and $|1\rangle$ before it totally leaks to the continuum. Hence there are Rabi-oscillations, though with a diminishingly amplitude. This is a "strong field" situation[2,3] (Fig. 2).

In the $q \ll 1$ sector the situation is quite different. Again, to make the point consider the $q = 0$ limit (Fig. 3.b), which implies that $W_{01} = 0$. Now the continuum is pumped directly from $|0\rangle$. The electron pumped into the 'flat' continuum "never comes back", which implies that Rabi oscillations of the $|0\rangle$- state population cannot set in, i.e., a "weak field" situation where the $|0\rangle$-state is drained out to the continuum. There is, however, an important difference between the $q \gg 1$ and $q \ll 1$ "weak field" regimes. In the former, γ_1 is an upper limit (to a good approximation) to the rate of depleting the $|0\rangle$-state population. In the latter, however, since the continuum is pumped directly the larger $W_{0\omega}$ is the faster $|0\rangle$ is depleted, i.e., no upper limit to the decay rate of $|0\rangle$.

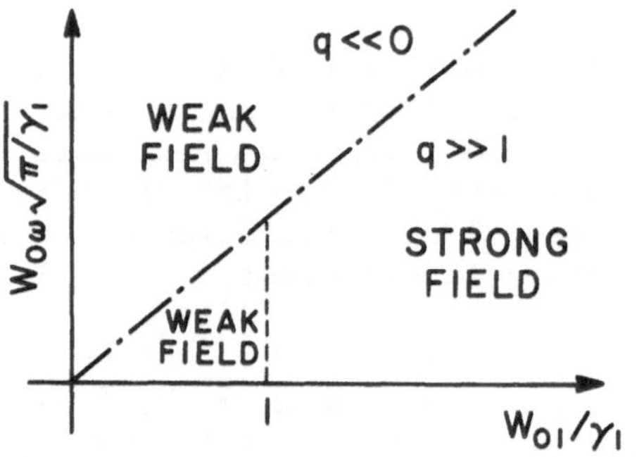

Fig. 2. Characterization of the regions in the $(W_{01}, W_{0\omega})$ space, see text.

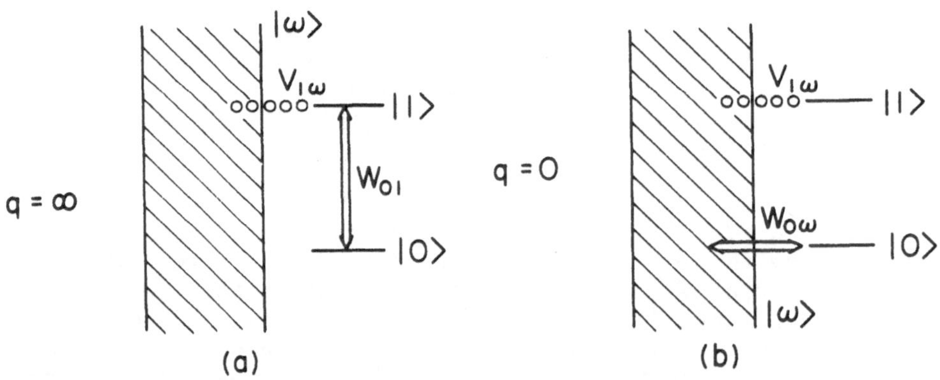

Fig. 3. Schematic picture of the couplings in the two interesting extremes of $q = \infty$, $q = 0$.

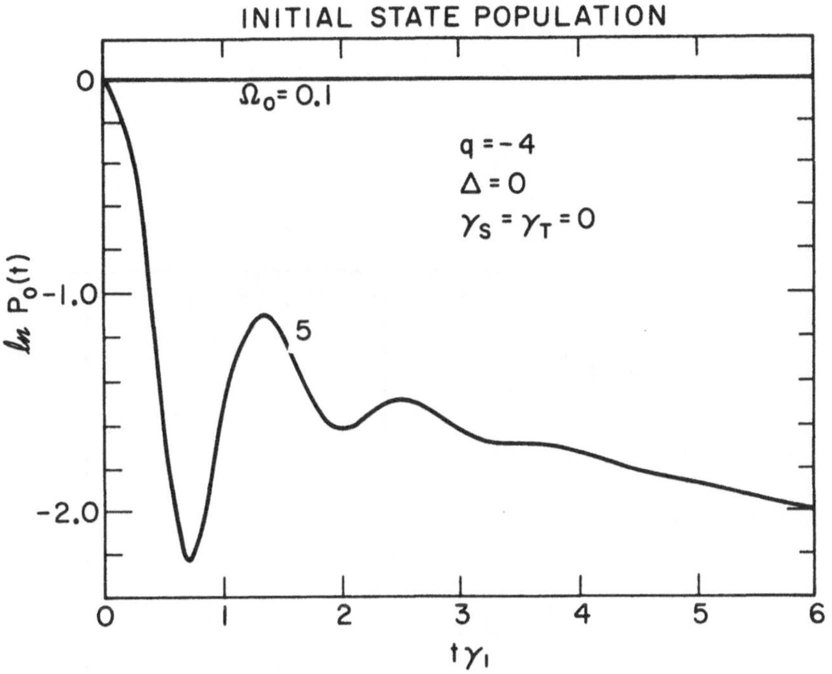

Fig. 4. Examples of $\ln P_0(t)$ in the regime $q \gg 1$, $\gamma_s = \gamma_T = 0$ and $W_{01} \ll \gamma_1$, $W_{01} \gg \gamma_1$. Note the transition from Rabi-oscillations to exponential decay.

The qualitative picture just described is reflected both in $P_o(t)$ and $S(\omega)$. Figures 4,5 pertain to the $q \gg 1$ sector, when both "weak-field" and "strong-field" regimes are expected. Concerning $P_o(t)$ (Figure 4) this is indeed the situation for two representative cases: When $\Omega_o \ll \gamma_1$, $\ln P_o(t)$ is a straight declining line, while for $\Omega_o \gg \gamma_1$ [when $q \gg 1$, $W_{01} \simeq \Omega_o/2$ eq. (3.8)] oscillations of frequency Ω_o are clearly seen. Fig. 5 shows the corresponding electron spectrum $S(\omega)$. In the weak field case only the "elastic" peak, around $\delta(\omega) = \omega - \omega_L \simeq 0$ can be discerned, with width $\ll \gamma_1$. However in the strong field case asymmetric peaks are observed which signal flux leaking during the "laps" the electron performed between $|0\rangle$ and $|1\rangle$ (the spectrum is a $t \to \infty$ limit). Note, however, that the positions of the

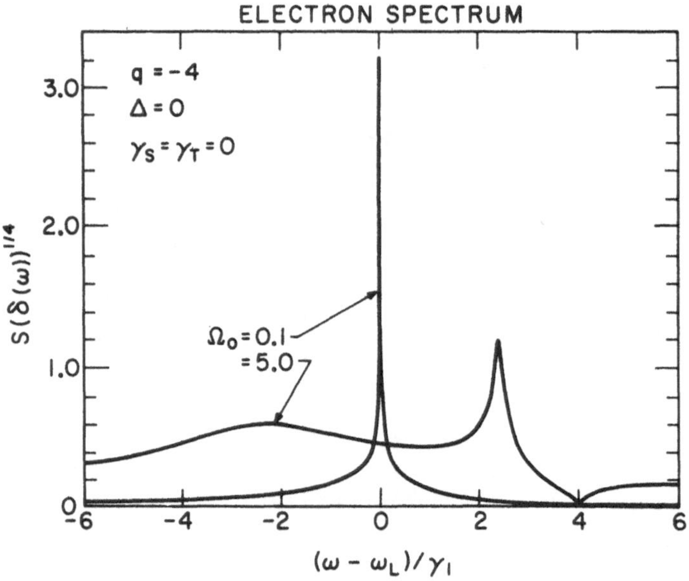

Fig. 5. Electron spectrum corresponding to the parameters of Fig. 4. The Fano zero is at $\delta(\omega) = \Delta - \gamma_1 q$.

peaks are very different from the "unperturbed" positions of $|0\rangle$, $|1\rangle$: $\delta(\omega) = 0, \Delta$. In fact they are much closer to the dressed states positions $.5(\Delta \pm \sqrt{\Delta^2 + \Omega_0^2})$.[2,3,9] This interpretation has been corroborated by considering many other examples as well and is consistent with the "strong field" picture. A deficiency of the present analysis is that little can be said with regard to the widths of the two resonances. This question is settled very naturally in the DRR, section 3.

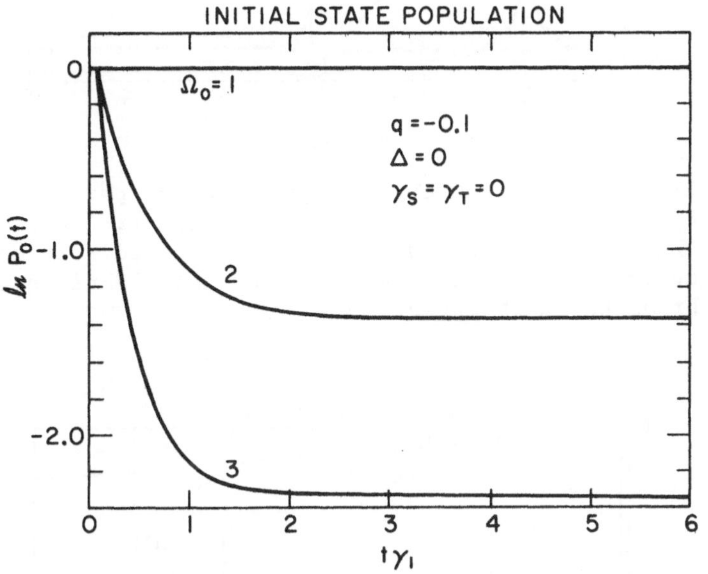

Fig. 6. An example of $\ln P_0(t)$ in the $q \ll 1$ sector, $\gamma_S = \gamma_T = 0$. The lack of Rabi-oscillations and the two-slopes structure is apparent. Interpretation given in text.

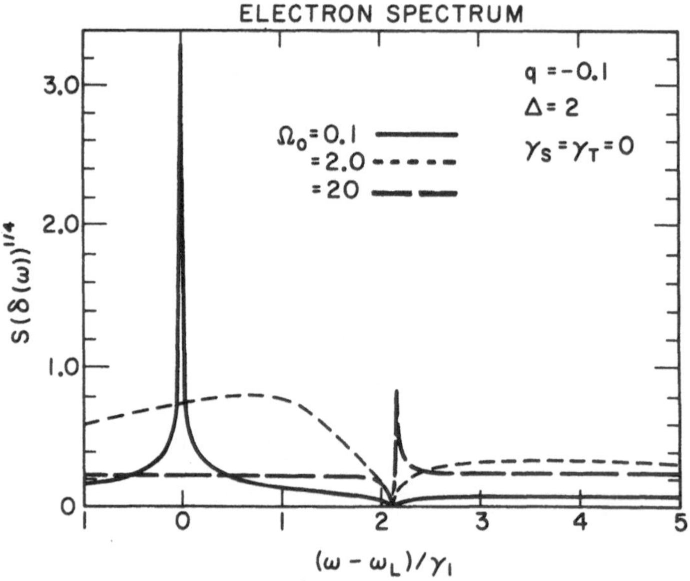

Fig. 7. Electron spectrum corresponding to the parameters in Fig. 6. For Ω_0 = .1, 2. The 'inelastic' peaks at $\delta(\omega) = \Delta$ is too narrow to be resolved. It does show up for Ω_0 = 20., at which point the elastic peak is too broad to be observed.

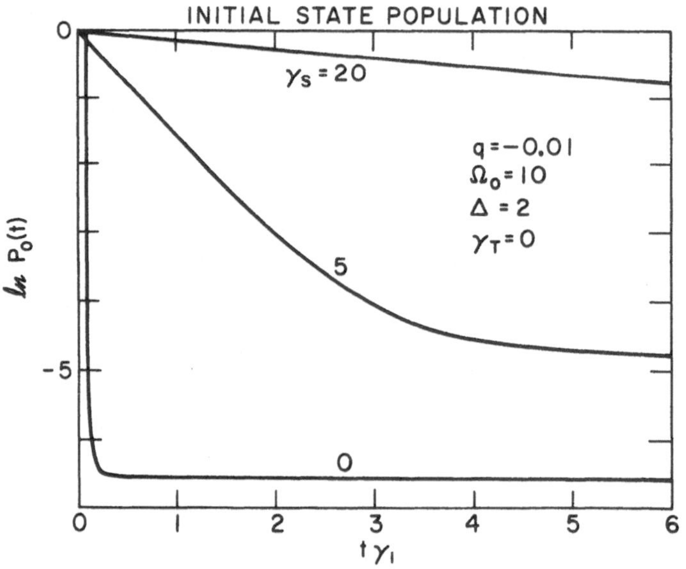

Fig. 8. The effect of spontaneous-decay channels on $\ell nP_0(t)$, in the $q \ll 1$ sector. Note that increasing γ_s decreases the "large" slope, with indiscernible effect on the second, small slope. Interpretation given in text.

Figures 6,7 exemplify the features in the $q \ll 1$ sector. Figure 6 shows the two-slope structure of $\ln P_0(t)$, the first one increases with Ω_0 [for $q \ll 1$, $W_{0\tilde{\omega}} \sim \Omega_0$, $W_{01} \simeq 0$. Eq. (3.8)], while the second slope is very close to zero and starts only after the population of $|0\rangle$ is substantially reduced. The interpretation of the two slopes is straightforward: The first slope represents the drainage of flux due to the $|0\rangle - |\tilde{\tilde{\omega}}\rangle$ transition which is direct. The second, almost vanishing slope, is due to drainage in the $|0\rangle - |1\rangle$ channel. Since this channel is suppressed and since the $|1\rangle$ - state is inefficiently coupled to the continuum for $q \ll 1$ (the Fano profile has a depression[11] of width γ_1 around $\delta(\omega) \simeq 0$), only a small amount of flux is depleted out in this channel, and only very slowly. Figure 7 shows a corresponding electron spectrum. Note the "elastic" peak, around $\delta(\omega) \simeq 0$, which is ever-broadened until it becomes indiscernible. The "inelastic" peak, around $\delta(\omega) \simeq \Delta$ is broadened as well, yet stays considerably narrower in comparison to the elastic peak (for $\Omega_0 = .1, 2.$ it is too narrow to be resolved). Asymmetric line shapes are seen to occur. The fact that the peaks are at $\delta(\omega) \cong 0, \Delta$ is consistent with the "dressed states" and "weak field" notions.

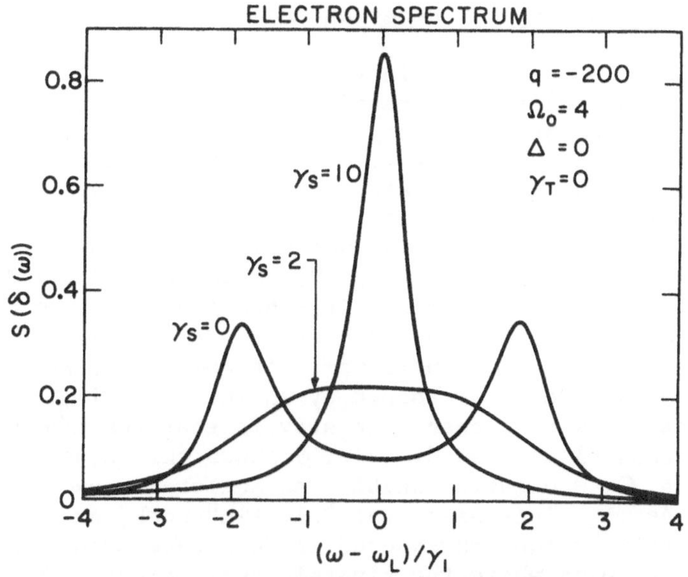

Fig. 9. The effect of spontaneous decay on the electron spectrum, in the particular case of Autler-Townes splitting for $\gamma_s = 0$. Note the "threshold" case for $\gamma_s = 2$, inbetween the Autler-Townes two peaks and the one, narrow, 'elastic' peak for $\gamma_s = 10$.

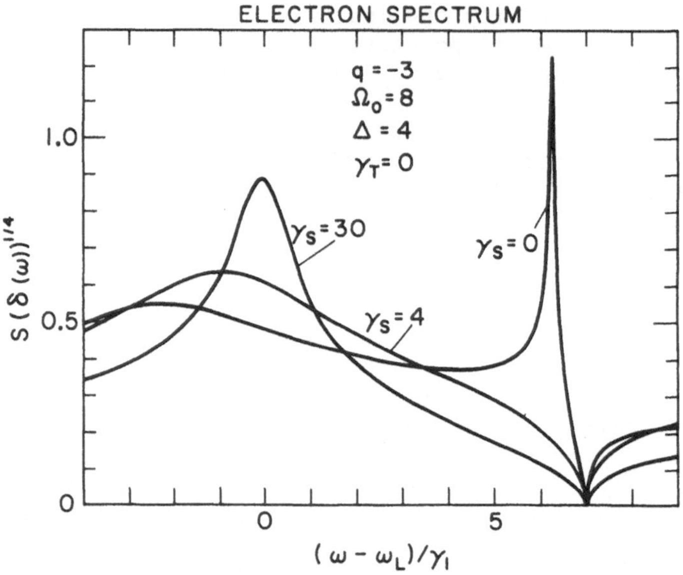

Fig. 10. Example for the "broadening-narrowing" effect of spontaneous-decays in a $q \gg 1$ case.

2.2 Effects of Spontaneous Decay and Phase-Jitter Relaxation

The other two parameters in (2.1) measure the strength of the two relaxation processes considered. The effect of spontaneous decay to the ground state is easy to assess. Since it is a decay to the ground state it tends to increase the fraction of time the system spends in the ground state. Hence $P_o(t)$ decays slower when $\gamma_s \neq 0$. This indeed is seen in Fig. 8 very clearly for the $q \ll 1$ sector. The effects on the second, almost vanishing slope are too small to be resolved. Fig. 9, is an example for the effects on the electron spectrum. Because increasing γ_s lengthens the lifetime of the $|0\rangle$ -state, the corresponding "elastic" peak becomes ever narrower. At the same time the "inelastic" peak is ever broadened by an increase of γ_s since the system has less of a chance to stay in that resonance now that a new decay channel is opened. Fig. 9 shows this happening in a case when the $\gamma_s = 0$ curve exhibits an Autler-Townes splitting. As γ_s increases, the "elastic" peak, around $\delta(\omega) \cong 0$ emerges. Note that the "threshold" in this case occurs at $\gamma_s = 2$. This is at the point where the Einstein coefficient $2\gamma_s$ equals the Rabi oscillation frequency Ω_o. A more typical situation is given in Fig. 10 where the narrowing-broadening role of γ_s is apparent.

The second kind of relaxation considered is the "phase jitter" mechanism,[3,18,19,20] which implies the attachment to the atomic dipole-moment a random, time dependent phase to simulate the effects of both soft elastic collisions with neighboring atoms in the target (collisional brodening) and phase fluctuations in the laser field itself.[2,3] Thus we insert in the Hamiltonian (1.4)

$$\tilde{\tilde{B}}(\omega) = |0\rangle\langle\tilde{\tilde{\omega}}| \Rightarrow |0\rangle\langle\tilde{\tilde{\omega}}|\, e^{i\psi(t)}$$

$$\langle\dot{\psi}(t)\;\dot{\psi}(t')\rangle = 2\gamma_T(t-t') \qquad (2.4)$$

The averaged effect of phase-jitter relaxation is to endow the laser with a band-width of magnitude γ_T . Consequently adding γ_T will invariably shorten the lifetime of $P_o(t)$ and broaden both resonances ('elastic', 'inelastic') in $S(\omega)$. This implies, in particular, that with regard to the 'elastic' peak, the "γ_s" and "γ_T" relaxations compete (i.e., narrowings vs. broadenings) while for the 'inelastic' peak they add up.[6] Fig. 11 demonstrates this γ_s, γ_T competetion.

Going back to the basic equation of motion (1.18), note that "γ_T" enters only in one place, i.e., in the equation for $\tilde{\tilde{B}}(\omega,t)$, side by side with "$-i\omega_L$" which also appears only there. This simple observation is embodied in the following "substitution rule": To include the "γ_T" effects, substitute in the equations of motion in the SDR

$$\pm i\omega_L \Rightarrow \pm i\omega_L - \gamma_T \quad \text{in} \quad \text{t-representation}$$

$$\pm i\omega_L \Rightarrow \pm i\omega_L + \gamma_T \quad \text{in} \quad \text{z-representation} \qquad (2.5)$$

The substitution rule (2.5) indicates quite clearly that phase-jitter relaxation is simulated by a finite-band laser beam.

2.3 Summary

To summarize, we have analyzed the effects of the "coherent" and "statistical" parameters in the model. The main message, as far as the coherent parameters (W_{01}, $W_{0\omega}$ or equivalently Ω_o, q) are concerned is that the spectrum can be understood in terms of two "dressed states" (corresponding to $|0\rangle$, $|1\rangle$) or rather the resonances that correspond to them. The behaviour of $\ln P_o(t)$ is universally either a straight line signalling a "weak field" regime, or oscillating-decaying behaviour in the "strong field" regime. With regard to the "statistical parameters", spontaneous decay to $|0\rangle$ (recycling) and "phase-jitter" relaxations have qualitatively different effects. The former lengthens the lifetime of $P_o(t)$ and narrows the "elastic" resonance, the latter vice versa.

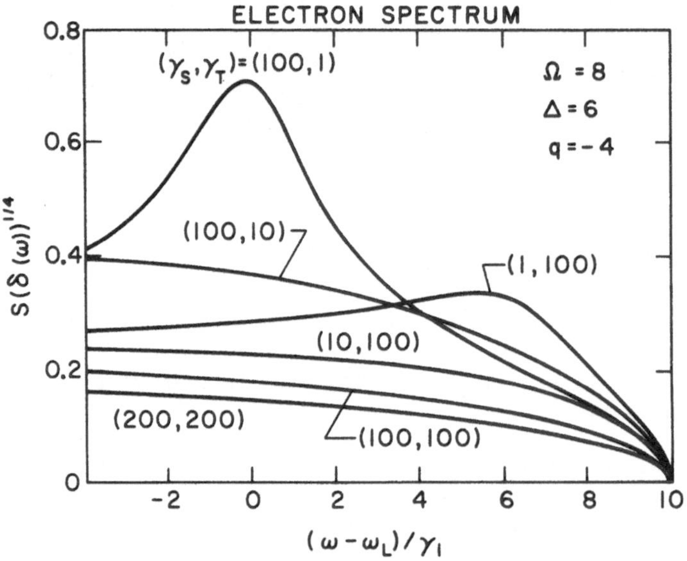

Fig. 11. The $\gamma_s - \gamma_T$ competing effects on the electron-spectrum. Phase-jitter relaxation of strength γ_T (Eq. (2.4)) imposes a minimal width γ_T in either of the two dressed resonance widths, in particular the elastic peak. Hence "γ_s" cannot narrow down the elastic peak beyond this value. A very large γ_T, such that $\gamma_T \gg \gamma_s, \gamma_1, \Omega_0$.. induces "redistribution".[5]

3. THE DRESSED RESONANCE REPRESENTATION (DRR)

As has been demonstrated in section 2 despite the complexity of the equations of motion (1.18) the electron spectrum can be understood in simple terms: There are two resonances in the spectrum whose positions can be identified with the two dressed state originating from $|0\rangle$, $|1\rangle$. In weak field situations, these positions coincide with the energies of $|0\rangle$, $|1\rangle$ properly shifted by the laser energy ω_L. In the strong field situation the splitting of the two resonances and shifts are that of the two dressed states. The idea underlying the DRR is to account for these features in the representation from the outset.[10,21,22]

In the SDR $|0\rangle$ and $|1\rangle$ are not treated on an equal footing. This asymmetry can be traced back to the initial "embedding" of the autoionizing state $|1\rangle$ (which defines $|\tilde{\omega}\rangle$). The "dressing", i.e., the shifts and distortions due to the radiation field are included via the equations of motion. Since the distortions can be substantial, while the equations of motion involve only the Fano (unperturbed) profile, no wonder the solution attains a complicated form. To reinstate the symmetry between the roles of $|0\rangle$, $|1\rangle$ a different route should be followed, schematically depicted in Fig. 12: First "dress" the $|0\rangle$, $|1\rangle$ states and then "embed" the two dressed states into the continuum as a 2-bound state 1-continuum Fano problem. Thus in the DRR the order of "embedding" and "dressing" is reversed.

There are several advantages to reformulating the model in the DRR. It obviates the role of the Dressed Resonances in the spectrum. As we shall see, it simplifies the equations of motion considerably and in particular the semi-classical limit is immediate to solve. It clarifies the relation between similar models which employ "embedding" either by a static interaction or by a radiation field. A bonus of this reformulation is that it offers a simple approximation scheme for the electron spectrum with an associated "small parameter" which is "1/q" for $q \gg 1$ and "q" for $q \ll 1$ sectors.

3.1 The Dressed Resonances and Semi-Classical Model

The dressed resonances are the result of embedding the dressed states into the continuum and as will be soon clear, they play a central role in understanding $P_o(t)$, $S(\omega)$. To introduce the concept consider the semi-classical two level atom in the rotating frame

$$\hat{H} = \hbar\Delta\,|1\rangle\langle 1| + W_{01}\,[\,|0\rangle\langle 1| + |1\rangle\langle 0|\,] \qquad (3.1)$$

The dressed states $|d_j\rangle$ are given by:[9]

$$\hat{h}\,|d_j\rangle = \hbar\omega_j\,|d_j\rangle \qquad (3.2a)$$

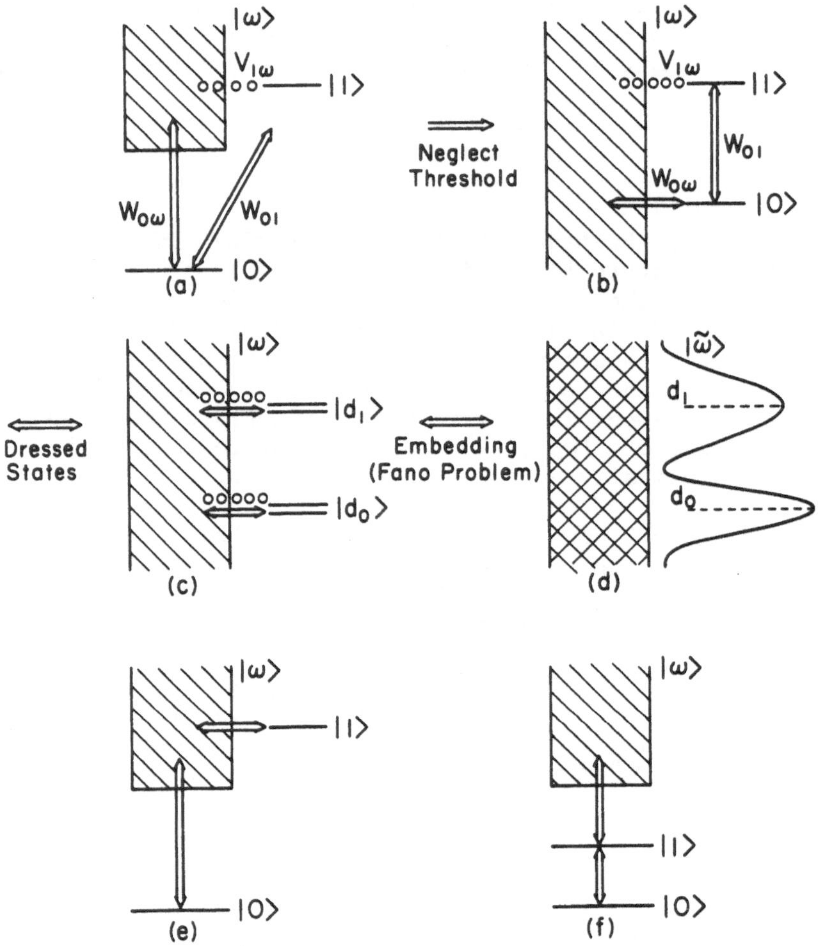

Fig. 12. Schematic presentation of the "dressing"-"embedding" steps in constructing the DRR(a - d). Notations as in Fig. 1, and $|d_i>$ denote the dressed states. Insets (e), (f) represent, in the present notation, two related models (Ref. 21, P.E. Coleman and P.L. Knight 1982), (Ref. 21, P.E. Coleman and P.L. Knight 1981), respectively.

and

$$|d_0\rangle = \cos\theta\,|0\rangle - \sin\theta\,|1\rangle \qquad \text{tg } 2\theta = \frac{2W_{01}}{\Delta}$$
$$|d_1\rangle = \sin\theta\,|0\rangle + \cos\theta\,|1\rangle \tag{3.2b}$$

The dressed <u>resonances</u> representation is defined in conjunction with a model when the continuum is added up (Fig. 1a):

$$\hat{H} = \hat{h} + \int d\omega\,\hbar\delta(\omega)|\omega\rangle\langle\omega| + \int d\omega\,\hbar W_{0\omega}\,[\,|0\rangle\langle\omega| + |\omega\rangle\langle 0|\,]$$
$$+ \int d\omega\,\hbar V_{1\omega}\,[\,|1\rangle\langle\omega| + |\omega\rangle\langle 1|\,] \tag{3.3}$$

The eigenvalues of $\hat{H}$ are easily solved by applying the Fano method[11] ($\delta(\omega) \equiv \omega - \omega_L$):

$$\hat{H}\,|\tilde{\omega}\rangle = \hbar\,\delta(\omega)\,|\tilde{\omega}\rangle$$
$$|\tilde{\omega}_0\rangle = a_0(\omega_0)|d_0\rangle + a_1(\omega_0)\,|d_1\rangle + \int d\omega\; b_{\omega_0}(\omega)\,|\omega\rangle \tag{3.4}$$

where

$$a_0(\omega_0) = \sqrt{\frac{\tilde{\gamma}_0}{\pi}}\;\frac{(\delta(\omega_0) - \omega_{d_1})}{p(\omega_0)} \qquad a_1(\omega_0) = \sqrt{\frac{\tilde{\gamma}_1}{\pi}}\;\frac{(\delta(\omega_0) - \omega_{d_0})}{p(\omega_0)}$$
$$\tilde{\gamma}_j = \pi\,|\langle\omega|\hat{V}|d_j\rangle|^2 \qquad j = 0, 1 \tag{3.5}$$

and the second order secular equation is

$$p(\omega) = (\omega - \omega_{d0} + i\tilde{\gamma}_0)\,(\omega - \omega_{d1} + i\tilde{\gamma}_1) + \tilde{\gamma}_0\tilde{\gamma}_1$$
$$\equiv (\omega - \Gamma_0)\,(\omega - \Gamma_1) \tag{3.6}$$

The usefulness and interpretation of the DRR are demonstrated by computing the electron spectrum

$$S(\omega) = \lim_{t\to\infty} \langle\tilde{\omega}|\hat{Q}_0\,\hat{\rho}(t)\,\hat{Q}_0|\tilde{\omega}\rangle = |\langle 0|\tilde{\omega}\rangle|^2 \lim_{t\to\infty}[1 - P_0(t)]$$
$$= \frac{N\,(\delta(\omega) - \Delta + \gamma_1 q)^2}{|p(\omega)|^2} \tag{3.7}$$

where $\hat{Q}_0 = \mathbb{1} - |0\rangle\langle 0|$ is a projection operator to the $|\tilde{\omega}\rangle$ continuum space and $\hat{\rho}(t) = \exp(\hat{H}t/ih)\ |0\rangle\langle 0|\ \exp[-\hat{H}t/ih]$. The last line in (3.7) is derived using eqs. (3.4),(3.5) and the relations

$$W_{01} = \frac{\Omega_0}{2}\frac{|q|}{\sqrt{q^2+1}} \qquad W_{0\omega} = \frac{\Omega_0}{\sqrt{4\pi\gamma_1}}\ \frac{\mathrm{sgn}(q)}{\sqrt{q^2+1}} \tag{3.8}$$

The simple expression (3.7) says it all: The complex zeros of $p(\omega)$, i.e., Γ_0, Γ_1 mark the maxima in $S(\omega)$. The lineshapes are skewed, however, by the numerator, which is expected since it reflects the existence of the Fano zero. Inverting the argument, we could have guessed the numerator of (3.7) and the only non trivial entry is the secular equation (of second order) in the denominator. Note that $p(\omega)$ provides <u>both</u> the positions and widths of the dressed resonances.

The evaluation of $P_0(t)$ is equally straightforward

$$P_0(t) = \mathrm{tr}[\hat{\rho}(t)\ |0\rangle\langle 0|\ ] = |J(t)|^2$$
$$J(t) = \int d\omega\ e^{-i\omega t}\ |\langle 0|\tilde{\omega}\rangle|^2 = \sum_{j=0}^{1} R_j\ e^{-i\Gamma_j t} \tag{3.9}$$

underscoring that the spectrum and $P_0(t)$ are intimately related. As we have observed the behaviour of $P_0(t)$ is always oscillating (for weak field, period $\to \infty$) with a decaying envelope. The interesting parameters are therefore the Rabi oscillation frequency ω_R, and the slope of the decaying envelope of $\ln P_0(t)$. Expression (3.9) suggests that we should identify:

$$\omega_R = \mathrm{Real}\ (\Gamma_0 - \Gamma_1^*)$$
$$\Gamma_S = \left|\mathrm{Min}\left\{\mathrm{Im}\ (\Gamma_0^* - \Gamma_0),\ \mathrm{Im}\ (\Gamma_1 - \Gamma_1^*),\ \mathrm{Im}\ (\Gamma_0 - \Gamma_1^*)\right\}\right| \tag{3.10}$$

Correspondingly a "strong" or "weak" field situation occurs whenever $\omega_R \gg \Gamma_S$, or $\omega_R \ll \Gamma_S$ respectively.

We see therefore that in the DRR, the calculation of $S(\omega)$ and $P_0(t)$ in the semi-classical model is immediate. Furthermore, all the gross features (position and widths of peaks, peridicity, decate rate) can be extracted from the two complex zeros of the second-order secular equation $p(\omega)$.

3.2 The Full Problem in the DRR

The next step is to transcribe the full problem - stimulated and spontaneous decay - to the DRR representation. This is straightforward starting from the unperturbed representation, Fig. (1.a). In this instance the appropriate complete set of atomic operators is

$$C(\omega, \omega') = |\tilde{\omega}\rangle\langle\tilde{\omega}'| \qquad (3.11)$$

and the equations of motion are:[10]

$$\frac{d}{dt} C(\omega_0, \omega_0'; t) = i(\omega_0 - \omega_0') \, C(\omega_0, \omega_0'; t) + \langle 0|\tilde{\omega}_0\rangle\langle\tilde{\omega}_0'|0\rangle M(t)$$
$$- F^*(\omega_0') \int d\omega' \, C^*(\omega_0, \omega'; t) \, R(\omega') - F(\omega_0) \int d\omega' R^*(\omega') C(\omega', \omega_0'; t) \qquad (3.12)$$

where the "perturbed" profiles are given by

$$F(\omega) = F_B(\omega) + F_C(\omega) \qquad R(\omega) = R_B(\omega) + R_C(\omega) \qquad (3.13a)$$

and

$$F_B(\omega_0) = W_{01} \langle 1|\tilde{\omega}_0\rangle \qquad R_B(\omega_0) = Q(\omega_1) \; F_B(\omega_0)$$
$$F_C(\omega_0) = W_{0\omega} \int d\omega \langle \omega_0|\tilde{\omega}\rangle \qquad R_C(\omega_0) = W_{0\omega} \int d\omega \, Q(\omega) \langle \omega_0|\tilde{\omega}\rangle \qquad (3.13b)$$

The precise definition of $M(t)$ is unimportant for the present purposes. It suffices to note that it is such that unitarity is conserved, i.e.

$$\frac{d}{dt} \left[\int d\omega \, C(\omega, \omega; t) \right] = 0 \qquad (3.14)$$

The simplicity of the equations of motion (3.12) when compared to its counterpart in the SDR, eq. (1.18) is gratifying. We can solve (3.12) exactly using the same method the third equation in (1.18) was "solved" for $C(\omega, \omega'; t)$, since both have the same structure. The crucial difference, however, is that in the DRR the profile-function $F(\omega)$ is the "perturbed" profile [eq. (3.13)], rather than Fano's unperturbed profile $\hat{F}(\omega)$ in (1.18). More important, note that decomposition (3.13) offers an approximation scheme: Since $F_B(\omega) \sim W_{01}$ and $F_C(\omega) \sim W_{0\omega}$, and, by virtue of (3.8), for $q \gg 1$ $W_{01} \gg W_{0\omega}$ while for $q \ll 1$ $W_{01} \ll W_{0\omega}$, two expansions emerge: For the $q \gg 1$ sector we can expand around $q = \infty$ in powers of "1/q", i.e., $F_B(\omega)$ is the leading term and $F_C(\omega)$ is a (1/q)-correction. Vice versa for $q \ll 1$, it is possible to expand around the $q = 0$ in powers of q by taking $F_C(\omega)$ as the leading term and $F_B(\omega)$ as a correction of order "q" term. The need for different expansions in the $q \gg 1$ and $q \ll 1$ sectors is in keeping with the different "physics" in these two sections, as discussed in section (2.1).

It can be shown[10] that the expression for the electron spectrum is of the form

$$S(\omega) = N \left| \frac{P(\omega) \langle 0 | \tilde{\omega} \rangle}{P(\omega)} \right|^2 \tag{3.15}$$

where

$$1 + iQ(\omega_1) \int d\omega' \frac{|F(\omega')|^2}{iz + \omega_c - \omega'} \equiv \frac{P(\omega_0, z)}{(iz+\omega_0-\Gamma_0)(iz+\omega_0-\Gamma_1)} \tag{3.16}$$

and N is a normalization factor. Hence, by approximating $F(\omega)$ in the manner indicated above, a (q, 1/q) expansion of the secular equation ensues. Using (3.13) and the expressions in section (3.1) the leading terms are[10]

$$iQ(\omega_1) \int d\omega' \frac{|F(\omega')|^2}{iz + \omega_0 - \omega'} = C_2 + \frac{C_1^{(0)} T_0^{(0)}}{iz+\omega_0-\Gamma_0} + \frac{C_1^{(0)} T_1^{(0)}}{iz+\omega_0-\Gamma_1} \tag{3.17}$$

where for $q = \infty$

$$C_2 = 0 \qquad C^{(0)}{}_1 = 2\gamma_1\gamma_s$$

$$T_j^{(0)} = \frac{\Gamma_j^2}{(\Gamma_j - \Gamma_k) P^*(\Gamma_j)} \qquad k \neq j, \quad j = 0,1 \tag{3.18a}$$

and for $q = 0$

$$C_2 = \gamma_s/\gamma_1 \qquad C^{(0)}{}_1 = 2\gamma_s/\gamma_1$$

$$T_j^{(0)} = \frac{[\tilde{\gamma}_0(\Gamma_j - \omega_{d_1}) + \tilde{\gamma}_1(\Gamma_j - \omega_{d_2})]}{\Gamma_j - \Gamma_k} \left\{ \frac{\tilde{\gamma}_0(\Gamma_j - \omega_{d_1}) + \tilde{\gamma}_1(\Gamma_j - \omega_{d_2})}{P^*(\Gamma_j)} - i \right\}$$

$$k \neq j. \quad j = 0,1 \tag{3.18b}$$

A comparison between the exact and approximate spectrum (to first order) is presented elsewhere,[10] vindicating the (q, 1/q) expansion.

The relative ease in obtaining $S(\omega)$ is due both to the use of the DRR equations of motion <u>and</u> that the evaluation of $S(\omega)$ requires the knowledge of $C(\omega, \omega;\ z=0)$ only. The calculation of $P_0(t)$ calls for the solution over the entire z-range and is not attempted here.

3.3 Conclusion

We have studied the electron spectrum and time-dependence of initial-state population in the context of a simple soluble model for laser-induced autoionization. To include spontaneous decay back to the initial state (recycling), the radiation field is quantized. In addition the model accommodates phase-jitter relaxation to account for fluctuations in the laser-beam and collisions with neighboring atoms.

The calculated results are easily understood in terms of underlying dressed resonances - the resonances generated by embedding the dressed states into the continuum of the atom. The effects of recycling and phase-jitter relaxations are also interpretable by realizing that the former is a decay to the initial state while the latter endows the laser field with an effective band width. These observations provide the motivation for reformulating the problem in a representation that incorporates the dressed resonances from the outset. This representation, the "Dressed Resonances Representation" (DRR) has been used successfully in the context of related models.[21,22]

In the DRR the equations of motion simplify considerably and the semi-classical limit of the model (the radiation field is treated as a classical field) is simple and transparent. An extra bonus is the identification of an approximation scheme (for the secular equation) in terms of a well defined "small parameter". The approximation has been checked and vindicated.[10]

REFERENCES

1. P. Lambropoulos and P. Zoller, Phys. Rev. A 24, 379 (1981); App. Optics 18, 3926 (1980).
2. K. Rzazewski and J.H. Eberly, Phys. Rev. Lett. 47, 408 (1981).
3. K. Rzazewski and J.H. Eberly, Phys. Rev. A 27, 2026 (1983).
4. G.S. Agarwal, S.L. Haan, K. Burnett and J. Cooper, Phys. Rev. Lett. 26, 1164 (1982); G.S. Agarwal, S.L. Haan, K. Burnett and J. Cooper, Phys. Rev. A 26, 2277 (1982); S.L. Haan and G.S. Agarwal, Proceedings of the Sixth International Conference on Spectral Line Shapes, K. Burnett, Editor (de Gruyter, Berlin 1982).
5. J.H. Eberly, K. Rzazewski and D. Agassi, Phys. Rev. Lett. 49, 693 (1982).
6. G.S. Agarwal and D. Agassi, Phys. Rev. A 27, 2254 (1983); D. Agassi and J.H. Eberly, Phys. Rev. A (in press).
7. M. Lewenstein, J.W. Haus and K. Rzazewski, Phys. Rev. Lett. 50, 417 (1983); J.W. Haus, M. Lewenstein and K. Rzazewski, Phys. Rev. A (in press); J.W. Haus, M. Lewenstein and K. Rzazewski (preprint 1983).

8. C. Cohen-Tannoudji and S. Reynald, Proceedings of the International Conference in Multiphoton Processes, J.H. Eberly and P. Lambropoulos, Editors (John Wiley & Sons, New York 1977); J. Phys. B 10, 345 (1977).
9. E. Courtens and A. Szoke, Phys. Rev. A 15, 1588 (1977).
10. D. Agassi and J.H. Eberly, in preparation.
11. U. Fano, Phys. Rev. 124, 1866 (1961).
12. The details of this coupling depend on the nature of the autoionizing state, e.g., S. Feneuille, S. Liberman, J. Pinard and A. Taleb, Phys. Rev. Lett. 42, 1404 (1979); W.E. Cooke, T.F. Gallagher, S.A. Edelstein and R.M. Hill, Phys. Rev. 40, 178 (1978).
13. See for instance L.Allen and J.H. Eberly, Optical Resonance and Two Level Atoms (John Wiley & Sons, New York 1975).
14. R. Louden, The Quantum Theory of Light (Calderon Press, Oxford 1979).
15. J.R. Ackerhalt and J.H. Eberly, Phys. Rev. D 10, 3350 (1974); P.W. Milonni and W.A. Smith, Phys. Rev. A 11, 814 (1975); P.W. Milonni, Phys. Rep. C 25, 1, (1976).
16. J.D. Cresser, Ph.D. Thesis (unpublished 1979).
17. G.S. Agarwal, Quantum Optics, Springer Tracts in Modern Physics, Vol. 70 (Springer, Berlin 1974).
18. G.S. Agarwal, Phys. Rev. Lett. 37, 1383 (1976); Phys. Rev. A 18, 1490 (1978).
19. J.H. Eberly, Phys. Rev. Lett. 37, 1387 (1976).
20. B.J. Dalton and P.L. Knight, J. Phys. B 15, 3997 (1982).
21. P.E. Coleman and P.L. Knight, J. Phys. B 14, 2139 (1981); P.E. Coleman and P.L. Knight, J. Phys. B 15, L235 (1982) and corrigendum, J. Phys. B 15, 1957 (1982); P.M. Radmore and P.L. Knight, J. Phys. B 15, 561 (1982).
22. M. Crance and L. Armstrong, J. Phys. B 15, 3199 (1982); L. Armstrong, invited talk D.E.A.P. Meeting, Boulder, Colorado, 1983.

COOPERATIVE EFFECTS IN ONE DIMENSION

V. Benza and E. Montaldi

Dipartimento di Fisica, Università di Milano
and Istituto Nazionale di Fisica Nucleare, Sezione di Milano
Via Celoria 16, 20133 Milano, Italy

INTRODUCTION

In this note we give an account of the results obtained in studying a one-dimensional lattice model of atoms coupled to a system of harmonic oscillators.

Our main interest being the occurrence of cooperative phenomena produced by the effective self-coupling between the atoms, we will analyze the model both as a thermodynamic system and as an isolated microscopic quantum mechanical system. From a phenomenological point of view, this last attitude is supported by the recently developed technique of trapping and cooling of neutral atoms[1], allowing direct detection of few radiators. In particular, observation of cooperative effects in systems of few tens of atoms has been made possible, as shown by Haroche and coworkers[2], who exhibited coherent emission and absorption from Rydberg atoms.

While the thermodynamic description will follow the standard lines of statistical mechanics, the quantum mechanical treatment will be performed in the framework of real space renormalization group.

Although our model provides a basis to describe a fairly large class of interactions of atoms with a Bose field (discretized over a lattice as a set of harmonic oscillators), we will restrict ourselves to the specific case of coupling with the radiation field.

THE MODEL

We consider a one-dimensional crystal lattice of atoms having spacing L interacting with the radiation field; regarding the atoms as pointlike objects at fixed positions, we can neglect the field over wavelengths of the order of the Bohr radius a_0, so that the coupling between radiation and matter can be treated in the dipole approximation.

We describe each atom as a system of 2K equally spaced levels, with associated angular momentum operators r_j (j = 1,2,3); the corresponding Hamiltonian is $\hbar\omega r_3$, where $\hbar\omega$ is the energy difference between contiguous levels.

We model the e.m. field by a lattice of oscillators with harmonic nearest neighbour couplings, the lattice spacing being Λ^{-1}.

We will show that the one-dimensional case already exhibits a rich phenomenology, essentially due to the fact that a nonlocal effective interaction between the atoms arises through their coupling with the field.

The Hamiltonian of the free e.m. field, which we assume to have a single direction of polarization, is

$$H_F = \Lambda \sum_{n=1}^{N'} \left[2\pi c^2 p_n^2 + \frac{1}{8\pi}(q_{n+1} - q_n)^2 \right] \qquad (1)$$

where N' is the number of oscillators, and the operators p_n and q_n satisfy the canonical commutation rules

$$[q_n , p_{n'}] = i\hbar\delta_{nn'} .$$

Of course, as Λ and $N' \to \infty$ with $\Lambda^{-1} N'$ finite, q goes over into the potential vector A and Λp into $(4\pi c^2)^{-1}\dot{A}$.

In terms of normal modes, eq. (1) becomes

$$H_F = \Lambda \sum_{n=1}^{N'} (2\pi c^2 P_n^2 + \frac{1}{4\pi}\lambda_n Q_n^2) , \qquad (2)$$

$$\lambda_n = 2\sin^2 \frac{n\pi}{2(N'+1)} , \quad Q_n = \left(\frac{2}{N'+1}\right)^{\frac{1}{2}} \sum_{m=1}^{N'} q_m \sin\frac{mn\pi}{N'+1}$$

or also

$$H_F = \sum_{n=1}^{N'} \hbar\nu_n (a_n^\dagger a_n + \frac{1}{2}) , \quad \nu_n = c\Lambda(2\lambda_n)^{\frac{1}{2}} \qquad (3)$$

where

$$a_n = (2\hbar)^{-\frac{1}{2}}(\alpha_n^{-\frac{1}{2}} Q_n + i\alpha_n^{\frac{1}{2}} P_n)$$

$$a_n^{\dagger} = (2\hbar)^{-\frac{1}{2}}(\alpha_n^{-\frac{1}{2}} Q_n - i\alpha_n^{\frac{1}{2}} P_n), \quad \alpha_n = \pi c\left(\frac{8}{\lambda_n}\right)^{\frac{1}{2}}.$$

Neglecting the field dynamics over wavelengths shorter than the interatomic spacing, we can associate to each site of a common lattice an atomic variable and a field variable, so that the total Hamiltonian is

$$H = \hbar\omega\sum_{n=1}^{N} r_{3,n} + \Lambda\sum_{n=1}^{N}\left[2\pi c^2 p_n^2 + \frac{1}{8\pi}(q_{n+1} - q_n)^2\right] + g\sum_{n=1}^{N} r_{2,n} q_n$$

$$[r_{i,n}, r_{j,n'}] = i\varepsilon_{ijl}\, r_{l,n}\, \delta_{nn'} . \qquad (4)$$

Searching for the occurrence of cooperative effects, we will consider in the sequel the case in which the number of atoms is high over an atomic wavelength $2\pi c/\omega$; this is just the condition under which Dicke[3] originally foresaw superradiance, furthermore it has been recently realized by Haroche et al., using transitions between Rydberg levels with atomic wavelengths of the order of millimeters.

Accordingly, we require the inequality $\omega \ll c\Lambda \ll c/a_0$, where the r.h.s. ensures the dipole approximation to be valid.

QUANTUM MECHANICAL TREATMENT.

We begin by considering an isolated quantum mechanical system: in this context, we search for coherence properties of the atomic states spontaneously produced by the effective interatomic coupling. As we will see, these coherence properties emerge from the structure of the ground state, which we are going to analyze by constructing a sequence of effective Hamiltonians pertaining to different length scales; at each stage of the procedure, the spectrum of the new Hamiltonian is required to minimize the spectrum of the former one.
Such a strategy follows the lines of the Kadanoff – Wilson renormalization group approach in the real space, in a quantum mechanical version recently

proposed by Fradkin and Raby[4], particularly suitable for our purposes.

We start by writing H as a sum of "cell" ($\mathfrak{h}^*$) and "site" ($\mathfrak{h}$) Hamiltonians (see Fig. 1).

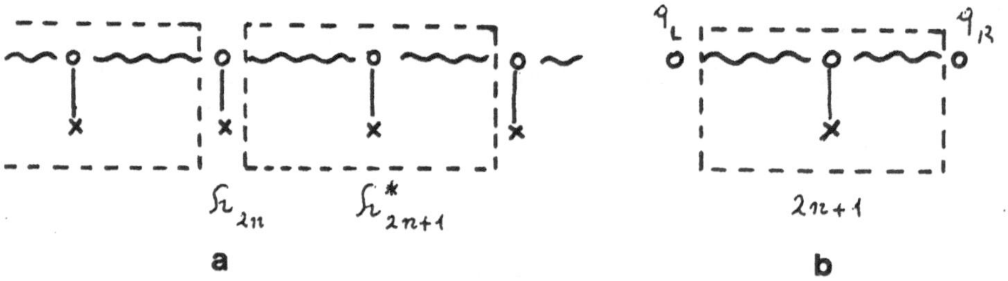

Fig. 1 : a) The lattice model : the atoms are indicated by x, the oscillators by o. Within dashed lines the odd sites and the corresponding coupling with the nearest-neighbour oscillators: such subsystems are described by $\mathfrak{h}^*$, while the even sites are described by $\mathfrak{h}$.
b) By q_L and q_R we indicate the oscillators coupled to the (2n+1)-th site; $q_{ext} \equiv \frac{1}{2}(q_L + q_R)$ (see eq. (5)).

$$H = \sum_n \mathfrak{h}_{2n} + \sum_n \mathfrak{h}^*_{2n+1}$$

$$\mathfrak{h} = \hbar\omega\tau_3 + \Lambda\left(2\pi c^2 p^2 + \frac{1}{4\pi} q^2\right) + g\tau_2 q$$

$$\mathfrak{h}^* = \mathfrak{h} - \frac{\Lambda}{2\pi} q\, q_{ext} \tag{5}$$

Observing that $\mathfrak{h}^*$ can be written as

$$\mathfrak{h}^* = \hbar\omega\tau_3 + H_{osc} - \frac{\Lambda}{4\pi} q_{ext}^2 + g\tau_2 q_{ext} - \frac{\pi}{\Lambda} g^2 \tau_2^2$$

$$H_{osc} = \Lambda\left[2\pi c^2 p^2 + \frac{1}{4\pi}\left(q - q_{ext} + \frac{2\pi}{\Lambda} g\tau_2\right)^2\right] \tag{6}$$

we replace H_{osc} with its lowest eigenvalue $\hbar c\Lambda/\sqrt{2}$; in so doing, we obtain a new Hamiltonian $\mathfrak{h}^{*\prime}$ whose spectrum minimizes the original one. It can be shown [5a,b] that the replacement of $\mathfrak{h}^*$ with $\mathfrak{h}^{*\prime}$ amounts to the exact resummation of the leading terms of both the perturbative series in g (weak coupling case) and in $\hbar\omega$ (strong coupling case); the basic assumption needed to get this result is the already quoted inequality

$\omega \ll c\Lambda$, implying that there is a large number of atoms over an atomic wavelength.

Dropping some technical details, for which we refer to Ref. 5a, we have thus the renormalization transformation (see Fig. 2)

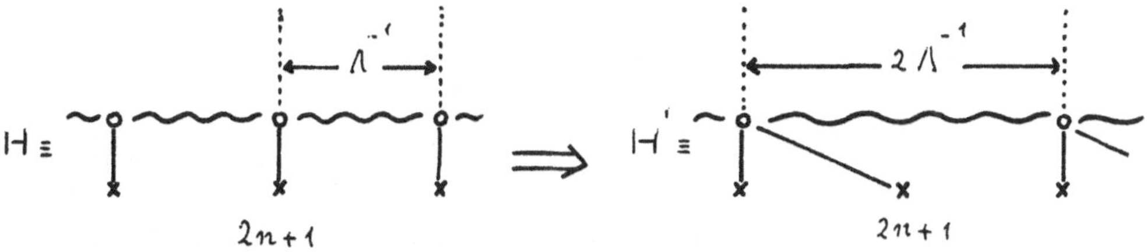

Fig. 2: To illustrate the renormalization transformation: the new Hamiltonian H' is associated to a lattice having spacing $2\Lambda^{-1}$. In the odd sites the oscillators have been frozen, while the atomic variables associated to 2n+1 and 2n group together to give clusters of double angular momentum ($r^{(1)}_{i,n_1} \equiv r_{i,2n} + r_{i,2n+1}$).

$$H \to H' = \frac{N}{2}\hbar c\Lambda/\sqrt{2} - \frac{\pi}{\Lambda} g^2 \sum_n r^2_{2,2n+1} + \sum_{n_1} h^{(1)}_{2n_1} + \sum_{n_1} h^{*(1)}_{2n_1+1} \qquad (n_1 \equiv 2n)$$

$$h^{(1)} = \hbar\omega r^{(1)}_3 + 2\pi c^2 \Lambda p^2 + \frac{1}{2}\cdot\frac{\Lambda}{4\pi} q^2 + g r^{(1)}_2 q$$

$$h^{*(1)} = h^{(1)} - \frac{1}{2}\cdot\frac{\Lambda}{2\pi} q\, q_{ext} \qquad (7)$$

The new Hamiltonian H', apart from the zero point energy of the odd site oscillators ($\frac{N}{2}\hbar c\Lambda/\sqrt{2}$) and the term associated with the polarization of the odd site atoms ($-\frac{\pi}{\Lambda}g^2\sum_n r^2_{2,2n+1}$), has the same form as the original one, provided that the initial spins are replaced by clusters of two spins, and the oscillator frequencies are rescaled by the factor $1/\sqrt{2}$.

Hence the transformation can be iterated obtaining, after s steps

$$H \to H^{(s)} = \frac{N}{2}\hbar c\Lambda/\sqrt{2}\sum_{i=0}^{s-1} 2^{-\frac{3i}{2}} - \frac{\pi}{\Lambda} g^2 P^2 + \sum_{n_s} \Omega^{(s)}_{2n_s} + \sum_{n_s} \Omega^{*(s)}_{2n_s+1}$$

$$P^2 = \sum_n (r_{2,2n+1})^2 + 2\sum_{n_1}(r^{(1)}_{2,2n_1+1})^2 + \dots + 2^{s-1}\sum_{n_{s-1}}(r^{(s-1)}_{2,2n_{s-1}+1})^2$$

$$n_i \equiv 2n_{i-1}\ , \quad i = 1, 2, \dots, s \ ; \quad r^{(i)}_{\ell, n_i} \equiv r^{(i-1)}_{\ell, 2n_{i-1}} + r^{(i-1)}_{\ell, 2n_{i-1}+1} \quad (8)$$

where $\Omega^{(s)}$ and $\Omega^{*(s)}$ have the same operator form as Ω and Ω^* with correspondingly rescaled coefficients, and $-\frac{\pi}{\Lambda} g^2 P^2$ is associated to the polarization of clusters of spins of increasing size. Of course, at the s – th iteration, the condition $\omega \ll c\Lambda$ must be replaced by $\omega \ll c\Lambda/2^s$.

We now proceed to evaluate the ground state of the system observing that, while the unperturbed atomic energy $\hbar\omega\sum_n r_{3,n}$ is left unaffected by the renormalization procedure, the contribution of the polarization term becomes more and more relevant as the number of iterations is increased.

In order to compare the relative weight of the two terms, we use as trial states the eigenstates $|j, m\rangle$ of the total atomic angular momentum $R_\ell = \sum_n r_{\ell, n}$, satisfying

$$R^2 |j, m\rangle = j(j+1)|j, m\rangle$$

$$R_3 |j, m\rangle = m|j, m\rangle$$

We first keep m fixed and vary with respect to j. In so doing, the unperturbed atomic energy does not vary, so that the approach to the minimal expectation value of $H^{(s)}$ requires P^2 maximal. To this aim, one has to maximize separately the contributions to P^2 associated to clusters of spins of different size; since this implies the highest symmetry for the atomic states, the maximal attainable value of j is needed. Thus, the atoms are spontaneously driven in a cooperative state of macroscopic angular momentum.

As a matter of fact, by finally varying with respect to m, it is easy to find a threshold in the number of iterations, say $\bar{s}$, beyond which the polarization contribution becomes comparable with the unperturbed atomic energy. Precisely, one has [5a]

$$\hbar\omega \simeq \frac{\pi}{4\Lambda} g^2 \, 2^{\bar{j}} \tag{9}$$

Hence, if N is large enough, once this condition is satisfied, the system goes over to a strong coupling regime, where the ground state tends to maximize R_2 instead of R_3. Since the renormalization procedure is still valid in such a regime [5b], it can be iterated until the rescaled lattice spacing is of the order of c/ω ; at this point, as already stated, the validity condition of the transformation fails. The final situation is thus as follows: if the length of the sample $N\Lambda^{-1}$ does not exceed $2\pi c/\omega$, then the relevant part of the effective Hamiltonian reduces to $\hbar\omega R_3 - \frac{\pi}{\Lambda} g^2 P^2$, where P^2 includes clusters of every size up to N. On the other hand, if $N\Lambda^{-1}$ exceeds $2\pi c/\omega$, the sample splits in clusters having length of the order $2\pi c/\omega$; furthermore, for each cluster, a single oscillator survives, interacting with its nearest neighbours over the same length.

In concluding this section, we remark that our effective cluster Hamiltonian has the same form as the Hamiltonian of the hierarchical model, originally introduced by Dyson [6] to show that even in one dimension phase transitions can occur, provided that a nonlocal coupling between spins is taken into account (see Fig. 3)

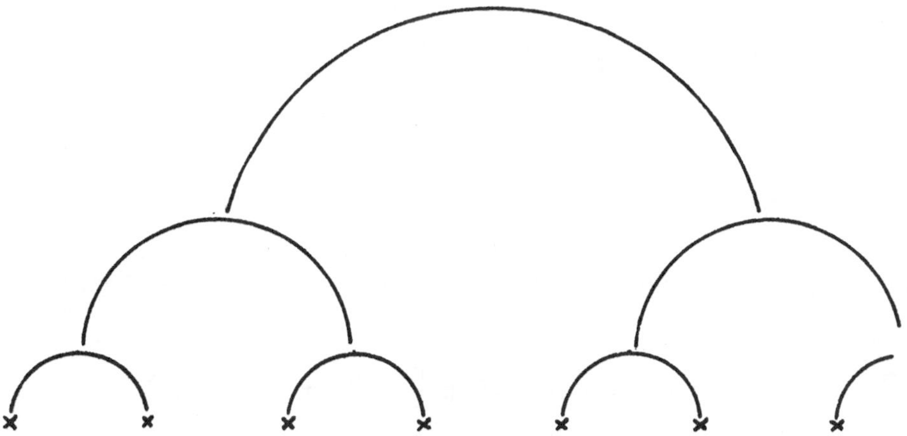

Fig. 3: The hierarchical structure of the interaction between spins in the Dyson's Hamiltonian, up to the third level. The interaction at the k-th level involves clusters of 2^k spins.

We note that, starting from the local Hamiltonian (4), the renormalization procedure generates in a natural way the hierarchical structure of the coupling, the key feature being the bilinear form of the interaction term $g\tau_2 q$.

CLASSICAL STATISTICAL MECHANICS OF THE MODEL

The threshold condition (9) has been derived in a purely quantum mechanical context at zero temperature. In order to get further insight into the physics of the model, we now discuss its classical counterpart at finite temperature; the operators $\tau_{i,n}$ become now the components of a vector of fixed length $\frac{S}{2}$, so that the Hamiltonian is

$$\mathcal{H}_N = \frac{1}{2}\hbar\omega S \sum_{n=1}^{N} \cos\Theta_n + \Lambda \sum_{n=1}^{N} \left[2\pi\alpha^2 p_n^2 + \frac{1}{8\pi}(q_{n+1} - q_n)^2\right] + \frac{1}{2} g S \sum_{n=1}^{N} q_n \sin\Theta_n \cos\phi_n \tag{10}$$

The partition function

$$Z_N = \int d\omega_1 \dots d\omega_N \int_{-\infty}^{\infty} dp_1 \dots dp_N \, dq_1 \dots dq_N \exp[-\beta\mathcal{H}_N]$$

$$(d\omega_i \equiv \sin\Theta_i \, d\Theta_i \, d\phi_i) \tag{11}$$

upon integration over the momenta p_i's and over the angles, becomes

$$Z_N = C_N \int_{-\infty}^{\infty} dx_1 \dots dx_N \exp[-(x_1^2 + \dots + x_N^2)] \cdot \prod_{m=1}^{N-1} \cosh(x_m x_{m+1}) \prod_{n=1}^{N} \mathcal{G}(x_n^2),$$

$$C_N = \left(\frac{2^{\frac{5}{2}}\pi^{\frac{3}{2}}}{\beta\kappa\Lambda}\right)^N, \quad \varphi(\xi) = \frac{\sinh\,(A^2 + B^2\xi)^{\frac{1}{2}}}{(A^2 + B^2\xi)^{\frac{1}{2}}}, \quad A = \frac{1}{2}\beta\hbar\omega S, \quad B = \left(\frac{\pi\beta}{\Lambda}\right)^{\frac{1}{2}} g S \tag{12}$$

where $x_i = \left(\frac{\beta\Lambda}{4\pi}\right)^{\frac{1}{2}} q_i$. We are interested in the behaviour of Z_N for large N at fixed density Λ.

As we will see, the function $\varphi(\xi)$, including the effect of the coupling with the atoms, plays an essential role in determining the most probable configuration of the oscillator amplitudes; of course, in the absence of such a coupling, this configuration simply reduces to $q_\kappa \equiv 0$.

The occurrence of a phase transition will be exhibited by deriving a threshold condition under which the mean values of the oscillator amplitudes are different from zero.

Since we are going to perform an asymptotic evaluation of Z_N through the Laplace method, it is convenient to convert the integral (12) into an integral over a single variable.

By putting

$$t = \sum_{r=1}^{N} x_r^2, \quad t_i = x_i^2 \Big/ \sum_{r=1}^{N} x_r^2, \quad i = 2, 3, \ldots, N \tag{13}$$

it can be shown[7] that Z_N takes the form

$$Z_N \simeq C_N \frac{\pi^{\frac{N}{2}}}{\Gamma(\frac{N}{2})} \int_0^\infty dt\, t^{\frac{N}{2}-1} e^{-t} \left[\varphi\left(\frac{t}{N^2}\right)\right]^N \Phi_2\left(\frac{1}{2}, \ldots, \frac{1}{2}; \frac{N}{2}; \xi_1 t, \ldots, \xi_N t\right)$$

$$\xi_i = \cos\frac{\pi i}{N+1} \tag{14}$$

where Φ_2 is a hypergeometric series of several arguments. It gives the leading contribution of the integration over $t_2, \ldots, t_N$, which occurs[5b] for $t_i = \frac{2}{N+1}\sin^2\frac{\pi i}{N+1}$.

We now observe that, for values of t up to order of unity, the integrand behaves as $t^{\frac{N}{2}-1}$, while for large t we must determine how the factor $\varphi^N \Phi_2$ competes with e^{-t}. To this aim, we approximate Φ_2 by its asymptotic expression[7]

$$\Phi_2 \sim \frac{\Gamma(\frac{N}{2})2^{\frac{N}{2}}}{\sqrt{\pi(N+1)}} \sin\left(\frac{\pi}{N+1}\right) t^{-\frac{1}{2}(N-1)} e^{t\cos\frac{\pi}{N+1}} \tag{15}$$

Taking into account eq. (15), a factor exp $[-(1-\cos\frac{\pi}{N+1})t] \simeq \exp(-\frac{\pi^2}{2N^2}t)$ occurs in the integrand; by letting $u = \pi^2 t/2N^3$ we are thus led, apart from a constant involving N, to consider the integral

$$\int_0^\infty du\, u^{-\frac{1}{2}} e^{N\Psi(u)} , \qquad \Psi(u) = -u + \ln \mathcal{G}\left(\frac{2N}{\pi^2}u\right) \tag{16}$$

The stationarity condition $\Psi'(u) = 0$ implies

$$m\eta = L\left(\frac{1}{2}\beta\hbar\omega S\eta\right)$$

$$\left(L(\xi) = \coth\xi - \frac{1}{\xi}\right)$$

$$\eta = \left[1 + \frac{8\kappa}{\beta\nu}\left(\frac{\partial}{\hbar\omega}\right)^2 u\right]^{\frac{1}{2}} , \qquad m = \frac{\hbar\omega\nu}{2S\kappa y^2} \tag{17}$$

where $\nu = \frac{\pi\kappa\Lambda}{N} \simeq \nu_1 = 2\kappa\Lambda \sin\frac{\pi}{2(N+1)}$ (compare eq. (3)) is the lowest oscillator eigenfrequency occurring in the sample.

Since $L(\xi) < 1$ and $\eta > 1$, condition (17) cannot be satisfied unless

$$y^2 > \frac{\hbar\omega\nu}{2S\kappa} \tag{18}$$

(see Fig. 4).

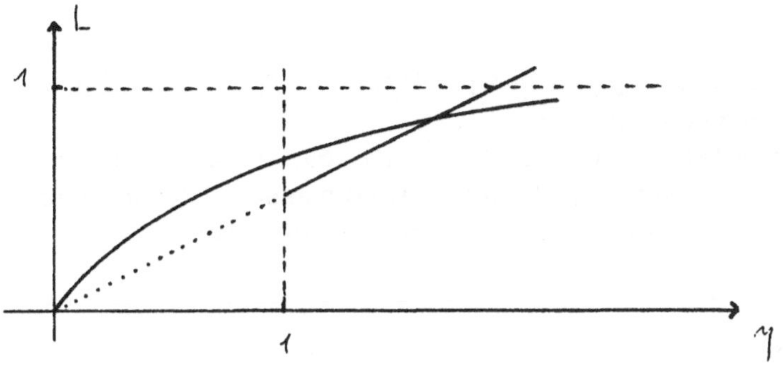

Fig. 4: Graphical solution of the stationarity equation (17). The straight line $m\eta$ intersects L provided that $m < 1$ (see condition (18)).

For values of g below threshold, the steady state of the system is given by $q_i = p_i = 0$ and $\Theta_i = \pi$. On the other hand, $L(\xi)$ being an increasing function, condition (17) taken at $\eta = 1$ for fixed m determines a critical value ($\beta = \beta_c$). For $\beta < \beta_c$ the steady state is the same as before, while for $\beta > \beta_c$ the equilibrium configuration of the oscillators is given by the lowest eigenmode. In particular, we can evaluate the energy stored in the oscillator sites in the limit $T \to 0$. The stationarity condition (17), for fixed m, reduces then to $m\eta = 1$, whence

$$\frac{\mu}{\beta} \simeq \frac{g^2 S^2 N}{2\pi\Lambda}$$

Recalling (see below eq. (14)) that the leading contribution to the partition function, for every β , arises at $t_i = \frac{2}{N+1}\sin^2\frac{\pi i}{N+1}$ ($i = 2, \dots, N$), we have, taking into account that $t = 2N^3\mu/\pi^2$

$$\eta_i^2 = \frac{4\pi}{\beta\Lambda}x_i^2 = \frac{4\pi}{\beta\Lambda}t\,t_i = \frac{8N^3}{\pi\Lambda}\cdot\frac{\mu}{\beta}\cdot\frac{2}{N+1}\sin^2\frac{\pi i}{N+1} \tag{19}$$

Thus, as $T \to 0$, the oscillator energy becomes (compare eq. (2))

$$E \simeq \frac{\Lambda}{4\pi}\lambda_1\varphi_1^2 \simeq \frac{g^2 S^2 N^2}{2\pi\Lambda} \tag{20}$$

Hence, the estimated value of the energy per site ($g^2 N/\Lambda$) agrees with the one previously given in eq. (9) for the macroscopic polarization in the quantum mechanical case.

As a matter of fact, E has the N^2 - behaviour typical of the field produced by N phased radiators; physically, it corresponds to the binding energy necessary to generate the macroscopic atomic dipole.

CONCLUSIONS

Starting from our model, which is the first lattice treatment in the framework of quantum optics, we have been able to exhibit the spontaneous build-up of macroscopic polarization in the atomic system.

Our approach is based on letting the lattice, over which the Bose field is discretized, to coincide with the atomic chain. As a first consequence of this assumption, the atomic sample extends over the same region occupied by the field, thus fully taking into account nonlinear propagation effects. Secondly, as already mentioned, wavelengths shorter than the interatomic spacing are neglected. Furthermore, we stress that the spectrum of the free Bose field is completely determined by the lattice.

In the quantum mechanical case, proceeding along the lines of the real space renormalization group, a threshold for the number N of lattice sites necessary to reach the cooperative regime has been derived (see eq. (9)).

In the classical case, by evaluating the partition function for large N, a second order phase transition has been demonstrated, the result at T = 0 being in full agreement with the quantum mechanical one.

A final remark on previous works[8], concerning superradiance as a phase transition starting from the Dicke Hamiltonian, is now in order. Even the multimode generalizations of the original Dicke model are essentially mean field theories, because there the atoms are described in terms of collective operators which do not take into account any spatial structure. Our approach, on the contrary, fully includes the spatial resolution, as explicitly shown by the interaction term $g \sum_{n=1}^{N} \tau_{1,n} q_n$; as already remarked, just this form of the coupling is responsible for the hierarchical structure of the nonlocal effective Hamiltonian.

Of course, a deeper understanding of the occurrence of macroscopic polarization requires an analysis in higher dimensions.

REFERENCES

1. V.S.Letokhov and V.G.Minogin: Phys.Rep. 73, 1 (1981) and references therein.
2. J.M.Raimond, P.Goy, M.Gross, C.Fabre and S.Haroche: Phys. Rev. Lett. 49, 117 and 1924 (1982).
3. R.H.Dicke: Phys. Rev. 93, 99 (1954).
4. E.Fradkin and S.Raby: Phys. Rev. D20, 2566 (1979).
5. a) V.Benza, E.Montaldi and M.Ciftan: Lett. Nuovo Cimento 36, 193 (1983); b) V.Benza and E.Montaldi, in preparation.
6. F.J.Dyson: Commun. Math. Phys. 12, 91 (1969).
7. V.Benza and E.Montaldi: Phys. Lett. 97A, 231 (1983).
8. K.Hepp and E.H.Lieb: Ann. Phys. 76, 360 (1973); Y.K.Wang and F.T.Hioe: Phys. Rev. A7, 831 (1973); G.M. Zaslavskii, Y.A.Kudenko and A.P.Slivinskii: Teor.Mat. Fiz. 33, 95 (1977); C.C.Sung and C.M.Bowden: J.Phys. A12, 2273 (1979); M.Kimura: Progr.Theor. Phys. 65, 437 (1981).

THEORY OF TIME-DEPENDENT SPECTRAL OBSERVATIONS IN QED AND QUANTUM OPTICS

J.H. Eberly

Department of Physics and Astronomy
University of Rochester
Rochester, New York 14627 USA

1. INTRODUCTION

The subject of measurement in quantum theory is more than 50 years old, and there are well-known controversies regarding hidden variables that are still lively today. The subject of spectra is twice as old, but not as well known. It is also the center of recent activity. In these lectures I will deal only with the older subject.

Spectra became widely interesting to theoretical physicists only at the end of the last century, when attention was attracted to the existence of a wide variety of natural phenomena that have no obvious beginning or end, that fluctuate indefinitely about a fixed mean.[1] Among these phenomena that were studied by physicists and mathematicians were ocean tides, sun spots, the earth's magnetic field, commodity prices, interest rates, and white light. Such random processes, without an obvious time origin, are termed stationary. By their nature they are not square integrable, and so they do not possess Fourier transforms in the traditional sense. It was an important issue how to define properly the spectrum of such a process.

This issue was settled by Norbert Wiener, with his development of generalized harmonic analysis in 1930. Wiener showed that the Fourier transform of the autocorrelation function of a stationary random process does exist. [1] This Fourier integral relationship:

$$S = 2 \,\mathrm{Re} \int_0^\infty d\tau \; e^{-i\omega\tau} \; \langle V^*(\tau)\, V(0) \rangle \tag{1.1}$$

is usually called the Wiener-Khintchine Theorem, and S is the "power spectrum" of the random process.

In this equation, and throughout these lectures, the angular brackets denote ensemble averages. In classical physics it may be a question to decide how to describe the implied ensemble. This question is not easier to answer in quantum physics. It is not a question that can be dealt with here, and we will interpret $\langle ... \rangle$ to mean $Tr[(...)\rho]$, where ρ is the usual density matrix.

Wiener's solution can be understood as a purely mathematical success. That is, Wiener discovered the sense in which endless natural signals (stationary random processes) have Fourier spectra, as do natural signals that begin and end at finite times. In the familiar finite-signal case, the spectrum is simply the signal's Fourier transform (really, its absolute square). In the infinite-signal case, the Wiener spectrum is again a Fourier integral, but of the signal's auto-correlation function. From a physical point of view, however, Wiener's solution to the spectrum problem is not beyond question. One can ask what part of Wiener's solution is more than mathematics? What part survives if one or another of Wiener's assumptions breaks down? How is Wiener's spectrum to be recorded in the laboratory? Should one expect Wiener's spectrum to be independent of time on the basis of a rough physical argument, or is it a mathematical property alone? In the next section we will sketch answers that have recently been given to these questions - at the same time we will be forced to adopt a new "universal" definition of spectrum, one that contains the Wiener definition as a particular limit.

2. OBSERVATION OF SPECTRA

The main elements of a spectral observation are sketched in Figure 1. We show a light scattering experiment. The measured signal in this case is the light emitted from a collection of atoms. First the atoms are excited by some external agency, in this case a laser pulse. To resolve the spectral components of the signal a filter is necessary, and to register the filtered signal a photo-detector is needed. Convenience dictates the exact form adopted by the experimenter for the filter and detector, but they commonly have three characteristics that permit an idealized but fairly general calculation to be made of the observed spectral intensity.

The <u>filter</u> is typically both <u>linear</u> and <u>causal</u>. Thus the complex electric field (analytic signal) that passes through it

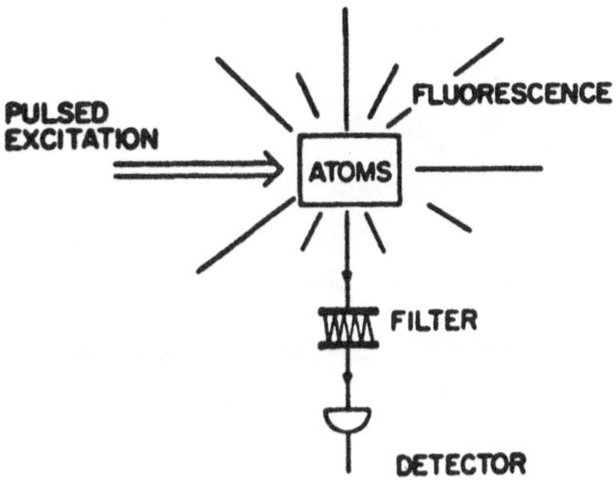

Fig. 1. Elements of Time-Dependent Spectral Measurement

and reaches the detector is related to the field emitted by the atoms through an integral transform:

$$E_D(t) = \int_{-\infty}^{t} H_F(t-t')\, E_A(t')dt'. \qquad (2.1)$$

Here the specific device chosen as the filter will determine the function H_F. The detector is typically a so-called square-law detector. That is, it responds directly to the square of the signal that reaches it. Thus the observed spectral intensity is simply $\mathcal{S} = E_D^2$. Of course, if the original emission process was stochastic, or if we are dealing with a signal for detection that is statistical, as is inevitably the case in a quantum treatment, then we write $\mathcal{S} = \langle E_D^2 \rangle$. What is more, the detection process almost always works by raising the energy of a detector atom, in which case the square law must be normally ordered.[2] This gives us our final working relation for the observed spectral intensity:

$$\mathcal{S} = \langle E_D^*(t)\, E_D(t) \rangle. \qquad (2.2)$$

Formula (2.2) for spectral intensity is not obviously related to S of (1.1). At first sight it appears that they cannot be related, as S is a function of frequency, not time, and $\mathcal{S}$ is a function of time, not frequency. Whatever connection exists must be found by identifying $E_A(t)$ with $V(t)$. These are the primary signals whose spectra are being calculated in each case. Obviously, in the case of $\mathcal{S}$, the filter must play a crucial physical role. Thus we consider the nature of the filter's transfer function $H_F(t)$ in a representative example.

A plane parallel Fabry-Perot etalon is a simple optical device with high frequency selectivity. Its theory is well understood,[3] and the Fourier transform of $H_F(t)$ is given by Airy's formula:

$$\tilde{h}_F(\omega) = \frac{(1-r^2)\, e^{i\omega d/c}}{1-r^2\, e^{2i\omega d/c}} \,. \tag{2.3}$$

Here r is the reflectivity of the plates and d is their distance of separation. A rough sketch of $|\tilde{h}(\omega)|^2$ is shown in Figure 2, in which the curves' minima are given by

$$M = \frac{1-r^2}{1+r^2} \,, \tag{2.4}$$

and where

$$\omega_n = \frac{n\pi c}{d} \,. \tag{2.5}$$

If $r \approx 1$, and $M \simeq 0$, the peaks in Fig. 2 are well isolated, and we can make a very simple and useful approximation to $\tilde{h}(\omega)$ in the neighborhood of one of the peaks:

$$\tilde{h}_F(\omega) \approx \frac{1}{\omega-\omega' + i\gamma'} \,, \tag{2.6}$$

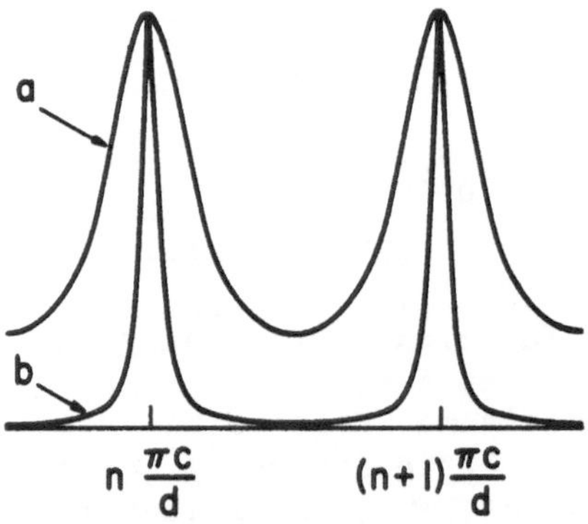

Fig. 2. Fabry-Perot interferometer spectral transmission curves, showing near-Lorentzian line shape in neighborhoods of transmission maxima for large reflectivity: (a) r = 0.60, (b) r = 0.90.

where $\omega' = \omega_n$ for any n, and

$$\gamma' = \frac{1-r^2}{r^2} \frac{c}{2d} \approx 0 . \tag{2.7}$$

We will use the prime to designate Fabry-Perot parameters. In other words, $\tilde{h}_F(\omega)$ is closely approximated by a narrow Lorentzian function, if the plate reflectivity is near unity. In this limit $H_F(t)$ is an exponential function:

$$H_F(t) = e^{-i\omega' t} e^{-\gamma' |t|} . \tag{2.8}$$

Note that, according to (2.1) and (2.8), an impulsive input field $E_A(t) \sim \delta(t)$ is stretched out to $E_D(t) \sim \exp[-(\gamma' + i\omega')t]$ by passing through the filter. Thus $\tau' = 1/\gamma'$ can be called the filter lifetime.

The combination of (2.1), (2.2), and (2.8) allows the spectral intensity at the detector to be written

$$S = 2\gamma' \int_{-\infty}^{t} dt_1 \int_{-\infty}^{t} dt_2 \; e^{-(\gamma' - i\omega')(t-t_1)} e^{-(\gamma' + i\omega')(t-t_2)}$$

$$\times \; \langle E_A^*(t_1) \, E_A^*(t_2) \rangle . \tag{2.9}$$

This is a normalized and idealized version of what is typically measured in an experiment.[4] It depends on an auto-correlation function recognizably analogous to the one appearing in S in (1.1), but not restricted to be stationary or to have any other mathematical property. Although (2.9) looks like a generalized Fourier integral, it is a double integral, unlike (1.1), and the exponential kernel arose from an approximate form of $H_F(t)$, not from any deep theoretical principle. However, as we show below, it is just the ordinary experimental foundation of (2.9) that gives (1.1) its physical legitimacy. At the same time, the degree of approximation used here in obtaining (2.9) should be kept in mind, because it indicates the limited degree to which the Wiener formula has direct physical relevance.

Let us now explore more closely the relation of (1.1) and (2.9). First we rewrite (2.9) as

$$S = 4\gamma' \; \mathrm{Re} \int_{-\infty}^{t} dt_1 \int_{-\infty}^{t_1} dt_2 \; e^{-(\gamma' + i\omega')(t-t_1)} e^{-(\gamma' - i\omega')(t-t_2)}$$

$$\times \; \langle E_A^*(t_1) \, E_A(t_2) \rangle .$$

Next we introduce a physical assumption, namely that $E_A(t)$ is activated at $t = 0$ and becomes stationary after some time elapses, say T_S. That is, suppose

$$\langle E_A^*(t+\tau)\, E_A(t)\rangle = \theta(t)\,\theta(t+\tau)\,\langle E_A^*(\tau)\, E_A(0)\rangle , \tag{2.10}$$

for all values of τ and for $t \geq T_S > 0$. We find, under this condition,

$$S = 4\,\gamma'\ \mathrm{Re} \int d\tau\ e^{(\gamma'-i\omega')\tau} \int_0^{t-\tau} dt_2\ e^{-2\gamma'(t-t_2)}$$
$$\times\ \langle E_A^*(t_2+\tau)\, E_A(t_2)\rangle . \tag{2.11}$$

Now we can break the t_2 integral into segments: $[0,T_S]$ and $[T_S,t-\tau]$. The first segment behaves like $\exp(-2\gamma' t)$ and can be ignored for $t \gg \tau' = 1/\gamma'$. We add this condition as a second physical assumption, and obtain

$$S = 4\,\gamma'\mathrm{Re} \int d\tau\ e^{(\gamma'-i\omega')\tau} \int_{T_S}^{t-\tau} dt_2\ e^{-2\gamma'(t-t_2)}$$
$$\times\ \theta(t_2)\theta(t_2+\tau)\ \langle E_A^*(\tau)\, E_A(0)\rangle$$
$$= 4\,\gamma'\ \mathrm{Re} \int d\tau\ \ e^{(\gamma'-i\omega')\tau}\ \langle E_A^*(\tau)\, E_A(0)\rangle$$
$$\times\ e^{-2\gamma' t} \left[\frac{e^{2\gamma'(t-\tau)} - e^{2\gamma' T_S}}{2\gamma'}\right]$$
$$= 2\mathrm{Re} \int d\tau\ \ e^{-(\gamma'+i\omega')\tau}\ \langle E_A^*(\tau)\, E_A(0)\rangle . \tag{2.12}$$

The last step requires another assumption that is slightly stronger than $t \gg \tau'$, namely $t-T_S \gg \tau'$. All of the τ integrals above run from 0 to t.

Comparison of (2.12) with (1.1) shows very close similarity, but not identity. One further physical assumption must be added. We assume that $E_A(t)$ is characterized by an autocorrelation time τ which is much shorter than τ'. Then the integrand is non-zero only for $\tau' \ll \tau$, in which interval $\exp(-\gamma'\tau) \approx 1$. Therefore, under all of these physical assumptions:

(i) $E_A(t)$ becomes stationary for $t - T_S > 0$,

(ii) $t - T_S \gg \tau$,

(iii) $\langle E_A^*(\tau)\, E_A(0)\rangle = 0$, for $\tau > \tau_c$, and

(iv) $\tau' \gg \tau_c$,

we can write

$$S = 2 \text{ Re} \int_0^\infty d\tau e^{-i\omega'\tau} \langle E_A^*(\tau)\, E_A(0)\rangle \quad , \tag{2.13}$$

which is just (1.1) if ω is interpreted as the filter's pass frequency ω'.

Thus, conditions (i) - (iv) provide the physical meaning of the Wiener power spectrum. They show when the Wiener spectrum (1.1) can be expected to have physical relevance. These conditions need not be invoked for (2.9), of course, which remains much more generally applicable to physical situations. We will describe two situations of experimental interest for which (1.1) is useless, in Secs. 3 and 4, and use (2.9) to calculate spectra.

The photodetection rate $\langle E_D^2(t)\rangle$ clearly depends on time, as well as on the filter's pass frequency ω', so the physical spectrum is a properly time-dependent spectrum.[4] It is positive definite by construction, and has none of the awkward properties of the mathematical generalizations of S that have been proposed by Page, Lampard, Silverman, Mark, and others in the past.[5]

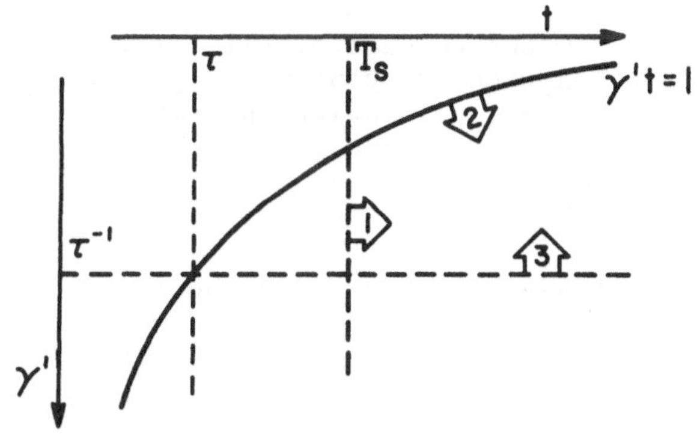

Fig. 3. Time-frequency phase space for time-dependent spectral measurement. Three arrows indicate the region of conventional experiments: (1) $t > T_S > \tau$, the variable to be observed is stationary; (2) $t > 1/\gamma'$, the filter is in steady state; (3) $\gamma' < 1/\tau$, the filter has good spectral resolution. Conventional definition (1.1) is restricted to the union of regions 1 and 2 and 3, but (2.9) is not.

Because the physical spectrum is an analytic function of both time and frequency, it permits, in principle, examination of arbitrarily narrow intervals of t and ω. It is not, however, exempt from what one conventionally understands to be time-bandwidth inequalities. This is apparent in conditions (i) - (iv) above and is illustrated in Figure 3.

3. PERFECT CAVITY SPECTROSCOPY

We emphasized in Sec. 2 that stationarity (time-translation invariance) is an essential property of a signal if it is to have a Wiener spectrum. No such property is required of a signal in order to have a well-defined physical spectrum.[4] In principle no natural signal is stationary. In practice most signals are stationary because their initial features are rapidly damped out. However, if the damping is especially slow it may well be possible, even necessary, to record the spectrum of a signal before it has become stationary.

Atoms are subject to a wide variety of interactions that have the effect of damping any radiative dipole oscillations. These damping interactions give rise to spectral line broadening. By isolating an atom more and more completely from its environment, spectroscopists have traditionally sought to achieve narrower spectral lines. We will now discuss a situation in which such isolation not only produces the ultimate in line narrowing, but also prevents the achievement of stationarity, and so alters the qualitative character of the spectrum.

In free space, in the absence of all collisions, an atom is still capable of relaxation by spontaneous photon emission. The rate of purely spontaneous relaxation is provided by the Einstein A coefficient:

$$A = \frac{4}{3}\,\frac{d^2\omega^3}{\hbar c^3}\,. \tag{3.1}$$

This number is basically the product of the atomic oscillator strength

$$f = \frac{2m}{\hbar}\, r^2\,\omega\;, \tag{3.2}$$

and the density of electromagnetic modes in free space:

$$\rho(\omega) = \frac{\omega^2}{\pi^2 c^3}\,. \tag{3.3}$$

Obviously A is smaller for transitions with weaker f values. Also, as Purcell pointed out almost 40 years ago,[6] A may be made smaller by reducing $\rho(\omega)$. In a cavity the density of states $\rho(\omega)$ can be much lower (below cutoff) or higher (near a cavity resonance) than the free space value (3.3). The ultimate is achieved below the cavity cutoff frequency ω_{co}, where $\rho = 0$ in the case of an ideal lossless cavity. In such a cavity, therefore, $A = 0$ and an atom whose transition frequency falls below ω_{co} cannot radiate. Its natural linewidth is zero.[7] Such an ideal cavity does not exist, but close approximations to it are possible to imagine, using cryogenic techniques to obtain extremely high cavity Q values ($Q > 10^7 - 10^9$). Figure 4 shows the character of $\rho(\omega)$ for a cavity, and the spectral regions that offer interesting experimental possibilities are marked. These occur when $\rho(\omega)$ deviates sharply from formula (3.3).

I will not undertake a description of the methods proposed for preparing atoms and cavities in order to realize a good approximation to the ideal conditions described above. They can be found elsewhere.[8] Present purposes are served by noting that experiments appear to be possible for which the Einstein A coefficient is anomalously very small (many orders smaller than the atomic f value would suggest). For such experiments a very simple model theory appear well-suited. This is the Jaynes-Cummings model,[9,10] which treats fully quantum electrodynamically the radiative dipole oscillations of a single two-level atom in a lossless cavity which supports a single mode of radiation.

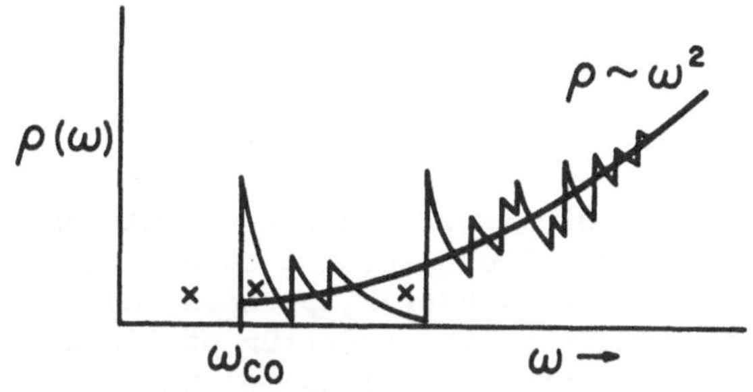

Fig. 4. Schematic representation of cavity density of states, showing prominent resonances near cut-off and gradual smoothing and convergence far above cut-off to the free-space ω^2 law. The crosses mark regions where cavity experiments may be most interesting because $\rho(\omega)$ deviates strongly from the free-space curve.

The Jaynes-Cummings model Hamiltonian is given by

$$\hat{H} = \hbar\,\omega_c\,\hat{a}_c^*\,a_c + \hbar\,\omega_{21}\,\hat{\sigma}_{22} + \hbar\,\lambda\,(\hat{a}_c\,\hat{\sigma}_{12} + \text{h.c.}). \tag{3.4}$$

Here $[\hat{a}_c, \hat{a}_c^*] = 1$, as expected of photon creation and destruction operators. The σ's are atomic transition-projection operators, given by $\hat{\sigma}_{ij} = |i\rangle\langle j|$, where $i,j = 1,2$. Their commutator properties are obvious.[11] The parameter λ serves as the dipole coupling constant:

$$\lambda = \sqrt{(2\pi\,\hbar\,\omega_c/V_c)}\;\hat{\varepsilon}_c\cdot\vec{d}_{21}\ , \tag{3.5}$$

where V_c is the cavity volume, etc. This Hamiltonian has been carefully studied [11] because it offers a moderately realistic caricature of actual atom-radiation interactions, but at the same time allows exact solutions for eigenvalues, eigenstates, and dynamics for all values of the coupling parameter λ, time t, resonance detuning $\Delta = \omega_{21} - \omega_c$, and initial conditions.

The Jaynes-Cummings Hamiltonian is Hermitean. The dynamical evolution implied by it is fully unitary, and shows no relaxation. It does permit spontaneous emission, however, in the sense that the quantum state of no photons and the atom excited is not stable. From this state the atom spontaneously emits a photon and drops to its lower level. (Note that this is not a stable state either.) Since spontaneously emitted photons are necessarily emitted into the same mode as all the other photons in the cavity, they cannot be distinguished kinematically (by different wave vector or frequency), but they can be distinguished dynamically. That is, one can say that the non-zero commutator $[\hat{a}_c, \hat{a}_c^*] = 1$ ensures quantum stochasticity of the emission process.

We assume, following Kubo for example, that the atomic dipole auto-correlation function is the basis for the emitted field auto-correlation function required in (1.1) or (2.9). That is, we assume

$$\langle E_A^*(t_1)\,E_A(t_2)\rangle = \kappa\,\langle\hat{\sigma}_{21}(t_1)\,\hat{\sigma}_{12}(t_2)\rangle\ , \tag{3.6}$$

where κ is a constant of proportionality. This assumption does not stand extremely critical examination in any case, and it is especially shaky here. It is better, perhaps, simply to say in all cases that one calculates the spectrum of dipole evolution. The exact Heisenberg operator solutions for the dipole operators $\hat{\sigma}_{12}$ and $\hat{\sigma}_{21}$ were found by Ackerhalt; [12] and Narozhny, et al., have computed the required correlation.[11]

$$D(t,\tau) = \langle\hat{\sigma}_{21}(t)\,\hat{\sigma}_{12}(t-\tau)\rangle\ . \tag{3.7a}$$

In the simplest case, when the expectation is computed in the field vacuum, with the atom initially excited, one finds: [13]

$$D_{+,0}(t,\tau) = (2\nu)^{-2} e^{i(\omega_c + \Delta/2)\tau}[(\nu+\Delta/2)^2 e^{i\nu\tau} + $$
$$+ (\nu - \Delta/2)^2 e^{-i\nu\tau} + 2\lambda^2 \cos\nu(2t-\tau)]. \qquad (3.7b)$$

Here ν is a kind of vacuum Rabi frequency:

$$\nu = [\lambda^2 + (\Delta/2)^2]^{\frac{1}{2}} , \qquad (3.8)$$

and the subscripts $+,0$ identify the initial atomic state (+ = excited) and field state (0 = no photons). In the Heisenberg picture only the initial state needs to be specified.

The correlation given in (3.7b) is not stationary. It obviously depends on t as well as on τ. A three-dimensional view of $D(t,\tau)$, with the fast cavity oscillation $\exp - i\omega\tau$ removed, is shown in Fig. 5. The Wiener spectrum is not defined, but the physical spectrum is easily computed. In the limit $\gamma t \gg 1$ one finds (now we drop the primes from ω' and γ' for simplicity):

$$S \to (4\nu)^{-2} [\delta^2 + \gamma^2]^{-1} \{(\nu + \Delta/2)^2 + \lambda^2 (\nu^2 + \gamma^2)^{-1}$$
$$\mathrm{Re}(\gamma - i\delta)(\gamma - i\nu) e^{-2i\nu t}\} + \nu \to -\nu . \qquad (3.9a)$$

The positions of the two spectral peaks are determined by δ:

$$\delta = \omega - (\omega_c + \nu + \Delta/2) . \qquad (3.9b)$$

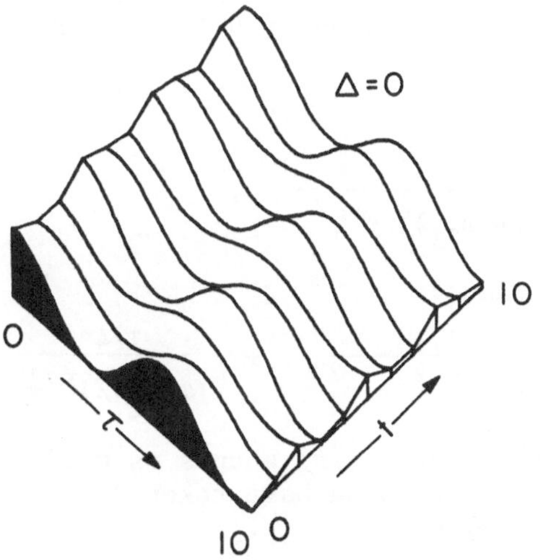

Fig. 5. Non-stationary Jaynes-Cummings dipole autocorrelation, with t and τ in units of λ^{-1}.

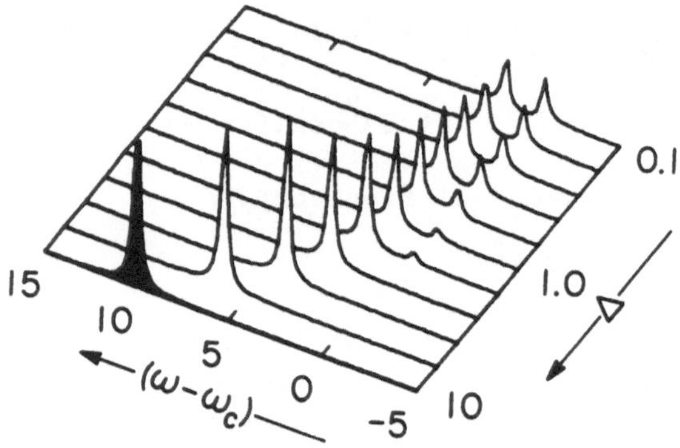

Fig. 6. Ideal cavity spontaneous emission spectra, showing the behavior predicted by (3.10a) for large Δ and that predicted by (3.10b) for small Δ. For intermediate values of Δ the low peak near $\omega = \omega_c$ is the analog of the elastic Rayleigh peak in free-space light scattering. Both axes are in units of λ . [From Ref. 13]

Note that even when $\gamma t >> 1$ there is a residual time dependence in S that does not damp out. It is significant at long times only if $\gamma \approx \lambda$.

Two limiting cases of this formula for the spectrum, far off resonance and exactly on resonance ($\Delta >> \lambda$ and $\Delta = 0$), are worth special attention:

(a) far off resonance:

$$S(\omega) = \frac{1}{(\omega-\omega_{21})^2 + \gamma^2} , \tag{3.10a}$$

(b) exactly on resonance:

$$S(\omega) = \frac{1/4}{(\omega-\omega_c-\lambda)^2 + \gamma^2} + \frac{1/4}{(\omega-\omega_c+\lambda)^2 + \gamma^2} . \tag{3.10b}$$

Figure 6 shows these spectral line shapes as well as intermediate cases. In all of these cases we have taken $\lambda >> \gamma$ so that the two peaks of (3.10b) are evident.

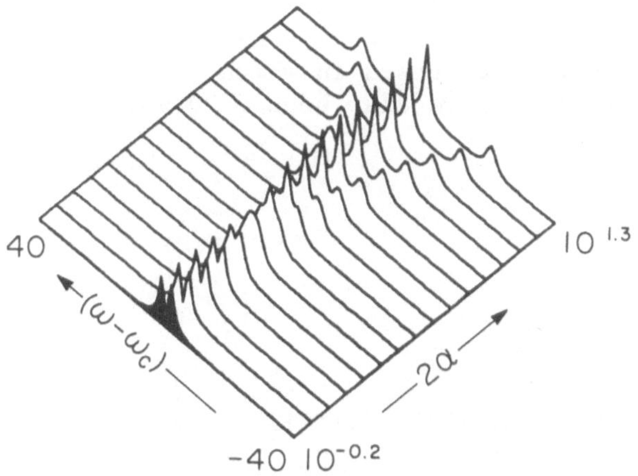

Fig. 7. Spectra of resonant ($\Delta = \omega_c - \omega_{21} = 0$) atomic emission in an ideal cavity initially excited in the coherent state $|\alpha\rangle$.

Several observations can be made about equations (3.10). First, the widths of the peaks are determined solely by γ, i.e., by the interferometer used to resolve the spectrum and not by the radiation process. This is expected since our model is lossless. In practice a finite cavity Q value will also contribute to the linewidth and perhaps make the dominant contribution. Similarly, the shape of the line is Lorentzian only because of approximation (2.6). Nevertheless, the calculated spectrum (3.10a) shows the principal feature projected [7] by Kleppner, an observable linewidth possibly much narrower than the "natural" limit, namely A. Second, the on-resonance spectrum shows peak splitting. This is a new feature [13] not previously predicted whose origin is in the close atom-cavity coupling.

It is possible, perhaps even highly probable in any realistic attempt to observe the spectra predicted in (3.10), that the cavity will contain photons initially as well as the excited atom. We have computed the physical spectrum for several such cases,[14] and show one set of examples in Fig. 7. The atom and cavity are taken exactly resonant with each other, and in addition the cavity mode is taken to be excited initially in a coherent state The expected number of photons initially in the cavity is thus $n = \alpha^*\alpha$. As Fig. 7 shows, a small value of α makes very little difference to the shape of the spectrum (compare the spectrum for $\Delta = 0.1$ in Fig. 6 with the spectrum for $2\alpha = 10^{-0.2}$ in Fig. 7). However, for values of α greater than 2 or 3, significant differences are evident, and the spectrum takes on the distinguishing three-peak form of the Mollow-Stroud spectrum [15] predicted and observed for two-level atoms in free space. Calculations of spectra for cases in which the cavity is initially in a photon number state have also been made and will be presented elsewhere.[14]

4. TRANSIENT LIGHT SCATTERING SPECTRA

The example of Sec. 3 was unsuited to the traditional Wiener definition of spectrum because of the ideal, completely loss-free nature of the model. In this section we will deal with a less ideal situation, but one for which the Wiener spectrum is still not appropriate, because we consider the spectral observation to be completed well before relaxation has led to stationarity. This is the case for transient light scattering.[16]

The simplest prototype of experiments in transient light scattering is shown in Fig. 1. If the pulsed excitation is shorter than any of the atom's dipole relaxation times, then the dipole auto-correlation cannot reach its stationary state during the experiment. In effect there are two transients in such an experiment: one associated with the turn-on and one with the turn-off of the exciting pulse. We will consider a somewhat simpler case by taking the excitation pulse to be quite long and smooth, allowing the atoms to reach steady state. We will consider the spectrum during times after the turn-off of the excitation. In this way only the turn-off transient will be significant. A calculation focused on the turn-on transients has also been carried out,[17] but the experimental realization is much less straightforward.

As in all cases, we rely on the proportionality given in (3.6) and evaluate the auto-correlation (3.7a), under the assumption of near resonance. That is, we consider only the emission from a specific two-level transition that is excited by a steady resonant laser intensity I(t) that turns <u>off</u> abruptly at a certain instant which we label $t = t_p$, as shown in Fig. 8a. The value of the needed correlation has been given elsewhere.[18] It is very complicated, as is the resulting physical spectrum, and we will not reproduce the formulas here.

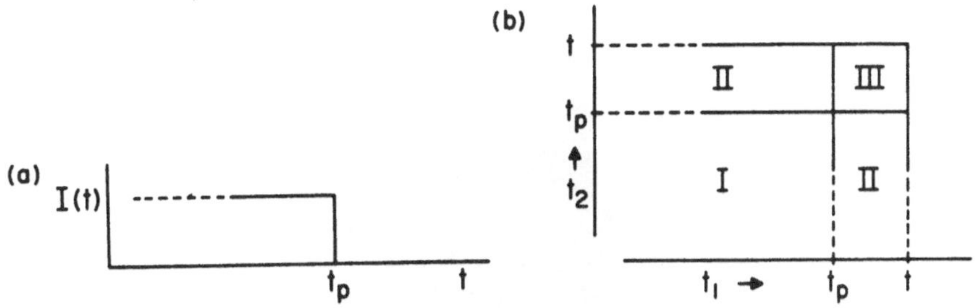

Fig. 8. Diagrams indicating pulse behavior. (a) laser intensity is constant until $t = t_p$. (b) regions of integration in Eq. (2.9): (I) entirely before turn-off, (III) entirely after turn-off, and (II) interference region.

What is more important in the present context is that the physical spectrum is well-defined for all times and the Wiener spectrum is not. In fact, the Wiener spectrum serves only as an initial condition for the physical spectrum in this case. Consider the double time integral given in (2.9). For transient spectra at times $t > t_p$ the integrations can be divided into three categories as shown in Fig. 8b. The contribution from region I, in which $-\infty \leq t_1 \leq t_p$ and $-\infty \leq t_2 \leq t_p$, gives the usual Wiener spectrum. This is because region I reflects only the effect of the constant excitation pulse. The region is large enough so that all correlations are certainly stationary. Region III, in which both t_1 and t_2 are greater than t_p (i.e., the excitation is off), reflects the effect of a steady excitation, but an excitation of zero amplitude. If there is light emitted by atoms in this time region it must be pure fluorescence, as no exciting light is incident to be scattered. It is clearly region II that is most interesting. It reflects the quantum interference of $\hat{\sigma}_{21}(t_1)$ during the "on" period of the excitation with $\hat{\sigma}_{12}(t_2)$ during the "off" period of the excitation (or vice versa). Obviously this interference exists only if $t > t_p$.

Figure 9 shows a set of transient spectra, as reported by Huang, Tanaś, and Eberly.[18] The spectra show clearly the effect of the Fabry-Perot interferometer as well as the spectral time dependence. It is clear on physical grounds that the spectrum must eventually become zero because there are no new quanta to be scattered at $t = t_p$. Of course the spectrum does not immediately go to zero at $t = t_p$ because both the atoms and the interferometer can store quanta for a finite time. What is perhaps unexpected is that these two independent storage mechanisms can interfere with each other. This is evident in the "waves" of spectral amplitude appearing in spectra (b) - (d) in Fig. 9.

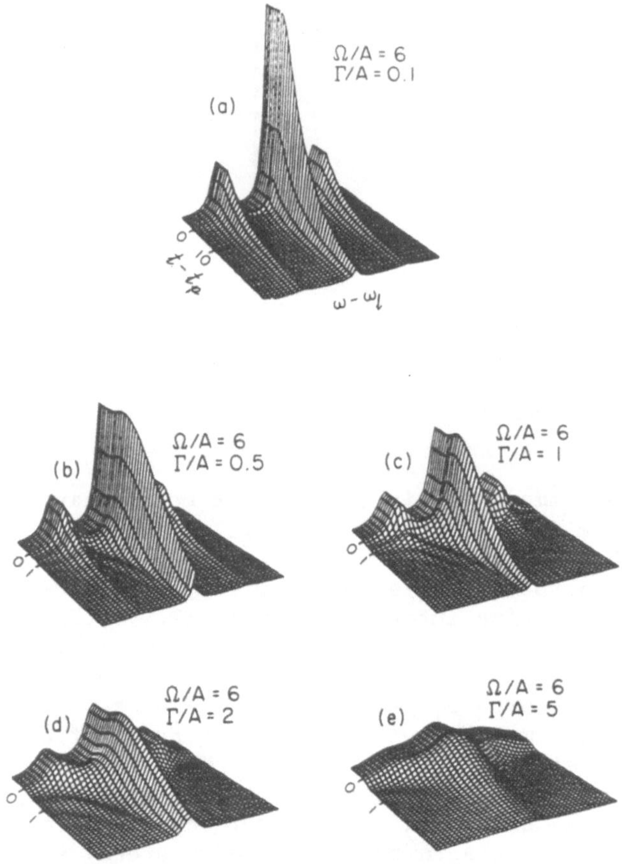

Fig. 9. Two-level-atom scattering spectra, showing steady state appropriate to Rabi frequency $\Omega = 6A$ prior to laser turn-off at $t - t_p = 0$, and transient spectra for $t > t_p$. Γ is filter bandwidth and A^{-1} is atomic lifetime (denoted γ and τ in text). [From Ref. 18].

Finally, another question altogether can be addressed, using these transient spectra. This is the question of "dressing". The use of combined atom-field "dressed" states has a long history in quantum optics. For example, in 1958 the original discussion [9] of the Jaynes-Cummings problem was made in terms of joint atom-field eigenstates, what are now called dressed states of the atom-field interaction. In one respect these dressed states are only a mathematical basis, and in another respect they have a direct physical reality, particularly when the atom-field interaction Hamiltonian is not negligible.

The question of interest is the speed with which physical dressing takes place. That is, how fast does the atom-field interaction cause the atom-field complex to be better described by the dressed states than by the originally relevant "bare" states? In the present case the time-dependent physical spectrum allows one to study this point, as follows. It is well known that long after laser turn-on the fully dressed system of two-level atom and single-mode exciting laser field exhibits a three-peaked light scattering spectrum,[15] whereas the bare system before laser turn-on exhibits a spectrum with only a single peak. Clearly it will also happen that, after the exciting laser is turned <u>off</u>, the three-peaked spectrum should revert to one-peaked form. This is obviously a time-dependent spectral change, ideally appropriate for the physical spectrum and impossible to handle with the Wiener spectrum.

The spectra already shown in Fig. 9 make some statements about the speed of dressing (actually, in the present example we are talking about undressing). It is clear that in some cases the side peaks vanish long before the central peak vanishes (i.e., the spectrum goes from 3-peaked form to 1-peaked form) and the atom becomes undressed. This can be analyzed in the following way.

In all cases the laser itself is turned off abruptly at $t = t_p$. The question is, how rapidly does the atom follow? This is a matter for the interferometer. If the Fabry-Perot has a very high reflectivity then it will store for a long time (and continue to pass on to the detector) light that displays the 3-peaked spectrum. But this light is not representative of the turned-off state. Thus we should use an interferometer with decreased reflectivity, one that will quickly pass to the detector the details of the response to the turn-off. In other words the ratio of interferometer bandwidth to emission bandwidth should be increased.

Figure 10 shows the results of Huang, Tanaś, and Eberly [18] in this respect. In part (a) the spectral intensity of the central peak is shown as a function of time, and in part (b) the same is shown for the side peaks. The central peak decays more rapidly as Γ/A increases, until $\Gamma/A \approx 5$, at which value an increase in Γ makes no appreciable increase in the decay. The side peaks behave differently. They also decay more rapidly until $\Gamma/A \approx 5$, but for larger Γ values they actually begin to decay less rapidly. What is the meaning of this, and what is the significance of $\Gamma/A \approx 5$?

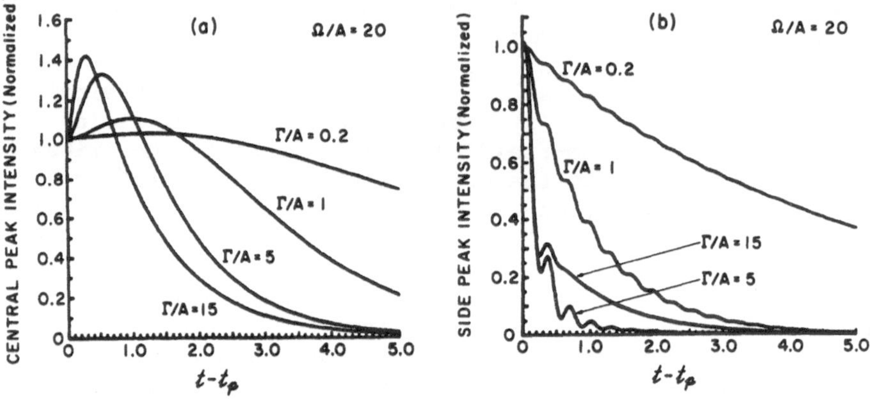

Fig. 10. (a) Central and (b) side peak heights of spectra similar to those of Fig. 9, except $\Omega = 20A$ and Γ ranges from 0.2A to 15A. Note that the central peak decays more rapidly for larger Γ, but that the side peak does the same only for Γ values up to $\Gamma = 5A$. The side peak decays *less* rapidly for $\Gamma = 15A$ because the filter resolution is then too poor to distinguish the side peaks from the center peak. This determines the speed of undressing. [From Ref. 18].

Both halves of this question have the same answer. The behavior noted occurs because, for sufficiently large Γ, the interferometer cannot resolve the side peaks separately from the central one. The splitting of the peaks is Ω, and in the examples shown, $\Omega = 20 \approx 5\pi$. In other words, when the spectrometer's time resolution Γ' becomes comparable to half a cycle of the splitting frequency $\pi/\Omega \approx 1/6$, then the existence of spectral splitting cannot be observed. This is precisely in the nature of a classic time-frequency uncertainty product. We conclude that undressing cannot be observed to occur faster than the reciprocal of the spectral splitting that identifies the dressed state.

ACKNOWLEDGMENTS: I am indebted to my colleagues and collaborators for their shared insights into time-dependent spectra. In connection with the topics mentioned in this review I thank particularly X.Y. Huang, R. Kornblith, J.J. Sanchez-Mondragon, and K. Wódkiewicz. The research reported here was partially supported by the U.S. Department of Energy and by the U.S. Office of Naval Research. Some further developments are reported by Dr. Wódkiewicz in another paper in this volume.

REFERENCES

1. See, for example, N. Wiener's summary in his classic paper: Acta Math. 55, 117 (1930).
2. Observations that do not imply normal ordering are discussed in I.C. Khoo and J.H. Eberly, Phys. Rev. A 14, 2174 (1976) and 18, 2184 (1978). Earlier references are also given.
3. See, for example, M. Born and E. Wolf, Principles of Optics (Pergamon Press, London), Fifth Edition, Sec. 7.6.
4. It was first discussed in J.H. Eberly and K. Wódkiewicz, J. Opt. Soc. Am. 67, 1252 (1977).
5. See Ref. 4 for many references to early attempts to define time-dependent spectra.
6. E.M. Purcell, Phys. Rev. 69, 681 (1946).
7. Experiments along these lines have been proposed recently: D. Kleppner, Phys. Rev. Letters 47, 233 (1981).
8. See, for example, the discussions given by Kleppner in Ref.7, and by H. Walther elsewhere in this volume, and by S. Haroche in Phys. Rev. Letters 51, 1175 (1983). Earlier references are also given in these papers.
9. E.T. Jaynes, Microwave Laboratory Report 502, Stanford University, 1958 (unpublished).
10. E.T. Jaynes and F.W. Cummings, Proc. I.E.E.E. 51, 89 (1963).
11. For recent work, and many earlier references, see N.B. Narozhny, J.J. Sanchez-Mondragon, and J.H. Eberly, Phys. Rev. A 23, 236 (1981).
12. J.R. Ackerhalt, Ph.D. Thesis, University of Rochester 1974 (unpublished). See also J.R. Ackerhalt and K. Rzażewski, Phys. Rev. A 12, 2549 (1975), Sec. IV.D.
13. J.J. Sanchez-Mondragon, N.B. Narozhny, and J.H. Eberly, Phys. Rev. Letters 51, 550 (1983).
14. J.J. Sanchez-Mondragon and J.H. Eberly (unpublished).
15. B.R. Mollow, Phys. Rev. 188, 1969 (1969), and F. Schuda, C.R. Stroud, Jr., and M. Hercher, J. Phys. B 7, L198 (1974).
16. References to experimental work, and an approximate theoretical discussion which uses an approach similar to that of the physical spectrum, have been given by E. Courtens and A. Szöke, Phys. Rev. A 15, 1588 (1977), and 17, 2119 (1978).
17. J.H. Eberly, C.V. Kunasz, and K. Wódkiewicz, J. Phys. B 13, 217 (1980).
18. X.Y. Huang, R. Tanaś, and J.H. Eberly, Phys. Rev. A 26, 892 (1982).

KELDYSH APPROXIMATION REVISITED

Howard R. Reiss

Arizona Research Laboratories
University of Arizona
Tucson, Arizona

and

Physics Department
The American University
Washington, D.C.

INTRODUCTION

The Keldysh approximation,[1] introduced in 1964, is important as the first intense-field approximation method designed for bound-state problems. The method is to treat photodetachment from a bound state by assuming that the dynamics of the detached electron are dominated by the applied field, and hence to describe the detached electron by a Volkov wave function. The Volkov wave function is an exact solution for a free, charged particle in a plane-wave electromagnetic field.

Analytical complexity created difficulties both for practical applications[2-4] of the Keldysh approximation, and for general investigations of intense-field phenomena. Recently, the Keldysh approximation has been reformulated[5] in an analytically simple way, which makes possible many new results and insights. These include: the correct low intensity limit; the basic intensity parameter (in addition to a secondary parameter found by Keldysh); the radius of convergence of perturbation theory; qualitative evidence of the failure of perturbation theory; and a novel and striking specific intensity effect.

ASPECTS OF THE ORIGINAL KELDYSH WORK

In the Keldysh paper,[1] the Volkov solution, derived in the Coulomb gauge, is transformed to a gauge with interaction potential $-e\vec{E}\cdot\vec{r}$ before it is applied. The ensuing formalism cannot be carried very far before further analytical approximations must be made. It is assumed very early in the paper that the photon number is large. This causes the method to give poor results[2-4] when applied in the perturbation theory limit, although the correct tunneling limit is obtained. The low intensity limit gives the correct intensity dependence, but the magnitude of the transition probability is much too small. The tunneling limit follows from allowing field frequency to approach zero as photon number goes to infinity. Apart from finding limits, the analytical forms obtained by Keldysh are too complicated to permit general investigation of nonperturbative effects.

FORMAL RESULTS OF THE REVISED KELDYSH APPROXIMATION

The Keldysh approximation can be cast[5] in an analytically much more tractable form than in the original work. A key step in this is to carry out the calculation entirely in Coulomb gauge.

An exact formal expression for the S-matrix (or transition amplitude) for the problem of photodetachment of an electron from a center of force is

$$\begin{aligned}(S-1)_{fi} &= -i \int dt_1 (\psi_{Af}, V_A \psi_{Bi})_{t_1} \\ &- \sum_j \int dt_1 \int dt_2\, \theta(t_2-t_1)(\psi_f, V_B \psi_{Aj})_{t_2} (\psi_{Aj}, V_A \psi_{Bi})_{t_1} \, . \end{aligned} \quad (1)$$

In this expression, ψ_A represents an exact solution for the electron in the presence only of the applied electromagnetic field, given by the interaction potential V_A; ψ_B is an exact solution for the electron in the presence only of the binding potential, given by V_B; the subscripts f and i refer to final and initial states, with ψ_f being a complete final-state solution; $\theta(t_2-t_1)$ is a standard step function; and the t subscripts on the Hilbert-space scalar products represent the time coordinate for all of the quantities in the product. Retention of only the first term in Eq. (1) corresponds to the Keldysh approximation. An order-of-magnitude comparison of a general term in the sum over j in Eq. (1) with the leading term gives a ratio which is independent of field intensity, but which is proportional to the ratio of the range of the potential to the wavelength of the applied field. Thus the validity of the Keldysh approximation depends upon the short range of the potential. This is in contrast to perturbation theory, which depends on the smallness of the field intensity. The requirement for a short-range potential means that the Keldysh method will not work for atomic ionization, with its long-range Coulomb attraction for the ionized electron. It does apply, however, to photodetachment of a singly charged negative ion.

With the interaction operator for the electromagnetic field given by the usual expression

$$V_A(t) = -e\vec{A}\cdot\vec{p}/m + e^2\vec{A}^2/2m ,$$

then the non-relativistic Volkov solution ψ_A satisfies the eigenvalue equation

$$V_A(t)\psi_A = V_A(\vec{p},t)\psi_A ,$$

where $V_A(\vec{p},t)$ is the momentum eigenvalue of the $V_A(t)$ operator. The S-matrix for the Keldysh approximation can then be written in general in the explicit form

$$(S-1)_{fi} = \frac{i}{V^{1/2}} \Phi_i(\vec{p}) \left(\frac{p^2}{2m} - E_i\right) \int_{-\infty}^{\infty} dt e^{i(p^2/2m-E_i)t}$$

$$\times \exp[i \int_{-\infty}^{t} d\tau \, V_A(\vec{p},\tau)] , \tag{2}$$

where

$$\Phi_i(\vec{p}) = \int d^3r e^{-i\vec{p}\cdot\vec{r}} \phi_i(\vec{r}) \tag{3}$$

is the Fourier Transform of the initial bound-state wave function. The convention that $\hbar=c=1$ is used throughout this work.

The well-defined closed form given in Eq. (2) can be made even more explicit if the applied field is taken to be monochromatic. For a circularly polarized field, the differential total transition probability per unit time which follows from Eq. (2) in the long-wavelength approximation is

$$\frac{dW}{d\Omega} = \frac{(2m^3\omega^5)^{1/2}}{(2\pi)^2} \sum_{n=n_0}^{\infty} (n-z)^2(n-z-\varepsilon)^{1/2} |\Phi_i(\vec{p})|^2 J_n^{\,2}(z^{1/2}\gamma) , \tag{4}$$

while for linear polarization of the field the result is

$$\frac{dW}{d\Omega} = \frac{(2m^3\omega^5)^{1/2}}{(2\pi)^2} \sum_{n=n_0}^{\infty} (n-z)^2(n-z-\varepsilon)^{1/2} |\Phi_i(\vec{p})|^2 J_n^{\,2}(z^{1/2}\alpha,-\tfrac{1}{2}z). \tag{5}$$

In Eqs. (4) and (5), ω is the circular frequency of the applied field; z is the intensity parameter

$$z = \frac{e^2a^2}{4m\omega} , \qquad (6)$$

where a is the amplitude of the Coulomb-gauge vector potential of the field; ε is the binding energy of the detached particle, expressed in units of ω, the field photon energy; and n_0 is the lowest photon order of interaction, specified by

$$n_o = [\varepsilon + z] , \qquad (7)$$

where the square bracket signifies the smallest integer containing the quantity in the bracket. The quantity γ in Eq. (4) is

$$\gamma = 2(n-z-\varepsilon)^{1/2} \sin\theta ,$$

where θ is the polar angle of $\vec{p}$ with respect to $\vec{k}$, the propagation vector of the field; and the quantity α in Eq. (5) is

$$\alpha = 8^{1/2}(n-z-\varepsilon)^{1/2} \cos\theta ,$$

where here θ is the polar angle of $\vec{p}$ with respect to the axis of linear polarization. Both Eqs. (4) and (5) are subject to the energy conservation condition

$$p^2/2m = \omega(n-\varepsilon-z).$$

One can see that Eqs. (4) and (5) are identical in form, with the only difference being that the ordinary Bessel function of Eq. (4) is replaced in Eq. (5) by a generalized Bessel function defined[5] by

$$J_n(u,v) = \sum_{k=-\infty}^{\infty} J_{n-2k}(u)\, J_k(v) . \qquad (8)$$

Equations (4) and (5) are totally explicit results upon specification of the initial-state momentum wave function of Eq. (3). Equations (4) and (5) are exact in a Keldysh sense. The only approximations that enter into them are the long-wavelength approximation and the basic Keldysh approximation of retaining only the leading term in Eq. (1).

The intensity parameter specified in Eq. (6) is the basic intensity parameter of the Keldysh problem. However, it was not found in Ref. 1.

LIMITING FORMS

From Eqs. (4) and (5), one can find general results for any photon order and any $\Phi_i(\vec{p})$ for the low-intensity or perturbative limit, given by $z<<1$. For example, for $n=1$ in the negative hydrogen ion problem, exactly the perturbation theory result is obtained[5] in this limit. A more stringent comparison is available if the Keldysh theory is applied to transitions from the valence band to the conduction band in a simple band-gap solid. In that case, for circularly polarized light, it is possible to establish perturbation theory results to all orders.[6] The $z<<1$ limit of Eq. (4) is then found to agree with the perturbative results to all orders.[6]

The high-intensity limit of Eqs. (4) and (5) can be ascertained in general, but the results are complicated. However, if in addition to $z>>1$, one also requires $\varepsilon>>1$ and $z_1>>1$ in Eq. (5), then the same Oppenheimer tunneling limit[7] found by Keldysh is again obtained, where z_1 is a secondary intensity parameter defined by

$$z_1 = 2z/\varepsilon = e^2a^2/2m\omega\varepsilon \; . \tag{9}$$

The intensity parameter of Eq. (9) is exactly the inverse square of the parameter found by Keldysh, who found only one intensity parameter. The reason is that large photon orders were assumed at an early stage in Ref. 1.

One limitation which must be noted in the $z>>1$ case is that the use of long-wavelength approximation in the original Volkov wave functions implies an upper limit on intensity[8] given by $z_f<<1$, where z_f is the standard free-electron intensity parameter

$$z_f = e^2a^2/2m^2 \; .$$

Since the ratios $z/z_f = m/2\omega$ and $z_1/z_f = m/\varepsilon\omega$ are both of the order of 10^6 for H^- in a CO_2 laser field, as an example, then it is perfectly possible to have $z>>1$, $z_1>>1$, $z_f<<1$ simultaneously satisfied.

CONVERGENCE OF PERTURBATION THEORY

One of the great benefits which attaches to the existence of a simple analytical form for the transition probability is that it can be examined for the singularities which limit the convergence of perturbation theory. Only the circular polarization case has been considered, since this is so much simpler than linear polarization.

The procedure is to treat a physical problem for which an analytical form for $\Phi_i(\vec{p})$ can be found, insert this into Eq. (4), and then find the singularity structure of $dW/d\Omega$ in the complex coupling constant plane. An expansion of $dW/d\Omega$ in powers of z is the perturbation expansion, since z is proportional to the fine-structure constant. The definition given in Eq. (6) can be rewritten as

$$z = \alpha_o \rho ƛ^2 \lambda_c \, , \tag{10}$$

where α_o is the fine structure constant, $ƛ$ is the field wavelength divided by 2π, λ_c is the electron Compton wavelength, and ρ is the photon density. The form given in Eq. (10) exhibits the physical content of the intensity parameter to be the product of the fine structure constant and the number of photons contained in a physical interaction volume $ƛ^2\lambda_c$.

The above procedure can be applied to H^-, since an analytical approximation is available for the ground-state wave function.[9] Equation (4) then takes the form

$$\frac{dW}{d\Omega} = \frac{f^2}{\pi}\, \omega\varepsilon^{1/2} \sum_{n=n_o}^{\infty} (n-z-\varepsilon)^{1/2}\, J_n{}^2(z^{1/2}\gamma) \, , \tag{11}$$

where f^2 is an empirically determined constant. It is readily shown that $J_n{}^2(z^{1/2}\gamma)$ is an entire function of z, and so the only singularities in the complex z plane are branch points at $n-z-\varepsilon=0$, and essential singularities whenever n_o as given in Eq. (7) increments to a different integer. The singularity nearest to the origin determines the radius of convergence, which is thus

$$z_R = [\varepsilon] - \varepsilon \, . \tag{12}$$

Therefore, one has always $0 \leq z_R < 1$. For H^- in 10.6 μm radiation, the binding energy of 0.75 eV gives $\varepsilon = 6.41$, and so $[\varepsilon]$ is 7. This gives

$$z_R = 0.59 \, . \tag{13}$$

One can then examine H^- results for $z \geq 0.59$ to explore nonperturbative behavior.

The legitimacy of a radius of convergence found for Eq. (11) can be challenged on two grounds: (i) $\Phi_i(\vec{p})$ was found from an approximate, not an exact, wave function; and (ii) the Keldysh approximation itself is an incomplete statement of the transition probability. However, the analytical wave function approximation that was used did not contribute to the limiting singularity, which was pre-existing in Eq. (4). A true momentum wave function might contribute other singularities, but it could not remove the one given in Eq. (12). With respect to the second objection above, it can be noted from Eq. (1) that the neglected higher-order terms in the S-matrix contain factors of V_B not present in the first, or Keldysh, term. These neglected higher-order terms thus have a different analytical structure than the first term, and so they can add extra singularities but they cannot remove any. Therefore the true radius of convergence might be less than that stated in Eqs. (12) and (13), but not greater.

NONPERTURBATIVE QUALITATIVE BEHAVIOR

If one considers again the example of H^- photodetachment by 10.6 μm radiation, the low-intensity lowest order process is $n_0=7$. If the log of the total transition probability per unit time is plotted against log z, perturbation theory predicts that the result should be a straight line with a slope of 7. When, in fact, Eq. (11) is integrated over solid angle and plotted in such fashion, the result departs only very slightly from a straight line of slope 7 even when plotted far beyond the convergence limit stated in Eq. (13). The effects of surpassing the perturbation theory limit are notable only by their absence. This agrees qualitatively with experimental results[10] in a somewhat different physical problem, where perturbative behavior persisted in high order multiphoton ionization of helium atoms well beyond the intensity where perturbation theory was expected to fail.

However, if one takes the result of Eq. (11) and examines it term by term in the sum over n, nonperturbative behavior emerges immediately. For example, at z=1, which is substantially more intense than the limit (13), the lowest order process is not 7, but $n_0=8$ from Eq. (7). The ninth- and tenth-order processes are then found to contribute substantially more to the transition amplitude than the eighth-order process, and even orders up to n=15 contribute more than the lowest-order process. This is very nonperturbative behavior, but it would be experimentally manifest only if energy discrimination were done on the ionized electrons so that the different orders could be distinguished. When all orders are summed, the result is very much like an extrapolation of low-intensity behavior in which only n=7 contributes.

A very striking intensity effect is predicted to occur in connection with a measurement which has not yet been attempted. If one calculates the linear polarizaton analogue of Eq. (11), performs the integration over solid angle, and plots log W against log z, it is again found that the curve looks very much like a straight line with a slope of 7 up to quite large values of z. However, in both the circular and linear polarization cases, there are some departures from straight-line behavior, and they are different in the two cases. If the ratio of the transition probabilities for circular and linear polarization is found, this ratio has a constant value of 1.04×10^{-3} in the perturbation theory limit. By the time the intensity has risen to the limit given in Eq. (13), the ratio has become more than ten times as large as the low-intensity value. This rises further to 40 times the low-intensity limit by the time $z=1$ is reached. This is a very striking intensity effect. Yet the necessary field strengths are physically attainable. At 10.6 μm, $z=0.59$ requires 6.6×10^9 W/cm^2.

REFERENCES

1. L. V. Keldysh, Zh. Eksp. Teor. Fiz. 47, 1945 (1964) [Sov. Phys.-JETP 20, 1307 (1965)].
2. N. G. Basov, A. Z. Grasyuk, I. G. Zubarev, V. A. Katulin, and O. N. Krokhin, Zh. Eksp. Teor. Fiz. 50, 551 (1966) [Sov. Phys.-JETP 23, 366 (1966)].
3. J. H. Yee, Phys. Rev. B 3, 355 (1971).
4. F. Adduci, I. M. Catalano, A. Cingolani, and A. Minafra, Phys. Rev. B 15, 926 (1977).
5. H. R. Reiss, Phys. Rev. A 22, 1786 (1980).
6. H. D. Jones and H. R. Reiss, Phys. Rev. B 16, 2466 (1977).
7. J. R. Oppenheimer, Phys. Rev. 31, 66 (1928).
8. H. R. Reiss, other contribution in this volume.
9. B. H. Armstrong, Phys. Rev. 131, 1132 (1963).
10. L. A. Lompre, G. Mainfray, C. Manus, S. Repoux, and J. Thebault, Phys. Rev. Lett. 36, 949 (1976).

COHERENCE AND SATURATION IN STRONG FIELD IONIZATION

Kazimierz Rzążewski

Institute for Theoretical Physics
Polish Academy of Sciences
02-668 Warsaw, Al.Lotników 32/46, Poland

1. INTRODUCTION

In a series of lectures, together with Don Agassi, we will review a progress made in the last 3 years in the study of strong field ionization. The development of this theory (and to a lesser extent its mathematical tools) was strongly influenced by the well known theory of bound-bound transitions based on the, so called, two-level atom model.[1]

We begin with a very brief discussion of a structure of the theory of bound-bound transitions.

With the advent of narrow band tunable lasers, the resonant or nearly resonant interaction of the atom with the monochromatic coherent light wave can be described adequately by a very simple model in which all the complicated level structure of the real atom is replaced by the two levels only.

There are two basically different regimes of the interaction: i) weak field regime, and ii) strong field regime.

ad. i) Weak field is not capable of altering the properties of the atom in a significant way. It is rather a probe of the properties of the atom itself. The interaction with such a field yields informations about atomic parameters like the precise energy spacing between the atomic levels, values of the dipole moments for various transitions and different relaxation constants. In the case of the weak field, all the relevant quantities can be computed by perturbative methods.

ad. ii) In this case one is probing the properties of the strongly coupled system composed of the atom and the field. Known atomic parameters enter the Hamiltonian describing the system. One should not treat the interaction between the atom and the field perturbatively in this case. A number of new phenomena were discovered this way. They played the central role in the quantum optics of the last two decades. Notions of Rabi oscillations, Mollow spectrum of resonance fluorescence or antibunching of photons are widely known.

Having the two regimes in mind, let us turn to a short discussion of a weak field ionization.

2. WEAK FIELD IONIZATION

The main quantity measured in the weak field ionization is the ionization rate R. It is the total number of photoelectrons released from the atom per second. It has a dimension of frequency. One can measure the ionization rate R as a function of laser frequency ω_L.

Perturbation theory applied to the transition from the discrete state to the continuum under the influence of the harmonic perturbation yields the Fermi's golden rule[2] for the ionization rate:

$$R(\omega_L) = \frac{2\pi}{\hbar} |\langle 0|\vec{d}\cdot\vec{E}|\omega\rangle|^2 \qquad (2.1)$$

where $\vec{E}$ is the amplitude of the laser light, $\vec{d}$ is the dipole moment operator, $|0\rangle$ denotes the initial bound state of energy $E_0>0$ and $|\omega\rangle$ is the generalized continuous spectrum eigenstate of energy $\hbar\omega$. The conservation of energy follows directly from the lowest order perturbation theory. Only the state $|\omega\rangle$ satisfying

$$\hbar\omega = \hbar\omega - E_0 \qquad (2.2)$$

can be reached. In other words, the momochromatic, weak laser produces momochromatic electrons.

In many textbooks the Fermi's golden rule contains another factor known as the density of states. This last factor is simply equal to one if we normalize the continuum states to the energy delta function:

$$\langle\omega|\omega'\rangle = \delta(\omega-\omega') \qquad (2.3)$$

Such normalization is assumed in this lecture.

The orthodox lowest order perturbative approach neglects a depletion of the initial state. This is clearly inadequate if even weak field is allowed to act on the atom sufficiently long.

The simple kinetic equation type argument* tells, that the population of the ground state $P(t)$ changes in time exponentially:

$$P(t) = e^{-Rt} \tag{2.4}$$

according to the most common quantum mechanical decay law. So the photoelectrons are emitted on a time scale $1/R$. According to the time-energy uncertainty principle they should have their spread in energies $\sim\hbar R$. In other words the weak field ionization rate is also a width of the long time photoelectron spectrum. This spectrum is power broadened.

We are in the position now to tell what is a strong field in the context of single photon ionization. It is heuristically clear that the laser field can be regarded as strong if the width of the photoelectron spectrum R is comparable with the characteristic interval of the laser frequencies ω_L, over which $R(\omega_L)$ itself changes significantly. Therefore, for the applications of the theory to be developed here we should look for the narrow structures in the $R(\omega_L)$.

We conclude this section with the formulation of the simplest Hamiltonian used for the strong field ionization. Such a Hamiltonian is obviously suggested by the analogy with the two-level atom case. One can derive it from the standard dipole coupling Hamiltonian via a chain of approximations similar to those invented for two-level atoms.[1] The Hamiltonian reads:

$$H = E_0|0\rangle\langle 0| + \hbar\int \omega|\omega\rangle\langle\omega|d\omega + \hbar\int d_\omega \cdot$$

$$\cdot[\Omega(\omega)e^{i\omega_L t}|0\rangle\langle\omega| + \Omega^*(\omega)e^{-i\omega_L t}|\omega\rangle\langle 0|] \tag{2.5}$$

The interaction here is described by the function $\Omega(\omega)$ which tells how strongly different points of the continuous spectrum are coupled to the bound state.

The contact with the informations obtained from the weak field ionization is achieved by equating the weak field (measured experimentally) ionization rate with the one computed via Fermi's golden rule from the Hamiltonian (2.5):

$$R(\omega_L) = 2\pi|\Omega(\omega_L)|^2 \tag{2.6}$$

From (2.6) we see that the weak field ionization rate determines the coupling function $\Omega(\omega)$ up to a phase factor.

---

*We will derive this result formally in Section 5.

3. AUTOIONIZATION RESONANCES

The simplest ionization transition discussed in all the textbooks on quantum mechanics is that from the ground state of hydrogen atom.

The simplest formula for the ionization rate gives:[2]

$$R(\omega_L) \sim E^2 \frac{\omega^{2/3}}{\left(1 + \frac{\hbar\omega}{|E_0|}\right)^6} \tag{3.1}$$

it is obtained with a plane-wave function instead of Coulomb scattering wave function in Fermi's golden rule. It is a smooth function which changes significantly over energy interval of the order of the ionization energy $|E_0|$. No hope to saturate this ionization profile.

Practically all the other atoms, however, have narrow resonances in the photo-ionization cross-section. The physical mechanism of such resonances can be easily explained on the helium atom. Its level structure can be described in the first approximation by a hydrogen-like Hamiltonian, neglecting the Coulomb interaction between the electrons. In this approximation a state with both electrons excited has the same energy as that having one electron in the ground state. If the electron-electron interaction is now recalled, its Hamiltonian will have nonvanishing matrix element between the doubly excited bound state and the continuum state. In effect the bound state becomes unstable, decays to the continuum. One of the electrons gets ionized at the expense of energy of the other electron. The phenomenon is called the autoionization.

The ionization by the light tuned near the autoionization region of the continuous spectrum of the atom produces a well known interference pattern, known as Fano profile[3] (see Fig. 1). It is the ionization to the autoionization resonance which will be a central theme of these lectures.

We will use the following parametrization[4] of the radiative matrix element $\Omega(\omega)$:

$$\Omega(\omega) = \frac{\Omega_0}{\sqrt{4\pi\gamma_0}} \left[\frac{\gamma_0}{\omega-\omega_e-i\gamma_0} - \frac{1}{q+i} \frac{i\gamma_1}{\omega-\omega_e-i\gamma_1} \right] \tag{3.2}$$

which only slightly differs from that used in the original paper of U. Fano. The resonance is located at $\hbar\omega_e$. Its width is γ_0, real parameter q is Fano's asymmetry parameter controlling the influence of the second Lorentzian term which is called background. The width of the background γ_1, is much greater than γ_0 and in most results

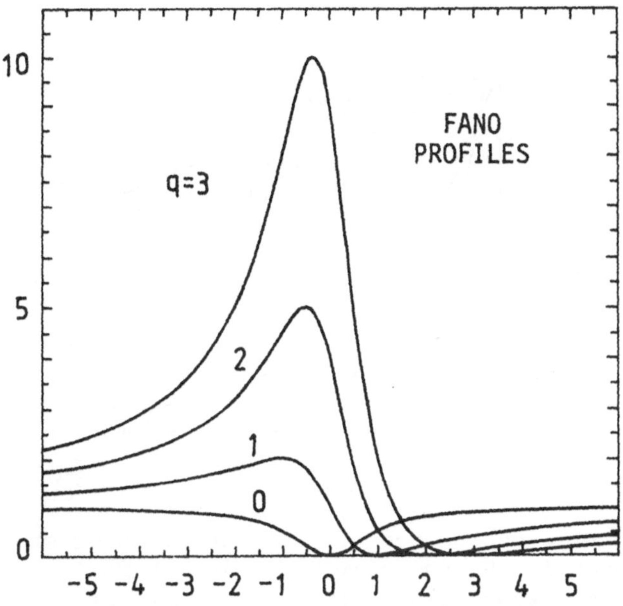

Fig. 1. Plots of $|\Omega(\omega)|^2$ for the autoionization resonance for several values of the asymmetry parameter q showing the Fano zeros at $\omega = -q\gamma_0$

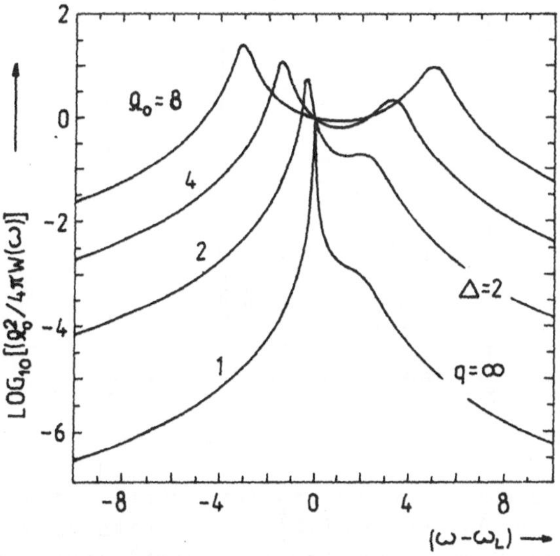

Fig. 2. Logarithmic plot of the electron spectrum, showing increasingly significant Autler-Townes splitting as Ω_0 increases from 1 to 8.

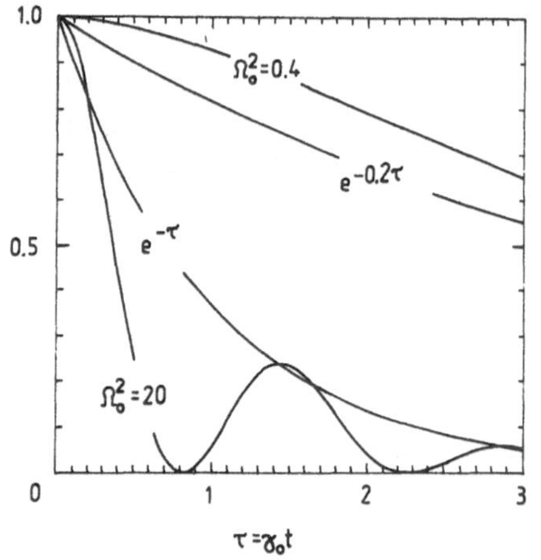

Fig. 3. Exact time dependence of the population of state $|0\rangle (q=\infty)$ is shown in the two curves labeled by Ω_0^2 (in units of γ_0). The two exponential curves show that for low Ω_0 the population leaks out at a rate approximated by $\Omega_0^2/2\gamma_0$ (which is the one-photon low-power ionization rate), and for high Ω_0 at the rate γ_0 (which is the width of the autoionizing resonance).

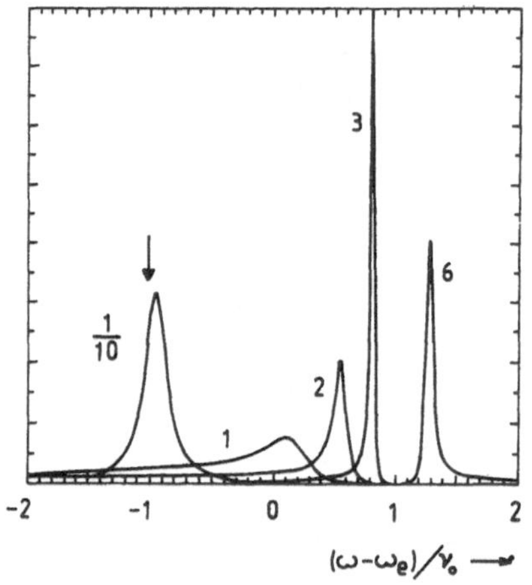

Fig. 4. Long-time electron spectra for several laser intensities, labeled by the value of $\Omega_0^2/4$ (in units of γ_0^2) for $q=1$. The arrow at -1 shows the (resonant) laser frequency. The position of the confluence is evident at +1 on the normalized energy axis.

we simply take a limit $\gamma_1 \to \infty$, so our background is effectively flat.

The overall strength of the coupling is controlled by the frequency Ω_0 which is proportional to the field strength of the applied light. By the analogy with the two-level atom we call it Rabi frequency.

Note, that in the limit $q \to \infty$ we obtain simply the resonance of the simplest, symmetric Lorentzian type.

The narrowest autoionization resonances resolved so far are not those of free, neutral atoms but are rather produced by an external influence. Highly excited Rydberg states of the neutral atoms placed in a static electric field get mixed with the continuum and ionize very much the way configuration interaction produces the phenomenon of autoionization.[5] Also static magnetic field has reportedly very similar effect.[6]

Our theory applies regardless of the origin of the resonance.

4. THRESHOLD BEHAVIOR

There is obviously another interesting region of ionization--vicinity of the ionization threshold. The study of the threshold behavior is rather complicated in the case of the neutral atoms. In this case a long-range Coulomb potential supports an infinite number of Rydberg states producing an essential singularity in the radiative matrix element at the threshold. Instead, the negative ion has only a finite number of bound states, there is an energy gap below the threshold and a simple power dependence describes the density of the scattering states.[7] The corresponding power dependence of the photodetachment cross-section has been verified experimentally.

The power of the energy appearing at the photodetachment threshold depends on the angular momentum of the excited continuum. Wigner's theory gives

$$|\Omega(\omega)|^2 \sim \omega^{\ell+1/2} \tag{4.1}$$

In the example of the s-wave continuum treated below, we chose the following model:

$$|\Omega(\omega)|^2 = \frac{A}{\pi}\frac{\sqrt{\beta\omega}}{\omega+\beta} \tag{4.2}$$

where β is a cut-off parameter. The parameter A is proportional to the intensity of the applied field and measures the over-all strength of the interaction.

It is clear that tuning the laser at the threshold is not a condition for a resonance. The usual argument for the rotating wave approximation assumed in the Hamiltonian (2.5) does not hold. We will comment on the corrections arising from the counter-rotating terms in the strong field photodetachment near the threshold at the end of the next Section.

5. THE SOLUTION OF THE MODEL

To solve for the time evolution of the system described by our simple Hamiltonian (2.5) we write down a time-dependent Schrödinger equation:

$$i\hbar \frac{\partial}{\partial t} |\psi\rangle = H|\psi\rangle \tag{5.1}$$

and look for its solution in the form:

$$|\psi\rangle = e^{-i\omega_L t} \left[\alpha(t)|0\rangle + \int d\omega \beta(\omega,t)|\omega\rangle\right] \tag{5.2}$$

Taking $E_0=0$ for convenience, the c - number equations for the amplitudes $\alpha(t)$ and $\beta(\omega,t)$ are

$$\dot{\alpha} = -i \int d\omega \Omega(\omega)\beta(\omega,t) \tag{5.3a}$$

$$\dot{\beta}(\omega,t) = -i(\omega-\omega_L)\beta(\omega,t)-i\Omega^*(\omega)\alpha(t) \tag{5.3b}$$

Assuming a step turn-on of the laser at $t=0$ with the obvious initial condition $\alpha(0)=1$, $\beta(\omega,0)=0$, we can solve the equations (5.3) in the Laplace transform domain in time variable.

This way, the time evolution of the probability amplitude $\alpha(t)$ is given as the following inverse Laplace transform:

$$\alpha(t) = \int_\Gamma \frac{e^{zt}}{\mathcal{H}(z)} \frac{dz}{2\pi i} \tag{5.4}$$

where the resolvent function is given by

$$\mathcal{H}(z) = z + i\omega_L + \int \frac{|\Omega(\omega)|^2}{z+i\omega} d\omega \tag{5.5}$$

and the contour of integration is placed to the right from all the singularities of the integrand. The physical branch of the multi-valued function (z) is obtained if the integration in (5.5) is performed with $\mathrm{Re}z>0$.

One can define a spectrum of outgoing electrons as

$$W(\omega) = \lim_{t\to\infty} |\beta(\omega,t)|^2 \tag{5.6}$$

If all the singularities of $\mathcal{H}(z)$ lie to the left of the imaginary axis, then $\alpha(t)\to 0$ as $t\to\infty$. In this case the photoelectron spectrum is normalized to unity and given by

$$W(\omega)= \lim_{\varepsilon\to 0+} \left|\frac{\Omega(\omega)}{\mathcal{H}(-i\omega+\varepsilon)}\right|^2 \tag{5.7}$$

and is available even if the explicit evaluation of the inverse Laplace transform (5.4) is difficult.

The properties of the weak field photoionization outlined in Section 2 are easily recovered from our present formulation.

In this case the dominant contribution to the time evolution of $\alpha(t)$ is expected to come from a zero of $\mathcal{H}(z)$ which, when the interaction $\Omega(\omega)$ tends to zero approaches the free evolution $z_0=-i\omega_L$.

Solving equation

$$\mathcal{H}(z)=0$$

perturbatively we obtain in the second order in $\Omega(\omega)$

$$z_1= -i\omega_L + i\int \frac{|\Omega(\omega)|^2}{\omega-\omega_L-i\varepsilon}\, d\omega \tag{5.8}$$

the $-i\varepsilon$ prescription for the singularity is obviously dictated by $\mathrm{Re}z>0$ requirement for the evolution to the future branch of $\mathcal{H}(z)$.

Using the well known formula from the theory of distributions:

$$\frac{1}{x-i\varepsilon} = P\,\frac{1}{x} + i\pi\delta(x) \tag{5.9}$$

we obtain the position of the pole:

$$z_1 = -i\left(\omega_L - p\int \frac{|\Omega(\omega)|^2}{\omega-\omega_L}\, d\omega\right) - \pi\, |\Omega(\omega_L)|^2 \tag{5.10}$$

With this pole giving rise to the only contribution to $\alpha(t)$ we get

$$P(t) = |\alpha(t)|^2 = e^{-2\pi|\Omega(\omega_L)|^2} \quad . \tag{5.11}$$

In full agreement with our approximation we may substitute

$$\mathcal{H}(z) \approx z - z_1$$

into (5.7) finding $R=2\pi|\Omega(\omega_L)|^2$ as a width of the spectrum. It all supports our discussion from Section 2. As a bonus we notice a small displacement of the position of the spectrum's maximum ω_L. This displacement is usually called the a.c. Stark shift. The approximation described above can be applied to a wide range of dynamical problems solved in the Laplace transform domain. It is called the single pole approximation.

Remarkably, our parametrization of the Fano profile (3.2), in the limit of the flat background taken after the integral defining $\mathcal{H}(z)$ is performed, leads to a very simple resolvent, which has only two complex zeros:

$$\mathcal{H}(z) = z + i\omega + \frac{\Omega_0^2}{4}\left[\frac{iq+1}{iq-1}\cdot\left[\frac{1}{z+i\omega_e+\gamma_0} + \frac{1}{1+q^2}\cdot\frac{1}{\gamma_0}\right]\right] \tag{5.12}$$

which can be computed as the roots of a quadratic equation. All the results in this case can be therefore calculated analytically.

Instead of quoting here rather complicated formulae (they can be found in Ref. 8) we will discuss the results graphically. We begin the discussion of the symmetric Fano profile or $q\to\infty$ case.

In general, there are two maxima in the photoelectron's spectrum.[4] Very clear interpretation is possible in two extreme cases:

i) $\Delta = \omega_e - \omega_L = 0$ - perfect resonance:

For the weak field both peaks overlap and we have by now familiar perturbative, power broadened situation. As Ω_0 increases, the form of the spectrum changes to symmetrically displaced two-peaked spectrum (Fig. 2). The distance between the peaks increases with Ω_0 and is approximately equal to Ω_0 for the very strong field. The phenomenon is a bound-free transition counterpart of a well known phenomenon called

Autler-Townes splitting. Another way of looking at the difference between the weak and strong field case is illustrated by Fig. 3. We plotted here the time evolution $P(t)$ for the weak and strong field. The distinction between single and double peak spectra corresponds to the switch from monotonic decay to the damped Rabi oscillations. The two exponential curves show that for low Ω_0 the population leaks out at a Fermi's golden rule rate $\Omega_0/2\gamma_0$, and for high Ω_0 at the rate γ_0 (which is the width of the autoionizing resonance).

ii) If $\Delta \neq 0$, for the weak field, the spectrum is composed of two peaks--one large (area ~1) centered at ω_L (a.c. Stark shifted) elastic, the other small, inelastic centered at atomic resonance frequency ω_e (area $\sim\Omega_0^2/4\Delta^2$). As Ω_0 increases the clear identification of elastic and inelastic peaks becomes difficult, they get more and more symmetric and the picture resembles the one discussed in the i) point (Fig. 2).

The appearance of the oscillatory mode in the evolution of $P(t)$ is a new feature. For sufficiently strong laser field the ionization process is no longer a simple monotonic decay but contains the repeated coherent continuum-to--discrete electronic recombinations.

Completely new features appear for the finite q or asymmetric Fano profiles. There are still two peaks, but there is also a nontrivial ω dependence of the numerator of (5.7). It makes the photoelectron spectrum vanish always at the Fano minimum. The splitted spectrum is very asymmetric (Fig. 4). As the Rabi frequency Ω_0 increases and one of the Autler-Townes peaks approaches the Fano minimum a remarkable line narrowing occurs. This narrowing reverses for still higher powers although it goes to a finite point as $\Omega_0\to\infty$.

We call the critical line narrowing discovered here a confluence of coherences. It occurs when the Fano minimum, a result of the coherent configuration Coulombic interaction, coincides with the position of one of the peaks--itself a result of the coherent radiative interaction.

The critical value of the Rabi frequency is given by the formula

$$\Omega_0^2 = 4\gamma_0(1+q^2)(\gamma_0-\Delta/q) \tag{5.13}$$

The critical line narrowing via time-energy uncertainty relation means that a fraction of the population of the ground state gets trapped under confluence condition. This is illustrated by Fig. 5. All of the properties mentioned here are well illustrated by Fig. 6 and Fig. 7. Here we have a map of trajectories traced by the zeros of the resolvent $\mathcal{H}(z)$ as Ω_0 increases from 0 to ∞ for infinite and finite q respectively. To help an imagination of a reader, we recall that the imaginary part of the zero is a characteristic frequency while its real part is a reciprocal of a characteristic decay time or a width of the spectral peak.

The discussed so far regularities are derived here in the very idealised model. There is a number of incoherences to be accounted for in a more realistic approach. They will be discussed in detail by Don Agassi. In our study of strong field autoionization we have been consequently extending all of the ω- integrations to $-\infty$ thus neglecting the existence of the ionization threshold.

The remainder of this Section will be devoted to a brief discussion of the threshold peculiarities on the simple example of the photodetachment[9] mentioned in Section 4.

The single pole approximation breaks down close to the threshold.

There exists a certain positive ω_c given by

$$\omega_c = \int_0^\infty \frac{|\Omega(\omega)|^2}{\omega} d\omega \tag{5.14}$$

defining the so-called dynamical threshold. Below that frequency there exists a purely imaginary pole of the resolvent contributing a purely oscillatory term to $\alpha(t)$. In fact the residue at the imaginary zero tends rapidly to unity as ω_L decreases. This obviously means that the energy of a single photon is too small to cause the photodetachment.

Besides the complex pole above ω_c and the imaginary pole below ω_c there is always a contribution coming from the cut of $\mathcal{H}(z)$.

This cut contribution is dominant in the immediate vicinity of the dynamical threshold giving rise to the nonexponential decay for both short and long times (Fig. 8).

As mentioned at the end of Section 4, the rotating wave approximation is not well justified for the photodetachment near the threshold. It was shown however by J. Javanainen with the help of a continued fractions method,[23] that the inclusion of the counter-rotating terms in the Hamiltonian results (to a very good approximation) in a significant modification of the expression for the dynamical threshold ω_c only.

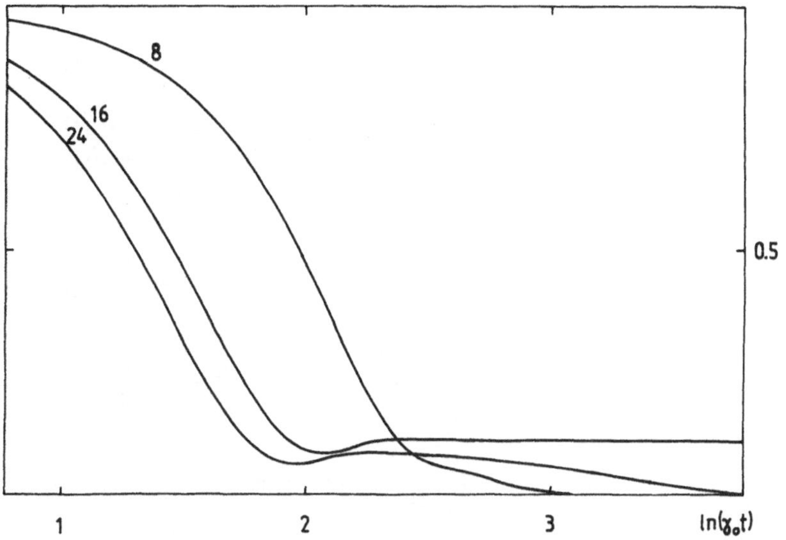

Fig. 5. Exact time dependence of the population of state $|0\rangle$ ($q=1$) is plotted versus $ln(\gamma_0 t)$. The curves are labeled by the value of Ω_0^2. The population trapping at confluence intensity $\Omega_0^2 = 16$ is evident.

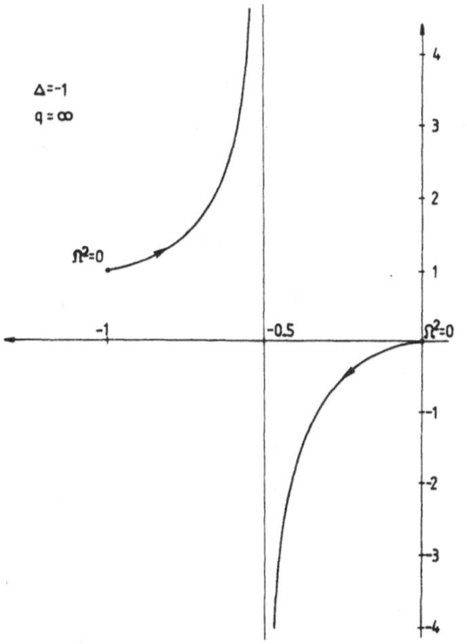

Fig. 6. Trajectories traced by the zeros of the resolvent (z) for the symmetric Fano profile as the Rabi frequency Ω_0 changes from 0 to ∞.

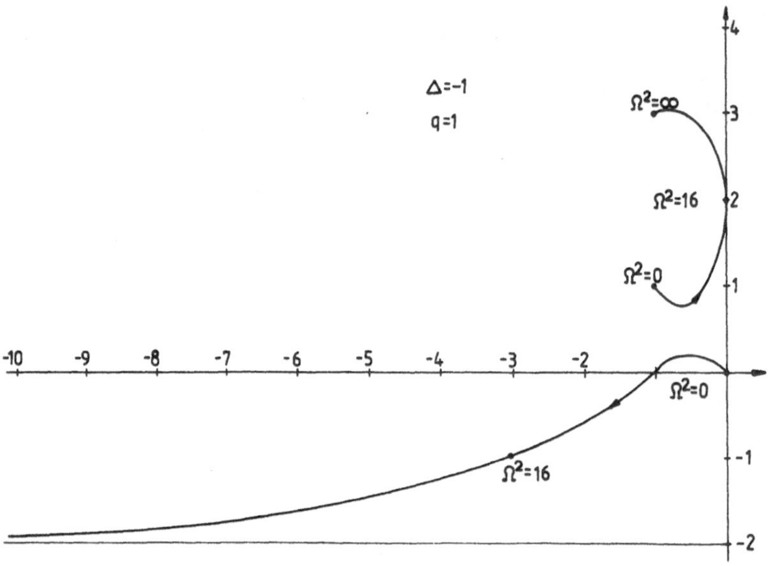

Fig. 7. Same as Fig. 6 but for the asymmetric Fano profile. Note the confluence point at $\Omega_0^2 = 16$.

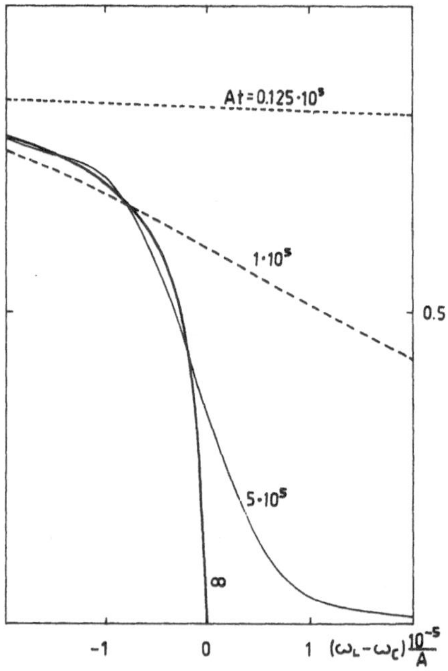

Fig. 8. Probability of remaining in the ground state $|\alpha(t)|^2$ plotted against laser frequency $10^{-5}(\omega - \omega_c)$ A varied in the vicinity of of thedynamical thereshold for different times: $At = .125 \cdot 10^5$, $1 \cdot 10^5$, $5 \cdot 10^5$, ∞; $\beta = 10^6$A.

6. THE RESONANCE FLUORESCENCE SPECTRUM ASSOCIATED WITH STRONG FIELD IONIZATION

Interaction of an atomic system driven by a strong monochromatic field with a full quantized electromagnetic field is one of the fundamental problems of quantum optics.

Besides the stimulated transitions caused by the external driving field the atom undergoes also some spontaneous transitions. Emitted photons have interesting spectral and statistical properties. In the case of a two-level atom, the system tends to a stationary state in which the spontaneous photons are emitted at constant rate. The notion of the power spectrum is then very useful. This spectrum has a familiar three peak structure[10] and has been calculated by many authors and also observed experimentally.

We will describe in this Section the basic results concerning photon-spectrum coming from spontaneous recombination in the strong field ionization.[11] There is an important difference between the two-level atom and the present problem. The ionization process is by its very nature a transient phenomenon. As time tends to infinity, the ionized electron leaves the interaction region and has no chance of the spontaneous recombination. Hence, the power spectrum tends to zero for long time. Of course, we have rather complicated notion of the, so called, physical spectrum[12] at our disposal. This allows us to follow the time evolution of the transient. We chose a somewhat different approach. Instead of talking about the rate at which photons of given frequency are emitted* we rather compute the energy spectrum or the total number of photons emitted to the particular mode of the electromagnetic field. This way, in spite of its being transient, we can study the process in some detail without computing time-dependent quantities.

Additional features related to the finite interaction time and the impact on the spectrum of the off diagonal relaxation will be covered by the separate lecture.

The study of the photon spectrum in the laser induced autoionization is interesting since it is yet another fundamental, quantum electrodynamic, strong field problem which can be solved explicitly.

It is also interesting in the context of the novel features of the strong field electron's spectrum reported in this lectures. These new features are, and for some time will be, difficult to observe directly, because the spectral resolution of electrons is very limited.

---

*The difficulty associated with time dependent rate versus infinite frequency resolution are probably clear to the reader. They are resolved by the notion of the "physical spectrum."[12]

Some other authors[13,14] suggest a double resonance type experiment to observe these regularities.

Spontaneously emitted photons present an alternative way to test our theoretical predictions.

The energy spectrum of scattered photons, averaged over solid angles, $\Omega_{\vec{k}}$, is given in the long-time limit by

$$S(k) = \lim_{t\to\infty} k^2 \sum_{\mu} \int d^2\Omega_{\vec{k}} \langle 0|a^+_{\vec{k}\mu}(t) a_{\vec{k}\mu}(t)|0\rangle \tag{6.1}$$

where the initial state of the system $|0\rangle$ requires that the atom be in the ground state and the electromagnetic field be in a coherent state describing a linearly polarized monochromatic electric field at frequency ω_L .

The calculation of $S(k)$ requires the solution of integro-differential equations for the coupled atom-field moments. The methods are very similar to those already discussed in the lectures of Don Agassi. The form of our matrix element $\Omega(\omega)$ makes all the relevant kernels of the Fredholm integral equations separable.

The results summarized here can be found in Refs.10 and 15. Consider first the case of a symmetric Fano profile ($q\to\infty$). For the weak field we have a single peaked spectrum which is power broadened. Note that the peak narrows as the spontaneous width grows. This is so because the spontaneous emission increases the lifetime for ionization by returning the atom to the ground state.

For $\Omega_0 >> \gamma_0, \gamma_s$, the photon spectrum has three peaks as shown in Fig. 9. If $\Delta=0$, the side peaks are displayed by $\pm\Omega_0$ and have widths $\gamma_0 \pm {}^3/_2\gamma_s$. Thus the central peak has a width γ_0 . The ratio of the heights is $(2\gamma_0+\gamma_s)/(\gamma_0+\gamma_s)$ so it is close to 2 for $\gamma_s << \gamma_0$.

For finite q completely new features occur. We plot the photon spectrum for $q=5$, $\Delta=0$ in Fig. 10 for two values of γ_s, as discussed in the figure caption. Again, for small values of the Rabi frequency, only one peak is present. Increasing the Rabi frequency results in a splitting of the central peak into a three-peak spectrum. However, the spectrum is asymmetric and the widths of the side peaks are broadened by further increase of the Rabi frequency. frequency.

The condition for the critical Rabi frequency producing the confluence is $\Omega_0^2 = 100$ in this case. At the confluence the central peak gets very narrow and its width is determined by the value of γ_s.

One can analyze also the total number of emitted photons. Generally, the total number of scattered photons per ionized atom is of the order γ_s/γ_0 . Again, an important exception is seen in the neighborhood of the confluence. In Fig. 11 we plot the integrated photon spectrum versus Rabi frequency for several values of detuning, $q = 5$. The curves show a dramatic enhancement in the total number of scattered photons at the confluence Ω_C .

By appropriate choice of the detuning, so that Ω_C is of order γ_0 , several orders of magnitude more photons should be emitted above the value γ_s/γ_0 . One can easily understand this fact. As we remember, at the confluence a fraction of the population of the ground state remains trapped in the state which is a mixture of the ground state and the continuum. It is obvious, that this state has nonzero dipole moment. Hence it radiates photons. The feedback makes the lifetime of the trapped state finite but long.

An experiment designed to count all the emitted photons would present an interesting test of our results.

The spontaneously emitted photons can serve to monitor the threshold peculiarities in the strong field photodetachment discussed in Section 5. With analytic properties of the radiative matrix element of the photodetachment more complicated than for Fano profile, the spectral properties of the recombination photons are rather difficult to calculate.

As noticed by Agarwal et al.[16] in the context of autoionization, much simpler is a task of computing a spectrum of the photons emitted to the third level. Within the rotating wave approximation, the atom can produce at most one photon and one can solve the problem through the time dependent Schrödinger equation. The state vector $\psi(t)$ can be represented as:

$$\psi(t)\rangle = \alpha(t)|0\rangle + \int_{\mathcal{D}}^{\infty} \beta(\omega,t)|\omega\rangle + \sum_{\mu}\int d^3k f_{\mu}(\vec{k},t) a^{+}_{\vec{k}\mu}|2\rangle \tag{6.2}$$

and the long time photon-number spectrum can be computed as

$$S(k) = \lim_{t\to\infty} \sum_{\mu} \int d^2\Omega_{\vec{k}} \, |f_{\mu}(\vec{k},t)|^2 \tag{6.3}$$

A typical plot is shown[17] on Fig. 12. A departure from a standard Lorentzian form of the spectrum is clearly seen if the laser is tuned very close to the dynamical threshold. A total number of emitted photons is very close to unity for the laser tuned below the dynamical threshold. We are in a Raman scattering

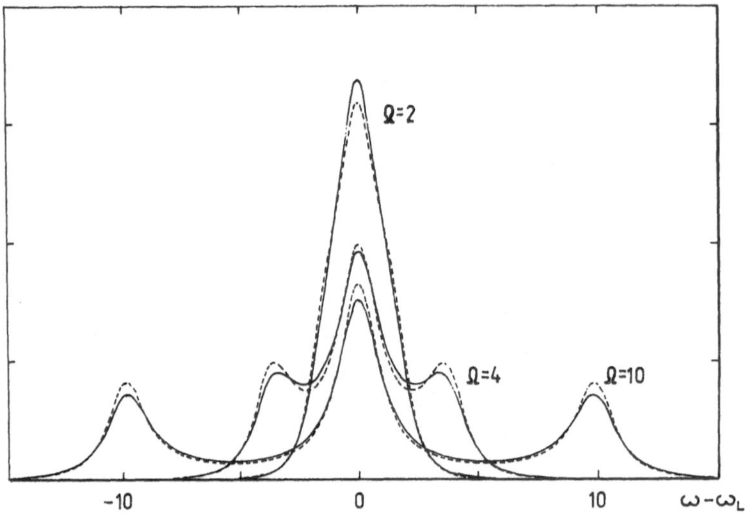

Fig. 9. The photon spectrum for the symmetric Fano profile, $\Delta = 0$ and increasing Ω_0 . The measured in γ_θ. Note that for a weak field the peak is narrowed, whereas for a strong field the peaks are reduced.

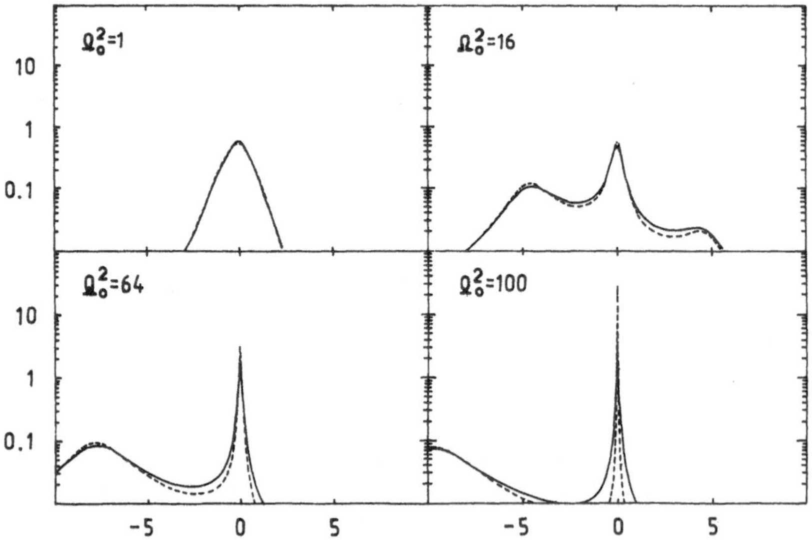

Fig. 10. A semilogarithmic plot of the photon spectrum for q=5, Δ = 0 and 4 different Rabi frequencies. The dashed curves are are the results for γ_s =.1. The confluence is a Ω_0^2 =. 100.

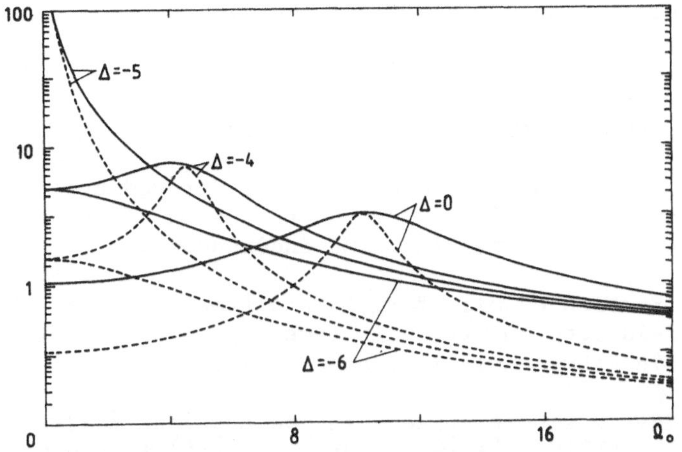

Fig. 11. A semilogarithmic plot of the total number of scattered photons versus the Rabi frequency for $q = 5$ and four values of the detuning. The spontaneous emission width $\gamma_s = .01$, while the solid curves represent $\gamma_s = .1$. The confluence is at $\Omega_0^2 = 100$.

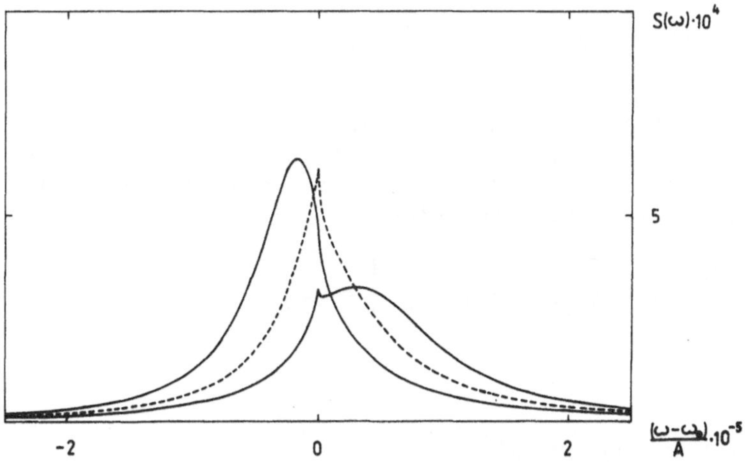

Fig. 12. The photon spectrum from the near threshold strong field photodetachment. The laser is tuned $-3\cdot10^{-6}$A below the dynamical threshold (dashed curve) and $+5.10^{-6}$A above the threshold. The scale on x-axis is 10^{-5}A ; γ_s=5·A , β =10^6

regime in this case. This total number of photons quickly tends to zero above the threshold, when the ionization channel competes effectively.

7. STRONG FIELD IONIZATION BY SMOOTH PULSES

All the results presented so far were derived under somewhat unrealistic assumption of a sudden switching-on of the strong laser signal, which otherwise remains time independent.

In this Section we will formulate a framework for a study of the strong field autoionization in time dependent fields. Some exactly soluble examples will be also given.[18]

We consider a smooth pulse with the envelope $E(t)=\epsilon_0 \cdot f(t)$ where ϵ_0 is its typical field strength and $f(t)$ is its dimensionless shape. The modified equations for the amplitudes $\alpha(t)$ and $\beta(\omega,t)$ read:

$$\dot{\alpha} = -i \quad d\omega\Omega(\omega) f(t) \beta(\omega,t) \tag{7.1a}$$

$$\dot{\beta} = -i(\omega-\omega_L)\beta(\omega,t) - i\Omega^*(\omega) f(t)\alpha(t) \tag{7.1b}$$

The typical field strength ϵ_0 has been included in the definition of $\Omega(\omega)$

We can eliminate the amplitude $\beta(\omega,t)$ assuming that $\beta(\omega,-\infty)$ and $f(t) \underset{t\to-\infty}{\longrightarrow} 0$:

$$\dot{\alpha} = -\int d\omega |\Omega(\omega)|^2 f(t) \int_{-\infty}^{t} f(\tau) e^{i(\omega-\omega_L)(\tau-t)} \alpha(\tau) \tag{7.2}$$

For a general $\Omega(\omega)$ it is not possible to convert integral equation (7.2) into differential equation. But for our parametrization of the Fano profile, one can derive (in $\gamma_1\to\infty$ limit) the following second order differential equation:

$$\ddot{\alpha} + \left[\xi - \frac{\dot{f}}{f} + bf^2\right]\dot{\alpha} + \left[Af + b\xi f + b\dot{f}\right] f\,\alpha = 0 \tag{7.3}$$

where

$$\xi = \gamma_0 + (\omega_e - \omega_L) = \gamma_0 + i\Delta$$

$$A = \frac{\Omega_0^2}{4}\;\frac{q-i}{q+i}$$

$$b = \frac{\Omega_0^2}{4}\;\frac{1}{\gamma_0}\;\frac{1}{1+q^2}$$

If the laser pulse originates at $t = -\infty$, the equation (7.3) should be supplemented with boundary conditions $\alpha(-\infty)=1$ and $\dot{\alpha}(-\infty)=0$.

The equation (7.3) forms a convenient starting point for a numerical study.

In this lecture, however, we will present a rather special exactly soluble case of a hyperbolic secant pulse:

Hyperbolic secant pulses play an important role in the theory of a coherent propagation of short light pulses through a continuous medium of two-level atoms.(1)

For the symmetric Fano profile, $q\to\infty$, one can express the amplitude $\alpha(t)$ in terms of the hypergeometric function:

$$\alpha(t) = F\left(\frac{\Omega_0}{2\gamma},\; -\frac{\Omega_0}{2\gamma}\,\Big|\,\frac{\gamma+\xi}{2\gamma}\,\Big|\,\frac{th\gamma t+1}{2}\right) \tag{7.4}$$

This function reduces to a polynomial if $\Omega_0/2\gamma = n$ is an integer. This "quantization condition" translates into the $2\pi n$ area of the pulse. The area of the pulse is defined as the integral over time of the time dependent Rabi frequency $\Omega_0\cdot f(t)$.

The polynomial in question is a Jacobi polynomial:

$$\alpha_n = J_n\left(0,\; \frac{\xi+\gamma}{2\gamma}\;\Big|\;\frac{th\gamma t+1}{2}\right) \tag{7.5}$$

Using the well known expression for the behavior of the hypergeometric function $F(a,b|c|z)$ near the singular point $z=1$ $(t=+\infty)$ we can derive the general formula for the probability $P(\infty) = \lim_{t\to\infty}|\alpha|^2$ of the atom to remain in its ground state after the pulse has passed:

$$P(\infty) = \left|\frac{\left[\Gamma\left(\frac{\xi+\gamma}{2\gamma}\right)\right]^2}{\Gamma\left(\frac{\xi+\gamma-\Omega_0}{2\gamma}\right)\Gamma\left(\frac{\xi+\gamma+\Omega_0}{2\gamma}\right)}\right|^2 \tag{7.6}$$

This formula shows, that the $2\pi n$ pulses are physically distinguished. For the detuning $\Delta = 0$, the atom is ionized completely by the $2\pi n$ pulse with sufficiently large n .

For example, for $\Delta = 0$, $\gamma = \gamma_0$, the formula (12) simplifies to:

$$P(\infty) = \frac{\sin^2 \frac{\Omega_0 \pi}{2\gamma}}{\left(\frac{\Omega_0 \pi}{2\gamma}\right)^2} \tag{7.7}$$

which is zero for all the $2\pi n$ pulses. On Fig. 13 we plotted the probability $P(t)$ for the pulses of different areas at resonance.

In the case of the $2\pi n$ pulses it is possible to derive an explicit formula for the photoelectron spectrum defined as:

$$W_n(\omega) = \lim_{t \to \infty} \left| \beta(\omega, t) \right|^2 \tag{7.8}$$

The main feature of this spectrum is its multipeaked structure. $W_n(\omega)$ can have up to n-maxima. This fact comes as a surprise. For a step turn-on of the constant laser signal one gets single peaked spectrum for weak field and two-peaked spectrum above a certain threshold Rabi frequency.

In the case of the smooth pulse of the kind discussed here, an instantaneous Rabi frequency varies between zero and its maximal value Ω_0 . The appearance of the multipeaked spectrum is therefore a new coherent phenomenon in the strong field bound-free transition. Fig. 14 shows the photoelectron spectra for the hyperbolic secant pulses of $2\pi n$ areas. The number of peaks is equal to n .

8. CONCLUSIONS

In the present series of lectures we have presented rather detailed analyses of the strong field ionization. We predict numerous new effects which are nonperturbative in nature.

There are many other recent theoretical papers closely related to our research outlined here.[13-22] They usually deal with somewhat more levels, more continua, more lasers. What is lacking is an experimental verification of all those predictions.

I end therefore with a plea for more experiments on strong field ionization.

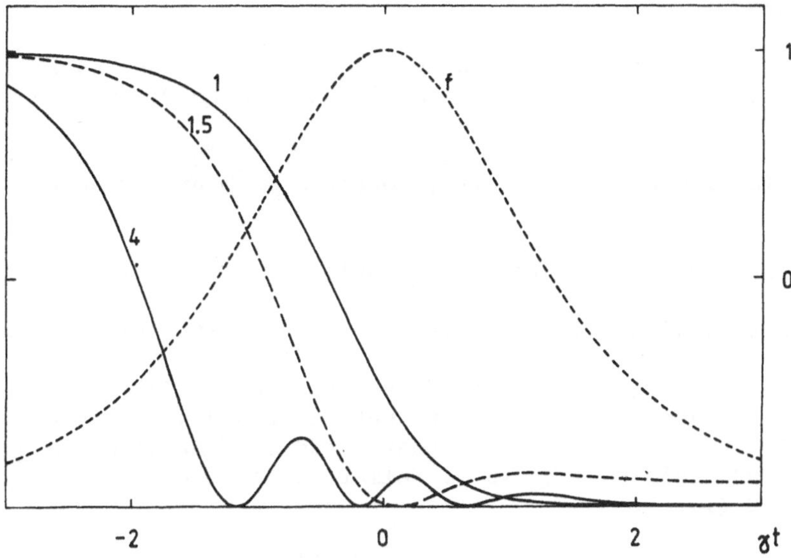

Figure 13

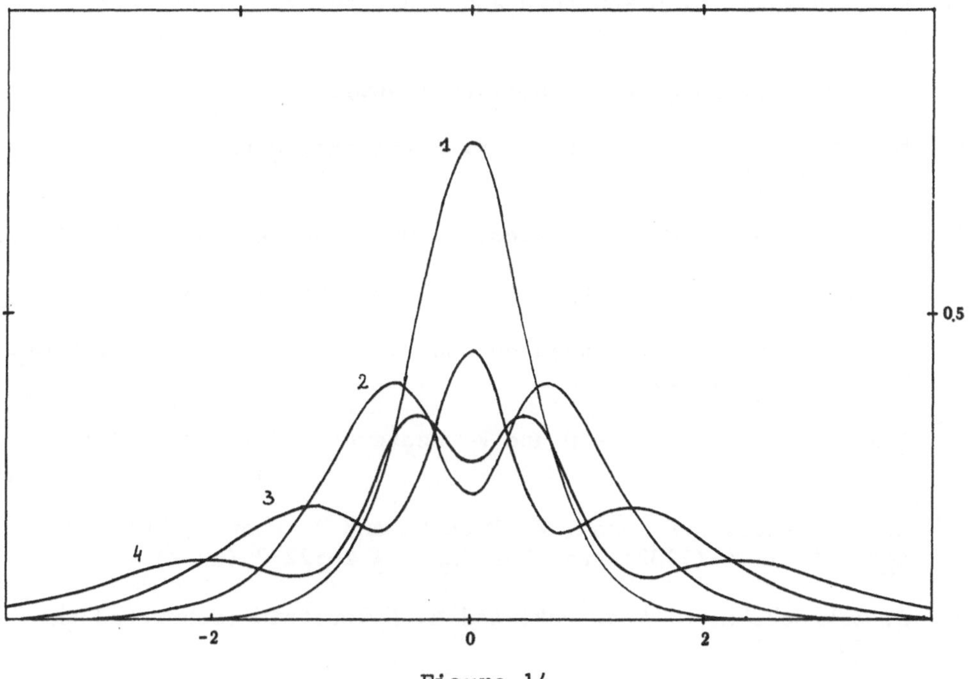

Figure 14

REFERENCES

1. For a review see for instance: L. Allen and J.H. Eberly, "Optical Resonance and Two-Level Atoms, John Wiley & Sons, New York (1975).

2. See for instance: L. Schiff, "Quantum Mechanics," McGraw-Hill, New York (1968).

3. U. Fano, Phys. Rev. 124:1866 (1961).

4. K. Rzążewski and J.H. Eberly, Phys. Rev. Lett. 47:408 (1981).

5. S. Feneuille, S. Liberman, J. Pinard, and A. Taleb, Phys. Rev. Lett.42:1404 (1979). An alternative method has been proposed by L. Armstrong, Jr., B.L. Beers, and S. Feneuille, Phys. Rev. A 12:1903 (1975) and by Yu.I. Heller and A.K. Popov, Opt. Commun. 18:449 (1976).

6. H. Friedrich, Phys. Rev. A 26:1827 (1982).

7. E.P. Wigner, Phys. Rev. 73:1002 (1948).

8. K. Rzążewski and J.H. Eberly, Phys. Rev. A 27:2026 (1983).

9. K. Rzążewski, M. Lewenstein and J.H. Eberly, J. Phys. B 15:L661 (1982).

10. B.R. Mollow, Phys. Rev. 188:1969 (1969).

11. M. Lewenstein, J.W. Haus and K. Rzazewski, Phys. Rev. Lett. 50:417 (1983).

12. J.H. Eberly and K. Wódkiewicz, J. Opt. Soc. Am. 67:1252 (1977).

13. P. Lambropoulous and P. Zoller, Phys. Rev. A 24:379 (1981).

14. A.I. Andryushin, A.E. Kazakov and M.V. Federov, Zh. Exp. Teor. Fiz. 82:91 (1982).

15. J.W. Haus, M. Lewenstein and K. Rzążewski, Phys. Rev. A (in press).

16. G.S. Agarwal, S.L. Haan, K. Burnett and J. Cooper, Phys. Rev. Lett. 48:1164 (1982) and Phys. Rev. A 26:2277 (1982).

17. J. Zakrzewski, K. Rzążewski and M. Lewenstein (submitted to J. Phys. B).

18. K. Rzążewski (submitted to Phys. Rev. A).

19. P.E. Coleman and P.L. Knight, J. Phys. B 15:L235 (1982).

20. Z. Białynicka-Birula, Phys. Rev. A (in press).

21. M. Crance and L. Armstrong, J. Phys. B 15:3199 (1982).

22. J. Zakrzewski, J. Phys. B (in press).

23. J. Javanainen, J. Phys. B. (in press).

QUANTUM FLUCTUATIONS AS CORRECTIONS TO SLOWLY VARYING QUANTITIES

Stig Stenholm

Research Institute for
Theoretical Physics
University of Helsinki
Helsinki, Finland

INTRODUCTION

One of the more intriguing questions in physics concerns the relation between quantum mechanics and classical mechanics, which is usually regarded as the limiting behaviour when Planck's constant $\hbar$ can be regarded as small in some sense. The usual semiclassical approximation of quantum theory does not in any natural way lead to classical trajectories and the expansion in $\hbar$ is rather singular.

In recent work on light pressure cooling (Stenholm 1983a) it has become clear that the approximation methods used rest essentially on an expansion in $\hbar$. Thus we have a case where the lowest order approximation is, indeed, found to correspond to the classical behaviour and corrections are due to quantum fluctuations. These emerge as a singular perturbation which must be handled with care. It has turned out that the velocity dispersion measure becomes anonomalously narrow, and this feature can be related to the sub-Poissonian nature of the spontaneously emitted light quanta. For a review see Stenholm (1983b). It thus appears interesting to understand the nature of the singular perturbation expansion and apply the same considerations to other problems of related structure.

In this work I present the basic situation in light pressure cooling in Sec. 2 as a motivation for the later, more formal treatment. This is presented in Sec. 3, and its relation to singular perturbation theory is discussed in Sec. 4. Section 5 treats a few applications: First the light pressure problem is cast into the present form. Next the onset of an instability due to spontaneous emission is reformulated in the presence of one strong laser. Finally the theory of quantum fluctuations in a single mode laser is discussed. In all these problems the classical solution can be seen to form the lowest approximation and the perturbative treatment describes fluctuations around this. Here we only outline the formulation of the problems; their detailed treatment is found in the references. In Sec. 6 there is a summary and discussion of our presentation.

2. THE APETIZER: LIGHT PRESSURE FORCE

We consider an atom of mass M travelling with the momentum $\vec{p}$. It contains two internal levels separated by the energy $\hbar\omega$. When a laser beam of frequency $\Omega=cq$ impinges on the atom, transitions take place when the Doppler shift along the laser axis compensates the detuning viz. for the velocity

$$v_0 = \frac{\omega-\Omega}{q} . \tag{1}$$

In the absorption process the atom also must take up the photon momentum $\hbar q$ in the direction of the laser beam. The situation is illustrated in Fig.1. A subsequent spontaneous emission event redistributes the outgoing momentum in a nearly isotropic way, and thus the average gain of velocity in the direction of the laser is

$$v_r = \frac{\hbar q}{M} . \tag{2}$$

Due to the randomness in the outgoing photon direction there appears, however, a spreading of the possible final states over a range

$$\Delta v = 2v_r . \tag{3}$$

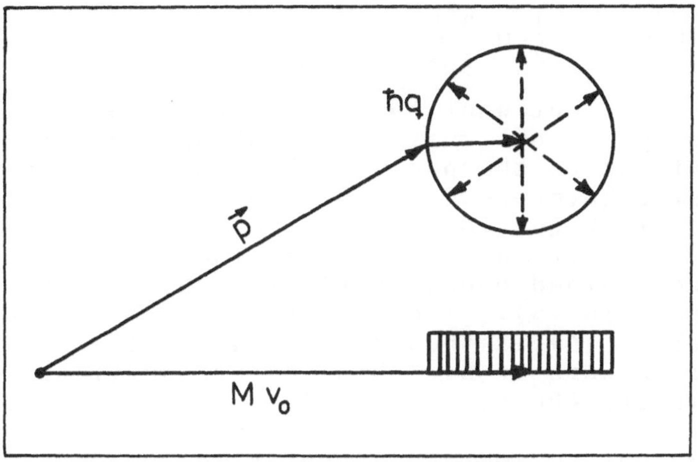

Figure 1.

For a strong laser field the atom is nearly immediately returned to the upper level and its average population probability is 1/2. With the spontaneous decay rate Γ the number of processes per unit time is $\Gamma/2$ and each changes the average atomic momentum by $\hbar q$. Hence the average light pressure force is

$$F \sim \frac{1}{2}\hbar\, q\, \Gamma\ . \tag{4}$$

This is a good estimate for the resonant light pressure force; more detailed reviews of the theory can be found in Kazantsev (1978) and Letokhov and Minogin (1981).

The spreading of the possible final states leads to a diffusive motion in momentum space, which from the theory of Brownian motion becomes

$$\frac{\langle \Delta p^2 \rangle}{t} \sim (\hbar q)^2\ \frac{1}{2}\ \Gamma \equiv D. \tag{5}$$

Combining (4) and (5) we can describe the development of the momentum distribution function W(p,t) by the Fokker-Planck equation

$$\frac{\partial W}{\partial t} + F\frac{\partial W}{\partial p} = \frac{1}{2} D \frac{\partial^2 W}{\partial p^2} . \qquad (6)$$

Here we do not want to discuss the implications of Eq. (6) but point out that the light pressure force (4) is proportional to the photon momentum i.e. to $\hbar$. It is a classical average force due to the incessant arrival of light quarta at the atom. The diffusion is due to the discrete nature of the single individual quantum process and consequently it contains the factor $\hbar^2$ as is seen in (5). The Fokker-Planck expansion used to derive (6) is thus a semiclassical expansion, formally in powers of $\hbar$. This feature survives in a more complete theory.

A new feature emerged recently in the discussion of photon momentum effects. Because each momentum kick of average magnitude $\hbar q$ is accompanied by the emission of one outgoing photon we can either count the photons or the momentum kicks. The distribution functions are the same. This fact was simultaneously realized for the free-electron laser, see Bambini and Stenholm (1979) or Stenholm and Bambini (1981), and the resonance fluorescence case by Mandel (1979) and taken up by Cook (1980b).

Starting from a state with initial atomic momentum p_0 and n_0 laser photons we can only go over into states of the type

$$|\psi\rangle = \sum_n c_n \, |p_0 + n\hbar q, \, n_0 - n\rangle , \qquad (7)$$

where n is the number of scattered photons. Thus both the momentum distribution W_{mom} and the photon distribution W_{ph} are given by the coefficients c_n according to

$$W_{mom}(p_0 + n\hbar q) = W_{ph}(n) = |c_n|^2 . \qquad (8)$$

For a more general state, a density matrix, the relation is less intuitively obvious but still true.

For the Brownian motion limit we can write

$$\langle p\rangle = \hbar q\langle n\rangle = Ft \qquad (9)$$

$$< \Delta p^2 > \; = \hbar^2 q^2 <\Delta n^2> = Dt \; . \qquad (10)$$

For purely uncorrelated spontaneous emission events the distribution is Poissonian and $<\Delta n^2> = <n>$. This suggested to Mandel (1979) the introduction of the normalized measure

$$Q = \frac{<\Delta n^2> - <n>}{<n>} \qquad (11)$$

for the deviation from Poissonian statistics. Introducing the relations (9) and (10) we find

$$Q = (D/\hbar qF) - 1 \; . \qquad (12)$$

When our estimates (4) and (5) are inserted Q is zero, indicating Poisson statistics. Our calculation is, however, too crude to allow this conclusion, and more detailed calculations are needed. For $Q>0$ the distribution is broader than a Poissonion and this indicates the bunching typical for bosons. The situation $Q<0$ suggests the interesting possibility of antibunching, which is foreign to both classical theories and simple-minded arguments based on Bose statistics. Its occurrence in resonance fluorenscence was suggested by Carmichael and Walls (1976a,b). From our treatment one can see that a decrease in Q must be due to a decrease in the diffusive spreading characterized by D.

To calculate the diffusion coefficient, we turn to a Wigner function representation of the density matrix, see e.g. Cook (1980a). The lower level population ρ_{11} at momentum p is coupled to the upper level population ρ_{22} at momentum $p+\hbar q$ by the strong laser field. The upper level decays on the average to the lower level at momentum $p+\hbar q$ wheras that level at p is filled directly from the same momentum. The situation is shown in Fig. 2. Thus the spontaneous emission couples the different momenta and an infinit set of coupled equations emerges. For the diagonal elements of the density matrix we write the equations

$$\frac{d}{dt}\rho_{22}(p+\hbar q) = -\Gamma \; \rho_{22}(p+\hbar q) + \ldots \qquad (13)$$

$$\frac{d}{dt}\rho_{11}(p) = \Gamma \; \rho_{22}(p) + \ldots \qquad , \qquad (14)$$

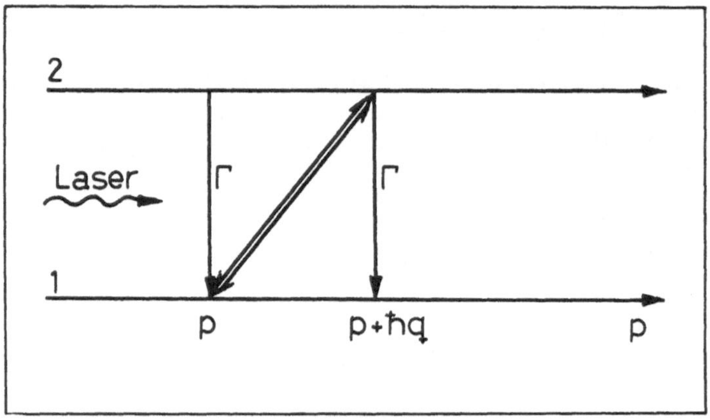

Figure 2

where the omitted terms are due to the strong laser field.

If we define the population as

$$W(p)= \rho_{22}(p+\hbar q) - \rho_{11}(p), \tag{15}$$

its equation of motion is from (13) and (14)

$$\frac{d}{dt} W(p)= \Gamma \left[\rho_{22}(p) - \rho_{22}(p+\hbar q)\right] = $$
$$=\left[-\Gamma \hbar q \frac{\partial}{\partial p} + \frac{1}{2}\hbar^2 q^2 \Gamma \frac{\partial^2}{\partial p^2}\right] \rho_{22}(p+\hbar q), \tag{16}$$

where we have expanded to second order in ($\hbar q$). The terms due to the laser field concel exactly. For a strong light field we know that half of the population appears in the upper state and if we set

$$\rho_{22}(p+\hbar q) \sim \frac{1}{2} W(p), \tag{17}$$

we directly obtain the Fokker-Planck equation (6) with the expressions (4) and (5) for the force and the diffusion coefficient. Thus the strong field limit of the theory leads to Poissonian statistics. This is due to the fact that a strong field flips the population rapidly between the upper and the lower states and thus allows a spontaneous emission event at almost any instant. No correlation between successive photon emission processes is possible.

In a more exact treatment the detailed calculation of the diffusion coefficient becomes more involved, see Gordon and Ashkin (1980) and Cook (1980a). Due to the drift of the atomic center of mass and the dipole moment there appears an anomalous contribution to the diffusion coefficient which can decrease its value, see Stenholm (1983a). This part is shown by Cook (1980b) to relate to photon statistics. Its experimental verification has recently been reported by Short and Mandel (1983).

In the following sections we shall formulate a very abstract approach to the calculations carried out in these works, relate these to other forms of perturbation theory, and give some applications.

3. PERTURBATION TREATMENT OF SLOWLY VARYING QUANTITIES

We are here considering the equation of motion of an abstract vector A developing according to the linear evolution equation

$$\frac{dA}{dt} = [M+\lambda N+\lambda^2 K]\ A, \tag{18}$$

where λ is some small expansion parameter. In our example in the previous section A was the density matrix for the two-level system. The linear operators are then super-operators, but in this finite-dimensional case we can easily take the four independent components of the density matrix $\{\rho_{21}, \rho_{22}, \rho_{11}, \rho_{12}\}$ to constitute the four components of a vector A. In the previous example the formal expansion parameter is the operator

$$\lambda = \hbar q \frac{\partial}{\partial p}\ , \tag{19}$$

see Eq. (16).

In the cases we are going to consider we assume the existence of a steady state when $\lambda=0$. We take

$$\lim_{t\to\infty} A(t) = \lim_{t\to\infty} e^{Mt} A(0) = a^0\ . \tag{20}$$

This assumes that M is a nonhermitean operator with dissipative terms i.e. ReM<0. Let us denote their magnitude by the decay rate Γ. From (20) it follows that a^0 is a steady state of (18) when $\lambda=0$ and

$$Ma^0 = 0. \tag{21}$$

Because M is nonhermitian the left-eigenvectors differ from the right-ones. We assume the existence of the element $\overline{a^0}$ such that

$$\overline{a^0}M = 0 \tag{22}$$

and $$(\overline{a^0}, a^0) = 1. \tag{23}$$

When this exists we can define the projections

$$P = a^0 \otimes \overline{a^0}; \; PA \equiv a^0(\overline{a^0}, A) \tag{24}$$

and

$$Q = 1 - P. \tag{25}$$

Here we remark that the space onto which P projects need not be one-dimensional. There may be more than one steady state, depending on the actual initial state.

Now the quantity PA is conserved when $\lambda=0$, and hence it denotes a slowly varying component, which can change only due to the presence of the perturbation terms in (18). Its equation of motion follows directly from (18), (21) and (22)

$$\frac{d}{dt} PA = \lambda PNPA + \lambda PNQA + \lambda^2 PKPA + \ldots \quad . \tag{26}$$

To the indicated order in λ we must also know the component QA in a consistent way. Its equation of motion is

$$\frac{d}{dt} QA = [QMQ + \lambda QNQ] \; QA + \lambda QNPA + \ldots \quad . \tag{27}$$

The order of magnitude of the rate of change here is given by QMQ and this we denoted by Γ earlier. In a time of order Γ^{-1} the component QA has already reached its steady state value in the presence of a given component PA. Just this situation favors the adiabatic elimination of the fast component QA, and allows us to follow the consequent drift of the slow component PA towards its eventual equilibrium value; see the discussion in

Haken (1977). The expansion parameter is (λ/Γ), and we can consistently put the time derivative in Eq. (27) equal to zero, if we keep only terms to order λ. Then the rate of change of PA is correct to order λ^2.

From Eq. (27) we thus find

$$QA = -\lambda\left[QMQ\right]^{-1} QNPA + O(\lambda^2) \tag{29}$$

and

$$\frac{d}{dt} PA = \lambda PNPA + \lambda^2\left[PKP - PNQ(QMQ)^{-1}QNP\right]PA. \tag{30}$$

This is, of course, the ordinary partitioning result combined with our case of a slowly varying component PA. When λ is the derivative operator (19) the ensuing equation is of the Fokker-Planck type.

The term containing the perturbation N to the second power is the anomalous contribution to the diffusion because it is not necessarily positive definite. If λ is the derivative operator it should, in the general case, be written as

$$\begin{aligned} &-\lambda PNQ\ \lambda(QMQ)^{-1}QNPA \\ &\quad = -\lambda^2\ PNQ(QMQ)^{-1}QNPA + \lambda\ (\lambda PNQ)(QMQ)^{-1}QNPA\ , \end{aligned} \tag{31}$$

where the term (λPNQ) denotes the derivative of the operator PNQ. The additional term adds to the term of (30) which is linear in λ a second order contribution. For consistency reasons we believe that this has to be neglected; see our discussion in Javanainen and Stenholm (1980a and 1981).

A few remarks on the present method are here needed. Partitioning techniques are, of curse, well known. New here is the application to a subspace with conserved vectors, and the inclusion of an adiabatic argument. The approach is related to operator versions of degenerate state perturbation theory, see Bogoliubov (1967), Bloch (1958) and Titulaer (1978 and 1980 a,b). In nonequilibrium systems, the aspect of degeneracy has been stressed by Haken (1975) and Dohm (1976). This treatment rests on the fact that the lowest order problem admits an infinity of solutions; the perturbation fixes the ambiguity and provides the solution to the dynamic problem.

We have also carried through the above derivation using the formalism of "multiple time scale perturbation theory" (see Nayfeh, 1973) and the result is the same. Recently the adiabatic elimination procedure in laser problems has been reconsidered by F. Haake; see Haake and Lewenstein (1983) and references therein.

4. CONNECTION WITH SINGULAR PERTURBATION PROBLEMS

The zero eigenvalue corresponding to the stationary state manifests itself in the fact that an adiabatic elimination to given order in λ can be carried out only when the time derivative d/dt is taken into account self-consistently to that order. This procedure is well used in physics and we shall here summarize the formalism to see its relation with our approach in the previous section.

The problem is again defined by Eqs. (18)-(20). We now make an ansatz for the solution in the form

$$A = a\alpha, \tag{32}$$

where we require

$$\frac{d}{dt} a = 0, (\overline{a^0}, a) = 1 \tag{33}$$

and attempt an expansion of the vector a in a power series

$$a = a^0 + \lambda a^1 + \ldots \quad . \tag{34}$$

From (18) we have exactly

$$\frac{d\alpha}{dt} = \lambda(\overline{a^0}, Na)\alpha + \lambda^2(\overline{a^0}, Ka)\alpha \quad . \tag{35}$$

To order λ^0 we have α= constant and $a = a^0$ so that $Ma^0 = 0$. To order λ^1 we need a to order λ^0 only and find directly

$$\frac{d\alpha}{dt} = \lambda(\overline{a^0}, Na^0)\alpha \tag{36}$$

in agreement with (30). To obtain the solution to order λ^2 we must solve for a^1 and then the time derivative of A enters according to

$$\frac{dA}{dt} = a\alpha^0 = (a^0 + \lambda a^1)\ |\lambda(\overline{a^0}, Na^0)\alpha|$$
$$= \lambda a^0(\overline{a^0}, Na^0)\alpha + O(\lambda^2). \qquad (37)$$

Combining this with the right-hand side of (18) to order λ^1 we find

$$a^0(\overline{a^0}, Na^0)\alpha = Na^0 + Ma^1 \qquad (38)$$

giving for the unknown a^1 the equation

$$Ma^1 = -(1-a^0 \otimes \overline{a^0})\ Na^0\alpha$$
$$= -QNa^0\alpha\ . \qquad (39)$$

In the subspace defined by Q,M can be inverted and we find

$$a = a^0 - \lambda M^{-1}QNa^0. \qquad (40)$$

At this stage we, however, have to preserve the normalization condition (33) to the consistent order and hence multiply (40) by a factor

$$C = 1+c\lambda, \qquad (41)$$

where c is determined to give

$$(\overline{a^0}, a) = C[(\overline{a^0}, a^0) - \lambda(\overline{a^0}, M^{-1}QNa^0)]$$
$$= 1+\lambda(c-(\overline{a^0}, M^{-1}QNa^0)) = 1 \qquad (42)$$

giving

$$c = (\overline{a^0}, M^{-1}QNa^0)\ . \qquad (43)$$

We find

$$a = [1+\lambda(\overline{a^0}, M^{-1}QNa^0)]\ [a^0-\lambda M^{-1}QNa^0]$$
$$= a^0 - \lambda[(1-a^0 x \overline{a^0})\ M^{-1}QNa^0]$$
$$= a^0 - \lambda QM^{-1}QNa^0. \qquad (44)$$

This shows that we have to take the unique solution of (39) which exists in the subspace defined by Q, i.e. we must identify $QM^{-1}Q$ with $(QMQ)^{-1}$, and the component a^1 has to be chosen such that $Pa^1 = 0$. These considerations are obvious but now introducing (44) into (35) we find to second order in λ the equation

$$\frac{d\alpha}{dt} = \lambda(\overline{a^0}, Na^0) + \lambda^2\left[(\overline{a^0}, Ka^0) - (\overline{a^0}, NQM^{-1}QNa^0)\right]\alpha, \tag{45}$$

which is exactly the result (30) of the previous Section. In the operator $OM^{-1}Q$ the first Q comes from the normalization condition and the second one from the slowly varying component of the time derivative. Because M^{-1} is singular the present approach has to be applied with great care, the zero eigenvalue is shifted to order λ and this must be included consistently in the time derivatives.

The procedure outlined above is a version of the Chapman-Enskog perturbation theory used in transport problems, see Tituleer (1980a). The approach by Minogin (1980 and 1981) is based on this expansion method. It has been of wide use in statistical physics, and can be applied to problems where the time derivative contributes to the small terms. It can be formally presented as a "multiple time scale" procedure.

5. PHYSICAL APPLICATIONS

In Sec. 2 we introduced the basic physical ideas of light pressure cooling. The formalism developed in Secs. 3 and 4 applies directly to this case when the 2x2 density matrix is taken as the vector A, the semiclassical strong field problem is taken as the operator M, and the spontaneous emission terms like those in Eq. (16) are taken as the perturbation. In fact, the recent work Stenholm (1983a) follows exactly these lines.

The formal expansion parameter λ in (19) is an operator. The rate of change of the momentum variable is given by the Doppler detuning in the Lorentzian depending on $(\omega\text{-}qv)$ and hence

$$\lambda \sim \hbar q \frac{\partial}{\partial p} \sim \frac{\hbar q^2}{M\Gamma} \sim \frac{\varepsilon}{\Gamma}, \tag{45}$$

where ε is the photon recoil energy. This dimensionless expansion parameter was used by Javanainen and Stenholm (1980a) and Letokhov and Minogin (1981).

The derivation of Fokker-Planck equations from the Wigner function was initiated by Kazantsev (1974) and discussed recently by Javanainen and Stenholm (1980b) and Cook (1980a).

The previous example concerned spontaneous emission recoil in resonance fluorenscence. Situated in a strong field an atom may also spontaneously start emitting at another frequency. Especially an atom in a cavity can be forced to emit into a narrow eigenmode; the operation becomes unstable. This case has been recently investigated by Zubairy et al. (1983 and work to be pubsished). It is also possible to cast their problem into the present formalism. The semiclassical problem forms the unperturbed part, and the other quantized modes are treated as a perturbation. For the reduced density operator in the space of the weak modes an operator equation of motion can be derived directly. The details of this will be published elsewhere.

Another case of interest emerges in an operating laser. For strong fields the semiclassical description suffices (see Sargent et al. 1974); but the quantum nature of the light can be seen as unavoidable fluctuations. They determine the ultimate line width possible.

The quantum fluctuations can here be treated as a perturbation if we use the coherent states $|z\rangle$ of Glauber (1963). The photon creation operator $b^{\dagger}$ acting on these states gives

$$b^{\dagger}|z\rangle\langle z| = (z^{*} + \frac{\partial}{\partial z})\ |z\rangle\langle z| \ . \tag{46}$$

When the derivative terms are omitted, the semiclassical problem emerges. If they are included and treated as a perturbation, we find a situation very similar to the one in laser cooling. To lowest order the probability function $P(z) = \mathrm{Tr}_{atom}\ \rho(z)$ is conserved, but an adiabatic elimination of the internal atom variables leads to a Fokker-Planck equation for $P(z)$.

The approach discussed was pioneered by Kazantsev and Surdutovich (1969) and discussed by Stenholm (1973).

It provides a convenient alternative to conventional theories, and it fits the present framework well.

6. CONCLUSIONS

We have found, that we can often formulate a problem so that the main part is a classically understood one and the quantum effects are perturbative fringes around this. When the classical problem contains some conserved quantity (subspace) there are close connections between the singular nature of the perturbation expansion and interesting physical effects like the anomalous diffusion and the antibunching.

Correctly formulated in terms of a well defined expansion parameter the problem contains no dilemma of the Ito-Stratonovich type. The consistent expansion of momentum gives e.g. for the light pressure case

$$\langle p \rangle \sim Ft \propto \lambda \tag{47}$$

$$\langle \Delta p^2 \rangle \sim Dt \propto \lambda^2 \ . \tag{48}$$

If we write the momentum as

$$\begin{aligned} \langle (p \pm \Delta p)^2 \rangle &= \langle p \rangle^2 + \langle \Delta p^2 \rangle \\ &= F^2 t^2 + Dt \ , \end{aligned} \tag{49}$$

we find that this is consistent to order λ^2; the addition of terms of order λ^2 to the force F would give inconsistent terms of order λ^3. If we add such terms to the Fokker-Planck expansion we have third derivative terms, which are known to be inconsistent with the probability interpretation of the theory (Pawula, 1967). Physically we can say that no quantum corrections are allowed in the expression for the classical force.

The theory is found to relate closely to several perturbative schemes used in singular perturbation problems:

- degenerate perturbation theory
- Chapman-Enskog methods
- multiple-time scale perturbations.

As the example of laser cooling shows us, it is not clear that the physical implications of these relations have been fully grasped yet.

REFERENCES

Bambini, A., and Stenholm, S., 1979, Quantum description of free-electrons in the laser, Opt. Comm., 30:391.

Bloch, C., 1958, Sur la théorie des perturbations des états lies, Nuclear Phys., 6:329.

Bogoliubov, N.N., 1967, Part 4.5, Perturbation theory for a degenerate level, in: "Lectures on Quantum Statistics, Vol. 1", Gordon and Breach, New York.

Carmichael, H.J., and Walls, D.F., 1976a, Proposal for the measurement of the resonant Stark effect by photon correlation techniques, J. Phys. B9:L43.

Carmichael, H.J. and Walls, D.F., 1976b. A quantum-mechanical master equation treatment of the dynamic Stark effect, J. Phys., B9:1199.

Cohen-Tannoudji, C., 1977, Atoms in strong resonant fields, in: "Frontiers in Laser Spectroscopy", R. Bailian, S. Haroche and S. Liberman eds., North Holland, Amsterdam.

Cook, R.J., 1980a, Theory of resonance radiation pressure, Phys. Rev., A22:1078.

Cook, R.J., 1980b, Photon statistics in resonance fluorescence from laser deflection of an atomic beam, Opt. Comm., 35:347.

Dohm, V., 1976, Exact steady-state solution of the quantum-mechanical single-mode laser model, Phys.Rev., A14:393.

Glauber, R.J., 1963, Coherent and incoheret states of the radiation field, Phys. Rev., 131:2766.

Gordon, J.P., and Ashkin, A., 1980, Motion of atoms in a radiation trap, Phys. Rev., A21:1606.

Haake, F., and Lewenstein, M., 1983, Adibatic expansion for the single-mode laser, Phys. Rev., A27:1013.

Haken, H., 1975, Cooperative phenomena in systems far from thermal equilibrium and nonphysical systems, Rev. Mod. Phys., 47:67.

Haken H., 1977, Chapter 7, Self-organization, in: "Synergetics", Springer-Verleg, Heidelberg.

Javanainen, J., and Stenholm, S., 1980a, Broad band resonant light pressure I: Basic equations, Appl. Phys., 21:35.

Javanainen, J., and Stenholm, S., 1980b, Laser cooling of trapped particles I: The heavy particle limit, Appl. Phys. 21:283.

Javanainen, J., and Stenholm, S., 1981, Laser cooling of trapped particles II: The fast particle limit, Appl. Phys., 24:71.

Kazantsev, A.P., 1974, Recoil effect in a strong resonant field, Sov.Phys. JETP, 40:825.

Kazantsev, A.P., 1978, Resonant light pressure, Sov. Phys. Uspehki, 21:58.

Kazantsev, A.P., and Surdutovich, G.I., 1969, The quantum theory of the laser, Sov. Phys. JETP, 29:1075.

Letokhov, V.S., and Minogin, V.G., 1981, Laser radiation pressure on free atoms, Phys. Reps., 73:1.

Mandel, L., 1979, Sub-Poissonian photon statistics in resonance fluorescence, Opt. Lett., 4:205.

Minogin, V.G., 1980, Kinetic equation for atoms interacting with laser radiation, Sov. Phys. JETP, 52:1032.

Minogin, V.G., 1981, Kinetic theory of the scattering of atoms by a resonant standing light wave, Sov.Phys. JETP, 53:1164.

Nayfeh, H., 1973, "Perturbation Methods", J. Wiley, New York.

Pawula, R.F., 1967, Approximation of the linear Boltzmann equation by the Fokker-Planck equation, Phys.Rev., 162:186.

Sargent III, M., Scully, M.O., and Lamb, Jr., W.E., 1974, "Laser Physics", Addison Wesley, New York.

Short, R., and Mandel, L., 1983, Observation of sub-Poissonian photon statistics, Phys. Rev.Lett., 51:384.

Stenholm, S., 1973, Quantum theory of electromagnetic fields interacting with atoms and molecules, Phys. Reps., 6:1.

Stenholm, S., 1983a, Distribution of photons and atomic momentum in resonance fluorescence, Phys. Rev., A27:2513.

Stenholm, S., 1983b, Physical applications of photon momentum, Invited talk at SICOLS '83, Interlaken, Switzerland.

Stenholm, S., and Bambini, A., 1981, Single-particle theory of the free-electron laser in a moving frame, IEEE J.Q. Electronics, QE-17: 1363.

Titulaer, U.M., 1978, A systematic solution procedure for the Fokker-Planck equation of a Brownian particle in the high-friction case, Physica, 91A:321.

Titulaer, U.M., 1980a, The Chapman-Enskog procedure as a form of degenerate perturbation theory, Physica, 100A:234.

Titulaer, U.M., 1980b, Corrections to the Smoluchowski equation in the presence of hydrodynamic interactions, Physica, 100A:251.

Zubairy, M.S., Sargent III, M., and De Martini, F., 1983, Quantum theory of laser and optical-bistability instabilities, Opt.Lett., 8:76.

RADIATION INTERACTION OF RYDBERG ATOMS

Herbert Walther

Sektion Physik, Universität München and
Max-Planck-Institut für Quantenoptik
D-8046 Garching, Fed. Rep. Germany

INTRODUCTION - PROPERIES OF RYDBERG ATOMS

When a valence electron of an atom is excited into an orbit with sufficiently high principal quantum number n and therefore far from the ionic core, the properties of the atom appear hydrogenic. The energy of these highly excited levels is given by the Rydberg formula, and so the states are also called Rydberg states. The phenomenological quantum defect δ_ℓ depends on the angular momentum ℓ. For states of low ℓ, where the orbits of the classical Bohr-Sommerfeld theory are ellipses of high eccentricity, the penetration and polarization of the electron core by the valence electron lead to large quantum defects and strong departures from the hydrogenic behaviour. As ℓ increases, the orbits become more circular and the atom becomes more hydrogenic, δ_ℓ changing with ℓ^{-5}.

The energy changes among highly excited states are small compared with the large changes between the lower levels. Since smooth changes are characteristic of classical systems (in which energy changes are continuous), Rydberg atoms can be expected to show classical properties. In particular, according to Bohr's correspondence principle, the frequency of electromagnetic radiation emitted for transitions between neighbouring states approaches the frequency at which the electron circulates around the ionic core. This suggests that many properties of these atoms can be understood in simple classical terms. Nevertheless, some very surprising properties of Rydberg atoms have recently been found, which has led to a steady increase in the number of experiments being performed on these atoms. The interest in Rydberg states is manifold:

a) The outer electron is a very good probe for the interatomic potential; quantum defects due to the penetration and polarization of the electron core are therefore being investigated as well as fine-structure and hyperfine-structure splittings.

b) Radiation interaction is large and different from that of ground state atoms owing to the large matrix elements for transitions to neighbouring levels; radiation-induced effects overcome spontaneous emission. Therefore Rydberg atoms in high-n states become sensitive to black-body radiation, and maser emission with only a small absolute number of radiators can also be observed. Observation and study of these effects allows testing of fundamental theories on light-matter interaction, which is not possible with ordinary atoms.

c) Collisional interaction becomes very important owing to the size of the atoms; its influence shows strong dependence on the main quantum numbers. It is thus found, for example, for the collisional angular momentum mixing in the low-n region that the cross-section increase is proportional to the geometric size of the atoms, i.e. n^4. As the size of the Rydberg orbit increases further, the electron distribution becomes very diffuse and the cross-section decreases.

d) The binding energy of the outer electron of a Rydberg atom is very small. Therefore external electric and magnetic fields show a very large influence even at small field strengths. The observation of similar effects for ground state atoms would require fields which are not attainable in the laboratory.

Table I: Scaling laws for properties of Rydberg atoms

Energy:	$E_n = R/(n-\delta_\ell)^2 = R/n^{*2}$
	n^* effective quantum number, δ_ℓ quantum defect
	R Rydberg constant
Radius:	$\langle r\rangle \sim n^{*2}$
Lifetimes:	$\tau \sim n^{*3}$ (low angular momentum states)
	$\tau \sim n^{*5}$ (high angular momentum states)
Fine-structure interval:	$\Delta E \sim 1/n^{*3}$

In Table I the scaling laws for the properties of Rydberg atoms are compiled: The radius of the charge distribution of the valence electron scales as n^{*2} and for $n^* = 50$ the linear dimension of the atom is already comparable with the wavelength of light in the visible region and competes with the size of larger biomolecules.

The electric polarizability for the quadratic Stark effect increases as n^{*7} and the diamagnetic interaction as n^{*4}. This allows one to perform experiments at field strengths high enough to make the interaction energy in the external electric or magnetic field comparable with or larger than the Coulomb energy of the atom. For practical reasons the corresponding field strengths for ground state atoms cannot be reached in the laboratory. The study of highly excited atoms in external electric and magnetic fields is therefore interesting in itself. (For reviews see References 1-4.)

The sensitivity of Rydberg atoms to external electric fields also means that the atoms already ionize in rather weak fields. This opens the possibility of a very effective detection, as will be discussed later.

The large Rydberg atom orbitals are characterized by natural lifetimes much longer than the ones of less excited atoms. In the case of hydrogen Rydberg states, the dependence of the lifetime on n can be obtained by fully quantum mechanical radiation rate calculations involving hydrogenic coulombic wavefunctions. For Rydberg states of other species the lifetimes (and the other radiative parameters) scale not exactly as a power of n but rather as a power of n^*. The n^* scaling law can be determined using Bates and Damgaard type of calculations[5]. The lifetimes scale either with n^{*3} (when ℓ is small) or with n^{*5} (when $\ell \cong n$).

In the following we will give a simple classical picture for this scaling law. (In this discussion we will not discriminate between n and n^*.) The rate of spontaneous emission of radiation for a transition from a state n to n' is given by the Einstein A coefficient:

$$A_{n\to n'} = 16\pi^3\upsilon^3 \langle r_{nn'}\rangle^2/3\varepsilon_0 hc^3,$$

where υ is the transition frequency and $\langle r_{nn'}\rangle$ the matrix element of the electric dipole operator between the initial n and the final state n'. For the case $n' \ll n$ one has a small $\langle r_{nn'}\rangle$ owing to the small overlap of the radial wave functions for n and n' and, as will be shown below, $A_{n\to n'} \sim n^{-3}$. If n' is close to n, the energy difference $E_n - E_{n'} \sim n^{-3}$ and $\langle r_{nn'}\rangle^2 \sim n^4$, and so $A_{n\to n'}$ becomes proportional to n^{-5}. The magnitude of the Einstein coefficient $A_{n\to n'}$ still depends on the angular momentum ℓ. This can be understood by simple classical arguments: for low angular momentum states (core penetration) the lifetime τ can be deduced from the third Kepler law. Accordingly, the electron period T is given by T $\cong (n^2 a_0)^{3/2} \cong n^3$ (in the classical picture T must be proportional to τ since a transition to a lower orbit is always more probable when the electron approaches the core and undergoes maximum acceleration). For the case of high angular momentum orbitals, in the classical picure, the electron radiates continuously and lowers its radius. The acceleration of the electron is inversely proportional to the square of the radius of the orbit, and so the power of the emitted radiation scales as n^{-8}. The distance between neighbouring Rydberg levels changes with n^{-3}; this gives a characteristic time requirement of $n^{-3}/n^{-8} = n^5$, for each step, corresponding to the scaling of the lifetimes of states with large ℓ.

The square of the matrix element $\langle r_{nn'} \rangle$ ($n \cong n'$) scales with n^4, showing a rather high transition probability for induced transitions. Rydberg atoms therefore strongly absorb microwave or far-infrared radiation. As a consequence, black-body radiation may cause strong mixing of the states. This is especially the case for states with high angular momenta since the spontaneous lifetimes for these are much larger and the induced transitions can therefore be saturated more easily than for the lower ℓ states.

We now wish to discuss also the scaling laws related to black-body-induced effects. The induced transition rate due to black-body radiation is proportional to $\langle r_{nn'} \rangle^2 S_\upsilon$, where S_υ is the energy flux of the black-body radiation per unit band width and unit surface area. At low frequencies (Rayleigh-Jeans limit) S_υ changes as υ^2. Considering the distance between the Rydberg states to scale as n^3 (here again we perform the discussion with n instead of n^*), it is therefore found that S_υ is proportional to n^6. Since $\langle r_{nn'} \rangle^2 \sim n^4$, it follows that the induced transition rate behaves as n^{-2}. Important in experiments is the ratio between the induced transition rate and the spontaneous rate, which changes as n^{-3} for low ℓ and as n^{-5} for high ℓ. This means that for a given atom and a given temperature there exists an n, above which the black-body-induced rate overcomes the spontaneous rate.

The sensitivity of Rydberg atoms to black-body radiation can also be explained in the following terms. The black-body radiation energy density can be expressed in terms of the average number of photons per mode $\bar{n}$. For the Rayleigh-Jeans limit this gives $\bar{n} = kT/h\upsilon$. At 300 K it follows that $kT/h \cong 6 \times 10^{12}$ Hz. For the vacuum fluctuations, which lead to spontaneous emission, $\bar{n} = 1/2$; this means that for frequencies larger than kT/h, where $\bar{n} \ll 1$, no significant black-body influence can be observed. However, for a Rydberg state with a transition frequency to a neighbouring state at 10^{11} Hz, where $\bar{n} = 60$, the black-body-induced transition rates can be orders of magnitude larger than the spontaneous rates.

In addition to population changes induced by the black-body radiation, energy shifts of the atomic levels also occur. Their magnitudes depend on the match of the atomic frequencies with the black-body frequencies and the strength of the coupling of the Rydberg atom to the black-body radiation[14,15,16].

In absorption spectroscopy with classical light sources only those Rydberg states could be investigated which can be optically excited directly from the ground state. For spectra of atoms with one valence electron this means that only the ^2P series can be studied. The alternative method of populating Rydberg states in an electric discharge and observing the spectrally resolved fluorescence is not practical: At the required particle densities collisional deactivitation is much more probable than radiative decay, since it is a result of the large collisional cross-sections of Rydberg states and of the long lifetimes. However, in atomic beam experiments where one can reach collision-free conditions it is possible to use electron bombardment or charge exchange collisions to populate Rydberg states. One drawback of this excitation process, however, is that it is not state-selective.

Most of the limitations discussed above were overcome after the invention of frequency-tunable lasers, little more than a decade ago. This lead to a renaissance of the spectroscopy of highly excited atomic states. The use of lasers to populate high-lying atomic levels in one, two or three excitation steps considerably increased the number of atomic states accessible to experiments. In particular, states with the same parity as the ground state could be reached. For atoms with large ionization potentials it can be advantageous to combine the collisional excitation of metastable states and subsequent laser excitation to Rydberg states[11,17,18].

For high-resolution spectroscopy of Rydberg atoms a low atomic density is also required in order to avoid collisional broadening or a collisional shift. This excludes absorption measurements from the onset since there larger densities are required. The alternative method of observing the fluorescence, however, is not suitable either ($n \geqq 15$) since the n^3 dependence of the radiative lifetime implies a corresponding decrease of the fluorescence intensity. Most experiments, therefore, exploit collisional, photo or field ionization to detect Rydberg atoms.

Field ionization was first observed in hydrogen, where about 10^6 V/cm has to be applied to ionize the $n = 4$ states. In the range of $n = 30$ an electric field of about 300 V/cm is enough to reach the onset of field ionization. The superposition of the atomic Coulomb potential and the linear slope potential of the externaly applied field results in a potential structure having a saddle point. The simplest approach to field ionization is to say that states above the saddle point fully ionize and states below are stable. For a state with principal quantum number n there exists a critical field defined by the onset of field ionization. Using the simple potential picture, one obtains[1]

$$E_{crit} \sim 1/n^4$$

As long as spectroscopic information about the unperturbed Rydberg atom is wanted, the excitation of the Rydberg state and field ionization have to be separated in time. If an electric field ramp is applied, after a time lag with respect to pulsed laser excitation, Rydberg atoms in different n-states will ionize at different electric fields, i.e. at different times, and can thus be discriminated.

INFLUENCE OF BLACK-BODY RADIATION ON RYDBERG ATOMS

The influence of black-body radiation on Rydberg atoms was first demonstrated in lifetime measurements. For instance, Gallagher et al.[6] observed that the measured lifetimes of the 16p and 17p states of Na are three times shorter than expected; the shorter lifetime was supposed to be due to black-body interaction. Haroche et al.[7] found a population transfer to nearby levels which could not be explained by spontaneous decay. More direct evidence of interaction with black-body radiation was observed later[8,9,10].

The influence of black-body radiation is demonstrated in Fig. 1. For this measurement the 5s23f state of the Sr atom was excited by a pulsed dye laser[11]. The Rydberg atoms were detected by field ionization. For this purpose a field ramp was used so that the different Rydberg states were successively ionized starting with the levels closest to the ionization limit[4]. The field ramp was started for the first measurement 1μs after the laser excitation. The measurements performed with larger time delays 2,6 and 12 μs clearly show the increasing population change due to the strong interaction with black-body radiation (see also Ref. [8]).

A consequence of the long radiative decay time of Rydberg levels and the very large value of the electric dipole matrix elements is that the saturating power for transitions between closely lying Rydberg levels is very small. The corresponding saturating power fluxes are proportional to n^{-10} for low and to n^{-14} for high angular momentum states. A very vivid way of describing the behaviour is to express the saturation power flux in terms of number of photons per surface of the size λ^2 and per lifetime, (the size λ^2 corresponds to the resonant cross-section). For $n \cong 30$ one obtains 10^2 and 1 for low and high angular momentum states, respectively. This means that for high angular momentum

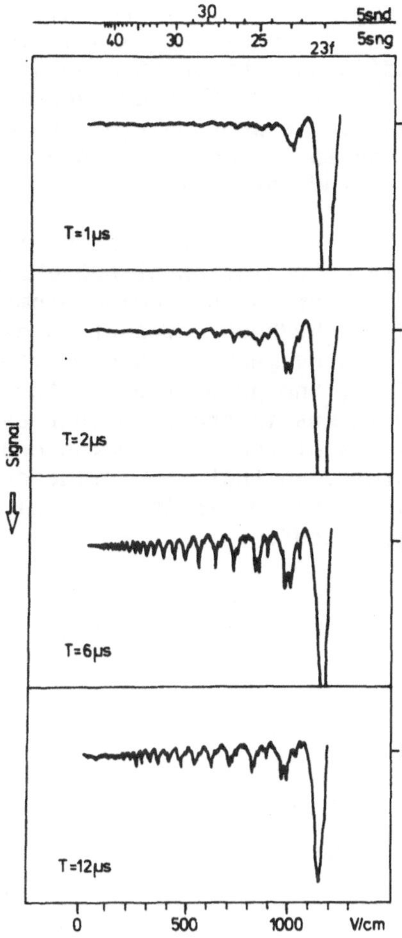

Fig. 1. Black-body induced transitions between Rydberg states.

states a single photon is required (in the chosen units) to saturate the transition to a neighbouring Rydberg level[12].

There are many applications of wide ranging importance for detectors in the submillimeter region, e.g. infrared and radio-astronomy, diagnostics of plasmas for nuclear fusion, stratospheric monitoring and materials research. The investigation of new principles for detectors is therefore as important as the further development and improvement of known detector principles. The ultimate sensitivity obtainable for detection of any radiation is of course reached when single photons can be monitored with a high probability and when the noise of the signal is only determined by the quantum noise of the radiation. The quantum noise limit of the radiation can be reached in the visible or near infrared spectral range since there the available photomultipliers allow the photocurrent of a single photon to be amplified to a value exceeding the noise of the dark current and the amplifier.

Ducas et al.[13] demonstrated the very sensitive detection of low-power far infrared laser radiation at 600 GHz by inducing transitions between Rydberg states of sodium atoms. To check the ultimate sensitivity of the Rydberg detector to microwave or far infrared radiation, two improvements of the previous experiments have to be effected. First, the population of the Rydberg atoms must be performed by cw lasers in order to increase the duty cycle and, in addition, the walls of the surrounding chamber have to be cooled to a low temperature, so that the influence of the thermal background radiation is minimized. In the following an experiment of this type will be described[9].

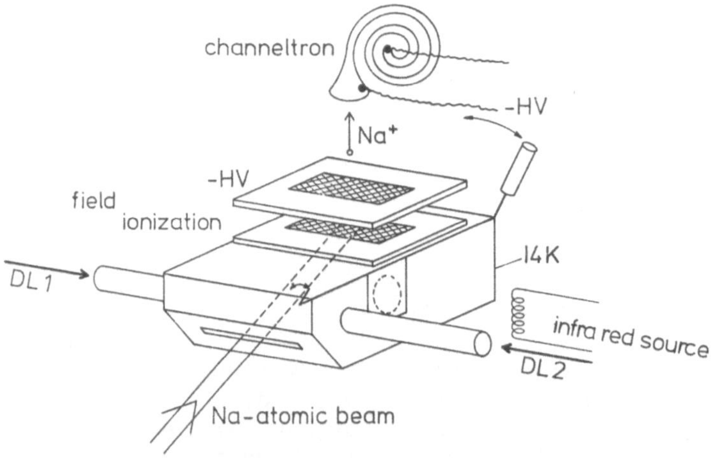

Fig. 2. Experimental set-up for the investigation of the interaction of black-body radiation with Rydberg atoms. The arrows marked by DL1 and DL2 indicate the dye laser beams for the first and second excitation step. For details see Reference 9.

Sodium atoms of an atomic beam were excited to high-lying ^2P-states in two steps via the $3^2P_{3/2}$ intermediate state. The $3^2S_{1/2}$, F = 2 → $3^2P_{3/2}$, F = 3 hyperfine transition (589 nm) is saturated with circularly polarized light by means of a single mode cw dye laser stabilized to this transition. A fraction of the atoms in the $3^2P_{3/2}$, F = 3 state is then excited to the n^2D state by means of a multimode dye laser whose cavity length is wobbled in order to obtain a more homogeneous intensity distribution over the laser line width of 100 GHz at a wavelength of about 4150 Å.

The interaction region is surrounded by a box cooled down to 14 K in order to keep the background of black-body radiation as low as possible (Fig. 2). The slit where the atomic beam enters the cooled box is covered with a wire mesh to reduce microwave and far infrared radiation emitted by the atomic beam oven. After a path of 20 mm the excited atoms leave the box through a second wire mesh which also acts as one plate of the capacitor used for field ionization. If an atom is field-ionized, the ion is accelerated and leaves the capacitor through the mesh in the negative plate and is detected by a channeltron multiplier.

A small flap is mounted at one side of the cooled box. By opening this flap, the Rydberg atoms can be exposed to the radiation of a heated wire. The microwave transitions induced in the Rydberg atoms by the thermal radiation of the wire are then monitored via field ionization. For the experiments described here the atoms were excited to the 22^2D state. The strength of the electric field was adjusted, so that atoms in the 22^2D state are not ionized. However, any transition induced to a higher state is detected through the ion signal.

By opening and closing the flap of the box the influence of the radiation from the heated wire could be investigated. Figure 3 shows the change of the ion count rate correlated with the opening and closing of the flap for different temperatures of the heated wire.

The lifetime of the 22^2D state is 10 μs. Only about 5 % of the initially excited Rydberg atoms therefore reach the field ionization region. The $\Delta n = 1$ transition to the 23^2P state is most likely to be induced. The 23^2P state has a lifetime of about 100 μs. All atoms excited by the microwave radiation therefore reach the field ionization region. The ratio between the rate of the detected ions and the rate of absorbed photons gives the quantum efficiency of the device of 3×10^{-3}.

As a result, the noise equivalent power (NEP) of the detector is 10^{-17} W/Hz$^{1/2}$, using an output bandwidth of 1 kHz. The NEP of this preliminary set-up favourably compares with the NEP of other

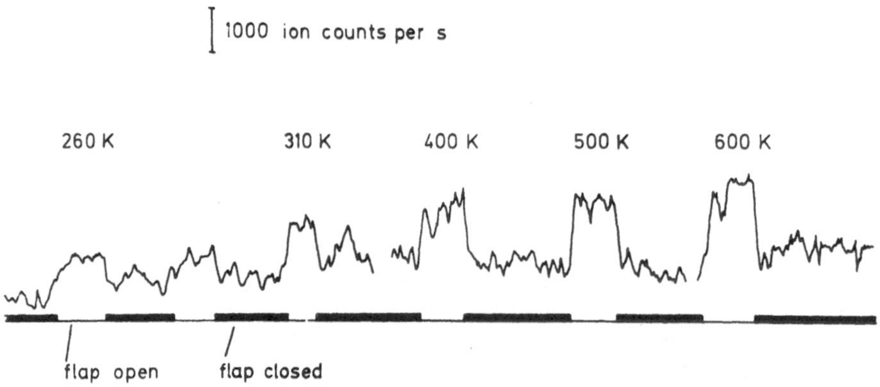

Fig. 3. Signal of the Rydberg detector. The background (not shown in the figure) is about five times the signal induced by the infrared source at 310 K and is due to the 14 K blackbody radiation emitted by the walls of the low temperature box.

detectors, which is at least one order of magnitude larger. The NEP was calculated by assuming a background noise which is equal to the signal at 310 K. The noise of the signal is not statistical, a major contribution comes from slow intensity and frequency fluctuations of the second laser indicating that a better NEP can be achieved by stabilizing the second laser.

Another considerable source of background signal is the blackbody radiation at 14 K, which is always present inside the cooled box. This background signal is about five times larger than the signal resulting from the radiation of the heated wire at a temperature of 300 K. If the box is cooled with liquid helium, this background can be reduced by a factor of six. With a stabilized single mode laser for the second excitation step it will thus be possible to obtain a NEP of 10^{-19} W/(Hz)$^{1/2}$. It has been demonstrated experimentally that the same value for the NEP can be obtained if coherent amplification of the microwave signal is performed in a Rydberg maser[38].

RADIATION INTERACTION OF RYDBERG ATOMS IN CAVITIES - A TEST SYSTEM FOR SIMPLE QUANTUM ELECTRODYNAMICAL EFFECTS

The invention of the maser has generated a great deal of interest in theoretical models describing the interaction of two-level atoms with a single mode of an electromagnetic field[19,20,21]. Although the first models treated a purely academic problem, modified versions were stimulated. These then led to an understanding of a major part of the experimentally observed phenomena, including the even larger variety of effects found after the laser was invented. In the experiments it was always necessary that large numbers of atoms and photons be present. This was due to the fact that small numbers of atoms could not be detected, and also that the interaction matrix elements between the radiation and atoms were small. A small number of photons in an experiment has the consequence that the atom-field evolution time indeed usually becomes much longer than other characteristic times such as the atomic relaxation, the atom-field interaction time, and the cavity mode damping time. The theories involving single electromagnetic modes and small photon occupation numbers are therefore not very realistic. They, however, predict a few very interesting and basic effects which are worthwhile studying experimentally. Among them are the following:

1. Modification of the spontaneous emission rate of a single atom in a resonant cavitiy.
2. Oscillatory energy exchange between a single atom and the cavity mode.
3. Disappearance and quantum revival of optical nutation induced on a single atom by a resonant field.

Rydberg atoms are very suitable for observing these effects for several reasons: They have a very strong coupling to the radiation field, as already mentioned; the transitions to neighbouring levels are in the region of millimetre waves, which allows one to build cavities with low-order modes that are reasonably large to ensure rather long interaction times; finally, the Rydberg atoms have long spontaneous emission times, and therefore only the interaction with the selected cavity mode is important and the coupling of the atoms to other cavity modes can be neglected. In the following several phenomena observable with Rydberg atoms are discussed in more detail.

Single Atom in Resonant Cavity - Modification of Spontaneous Emission Rates

The energy levels of the combined two-level atom and field system can be described in the dressed atom picture[22,44]. The lowest energy of the system is represented by $|g,0\rangle$ describing the atom in its ground state $|g\rangle$ with no photon in the cavity. The higher energy levels are separated by the energy of a photon. The states $|\pm n\rangle$

are a superposition of the states $|e,n\rangle$ (e stands for excited atomic state and n for the photon number) and $|g,n+1\rangle$ of the system without interaction between the cavity field and the atom. At resonances:

$$|\pm n\rangle = [|e,n\rangle \pm |g,n+1\rangle]/\sqrt{2}.$$

The energy separation between the levels $+n\rangle$ and $-n\rangle$ is $2h\,\Omega\sqrt{n+1}$; there is a small change proportional to $\sqrt{n}$ when the field strength is increased. The energy levels of the dressed atom taking the coupling with the field into account are shown in Fig. 4.

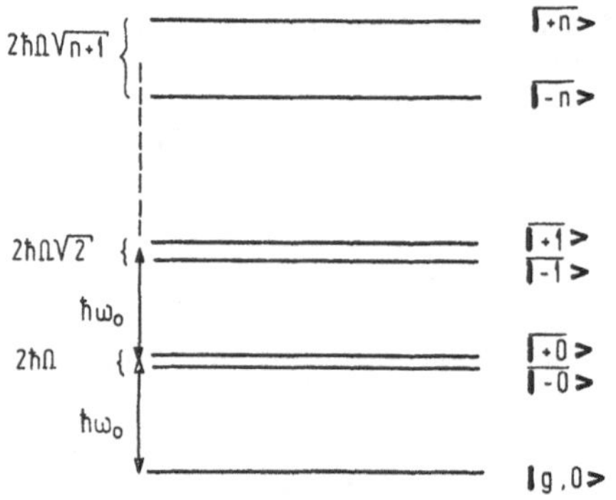

Fig. 4. Energy levels of a single two-level atom in the dressed atom description with resonant coupling to a cavity mode.

In a realistic description of the interaction the dissipative processes also have to be considered. Since Rydberg atoms have lifetimes longer than the atom-field interaction time, their relaxation can generally be neglected. However, the relaxation of the cavity field is important: the harmonic oscillator representing the field is coupled to a thermal reservoir at temperature T representing, for example, the cavity walls. The scheme shown in Fig. 5 gives the corresponding "coupling constants". The thermal equilibrium of the field mode is obtained in the characteristic time Q/ω, where Q is the quality factor of the cavity.

The behaviour of an atom entering an empty cavity (i.e., at

T = 0 K) in the excited state $|e\rangle$ depends on the relative size of Ω and ω/Q. If $\Omega > \omega/Q$ (small damping of the cavity), the probability of finding the atom in the state $|e\rangle$ undergoes a damped oscillation. This regime can be considered as a self-induced Rabi nutation in the field of the single photon emitted and reabsorbed by the atom. If $\Omega < \omega/Q$, the probability decreases exponentially at a rate $\Gamma_{cavity} = 4\,\Omega^2 Q/\omega$. There is a cavity-enhanced decay rate which is related

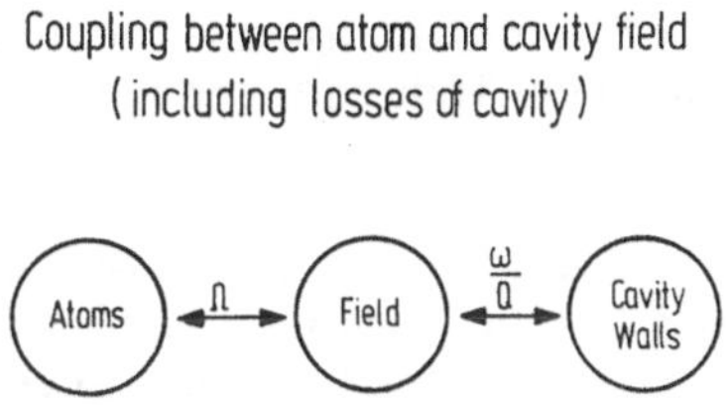

Fig. 5. Schematic description of the atom-single field mode system. The coupling is described by the one photon Rabi-frequency Ω and the characteristic damping time Q/ω of the cavity.

to the spontaneous rate in free space Γ_{spont} in the following way:

$$\Gamma_{cavity} = (3/4\pi^2) \cdot Q\lambda^3\, \Gamma_{spont} / V,$$

where V is the volume of the cavity and λ the wavelength of the radiation. This relation was predicted long ago by Purcell[23]. Physically, the cavity enhances the strength of the vacuum fluctuations at the resonance frequency; as a consequence the transition rate is increased. ($\Gamma_{cavity}/\Gamma_{spont}$ is obtained when the number of oscillator modes per unit frequency interval in a resonant cavity is divided by the corresponding value in free space.)

The opposite effect, the decrease of the decay rate, is obtained when the cavity is detuned. If the transition frequency of the atom lies below the fundamental frequency of the cavity, spontaneous emission is significantly inhibited. In an ideal case no mode is available for the photon and therefore spontaneous emission cannot occur[24].

To change the decay rate of an atom, in principle no resonator has to be present; any conducting surface near the radiator affects the mode density and, therefore, the radiation rate. Parallel-conducting planes can somewhat alter the emission rate but can only reduce the rate by a factor of 2 because of the existence of TEM modes which are independent of the separation.

To demonstrate experimentally the modification of the spontaneous decay rate, it is not necessary to go to single-atom densities in both cases. The experiments where the spontaneous emission is inhibited can also be performed with higher densities. However, in the opposite case, when the increase of the spontaneous rate is observed, a large number of excited atoms increases the field strength in the cavity and the induced transitions disturb the experiment.

The first experimental work on the inhibited spontaneous emission was done by Drexhage[25]. The fluorescence of a thin dye film near a mirror was investigated. Drexhage observed an alteration in the fluorescence lifetime arising from the interference of the molecular radiation with its surface image. An experiment with Rydberg atoms was recently performed by Vaidyanathan, Spencer and Kleppner[26]. They observed a wavelength-dependent cutoff in the absorption of black-body radiation by Rydberg atoms arising from a discontinuity in the density of modes between parallel-conducting plates. Absorption at a wavelength of 2/3 cm by atoms between planes 1/3 cm apart was measured at a temperature of 180 K. The discontinuity in the absorption rate occurred when the absorption wavelength was varied across the cutoff of the parallel-plate modes. The experiment was performed with Na atoms and the transition employed was 29d → 30p. For the tuning of the atomic resonance across the cutoff frequency a small electric field was applied to the parallel plates.

The first observation of enhanced atomic spontaneous emission in a resonant cavity was published by Goy, Raimond, Gross and Haroche[27]. Their experiment was performed with Rydberg atoms of Na excited in the 23S state in a niobium superconducting cavity resonant at 340 GHz. By taking advantage of the very strong electric dipole of these atoms and of the high Q value of the superconducting resonator cavity-tuning-dependent shortening of the lifetime was observed. This cooling, necessary for superconducting operation, also had the advantage of totally suppressing the black-body field effects ($n = 0$) required to test purely spontaneous emission effects in the cavity.

It was shown that the partial spontaneous emission probability

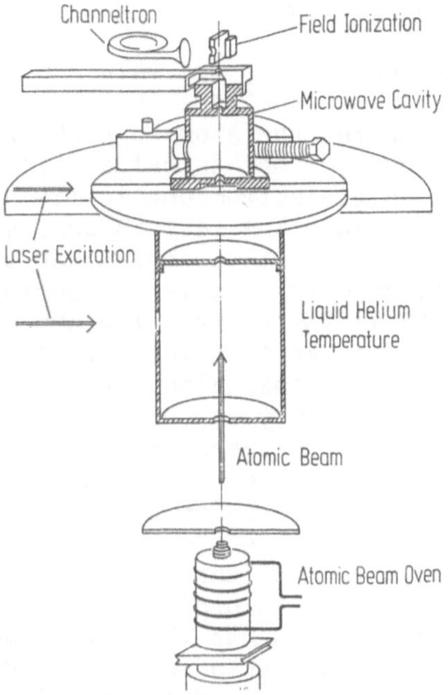

Fig. 6. Experimental set-up for the observation of single atom - single photon interaction. The Rydberg atoms are prepared in a specific velocity subgroup by a modulated stepwise excitation either with two laser beams or with modulated laser and microwave fields (the figure shows the position of two laser beams). The length of the cylindrical microwave cavity is about 20 mm. The parts are cooled to 2 K.

on the 23S → 22P transition in Na is increased from its free space value $\Gamma_{spont} = 150\ s^{-1}$ up to $\Gamma_{cavity} = 8 \times 10^4\ s^{-1}$. This enhanced rate is still 35 times smaller than the damping rate $\omega/Q = 2.8 \cdot 10^6 s^{-1}$

of the field in the cavity. This means that the photon emitted in the mode is absorbed in the mirrors much faster than the atoms decay. The atoms in the Rydberg states were detected by applying an electric field increasing in time for ionization. The two adjacent states were therefore subsequently ionized. The average number of atoms in the cavity was as low as 1.3.

With a tenfold increase in Q, the values of Γ_{cavity} and ω/Q would be of the same size, so that the emitted photon would be stored in the cavity long enough for the atom to reabsorb it. This would approach the regime of quantum mechanical oscillations between a two-level atom and a single electromagnetic field mode mentioned at the beginning of this section. The self-induced single-photon Rabi nutation is much more difficult to observe than the collective Rabi oscillation (which will be described later) because it occurs at a rate $\sqrt{N}$ times smaller (N is the number of atoms in the cavity) and thus requires the atom to be kept in the cavity for much longer times. Experiments to observe this single-atom -single-photon interaction are of present underway at ENS in Paris and in our laboratory. The set-up used in our laboratory (Meschede[28]) is shown in Fig. 6. The superconducting niobium cavity has a Q value of 5×10^9. In this experiment the atoms are velocity-selected during the stepwise excitation into the Rydberg states by using the modulated radiation of two lasers. Instead of the second laser a modulated microwave field can also be used in order to prepare a selected velocity subgroup. The latter method has the advantage that the first excitation step can already be performed into a Rydberg state with high main quantum number and long radiative lifetime. This reduces the losses due to spontaneous decay between the two excitation regions.

Experiments allowing observation of the single-atom - single-photon interaction are especially interesting in connection with the effects described in the next section.

Single Atom in Resonant Cavity - Disapearance and Revival of Optical Nutation

At temperatures T > 0 K the cavity also contains thermal photons. The effects described above therefore become more complicated since the atom evolves through an oscillatory or irreversibly damped transient regime towards a final state distribution corresponding to the thermal equilibrium. The transient behaviour is again dependent on whether there is weak ($\Omega \gg \omega/Q$) or strong damping ($\Omega \ll \omega/Q$) of the cavity. In the first case the transient regime can be described by a sum of elementary Rabi oscillations in a field in which the number of photons is a random quantity following the Bose-Einstein statistics. The distribution of Rabi frequencies results in an apparently random oscillation which for large n values very quickly collapses and then revives again (Faist et al.[29]; Meystre et al.[30]; Eberly et al.[31]; Knight and Radmore, et al.[32]. This behaviour is typical of a chaotic quantum field; a semi-classical description of a random Gaussian field does not give this result. This interesting phenomenon was always thought of as being incapable of experimental observation. However, the possibilities now opened up by Rydberg atoms bring us close to its realization. Superconducting cavities with Q values in the range between 10^9 and 10^{10} can be realized and, therefore, it should be possible to keep the damping small enough so that the oscillations are not washed out before their revival occurs.

N Atoms in Resonant Cavity - Collective Behaviour

The generalization of the single-atom effects described in the previous sections to N two-level atoms can be based on the ladder of equidistant non-degenerate states, the so-called symmetrical Dicke states[43]. Such states, where J + M atoms are excited in level $|e\rangle$ and J - M in level $|g\rangle$, are written formally

$$|JM\rangle = S\,|\underbrace{e,e,\ \ldots\ e}_{J+M};\underbrace{g,g,\ \ldots\ g}_{J-M}\rangle$$

where S is the symmetrization operator. (M = J and M = -J correspond to the totally excited and de-excited states, respectively.) The analysis of the atomic system by Dicke states is related to the atomic indiscernibility with respect to the single mode of the cavity. The N + 1 states $|JM\rangle$ describe situations in which strong correlations exist between the dipoles of different atoms resulting in a collective behaviour of the atoms in the cavity. Again the strong atom-to-field coupling of the Rydberg atoms is a big advantage, so that the experimental verification of the phenomena is much simpler than for "ordinary" atoms. The effects observed are cooperative features which cannot be interpreted in terms of an independent atom model: collective oscillations and superradiance, when the system is initially in the upper level, and collective absorption, when the system starts in the lower level.

For discussion of the phenomena two cases have again to be considered depending upon whether the Rabi frequency $\sqrt{N}\,\Omega$ is larger or smaller than the reciprocal of the cavity damping time ω/Q. In the case without black-body photons (T = 0 K), and $\omega/Q = 0$, and with all the atoms in the excited state, the spontaneous emission causes the atomic system to cascade through the ladder of eigenstates. The field strength in the cavity is increased and the photons are reabsorbed. The subsequent oscillations can be interpreted as a Rabi nutation in the field radiated by the atoms and stored in the cavity (Bonifacio and Preparata[33]; Scharf[34]). The oscillations show a rather complicated beating pattern for small N values[35].

For larger N values, the number of states to keep track of becomes prohibitive. Fortunately, the system can then be described in a classical way by using the concept of Bloch vector (see, for example, Allen and Eberly[20]; for the relation between the quantum mechanical and Bloch vector approaches see, for example, Bonifacio et al.[36]).

In the case of strong cavity damping the energy decays with a rate $T_R^{-1} = 4\,\Omega^2 NQ/\omega$ (Bonifacio and Preparata[33]). The value corresponds to $N \cdot \Gamma_{cavity}$ where Γ_{cavity} is the single-atom cavity-enhanced decay rate as discussed in the previous section.

The experimental observation of the above-mentioned effects was performed by Haroche and co-workers with an atomic beam of alkalis excited by pulsed lasers in the Rydberg states. Either the upper or the lower level of a millimetre-wave transition was in resonance with a mode of a cavity surrounding the atoms. The relatively long wavelength of the transitions allows all atoms to be excited in a region of constant field amplitude.

The Rydberg atoms are monitored by field ionization after the atoms have passed the cavity. In this way the number of atoms in the upper or the lower level of the microwave transitions was measured. In order to reconstruct the atomic evolution during the time the atoms spend in the cavity a small electrode producing an inhomogeneous electric field at a preset time t was inserted into the cavity. The Stark shift produced by this field suddenly brings the atoms out of the cavity resonance. A scheme of the experimental set-up is shown in Fig. 7. Therefore the atom-cavity coupling is interrupted and the detector measures the state the atom had at time t. By varying t, the dynamics of the atom-cavity interaction can be reconstructed.

Actual experiments were performed with cavities at T = 300 K whereas the theories deal with systems at T = 0 K. In fact, it can be shown that, as long as N is larger than the number of black-body photons in the cavity, the thermal-field contributions rapidly become negligible (Raimond et al.[37]). Black-body effects are relevant only at the onset of the emission, when the emitted field is still much smaller than the thermal one. With respect to fluctuations there is no difference since thermal and vacuum fields have the same statistical nature.

The experiments in a moderate Q cavity, typically $Q \approx 10^4$ (Moi et al.[38], Raimond et al.[37]), give the predicted cavity-assisted overdamped superradiance. This superradiant Rydberg "maser" is characterized by an extremely low inversion density threshold (N ~

10^4 atoms). The inverted medium emits a short burst of radiation and decays within a few hundred ns to the lower state of the transition (In the experiments mostly $nS_{1/2} \rightarrow (n-1)\, P_{1/2,\,3/2}$ or $nS_{1/2} \rightarrow (n-2)\, P_{1/2,\,3/2}$, $n \approx 30$). This maser emission was also detected by using Schottky heterodyne receivers (Moi et al.[38,39]). The latter detection technique is of course considerably less sensitive than the one based on atomic field ionization, which actually allows one to count the atoms which have radiated inside the cavity during a given time interval, and hence the emitted photons. Such a photon-counting type of experiment is quite novel in this part of the radiation spectrum.

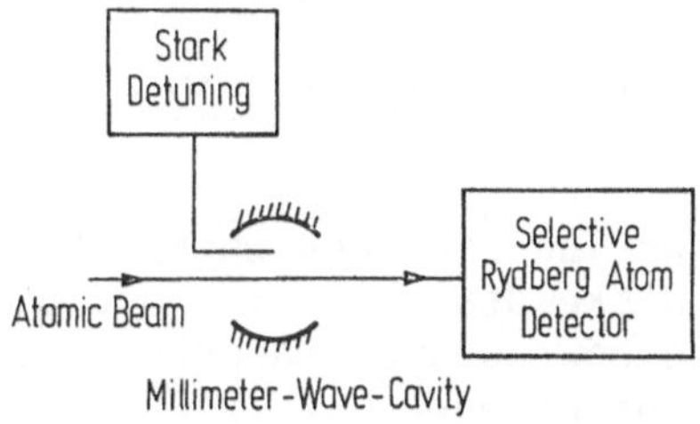

Fig. 7. Schematic of the Rydberg atom-cavity experimental set-up. The millimetre-wave cavity is made of copper spherical mirrors (for details and references see text).

Raimond et al.[37,40] succeeded in measuring the probability distribution $P(N,t)$ that N atoms have been deexcited at time t; the measured transition was $29\, S_{1/2} \rightarrow 28\, P_{1/2}$ at 162.4 GHz. At short times, $P(N,t)$ appears to obey a Bose-Einstein-type statistics, which is typical of a linearly amplified black-body field. At later times the amplification process becomes nonlinear and the distribution evolves into a broad bell-shaped Poisson-like function typical of a coherent process. These measurements represent the first direct and quantitative test of the theory in a system in which superradiance is not complicated by propagation and diffraction effects, as it normally is in the optical domain.

As discussed above, in the high-Q regime ($Q \approx 10^6$) one expects to observe an oscillatory exchange of energy between the atoms and the cavity field which can be described as a self-induced Rabi-nutation of the atomic system. The experimental observation was performed by Kaluzny et al.[41]. The transition investigated was 36 $S_{1/2} \rightarrow$ 35 $P_{1/2}$ of the Na atom. In order to remove the twofold degeneracy in the upper and lower levels and to study a true two-level atom transition, a small dc magnetic field was applied along the cavity axis and the cavity is tuned to resonance with 36 $S_{1/2}$, $m_J = + 1/2 \rightarrow$ 35 $P_{1/2}$, $m_J = 1/2$ transition at about 82 GHz.

The emission of the N atoms in the cavity occurs faster than it would in free space, essentially owing to the cavity enhancement effect. When N was sufficiently high ($N > 20\,000$), oscillations in the atomic population evolution become clearly observable. This collective self-nutation regime has also been discussed in the context of superradiance theories. It is then generally referred to as the "ringing" regime of superfluorescent emission. In the case of free-space superradiance this phenomenon has not yet been clearly observed since the simple Rabi nutation is then masked by multimode diffraction and propagation effects.

N Atoms in Resonant Cavity - Collective Absorption of Black-body Photons

In the previous section the case where the N atoms were initially in the excited Dicke state $|J, +J\rangle$ was discussed. In the following, the N atoms are now assumed to enter the cavity in the lowest state $|J, -J\rangle$; furthermore, it is assumed that the cavity field is in thermal equilibrium at a temperature $T \neq 0$ K.

The thermal photons represent a Bose-Einstein distribution with an average photon number $\bar{n} \neq 0$. As the time evolves, the atoms gain energy at the expense of the mode which is then supplemented by the thermal reservoir. The time constant for reaching thermal equilibrium depends on the values of N, $\bar{n}$ and ω/Q. Since the atomic energy diagram consists of non-degenerate equidistant levels with the same spacing as the field levels, the atoms will obviously reach an equilibrium described by a Boltzmann law quite similar to the Bose-Einstein distribution of the photon number in the field mode. (The only difference between the two distributions is that the number of levels for the atomic system is finite; this changes the normalization of the Boltzmann distribution.) As a consequence, the number of absorbed photons is limited to ΔN, no matter what value N has, and is equal to the average black-body photon number per mode (as soon as $N > \bar{n}$):

$$\Delta N = \bar{n} = [\exp(\hbar\omega/kT) - 1]^{-1},$$

which is close to $kT/\hbar\omega$ in the Rayleigh-Jeans limit.

The energy absorbed by N atoms in the cavity is not identical with the sum of energies that would be absorbed by N independent atoms. In this process the atomic sample evolves in a collective mode and behaves as a single quantum system exhibiting basic effects of Bose-Einstein statistics and Brownian motion (Raimond et al.[42]). A detailed study of the pulse-to-pulse random variations of ΔN around ΔN should allow one to probe the fluctuations of the cavity mode and to reconstruct their Bose-Einstein distribution. There is a connection between the absorption and emission of N atoms in a cavity: The atomic indiscernibility, which is responsible for superradiance when the system is initially excited, leads to a kind of "subabsorption" (Raimond et al.[42]) when it starts from its lower state.

The experimental demonstration of the effects just described was first performed by Raimond et al.[42]. The 30 $P_{1/2} \rightarrow 30\ S_{1/2}$ transition of Na was used in this experiment. The atoms were excited by a pulsed dye laser in a semifocal Fabry-Perot millimetre-wave cavity (~ 134 GHz) and detected by field ionization after the Na beam left the cavity. The cavity was coupled to a small, electrically heated tungsten wire whose temperature could be varied between 300 and 1600 K. For a resonant cavity an increase of the number of excited atoms is observed for small N values, but for $N \geq 5000$ ΔN reaches the limit expected from theory. The experiment demonstrated that, although the limiting value is independent of cavity characteristics, it changes with the radiation temperature. This could be demonstrated by increasing the temperature of the tungsten wire in the cavity. The experiment clearly demonstrated that the average number of excited atoms is exactly equal to the mean number $\bar{n}$ of black-body photons in the mode. The Rydberg atoms thus constitute an absolute radiation thermometer for the thermal field, the temperature being directly related to a particle number.

References

1. S. Feneuille, P. Jacquinot, Adv. Atom. Mol. Phys. 17:99 (1981).
2. D. Kleppner, in "Laser-Plasma Ineractions", Les Houches XXXVI (R. Balian, ed.) pp. 733-784, North-Holland, Amsterdam 1982.
3. D. Kleppner, M.G. Littman, M.L. Zimmerman, in "Rydberg States of Atoms and Molecules" (R.F. Stebbings and F.B. Dunning, eds.), pp. 73-116, Cambridge University Press, Cambridge 1983.
4. D. Delande, J.C. Gay, Comments At. Mol. Phys. 13:275 (1983).
5. D.R. Bates, A. Damgaard, Philos. Trans. Roy. Soc. London, 242: 101 (1949).
6. T. F. Gallagher, and W.E. Cooke, Phys. Rev. Lett. 42:835 (1979).
7. S. Haroche, C. Fabre, P. Goy, M. Gross, and J.M. Raymond, in: "Laser Spectroscopy IV", H. Walther, and K.W. Rothe, eds., Springer Series in Optical Sciences, Vol. 21, Springer Verlag, Berlin, Heidelberg, New York 1979.
8. E. J. Beiting, G.F. Hildebrandt, F.G. Kellert, G.W. Foltz, K.A. Smith, F.B. Dunning, and R.F. Stebbings, J. Chem. Phys. 70:3551 (1971).
9. H. Figger, G. Leuchs, R. Straubinger, H. Walther, Opt. Comm. 33:37 (1980).
10. P.R. Koch, H. Hieronymus, A.F.J. Van Raan, W. Raith, Physics Letters 75 A:273 (1980).
11. G. Rempe, Diplomarbeit, Ludwig-Maximilians Universität München, 1982.
12. S. Haroche, in: "Atomic Physics 7", D. Kleppner, and F.M. Pipkin, eds., Plenum Press, New York, London 1981, p. 141.
13. T.W. Ducas, W.P. Spencer, A.G. Vaidyanathan, W.H. Hamilton, and D. Kleppner, Appl. Phys. Lett. 35:382 (1979).
14. W.E. Cooke, T.F. Gallagher, Phys. Rev. A 21:588 (1980).
15. T.F. Gallagher, W. Sandner, K.A. Safinya, W.E. Cooke, Phys. Rev. A 23:2065 (1981).
16. J.W. Farley, W.H. Wing, Phys. Rev. A 23:2397 (1981).
17. R.F. Stebbings, C.J. Latimer, W.P. West, F.B. Dunning, T.B. Look, Phys. Rev. A 12:1453 (1975).
18. L. Barbier, R.J. Champeau, J. Phys. (Paris) 41:947 (1980).
19. E.T. Jaynes, F.W. Cummings , Proc. IEEE 51:89 (1963).
20. L. Allen, J.H. Eberly, "Optical Resonance and Two Level Atoms", Wiley, New York (1975).
21. P.L. Knight, P.W. Milonni, Phys. Rev. 66C:21 (1980).
22. S. Haroche, Ann. Phys. (Paris) 6:189 (1971).
23. E.M. Purcell, Phys. Rev. 69:681 (1946).
24. D. Kleppner, Phys. Rev. Lett. 47:233 (1981).
25. K.H. Drexhage, in "Progress in Optics" (E. Wolf, editor) Vol. 12, pp 165 North-Holland, Amsterdam (1974).
26. A. Vaidyanathan, W. Spencer, D. Kleppner, Phys. Rev. Lett. 47: 1592 (1981).

27. P. Goy, J.M. Raimond, M. Gross, S. Haroche, Phys. Rev. Lett. 50:1903 (1983).
28. D. Meschede, PhD thesis University of Munich, 1983
29. A. Faist, E. Geneux, P. Meystre, A. Quattropani, Helv. Phys. Acta 45:946 (1972).
30. P. Meystre, E. Geneux, A. Quattropani, A. Faist, Nuovo Cim. 25B:521 (1975).
31. J.H. Eberly, N.B. Narozhny, J.J. Sanchez-Mondragon, Phys. Rev. Lett. 44:1323 (1980).
32. P. Knight, P.M. Radmore, Phys. Rev. A 26:676 (1982).
33. R. Bonifacio, G. Preparata, Phys. Rev. A 2:336 (1970).
34. G. Scharf, Helv. Phys. Acta 43:806 (1970).
35. S. Haroche, in "New Trends in Atomic Physics", Proceedings of the Summer School Session XXXVIII, G. Grynberg, R. Stern, editors, North-Holland, Amsterdam 1982.
36. R. Bonifacio, D.M. Kim, M.O. Scully, Phys. Rev. 187:441 (1969).
37. J.M. Raimond, P. Goy, M. Gross, C. Fabre, S. Haroche, Phys. Rev. Lett. 49:1924 (1982).
38. L. Moi, P. Goy, M. Gross, J.M. Raimond, C. Fabre, S. Haroche, Phys. Rev. A 27:2043 and 2065 (1983).
39. L. Moi, C. Fabre, P. Goy, M. Gross, S. Haroche, P. Eucrenaz, G. Beaudin, B. Lazareff, Opt. Comm. 33:47 (1980).
40. J.M. Raimond, P. Goy, M. Gross, C. Fabre, S. Haroche, in "Laser Spectroscopy VI", H.P. Weber and W. Lüthy, editors, Springer Series in Optical Sciences, Vol. 40, pp 237 - 241, Springer, Berlin.
41. Y. Kaluzny, P. Goy, M. Gross, J.M. Raimond, S. Haroche, Phys. Rev. Lett. 51:1175 (1983).
42. J.M. Raimond, P. Goy, M. Gross, C. Fabre, S. Haroche, Phys. Rev. Lett. 49:117 (1982).
43. R.H. Dicke, Phys. Rev. 93:99 (1954).
44. S. Haroche, C. Fabre, J.M. Raimond, P. Goy, M. Gross, L. Moi, Joun. de Physique 43:C2 - 265 (1982).

27. T. Roy, [illegible] 50, 1807 (1983).
28. [illegible]
[illegible] 28, 546 (1972).
[illegible] (1975).
[illegible]

THE ROLE OF SCALAR PRODUCT AND WIGNER DISTRIBUTION IN OPTICAL AND QUANTUM MECHANICAL MEASUREMENTS

K. Wódkiewicz

Department of Physics and Astronomy
University of Rochester, Rochester, New York 14627, USA,
and
Institute of Theoretical Physics, Warsaw University
Warsaw 00-681, Poland.*

INTRODUCTION

The concept of phase-space measurement in Acoustics or Optics is much older than the now standard phase-space description of classical mechanics. A musical score offers perhaps the oldest and the simplest example of a phase space representation of an acoustical signal in the time and frequency domain.

The possibility and the convenience of a phase-space description of mechanical motions has been really appreciated in full only with the development of classical statistical mechanics where concepts like the phase-space distribution function or the Liouville theorem play a fundamental role.

With the advent of Quantum Mechanics, the basic notion of a probability amplitude has brought the Hilbert scalar product as an entirely new quantity into the physical description of dynamical systems. A phase-space description of Quantum Mechanics has been achieved with the introduction of the Wigner distribution functions.

On the other hand in Optics and Acoustics the notion of a spectrum for stationary or non-stationary signals has been, for a long time, the subject of very wide interest. After all, Fourier analysis, stationary and intrinsically non-stationary light fields are common notions to an electro-optical measurement.

---
* Permanent address.

In this paper we shall present a unified approach to the phase-space description of Optical and Quantum measurements. We shall see that from the operational point of view the notion of a time dependent spectrum of light and a joint measurement of position and momentum in Quantum Mechanics can be formulated in one common approach in which the scalar product, the Wigner function and the phase-space proximity are closely related to a realistic measuring process.

TIME DEPENDENT SPECTRUM OF LIGHT

The Time Dependent Spectrum of Light (TDSL) or the Physical Spectrum introduced and discussed in details in Ref. 1 (see also lectures of J.H. Eberly included in this volume) is defined by the following formula:

$$S(t,\omega) = \int dt_1 \int dt_2 J(t-t_1) e^{i\omega t_1} J^*(t-t_2) e^{-i\omega t_2} V^*(t_1) V(t_2). \quad (1)$$

In this definition, V(t) is a deterministic (we assume this for the simplicity of our arguments in this paper), complex amplitude of a spectrally analyzed electric field, measured by a detector with a setting frequency ω and with a causal filter response function J(t). For random signals V(t) an overall stochastic average of Eq.(1) should be additionally performed.

In order to obtain a scalar product form of Eq.(1) we introduce two operators, $\hat{T}$ and $\hat{\omega}$ defined by the following relations:

$$\hat{T}J(t) = tJ(t) \quad \text{and} \quad e^{t_o\hat{\omega}} J(t) = J(t+t_o) . \quad (2)$$

With the help of these definitions we can rewrite Eq.(1) for the TDSL in the following form:

$$S(t,\omega) = \left| \int dt_1 \; V^*(t_1) \; e^{i\omega(\hat{T}-t)} \; e^{-t\hat{\omega}} \; J(-t_1) \right|^2 . \quad (3)$$

The action of the operators $\hat{T}$ and $\hat{\omega}$ on the filter response function J(t) can be written in a compact form using the Dirac notation:

$$|t,\omega\rangle_J = e^{i\omega\hat{T}-t\hat{\omega}} |J\rangle = \hat{D}(t,\omega)|J\rangle . \quad (4)$$

With the help of this formula, the TDSL can be written as a scalar product or what is here equivalent, in the form of a matrix element of the operator $\hat{D}$:

$$S(t,\omega) = |\langle V|t,\omega\rangle_J|^2 = |\langle V|\hat{D}(t,\omega)|J\rangle|^2 . \quad (5)$$

The mathematical meaning of the $\hat{D}$ operation is clear from definition (4) and relations (3). This operator "shifts" the measured state $|V\rangle$ to the filter state $|J\rangle$ by amounts t and ω respectively in the time and frequency phase-space. From Eq. (5) follows also the classical uncertainty principle for the detected signals. From the obvious representation of the "time" and "frequency" operators, $\hat{\omega} = \frac{\partial}{\partial t}$ and $\hat{T} = t$, follows the fundamental relation:

$$[\hat{\omega}, \hat{T}] = 1 \quad (6)$$

limiting the accuracy of simultaneous time and frequency measurements, a fact well known in classical Optics.[2]

The scalar product form (5) of Eq. (1) suggests the interpretation of $S(t,\omega)$ as a proximity or a propensity in $t \times \omega$-space, of two systems V and J to have frequency and time differing by amounts t and ω, respectively.

SCALAR PRODUCT IN QUANTUM MECHANICS

A phase-space interpretation of the Quantum Mechanical scalar product has been proposed by Aharonov et al.[3] They have observed that for two systems described by states $|\phi\rangle$ and $|\psi\rangle$, the probability that the measurement of the relative positions and momenta will lead to $x_1 - x_2 = q$ and $p_1 - p_2 = p$ is equal to:

$$\left|\int dx\, \psi^*(x)\, \phi(x+q)\, e^{\frac{ipx}{\hbar}}\, e^{\frac{ipq}{2\hbar}}\right|^2 = \left|\langle\psi|q,p\rangle_\phi\right|^2 \quad (7)$$

where

$$|q,p\rangle_\phi = \exp\left(\frac{i}{\hbar}\, p\hat{x} + \frac{i}{\hbar}\, q\hat{p}\right)\, |\phi\rangle = \hat{D}(q,p)\,|\phi\rangle \quad (8)$$

with the standard $[\hat{x},\hat{p}] = i\hbar$. According to their interpretation the quantity

$$Pr(q,p) = \frac{1}{2\pi\hbar}\, \left|\langle\psi|q,p\rangle_\phi\right|^2 \quad (9)$$

could be regarded as a proximity or propensity in $q \times p$ phase-space of two systems $|\psi\rangle$ and $|\phi\rangle$ to have momentum and position differing by amounts q and p, respectively.

In Optics the TDSL has been obtained as a result of a realistic detection mechanism.[1] The natural question that then arises is whether or not it is possible to justify the interpretation of Eq. (9) on similar grounds.

QUANTUM PHASE SPACE DETECTOR

A detection scheme of position and momentum for a charged particle can be easily designed by simply scattering sharp laser pulses for example, on a beam of such particles.[4] These pulses with an electric field envelope $\mathcal{E}_q(x,t)$ centered around a detected position q and produced in δ-like impulses at time t_o interact with the charged particle via the standard dipole coupling $U_q(x,t) = -d \cdot \mathcal{E}_q(x,t)$. By analogy to the detection mechanism in Optics we can call the potential $U_q(x,t)$ the reference wave function or the state of the filter device and we shall denote it henceforth by $\psi_q^*(x)$.

In the Born approximation the wave function of the charged particle scattered by the laser pulse has the following form:

$$\phi_s(x,t) = \int dt' \int dx' K_o(xt;x't') \, \psi_q^*(x',t') \, \phi(x',t') \tag{10}$$

where K_o denotes the free Schrödinger propagator (i.e. unperturbed by light). Due to the pulse type interaction with the laser and in the far zone from the interaction retion (for $x \to \infty$ with $x = t \cdot p/m$, where p is the momentum of the moving particle) the probability of detection is equal to:[4]

$$| \phi_s(x,t_o)|^2 \; \alpha \left| \int dx_1 \, e^{\frac{-ipx_1}{\hbar}} \, \psi_q^*(x_1) \, \phi(x_1) \right|^2 . \tag{11}$$

This remarkably simple expression is equal up to a normalization factor to the phase-space propensity Pr(q,p) defined by Eq. (9). This means that as in Optics, the quantity Pr(q,p) can be obtained as a result of a realistic detection scheme.

WIGNER DISTRIBUTION FUNCTION

In 1932 Wigner introduced a phase space description of Quantum Mechanics with the help of the following function:[5]

$$W_\psi(q,p) = \int \frac{dx}{2\pi\hbar} \, \psi^*(q + \frac{x}{2}) \, \psi(q - \frac{x}{2}) \, e^{\frac{ipx}{\hbar}} . \tag{12}$$

It has been shown that the Wigner function defined by Eq. (12) can be obtained in a unique way from the following five general conditions [6]

i) W_ψ has to be a hermitian, real and bilinear function of ψ,
ii) Marginal averages of W_ψ correspond to position and momentum probability distributions,
iii) The correspondence between W_ψ and ψ is Galilei invariant,
iv) The correspondence between W_ψ and ψ is invariant with respect to space and time reflections,
v) There is a remarkable relation between the Wigner functions and the Hilbert space scalar product:

$$|\langle\psi|\phi\rangle|^2 = 2\pi\hbar \int dq \int dp \ W_\psi(q,p)\, W_\phi(q,p) \ . \tag{13}$$

As is well known, the fundamental drawback of such functions is the fact that conditions i - v, are incompatible with $W_\psi \geq 0$,[7] i.e., W_ψ cannot be regarded as a genuine probability distribution in phase-space. In fact a theorem proved by Hudson[8] and Piquet[9] says that W_ψ is positive for all values of q and p if and only if ψ is a Gaussian function.

In Optics the concept of Wigner function has been introduced by Ville in 1948.[10] For electrical or acoustical signals $V(t)$ he has defined the following time and frequency Wigner function:

$$W_V(t,\omega) = \int ds \ V^*(t + \tfrac{s}{2})\, V(t - \tfrac{s}{2})\, e^{i\omega s} . \tag{14}$$

The most important properties of the Optical Wigner function are listed below:

i) $\int \frac{d\omega}{2\pi} W_V(t,\omega) = |V(t)|^2$ and $\int dt \, W_V(t,\omega) = |\tilde{V}(\omega)|^2$,

ii) $W_{V_1 \cdot V_2}(t,\omega) = \int W_{V_1}(t,\omega-\omega')\, W_{V_2}(t,\omega')\, \frac{d\omega'}{2\pi}$,

iii) $W_{V_1 * V_2}(t,\omega) = \int W_{V_1}(t-t',\omega)\, W_{V_2}(t',\omega)\, dt'$,

iv) $\int\int W_{V_1}(t,\omega)\, W_{V_2}(t,\omega)\, \frac{dt\, d\omega}{2\pi} = \left| \int dt\, V_1^*(t)\, V_2(t) \right|^2$,

where $V_1 * V_2$ denotes the convolution of the optical signal and $\tilde{V}(\omega)$ is its Fourier transform.

Rewriting the definition (14) in the following form

$$W_V(t,\omega) = \int ds\, e^{i\omega t}\, e^{\frac{s}{2}\frac{\partial}{\partial t}}\, V^*(t)\, V(t-s) \tag{15}$$

we see that for a slowly varying in time t function $V^*(t)\ V(t-s)$, the approximation $\exp(\frac{s}{2}\frac{\partial}{\partial t}) \simeq 1$ is justified. In this limit the Wigner function becomes the instantaneous Page-Lampard power spectrum[11,12] introduced for non-stationary signals in 1952:

$$S_{PL}(t,\omega) \simeq W_V(t,\omega) \ . \tag{16}$$

The full discussion of the Page-Lampard spectrum and its relation to the TDSL given by Eq. (1) can be found in Ref. 1.

The important connection (16) indicates that under certain circumstances the optical Wigner function can have a close relation to a realistic time and frequency spectral measurement. In fact experiments of this type have been performed both for acoustical and optical waves.[13,14]

DETECTION, SCALAR PRODUCT AND THE WIGNER FUNCTION

In the previous sections we have given a general description of phase-space measurements based on detection mechanisms with realistic detectors or filters. These investigations lead in a natural way to the notion of the quantum propensity (9) or to the optical TDSL (5). A different mathematical description of quantum states or optical signals leads in a natural way to the Wigner functions (12) and (14). The natural question arises what is, if anything, the relation between these different phase-space functions.

For optical signals it has been shown in Ref. 15 that the TDSL is related to the Wigner function by the following relation:

$$S(t,\omega) = \iint dt' \frac{d\omega'}{2\pi}\, W_J(t-t', \omega-\omega')\, W_V(t',\omega') \tag{17}$$

i.e., it is an overlap function of the filter J and signal V Wigner functions. In Quantum Mechanics a very similar formula holds and as it has been shown in Ref. 16

$$Pr(q,p) = \iint dq'\, dp'\, W_\psi(q + q', p + p')\, W_\phi(q',p'). \tag{18}$$

These simple but very important relations indicate that it is possible to construct positive quantities $S(t,\omega)$ and $Pr(q,p)$ for all points of the appropriate phase space and for all signal and filter wave functions, with a well defined operational meaning based on realistic detection mechanisms. This shows that one can avoid the artificial smoothing procedures[17] used to obtain positive Wigner functions shifting the physical interpretation of realistic measurements from W_V and W_ψ to $S(t,\omega)$ and $Pr(q,p)$.

This has the natural consequence of regarding $Pr(q,p)$ as a genuine probability distribution over the phase space, but involving both the measured object and the detecting filter in the same way that $S(t,\omega)$ does. From the definition (9) we have

$$\iint dq\, dp\, Pr(q,p) = ||\psi||^2 \cdot ||\phi||^2 = 1 \qquad (19)$$

for normalized states ψ and ϕ. The marginal average of $Pr(q,p)$ has the form:

$$Pr(q) = \int dp\, Pr(q,p) = \int dq'\, \psi^*(q')\, \psi(q')\, \phi^*(q'+q)\, \phi(q'+q). \qquad (20)$$

This equation leads to the expectation value

$$\langle q\rangle = \int dq\, q Pr(q) = \langle\hat{x}\rangle_\psi - \langle\hat{x}\rangle_\phi \qquad (21)$$

where $\langle\hat{x}\rangle_\psi = \int dx\, \psi^*(x)\, x\, \psi(x)$. Formula (21) shows that $\langle q\rangle$ measures the relative position of the detected state $|\psi\rangle$ with respect to a reference given by the position of the filter state $|\phi\rangle$. For free evolution of the wave function we find that $Pr(q,p)$ satisfies a free Liouville equation:

$$\frac{\partial}{\partial t} Pr(q,p) + \frac{p}{m}\frac{\partial}{\partial q} Pr(q,p) = 0 . \qquad (22)$$

The advantage of using $Pr(q,p)$ instead of the scalar product $|\langle\psi|\phi\rangle|^2$ for the physical interpretation of measurements is obvious if any generalization of the Schrödinger equation to a nonlinear dynamical equation is performed. For a nonlinear wave equation the Hilbert space and its scalar product will lose its fundamental physical role in general. The derivation and the properties of $Pr(q,p)$ except for the connection (7) is basically independent from the Hilbert structure of wave mechanics. Nothing prevents obtaining more complicated relations between $Pr(q,p)$ and the proper wave functions while still conserving the entire physical interpretation of these propensities as genuine probability distributions in phase-space. As a possible abstract construction of such a propensity let us assume that the scattering process (10) can happen only in the second Born approximation. In this case we would obtain:

$$Pr(q,p) = \iint dq'\, dp'\, W_\psi^{(2)}(q+q',\, p+p')\, W_\phi(q',p')$$

with

$$W_\psi^{(2)}(q,p) = \int \frac{dx}{2\pi\hbar}\, \psi^{*2}\left(q+\frac{x}{2}\right) \psi^2\left(q-\frac{x}{2}\right) e^{\frac{ipx}{\hbar}} \quad \text{and}$$

an obvious non-Hilbertian structure between states $|\psi\rangle$ and $|\phi\rangle$ does emerge. Constructions of such quantities are under investigation.[4]

ACKNOWLEDGMENTS

The author would like to thank: I. Bialynicki-Birula, K.-H Brenner, J.D. Cresser, J.H. Eberly, F. Haake, A.W. Lohmann, M.O. Scully, B.W. Shore and S. Stenholm for many valuable discussions over an extended period of time about foundations of measurement theory in Quantum Mechanics and Optics. This research has been partially supported by the U.S. Department of Energy.

REFERENCES

1. J.H. Eberly and K. Wódkiewicz, J. Opt. Soc. Am. 67, 1252 (1977).
2. M. Born and E. Wolf, Principles of Optics, 5th ed. (Pergamon, Oxford, 1975).
3. Y. Aharonov, D.Z. Albert, C.K. Au, Phys. Rev. Lett. 47, 1029 (1982).
4. K. Wódkiewicz (to be published).
5. E.P. Wigner, Phys. Rev. 40, 749 (1932).
6. R.F. O'Connell and E.P. Wigner, Phys. Lett. 83A, 145 (1981).
7. E.P. Wigner in Perspectives in Quantum Theory, eds. W. Yourgrau and A. van der Merwe (Dover, New York, 1979).
8. R.L. Hudson, Rep. Math. Phys. 6, 249 (1974).
9. C. Piquet, C.R. Acad. Sc. Paris 279A, 107 (1974).
10. J. Ville, Câbles et Transm. 2 A, 1, 61 (1948).
11. C.H. Page, J. Appl. Phys. 23, 103 (1952).
12. D.G. Lampard, J. Appl. Phys. 25, 803 (1954).
13. H.O. Bartelt, K.-H. Brenner and A.W. Lohmann, Opt. Comm. 32, 1 (1980).
14. K.-H. Brenner and A.W. Lohmann, Opt. Comm. 42, 310 (1982).
15. K.-H. Brenner and K. Wódkiewicz, Opt. Comm. 43, 103 (1982).
16. R.F. O'Connell and A.K. Rajagopal, Phys. Rev. Lett. 48, 525 (1982).
17. See for example F. Soto and P. Claverie, Physica 109A, 133 (1981).

CONFIGURATION INTERACTION IN MULTIPHOTON IONIZATION

P. Zoller,* E. Matthias,† and S. J. Smith‡

Joint Institute for Laboratory Astrophysics
University of Colorado and National Bureau of Standards
Boulder, Colorado 80309

In the past decade we have gained a considerable understanding of multiphoton ionization (MPI) of alkali atoms, i.e., of single electron systems.[1] Presently the interest is turning toward the study of alkaline earth atoms which have two valence electrons outside closed shells.[1-8] The new aspect to be investigated in these systems is the effect of configuration interaction (CI) in MPI: alkaline earth atoms have many Rydberg series with different ionization thresholds corresponding to different excited core states of the inner of the two electrons, while the other one is in the Rydberg state. These doubly excited electron configurations mix due to the noncentral part of the Coulomb interaction with the Rydberg series of, for example, the singly excited states, leading to perturbation of the Rydberg states in the bound region and series of autoionizing resonances above the first ionization threshold.

Most dramatic effects in MPI of atoms with more than one valence electron are expected at the extremely high laser intensities ($\gtrsim 10^{10}$ W/cm^2) when the interaction of the electrons with the laser light becomes comparable to the strength of the CI in the atom.[2,3] The electron-light and electron-electron inter-

*1982-83 JILA Visiting Fellow, on leave from the University of Innsbruck, 6020 Innsbruck, Austria.
†1982-83 JILA Visiting Fellow, on leave from the Free University of Berlin, 1000 Berlin 33, West Germany.
‡Staff Member, Quantum Physics Division, National Bureau of Standards.

actions then compete with each other, i.e., the CI structure of the atom is modified in a strong external field, leading to a wealth of interesting interference effects which have been predicted theoretically but whose relation to present experiments[6] leaves for the moment some open questions.

Here we are concerned with some aspects of MPI in alkaline earth atoms in an intensity regime where CI still dominates the interaction with the light field. This situation corresponds to experiments performed with pulsed dye lasers using step-wise resonant excitation followed by ionization.[4,5,8] The development of a theoretical model for this case consists essentially of two steps. The first step is the description of the dynamics of the MPI process (including saturation, etc.) within a few-level system of the resonant states of the excitation sequence. This is a fairly standard problem in MPI well known from one-electron systems,[1] and amounts to parametrizing the ionization probability in terms of the Rabi frequencies of the resonant transitions, ionization and radiative decay widths of the resonant states, etc. The second and new problem in the context of CI is to express the energy dependence of these quantities (which are essentially dipole matrix elements) for a Rydberg series in the neighborhood of a bound interloper or autoionizing resonances in terms of CI parameters characterizing the state mixing. Multi-channel quantum defect theory[9] (MQDT) seems to be an ideal candidate for this purpose as it parametrizes the interaction of Rydberg series in terms of a few physically significant MQDT parameters which in many cases are known from spectroscopic work.

To illustrate the idea of using an MQDT parametrization to interpret experiments on resonant MPI (RMPI)[4,5,8] we qualitatively discuss below a recent three-photon ionization experiment with barium,[5] which studied state mixing in the intermediate bound states. Although our discussion concentrates on Ba, the aspects outlined below are of general relevance to RMPI of atoms with more than one valence electron.

In Ba the 5d7d doubly excited configuration causes strong state mixing in the 6snd Rydberg states near n = 26 (Fig. 1).[10-14] The RMPI was chosen to proceed via these 6snd $^{1,3}D_2$ states. Experimentally, the 6snd series ($19 \lesssim n \lesssim 30$) is convenient since the ionizing YAG-laser radiation reaches both the 6s and 5d continua of Ba^+ without exciting any autoionizing state.[15] Hence, two groups of photoelectrons are observed: <u>fast</u> ones (1.16 eV) originating from the 6s continuum and <u>slow</u> ones (0.46 eV) from the 5d continuum. The spectra in Fig. 2 were obtained by focusing the beams of two pulsed dye lasers (553.7 nm: $6s^2\ ^1S_0 \rightarrow 6s6p\ ^1P_1$ and 419-424 nm: $6s6p\ ^1P_1 \rightarrow 6snd\ ^{1,3}D_2$) and the YAG-laser radiation along the same axis onto a beam of Ba atoms in a field-free region. Typical power levels per pulse for the three successive

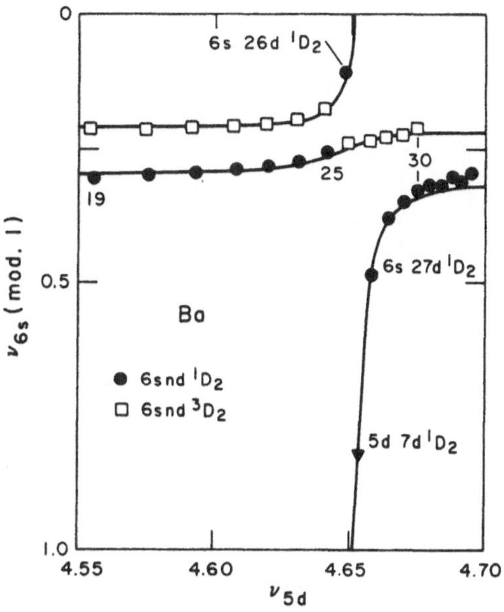

Fig. 1. Lu-Fano plot for the 5d7d perturbation in the $6snd^{1,3}D_2$ Rydberg series. Note that in our case this is essentially a plot of the quantum defect of the 6snd $^{1,3}D_2$ states versus energy. For explanation see text.

linearly polarized laser beams were 10^2 W (553.7 nm), 5 × 10^3 W (419-424 nm) and 2 MW (1064 nm) in a focal diameter of about 1 mm at the atom beam.

The influence of the $5d7d^1D_2$ perturber state (in jj-coupling[10] $5d_{5/2}7d_{3/2}$), located between n = 26 and 27, is obvious in the total ion yield (Fig. 2a) as well as in the spectra of fast (Fig. 2b) and slow (Fig. 2c) electrons. The simplest MQDT model assumes only the $6snd^{1,3}D_2$ states and the perturber to be mixed: $|6snd,J=2\rangle = Z_1|6snd^1D_2\rangle + Z_2|6snd^3D_2\rangle + Z_3|5d_{5/2}7d_{3/2},J=2\rangle$. The mixing coefficients Z_i follow from an MQDT analysis. In a Lu-Fano plot of the effective quantum numbers ν_{6s} (mod 1) versus ν_{5d} (mod 1), defined for each atomic level of energy E as

$$E = I_{6s} - \frac{Ry}{\nu_{6s}^2} = I_{5d} - \frac{Ry}{\nu_{5d}^2}$$

with I_{6s} and I_{5d} the ionization thresholds of $Ba^+(6s)$ and $Ba^+(5d)$ respectively, the energy levels lie on two separate branches[10,16] (Fig. 1). The continuous branch contains the $6snd^1D_2$ (n≲25) and $^3D_2$ (n≳26) states with ν_{6s} almost constant as a function of ν_{5d}. For these the admixture of the perturber state is small; instead there is strong singlet-triplet mixing. The $6snd^3D_2$ (n≲25) and

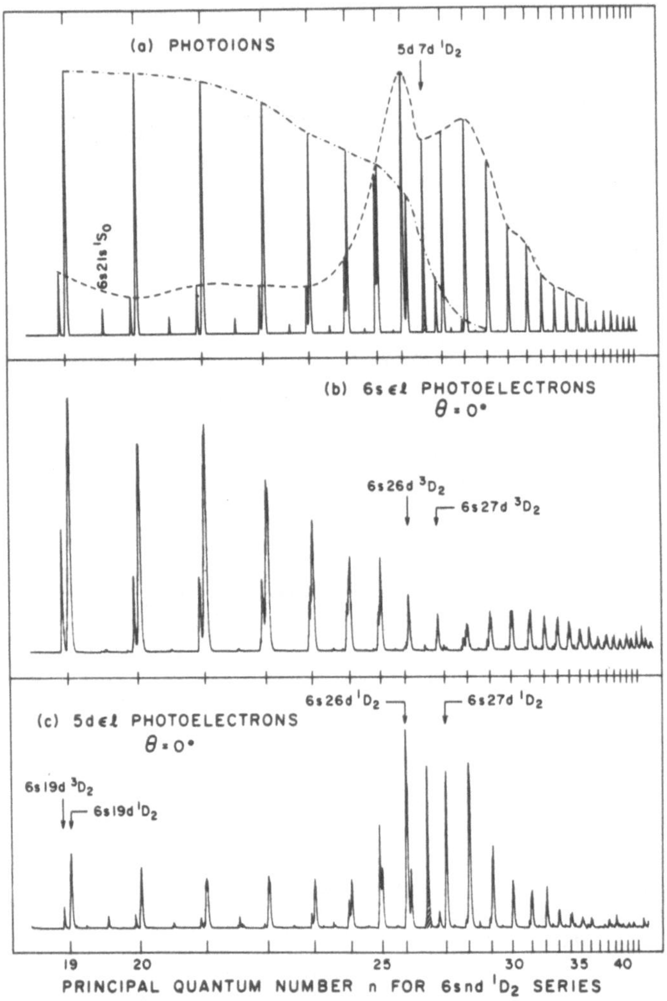

Fig. 2. a) Total yield of photoions for resonant three-photon ionization of Ba. b) and c) Angle-resolved and time-gated spectra of electrons originating from the 6s and 5d continua of Ba^+.

$^1D_2$ ($n \geq 26$) states, together with the 5d7d perturber, lie on the broken branch of the Lu-Fano plot. Along this branch ν_{6s} varies rapidly with ν_{5d} in the vicinity of the perturber, indicating that the $6s26d\,^1D_2$ and $6s27d\,^1D_2$ states contain large admixtures of the 5d7d configuration in addition to singlet-triplet mixing. This classification is identified in the ion spectrum in Fig. 2a where the continuous and broken branches are connected by dash-dot and dashed lines, respectively. The continuous branch terminates at n = 30 since triplet states were not resolved beyond this point.

Neglecting configuration interaction in the final state, the differential ionization rates from Rydberg states leaving behind a 6s or 5d core of Ba^+ are, in the usual notation,[1]

$$\frac{d\gamma_{6s}}{d\Omega} \propto \sum_{i=1,2} Z_i^2 \sum_{m_c m_s} |\langle 6s_{1/2}\ m_c;\vec{p}m_s|\vec{\mu}\vec{\varepsilon}|6snd\ ^{1,3}D_2, M_J=0\rangle|^2\ I_{YAG} \tag{1a}$$

$$\frac{d\gamma_{5d}}{d\Omega} \propto Z_3^2 \sum_{j=3/2}^{5/2} \sum_{m_c m_s} |\langle 5d_j m_c;\vec{p}m_s|\vec{\mu}\vec{\varepsilon}|5d7d\ ^1D_2, M_J=0\rangle|^2\ I_{YAG} \quad , \tag{1b}$$

assuming orthogonal radial wave functions. The sum includes angular quantum numbers of the ion core (m_c) and spin polarization (m_s). The ionization probability to the $Ba^+(5d)$ continuum is determined by the admixture of the perturber. Hence the dominant peaks in Fig. 2c stem from states on the broken branch of the Lu-Fano plot; the nearly constant background for $n \lesssim 24$ indicates the presence of nonresonant CI. In addition to the n-dependence of state mixing, the radial matrix elements involved in the excitation sequence may change over the perturbed region. They are almost constant along the continuous branch (apart from an overall decrease with n) and show for the broken branch a Fano profile as a function of energy near the perturber.

The intensities in Fig. 2 can be described by rate equations since the measurements satisfy broad bandwidth conditions.[1] If W is the excitation rate of the Rydberg states and κ their spontaneous decay rate, the ionization probabilities starting from the 6s6p state can be written

$$\frac{d}{d\Omega} P_{6s,5d} = \frac{W}{\kappa+2W}\left[1-e^{-(\kappa+2W)T}\right]e^{-\kappa\tau}$$
$$\times \frac{d/d\Omega(\gamma_{6s,5d})}{\gamma_{6s}+\gamma_{5d}+\kappa}\left[1-e^{-(\kappa+\gamma_{5d}+\gamma_{6s})T}\right] . \tag{2}$$

The first three factors describe the population remaining in the Rydberg states at time τ following excitation by a laser pulse of duration T; the last two represent the branching of the ionization signal. Since the $6s6p\,^1P_1$ state couples predominantly to the singlet part of the wave function, $W \propto Z_1^2 |R_{6p}^{nd}|^2 I_{blue}$. Thus, the contribution from states on the continuous branch of the Lu-Fano plot disappears near the pure[16] $6s30d\,^3D_2$ in both ion and fast-electron spectra [Fig. 2a and b]. Along the broken branch W reaches a maximum near the perturber state, because of an increase in the singlet admixture and a decrease in the radial matrix elements with n. The ionization signal is further enhanced by the higher ionization rate of the 7d electron [Eq. (1b)] compared to the Rydberg electron. In the perturbed region the ion yield is thus dominated by the Ba^+ 5d channel. The amplitude of the resonance in Fig. 2a and c is determined by the choice of laser polarization and by saturation of the ionization step. The observed double hump at n = 26 and 28 is explained by the reduction of radiative lifetimes caused by the 5d7d admixture,[11] which allows Rydberg states on the broken branch to decay during the $\tau \cong 30$ ns interval between exciting and ionizing radiation. Finally, the minimum near the perturber in Fig. 2b is caused by the large 5d7d admixture which, for the broken branch, diverts most of the Rydberg state population into the 5dε'ℓ' continuum.

Additional information about the coupling of the two valence electrons and the continuum wave functions can be obtained from the angular distribution of photoelectrons which from Eqs. (1) has the form[1]

$$W(\theta) = \sum_{\ell=0}^{N=3} A_{2\ell} P_{2\ell}(\cos\theta).$$

Here, θ is the angle between the electron momentum and the polarization vectors of the light. The coefficients $A_{2\ell}$ contain information about the structure of the states involved.[17,18] In Fig. 3 a few typical angular distributions illustrate the change in shape when passing from almost unperturbed 6snd states through the perturbed region. The angular distribution for 6s30d $^1D_2 \rightarrow$ 6sεℓ has again the same appearance as the one for 6s19d $^1D_2 \rightarrow$ 6sεℓ. To illustrate how to extract information from the angular distribution coefficients we will discuss A_6, which is not complicated by p-f interferences. For the normalized A_6 coefficient Eq. (1a) predicts

$$\frac{A_6}{A_0} = \frac{\frac{40}{77}\left(\frac{3}{2} Z_1^2 \, |R_s^f|^2 - Z_2^2 \, |R_t^f|^2\right)}{Z_1^2\left(\frac{4}{15} |R_s^p|^2 + \frac{9}{35} |R_s^f|^2\right) + Z_2^2\left(\frac{1}{5} |R_t^p|^2 + \frac{8}{35} |R_t^f|^2\right)} , \tag{3}$$

where $R_s^{p,f}$ and $R_t^{p,f}$ denote the radial matrix elements for singlet or triplet ionization. To calculate A_6, we assume $R_s \simeq R_t$ and find the ratio of radial matrix elements from quantum defect theory to be $|R^p|^2/|R^f|^2 \simeq 0.5$. The admixture coefficients were drawn from the MQDT analysis.[16] The result is plotted as a solid line in Fig. 4a together with values of A_6 obtained from the measured angular distributions and corrected for the finite solid angle. We observe overall agreement between theory and experiment. In particular, the data points for the nearly pure singlet and triplet states at n = 19 and 30 are well reproduced by the

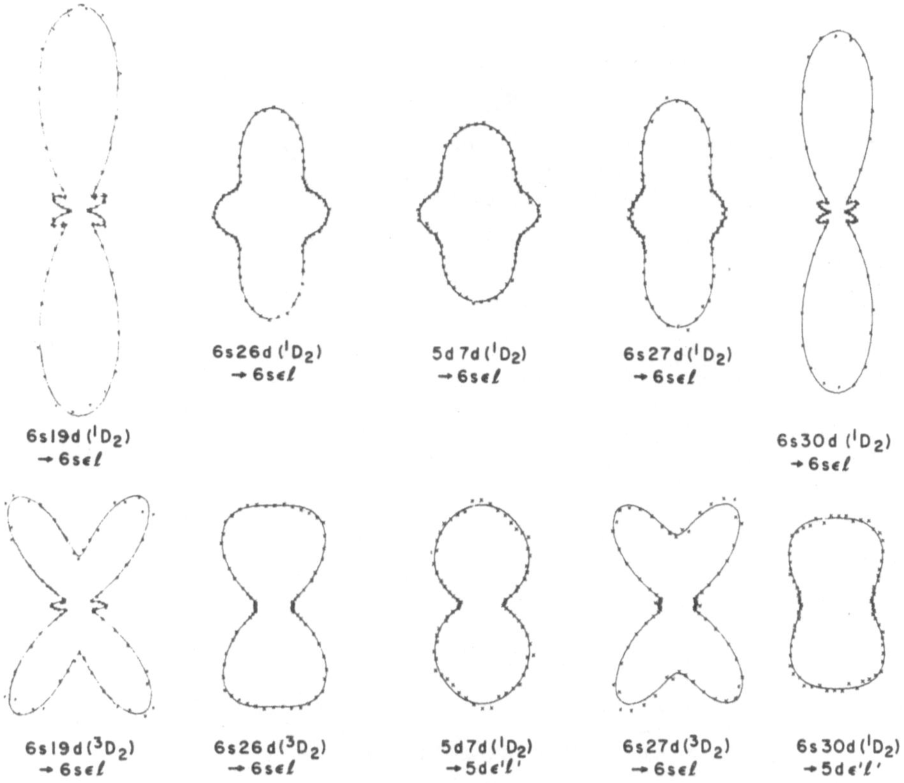

Fig. 3. Selected angular distributions of photoelectrons following RMPI of Ba. The θ = 0 axis points vertically; the linear polarizations of the three laser beams were aligned parallel. The solid lines are least-squares fits.

above approximations. The A_6 coefficients directly reflect the n-dependence of $Z_1{}^2$ shown in Fig. 4b. A change from singlet to triplet character leads to a change of sign for A_6 according to Eq. (3). Along the continuous branch of the Lu-Fano plot, the angular distributions consist of a superposition of pure singlet and triplet distributions. The situation is more complex for the broken branch near the perturber because the small probability for photoionization to the 6s continuum allows second order effects, such as final state CI, to become dominant. A typical angular distribution of 5dε'ℓ' electrons is illustrated in Fig. 3 for $5d7d\,^1D_2$. It consists mainly of an A_2 term, characteristic of an isotropic state, because A_6 vanishes for a pure $5d_{5/2}7d_{3/2}$ state[10] and the A_4 term is smeared out by the angular momentum exchange between the core and the ionized electron.

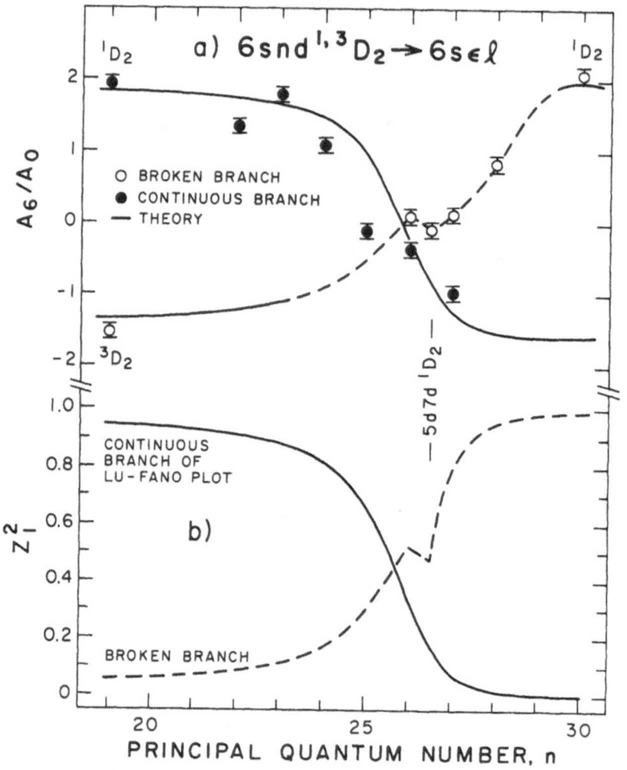

Fig. 4. a) Experimental and theoretical angular distribution coefficients A_6/A_0. The solid line was calculated using Eq. (3). The theoretical model does not apply to electrons from states on the broken branch near the perturber (dashed lines). b) Singlet admixture of $6snd^{1,3}D_2$ in Ba derived from an MQDT analysis (Ref. 10).

This work was supported in part by National Science Foundation Grants PHY82-00805 and INT81-20128 to the University of Colorado. E. Matthias and P. Zoller acknowlege the friendly hospitality of JILA.

References

1. For reviews on MPI see P. Lambropoulos, Adv. At. Mol. Phys. 12, 87 (1976); A. T. Georges and P. Lambropoulos, Adv. Electron. Electron Phys. 54, 191 (1980); G. Mainfray and C. Manus, Appl. Opt. 19, 3934 (1980); and various contributions in Multiphoton Ionization, edited by S. L. Chin and P. Lambropoulos, to be published.
2. For a discussion of strong field effects in autoionization see the papers in the present volume by D. Agassi, J. Haus, P.L. Knight, M. Lewenstein and K. Rzazewski.

3. Y.-S. Kim and P. Lambropoulos, Phys. Rev. Lett. 49, 1638 (1982).
4. W. Sandner, R. Kachru, K. A. Safinya, F. Gounand, W. E. Cooke and T. F. Gallagher, Phys. Rev. A 27, 1717 (1982).
5. E. Matthias, P. Zoller, D. S. Elliott, N. D. Piltch, S. J. Smith and G. Leuchs, Phys. Rev. Lett. 50, 1914 (1983).
6. D. Feldmann and K. H. Welge, J. Phys. B 15, 1651 (1982).
7. J. J. Wynne and J. P. Hermann, Opt. Lett. 4, 106 (1979); M. Aymar, P. Camus, and A. El Himdy, J. Phys. B 15, L759 (1982).
8. N. H. Tran, R. Kachru, and T. F. Gallagher, Phys. Rev. A 26, 3016 (1982); W. E. Cooke and S. A. Bhatti, Phys. Rev. A 26, 391 (1982), and references therein; C. L. Cromer, "Autoionization enhanced VUV generation in strontium," Ph.D. Thesis, University of Southern California (1982).
9. M. J. Seaton, Rep. Prog. Phys. 46, 167 (1983) and references therein.
10. M. Aymar and O. Robaux, J. Phys. B 12, 531 (1979).
11. K. Bhatia, P. Grafström, C. Levinson, H. Lundberg, L. Nilsson, and S. Svanberg, Z. Phys. A 303, 1 (1981); M. Aymar, R.-J. Champeau, C. Delsart, and J.-C. Keller, J. Phys. B 14, 4489 (1981).
12. P. Grafström, C. Levinson, H. Lundberg, S. Svanberg, P. Grundevik, L. Nilsson, and M. Aymar, Z. Phys. A308, 95 (1982).
13. M. L. Zimmerman, T. W. Ducas, M. G. Littman, and D. Kleppner, J. Phys. B 11, L11 (1978).
14. H. Rinneberg and J. Neukammer, Phys. Rev. Lett. 49, 124 (1982); P. Grafström, J. Zhan-Kui, G. Jönsson, S. Kröll, C. Levinson, H. Lundberg, and S. Svanberg, Z. Phys. A306, 281 (1982).
15. R. D. Hudson, V. L. Carter, and P. A. Young, Phys. Rev. A 2, 643 (1970).
16. H. Rinneberg and J. Neukammer, Phys. Rev. A 27, 1779 (1983).
17. J. A. Duncanson, Jr., M. P. Strand, A. Lindgard, and R. S. Berry, Phys. Rev. Lett. 37, 987 (1976); G. Leuchs, S. J. Smith, and H. Walther, in Laser Spectroscopy IV, Springer Series in Optical Sciences, Vol. 21, edited by H. Walther and K. W. Rothe (Heidelberg, 1979), p. 255; J. C. Hansen, J. A. Duncanson, Jr., R.-L. Chien, and R. S. Berry, Phys. Rev. A 21, 222 (1980), and references therein.
18. G. Leuchs, in Laser Physics, Proc. 3rd New Zealand Summer School, edited by D. F. Walls and J. D. Harvey (in press), and references therein.

NONLINEAR WAVE EQUATION FOR DECAYING STATES

M. D. Girardeau

Department of Physics and Institutes of Chemical Physics
and Theoretical Science,
University of Oregon, Eugene, OR 97403, U.S.A.

ABSTRACT

Normalizable wave functions $\mathcal{Q}$ describing initial wave packets for a decaying particle, as well as their complex resonance energies z, can be determined from the variational principle $\delta[z-\lambda\langle\mathcal{Q}|\mathcal{Q}\rangle]=0$ where z is the appropriate pole of $\langle\mathcal{Q}|(z-H)^{-1}|\mathcal{Q}\rangle$, λ is a Lagrange multiplier for the normalization constraint $\langle\mathcal{Q}|\mathcal{Q}\rangle=1$, and the variation is with respect to the functional form of $\mathcal{Q}$. This yields a nonhermitian, nonlinear eigenvalue - eigenfunction equation for $\mathcal{Q}$, λ, and hence z. In the limit of a stable state it reduces to the usual linear Schrodinger equation.

INTRODUCTION

The discussion will be quite brief, since the variational principle considered here has already appeared in print[1,2] as has an improved version[3] developed since the Symposium at which this material was presented.

In order to prevent misunderstanding, let me start by pointing out a few things that I am <u>not</u> trying to do: (a) I am not interested herein in calculating the time-dependent wave function $|\mathcal{Q}_\alpha(t)\rangle$ of a decaying state, given its form $|\mathcal{Q}_\alpha\rangle=|\mathcal{Q}_\alpha(0)\rangle$ at time zero. Methods for doing that are well-known but are of no use if one does <u>not</u> know the <u>initial</u> state $|\mathcal{Q}_\alpha\rangle$. (b) I am not interested here in finding scattering wave functions versus energy. These are not normalizable and cannot be used as <u>discrete</u> basis elements in calculations of various dynamical processes influenced by resonances (or more generally, by decaying states). (c) By the same token, I will not be concerned herein with non-normalizable formal eigenfunctions with complex energy eigenvalues, the Gamow-Siegert states[4,5].

I am interested in finding a general criterion for choice of discrete (normalizable) states representing transient decaying states. They could then be used as initial states $|\mathcal{Q}_\alpha\rangle$ in evaluation of time-dependent decaying states or as discrete basis elements in the time - independent approach to collision theory. This motivation is essentially the same as that of analytical continuation methods such as the complex rotation and complex stabilization methods[6-9]. In fact, it can be regarded as an extension of the complex stabilization method from the case of Rayleigh-Ritz parameter variation in a given finite basis to that of complete functional variation.

THE VARIATIONAL PRINCIPLE

In the complex stabilization method[8,9] one considers poles of the resolvent $(z-H)^{-1}$ in a finite-order matrix representation in terms of a given finite basis containing variational parameters. These poles describe "resonance trajectories" as each parameter is varied holding the others fixed. Each such trajectory corresponds to a one-parameter family of trial states. These trajectories generally "pause" in some region of the z-plane before moving away again, as each parameter is varied in turn. One tries to adjust the parameters so that the "pause" occurs at nearly the same z for various different trajectories (different parameters varied). The resonance value z_{res} thus determined has a certain amount of intrinsic "fuzziness", although the wave function itself is more precisely determined, in analogy with the usual Rayleigh-Ritz method.

In order to generalize this method to obtain a precise variational criterion, let us dispense with given parametrized bases and allow full functional variation, subject only to the constraint that the wave function be normalized. Let $\mathcal{Q}_\alpha(X)$ be the wave function describing the initial wave packet $\mathcal{Q}_\alpha(X)=\mathcal{Q}_\alpha(X,0)$ for some time-dependent decaying state $\mathcal{Q}_\alpha(X,t)$; here X stands for the coordinates of all the constituent particles of the given state. The persistence amplitude of this state is the Green's function

$$g_\alpha(t)=-i\langle\mathcal{Q}_\alpha(0)|\mathcal{Q}_\alpha(t)\rangle=-i\langle\mathcal{Q}_\alpha|e^{-iHt}|\mathcal{Q}_\alpha\rangle$$

which can be Laplace-transformed to give the complex-energy Green's function

$$\tilde{g}_\alpha(z)=\int_0^\infty g_\alpha(t)e^{izt}dt=\langle\mathcal{Q}_\alpha|(z-H)^{-1}|\mathcal{Q}_\alpha\rangle \quad .$$

The analytic continuation of $\tilde{g}_\alpha(z)$ across the continuum cut into the lower half plane will have a pole z_α characterizing the decay.

The position of this pole is clearly a functional of the initial wave function $\mathcal{Q}_\alpha(X)$, i.e., $z_\alpha = z_\alpha[\mathcal{Q}_\alpha]$. Our variational principle consists of the requirement that this pole be stationary under functional variation of the wave function, subject to a normalization contraint:

$$\frac{\delta}{\delta \mathcal{Q}_\alpha^*(X)}\left[z_\alpha - \lambda_\alpha \langle \mathcal{Q}_\alpha | \mathcal{Q}_\alpha \rangle\right] = 0 \; .$$

NONLINEAR WAVE EQUATION

To get a more explicit form of the variational equation introduce the complex self energy Σ_α (z) <u>defined</u> by

$$\mathcal{G}_\alpha(z) = [z - \mathcal{E}_\alpha - \Sigma_\alpha(z)]^{-1}$$

where $\mathcal{E}_\alpha = \langle \mathcal{Q}_\alpha | H | \mathcal{Q}_\alpha \rangle$. The pole z_α is determined by

$$z_\alpha - \mathcal{E}_\alpha - \Sigma_\alpha(z_\alpha) = 0.$$

Functional differentiation gives

$$\frac{\delta z_\alpha}{\delta \mathcal{Q}_\alpha^*(X)} - H\mathcal{Q}_\alpha(X) - \frac{\delta \Sigma_\alpha}{\delta \mathcal{Q}_\alpha^*(X)} - \Sigma_\alpha'(z_\alpha)\frac{\delta z_\alpha}{\delta \mathcal{Q}_\alpha^*(X)} = 0$$

Note that in general Σ_α is both a function of z and a functional of $\mathcal{Q}_\alpha$ and $\mathcal{Q}_\alpha^*$; Σ_α'(z) denotes its partial derivative with respect to z, holding $\mathcal{Q}_\alpha$ and $\mathcal{Q}_\alpha^*$ constant. Substitution into the variational equation gives

$$[1 - \Sigma_\alpha'(z_\alpha)]^{-1}\left[H\mathcal{Q}_\alpha(X) + \frac{\delta \Sigma_\alpha}{\delta \mathcal{Q}_\alpha^*(X)}(z_\alpha)\right] = \lambda_\alpha \mathcal{Q}_\alpha(X).$$

If $\mathcal{Q}_\alpha$ is chosen to be a bound state with eigenvalue $\mathcal{E}_\alpha$, then Σ_α vanishes and we have the usual linear Schrodinger equation, with $\lambda_\alpha = \mathcal{E}_\alpha$. However, in general this wave equation is both nonlinear (because z_α and Σ_α are functional of $\mathcal{Q}_\alpha$ and $\mathcal{Q}_\alpha^*$) and nonhermitian (because Σ_α is complex). It defines a <u>nonlinear eigenvalue problem</u> for $\mathcal{Q}_\alpha$ and λ_α, hence also z_α.

The self energy Σ_α can be conveniently evaluated in a Liouville space representation[10]. Numerical solution have been obtained for the case of one particle in one dimension tunnelling out of a double delta barrier V(X)= b $[\delta(X+\frac{1}{2}) + \delta(X-\frac{1}{2})]$. We shall not present them here since the same potential has been used to illustrate the nature of the solutions of an improved (but closely related) formulation of the variational principle[3]. The improvement consists of use of a biorthogonal inner product $\int \bar{\mathcal{Q}}_\alpha(X)\mathcal{Q}_\beta(X)dX$ with $\bar{\mathcal{Q}}_\alpha \neq \mathcal{Q}_\alpha^*$. This is more natural in view of the nonhermitian form of the eigenfunction-eigenvalue problem, and also facilitates comparison with previous variational formulations[11-13] in which biorthogonal expansions were used.

ACKNOWLEDGEMENT

This work was supported by the U.S. Office of Naval Research.

REFERENCES

1. M. D. Girardeau, in: "Recent Progress in Many-Body Theories," J. G. Zabolitzky, M. de Llano, M. Fortes, and J. W. Clark, eds. Lecture Notes in Physics Vol. 142, Springer-Verlag, Berlin (1981), pp. 355ff.
2. M. D. Girardeau, Bull. Am. Phys. Soc. 28:782 (1983).
3. C. F. Hart and M. D. Girardeau, Phys. Rev. Lett. 51:1725 (1983).
4. G. Gamow, Z. Phys. 51:204 (1928).
5. A. J. F. Siegert, Phys. Rev. 56:750 (1939).
6. J. Aguilar and J. M. Combes, Commun. Math. Phys. 22:269 (1971).
7. E. Balslev and J. M. Combes, Commun. Math. Phys. 22:280 (1971).
8. B. R. Junker, Phys. Rev. Lett. 44:1487 (1980).
9. B. R. Junker, in: "Advances in Atomic and Molecular Physics," D. Bates and B. Bederson, eds., Academic Press, New York (1982) Vol. 18, pp. 207ff.
10. M. D. Girardeau, Phys. Rev. A 28:1056 (1983).
    M. D. Girardeau and C. F. Hart, Phys. Rev. A 28:1072 (1983).
11. A. Herzenberg and F. Mandl, Proc. Roy. Soc. London A 274:253 (1963
12. J. N. Bardsley and B. R. Junker, J. Phys. B 5:L178 (1972).
13. R. A. Bain, J. N. Bardsley, B. R. Junker, and C.V. Sukumar, J. Phys. B 7:2189 (1974).

INTERFERENCE PHENOMENA IN UNSTABLE STATES

Laurent P. Lévy

AT&T Bell Laboratories
Room 1E - 236
600 Mountain Avenue
Murray Hill, New Jersey 07974

I. INTRODUCTION

The prediction of interference phenomena in microscopic systems has been one of the most outstanding successes of quantum mechanics. Interferences may involve spatial coherence of a quantum mechanical wavefunction such as encountered in Young's two-slit experiments[1], in neutron interferometer experiments[2], or in many real scattering processes. Alternatively, they may manifest themselves as temporal coherences between the different quantum states of the system, such as in quantum beats in atomic radiative processes.[3] In all cases, the system carries phase information which in turn determines the structure of the interference phenomena. In many physical examples, the phase information is not simply a scalar factor $e^{i\phi}$ but may often have a vectorial or spinorial character. In systems with spontaneously broken symmetry (^{4}He, ferromagnets, etc.), the phase or order parameter depends on a continuous variable and is a macroscopic observable.

In all cases, the study of interference phenomena reduces to the study of the dependence of the quantum mechanical phase on an external (or scattering) potential. For example, a radiative process may exhibit quantum beats when the emitted radiation is modulated by time-dependent oscillations. These oscillations can be viewed as the systematic phase drift between different states in a coherent superposition arising from field-induced energy differences. The external field alters the phase, which in turn induces a quantum beat.

A perhaps less familiar dynamical effect which affects the quantum mechanical phase is the damping of quantum mechanical states (or excitations) or, more precisely, the reaction of a quantum mechanical system to the coupling to an external bath (radiation field, phonon field). We know, for example, that a classical damped harmonic oscillator has on resonance a response 90° out of phase with respect to the driving field.

In a quantum system, the "radiative reaction" also introduces additional phase lags of the quantum mechanical amplitudes which depend on the precise spectrum and structure of the quantum system in the external fields which it experiences. This additional dependence of the radiative reaction or "self-energy" on external fields also modifies the dependence of the quantum mechanical phases on the applied field, and thus the interference phenomena is altered in a fundamental way by the presence of a radiative reaction. The phase lags may even allow some new quantum mechanical interferences between amplitudes which would otherwise be in quadrature because of the symmetry structure of the Hamiltonian. Thus, one finds new possible interference phenomena which have no parallel in conservative systems. Their origin may be understood from symmetry principles. The implications of time-reversal symmetry in unstable systems will be invoked for this purpose here. The plan of the paper is as follows: we first discuss the quantum mechanics of damped systems with special reference to the atomic physics applications which will be described in section IV. The role of time-reversal invariance in damped systems is analyzed in section II and the lower symmetry associated with the presence of damping will be shown to imply the existence of new interference phenomena. In section III we give a very simple example borrowed from the hydrogen atom, the E1-M1 interference observed in the quenching of the metastable 2S state in a combination of electric and magnetic field.[4] Finally, the last part is devoted to the description of two experiments which have demonstrated the presence of damping interferences, the first one performed in an induced transition in the metastable 2S shell of hydrogen[5], the second, on the anisotropic quenching of $^4He^+$ in an external electric field.[6]

II. UNSTABLE STATES IN QUANTUM MECHANICS

We consider a system (atom) described by a Hamiltonian H_a, with a spectrum of states $|i\rangle$ and energies E_i belonging to a Hilbert space $\mathcal{H}_a$. The radiation field couples the atom to the photon Hilbert space H_γ. Because of this coupling, the atomic Hamiltonian acquires a self-energy Σ and to lowest order,

$$\Sigma(\omega) = \frac{\alpha}{4\pi^2}\int \vec{\varepsilon}\,(\vec{k})\cdot\vec{v}\,\frac{1}{H_a+k-\omega}\,\vec{\varepsilon}(\vec{k})\cdot\vec{v}\;d^3k. \qquad \text{(II.1)}$$

The new spectrum may be obtained as the roots of the secular equation

$$H_a - \omega - \Sigma(\omega) = 0. \qquad \text{(II.2)}$$

If the matrix elements of the self-energy satisfy the relations

$$|\langle i|\Sigma(\omega)|j\rangle|^2/(E_j - E_i) \ll \langle i|\Sigma(\omega)|i\rangle \ll E_i, \tag{II.3}$$

one may reduce the problem to an effective Hamiltonian

$$H_{eff} = H_a - \Sigma \tag{II.4}$$

where

$$\langle i|\Sigma|i\rangle = \langle i|\Sigma(E_i)|i\rangle \quad \text{and} \quad \langle i|\Sigma|j\rangle = 0. \tag{II.5}$$

When this reduction is not possible, the conclusions we shall reach are little affected. Because the self-energy II.1 has a singularity along the real positive axis, the diagonal self-energy acquires an imaginary part and the effective Hamiltonian is no longer Hermitian. Thus, one is led to distinguish between the right eigenvectors

$$(H_{eff} - E_i')|i\rangle = 0, \tag{II.6}$$

which have the usual time dependence $|i\rangle \to e^{-iE_i't}|i\rangle$, exponentially damped in time and the left eigenvectors[7]

$$\langle\bar{j}|(H_{eff} - E_j') = 0. \tag{II.7}$$

The left eigenvectors $\langle\bar{j}|$ are not Hermitian conjugates of $|j\rangle$. They satisfy the orthogonality relation

$$\langle\bar{j}|i\rangle = \delta_{i,j}. \tag{II.8}$$

To clarify the role played by left and right eigenvectors, it is useful to write explicitly the form taken by Rayleigh-Schrödinger perturbative expansion for the perturbed Hamiltonian

$$H' = H_{eff} + V. \tag{II.9}$$

The new energy levels and right eigenvectors may be obtained with

$$E_i' = E_i + \langle\bar{i}|V|i\rangle, \tag{II.10a}$$

$$|i'\rangle = |i\rangle + \sum_k |k\rangle \frac{\langle\bar{k}|V|i\rangle}{E_i - E_k}, \tag{II.10b}$$

whereas the new left eigenvectors are given by

$$\langle i'| = \langle i| + \sum_k \frac{\langle\bar{i}|V|k\rangle}{E_i - E_k} \langle\bar{k}|. \tag{II.11}$$

We may, as an illustration, calculate the matrix elements of the dipole operator, $d = -e\vec{\varepsilon}_\lambda\cdot\vec{r}$ for the perturbed system,

$$\langle\bar{i}'|d|j'\rangle = \langle\bar{i}|d|j\rangle + \sum_{k\neq i}\left(\frac{\langle\bar{i}|d|k\rangle\langle\bar{k}|V|j\rangle}{E_i - E_k} + \frac{\langle\bar{i}|V|k\rangle\langle\bar{k}|d|j\rangle}{E_i - E_k}\right). \tag{II.12}$$

Note that the energy denominators have not been complex conjugated and that $\langle\bar{i}'|d|j'\rangle \neq \langle i'|d|j'\rangle$.[8]

III. TIME REVERSAL SYMMETRY

We consider the symmetry of a stable system under the reversal of the flow of time. We assume the system is governed by the Schrödinger equation

$$i\hbar \frac{d}{dt} \Psi(r,t) = H(r,t)\Psi(r,t), \tag{III.1}$$

where the Hamiltonian H may depend explicitly on space and time through the presence of an external electromagnetic field ($\Phi(r,t)$, $\vec{A}(r,t)$),

$$H(r,t) = \frac{1}{2m}(\vec{p} - \frac{e}{c}\vec{A}(r,t)) - \frac{Ze^2}{r} + U(r)\vec{\sigma}\cdot\vec{L} + e\Phi(r,t). \tag{III.2}$$

The invariance of the equation of motion under the reversal of time simply states the existence of a quantum state $\Psi(r,t'=t-T)$ which will evolve through the same intermediate states as Ψ, but backward in time

$$i\hbar \frac{d}{dt'} \Psi'(r,t') = H(r,t')\Psi(r,t'). \tag{III.3}$$

Thus, the existence of such symmetry implies that there is an operator θ such that $\Psi(r,t) = \theta\Psi(T-t,r)$ is a solution of eq. III.3.[8] Because the vector potential is generated by currents,

$$\vec{A}'(t) = \theta\vec{A}(t)\theta^{-1} = -\vec{A}(r,T-t), \tag{III.4}$$

while $\Phi'(t) = \Phi(r,T-t)$, since the scalar potential is induced by charges. The construction of the operator θ may be performed with the following theorems[9]: if we consider an arbitrary superposition of states $\Psi = \sum_i c_i\Phi_i$, its time reversed state will have the proper time development only if the operator θ is anti-unitary. Indeed we must have

$$e^{-iHt}\theta = \theta e^{iHt}$$

$$\text{or} \quad H\theta + \theta H = 0. \tag{III.5}$$

All anti-unitary operators may be decomposed as the product of a Hermitian operator U and the complex conjugation operator K

$$\theta = UK. \tag{III.6}$$

For the Hamiltonian III.2 of a one-electron atom, it is easy to verify that $U = i\sigma_y$ leads to the proper time reversed equation III.3.

In Dirac notation, the time reversed state of a state $|\alpha\rangle$

$$|\alpha\rangle = \sum_n c_n(t)\ |n\rangle \tag{III.7}$$

may be written as

$$|\alpha'\rangle = \sum_n c_n^*(T-t)\,|n'\rangle \tag{III.8}$$

where $\langle r|n'\rangle = i\sigma_y\ \langle r|n\rangle$, and for two arbitrary states, we obtain the general reciprocity relation

$$\langle\beta,t|\alpha,t\rangle = \langle\alpha,T-t|\beta,T-t\rangle. \tag{III.9}$$

The interchange of initial and final state is inherent to the time reversal symmetry which reverses the role of incoming and outgoing waves. When $\vec{A}\neq 0$, θ will reverse the direction of the magnetic field and the correspondence between Zeeman components and their time-reversed images is not straightforward. We introduce for this reason the symmetry $\theta_y = -i\sigma_y\theta$,[9] which is the time reversal operation followed by a rotation of π around the y axis. This leaves B unchanged and

$$\begin{aligned}\theta_y E_{x,z}(t,r) &= -E_{x,z}(T-t,r')\\ \theta_y E_y(t,r) &= E_y(T-t,r).\end{aligned} \tag{III.10}$$

For an unstable system in which the atomic and radiative sectors are coupled, θ_y remains a symmetry of the complete Hamiltonian and one easily verifies that the time-reversed states of the right eigen-vector of the effective Hamiltonian can be identified with the left eigenvectors. From

$$\begin{aligned}&(H - E_\alpha)|\alpha\rangle \rightarrow \langle m|(H - E_\alpha)|\alpha\rangle = 0,\\ &\rightarrow \langle\alpha'|(H - E_\alpha)|m'\rangle = 0, \text{ or } \quad \langle\alpha'|(H - E_\alpha) = 0.\end{aligned} \tag{III.11}$$

Thus we may identify $\langle\bar{\alpha}|$ with $\langle\alpha'|$. If U is the time evolution operator,

$$U|\alpha\rangle = \sum_\beta U_{\beta,\alpha}|\beta\rangle, \tag{III.12}$$

where $U_{\beta,\alpha}$ is a transition amplitude $U_{\beta,\alpha} = \langle\bar{\beta}|U|\alpha\rangle$. Since θ_y is a symmetry for the effective Hamiltonian,

$$\begin{aligned}\langle\bar{\beta}|U(\Phi',A')|\alpha\rangle &= \langle\beta'|U(\Phi',A')|\alpha\rangle = \langle\alpha'|U(\Phi,A)|\beta\rangle\\ &= \langle\bar{\alpha}|U(\phi,A)|\beta\rangle,\end{aligned} \tag{III.13}$$

which is the fundamental reciprocity relation for unstable states.

IV. E1-M1 INTERFERENCES IN THE QUENCHING OF THE METASTABLE 2S STATE OF HYDROGEN

The level crossing in the n=2 shell of atomic hydrogen in the vicinity of 550 Gauss is shown in Fig. 1. On crossing, the β (2S, $m_J = -1/2$) state and the e (2P, $m_J = 1/2$) states are separated only by the radiative width $-i\hbar(\Gamma_{2S}-\Gamma_{2P})/2$, and we therefore expect damping effects to be important. In an electric field the two levels are mixed and, to first order in perturbation theory, the Stark-perturbed β state is

$$|\beta_s\rangle = |\beta\rangle + \delta E^- \;|e\rangle \qquad \text{(IV.1)}$$

where

$$\delta = \frac{\sqrt{6}\, ea_0}{E_\beta - E_e + i\hbar\Gamma/2}\,, \qquad \text{(IV.2)}$$

and

$$E^- = (E_x - iE_y)/\sqrt{2}. \qquad \text{(IV.3)}$$

The 2S state can decay to the ground state by a relativistically allowed M1 process, the matrix element for which is given by

$$\langle \hat{k},\hat{\varepsilon}_\lambda | \times \langle 1S,\pm 1/2 | \hat{\varepsilon}_\lambda \cdot \vec{\alpha} e^{-i\vec{k}\cdot\vec{x}} | 2S,-1/2\rangle$$
$$= -I(k)\ \langle \pm 1/2 | (\vec{\sigma}\cdot\hat{\varepsilon})(\sigma\cdot k) | -1/2\rangle, \qquad \text{(IV.4)}$$

whereas the e state decays predominantly by an E1 process,

$$\langle \hat{k},\hat{\varepsilon}_\lambda | \times \langle 1S,\pm 1/2 | \hat{\varepsilon}_\lambda \cdot \vec{\alpha} e^{-i\vec{k}\cdot\vec{x}} | 2P,1/2\rangle = iJ(k)\ \langle \pm 1/2 | \vec{\sigma}\cdot\hat{\varepsilon}_\lambda | 1/2\ . \qquad \text{(IV.5)}$$

The radial integrals I(k) and J(k) are given in References 10 and 11. The transition amplitude of the Stark-perturbed state to the ground state is obtained by applying the perturbative expansion (II.12) for unstable states,

$$\langle \hat{k},\hat{\varepsilon}_\lambda | \times \langle 1S,\pm 1/2 | \hat{\varepsilon}_\lambda \cdot \vec{\alpha} e^{-\vec{k}\cdot\vec{x}} | \beta_s\rangle$$
$$= i(\delta E^- J(k) \langle \pm 1/2 | \vec{\sigma}\cdot\vec{\varepsilon}_\lambda | 1/2\rangle + iI(k) \langle \pm 1/2 | (\vec{\sigma}\cdot\hat{\varepsilon}_\lambda)(\vec{\sigma}\cdot\hat{k}) | 1/2\rangle). \qquad \text{(IV.6)}$$

The angular distribution of the emitted radiation is obtained by squaring the transition amplitude and tracing over the undetected final states (polarization and ground state components):

$$\Phi(k) = \gamma M + \gamma_\perp \vec{E}^2{}_\perp + e_\perp (\hat{k}\cdot\vec{E})_\perp + f_\perp(\hat{k},\vec{E},\hat{B}). \qquad \text{(IV.7)}$$

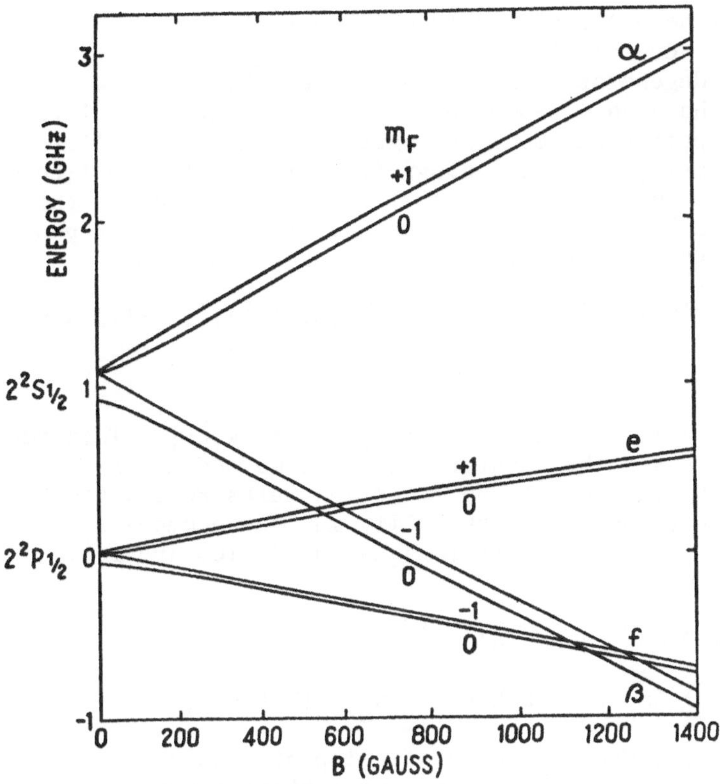

Fig. 1 Zeeman diagram for hydrogen at low fields, showing the hyperfine structure. The states are designated by the good quantum number m_F and the Greek letters α,β for S states and the Roman letters e,f for P states.

where $\gamma = |\delta|^2\Gamma$, $e = 2\sqrt{\gamma_M\Gamma}\ \mathrm{Im}(\delta)$, $f = 2\sqrt{\gamma_M\Gamma}\ \mathrm{Re}(\delta)$. Γ and γ_M are the E1 and M1 decay rates of the 2P and 2S states, respectively. If the θ_y operation is performed on the last two invariant scalars, we find $(\hat{k}\cdot\vec{E})_\perp \rightarrow -(\hat{k}\cdot\vec{E})_\perp$ whereas $(\hat{k},\vec{E},\hat{B})$ is even. We also note that $e_\perp$ is proportional to the lifetime of the 2P state, i.e., this interference term is possible only through the radiative reaction on the atom. The lineshapes of the invariant coefficients $e_\perp$ and $f_\perp$ as the magnetic field strength is varied are, respectively, resonant and dispersive.[4]

As seen in eq. IV.6, the E1 and M1 radiative matrix elements between the n=2 and n=1 shells are relatively imaginary. This is a consequence of rotational and time-reversal invariance. On these grounds, the corresponding amplitudes would not be expected to interfere. However, the Stark-induced amplitude acquires a phase lag through the large radiative reaction on the e state, which is maximum at the level crossing where the β and e states are separated only by their radiative width. A classical analogy with a damped harmonic oscillator (the e state) weakly coupled to a high Q oscillator (the β state) can be drawn: If the resonance frequency of the two oscillators become degenerate, the coordinate of the damped oscillator lags 90° behind the high Q resonator. The important role played by other symmetries (parity and Coulomb symmetries) does not allow one to carry the quasi-classical analogy further than the amplitude-phase relation for a damped oscillator. Finally, we clarify the role played by the θ_y symmetry in these interferences. For photon emission, the fundamental reciprocity relation takes the form

$$\langle 1S,\pm1/2|\times\langle\hat{k},\hat{\varepsilon}_\lambda|U(\Phi,\vec{A})|\beta_s\rangle = \langle\bar{\beta}_s|U(\Phi',\vec{A}')|1S,\pm1/2\rangle\times|\hat{k}',\hat{\varepsilon}_\lambda'\rangle, \tag{IV.8}$$

where $|\hat{k}',\hat{\varepsilon}_\lambda'\rangle$ is the θ_y reversed state of $\langle\hat{k},\hat{\varepsilon}_\lambda|$. This reciprocity relation expresses the identity of the time evolution of two distinct processes and, in general, no selection rules follow for damped systems. The transition probability may be decomposed in terms of even (I_n) and odd invariants (J_n) as the fields are replaced by their θ_y reversed images,

$$|\langle 1S,\pm1/2|\times\langle\hat{k},\hat{\varepsilon}_\lambda|U(\Phi,\vec{A})|\beta\rangle|^2 = \sum_n [a_n I_n(\vec{E},\vec{B},\hat{k},\hat{\varepsilon}_\lambda) + b_n J_n(\vec{E},\hat{B},\hat{k},\hat{\varepsilon}_\lambda)]. \tag{IV.9}$$

Substituting the fields and the final photon quantum numbers by their θ_y reversed images in (IV.9) and subtracting the original equation we find

$$|\langle 1S,\pm1/2|\times\langle\hat{k},\hat{\varepsilon}_\lambda|U(\Phi,\vec{A})|\beta\rangle|^2 - |\langle 1S,\pm1/2|\times\langle\hat{k}',\hat{\varepsilon}_\lambda'|U(\Phi',\vec{A}')|\beta\rangle|^2 = \sum_n b_n J_n(\vec{E},\hat{B},\hat{k},\hat{\varepsilon}_\lambda). \tag{IV.10}$$

For a damped system $U^\dagger \neq U$ and the reciprocity relation (IV.8) deso not require the θ_y odd invariant to vanish.[4] This may be viewed as a reduction of the symmetry ($U^\dagger = U$) of the original Hamiltonian to a lower symmetry and some terms previously excluded are no longer. It is interesting to note that the "selection rule" $J_n \equiv 0$ remains true for the diagonal matrix elements of the time evolution operator as a direct consequence of the general reciprocity relation (III.13). This excludes the possibility of having any θ_y-odd interference in the probability to remain in the initial state.

V. EXPERIMENTAL TESTS OF DAMPING INTERFERENCESS

A. Stark-induced transitions in H

In this experiment[5] the formally T-odd interferences are observed on a Stark-induced electric dipole transition, within the n=2 shell of atomic hydrogen, between two Zeeman sub-levels α_0 ($m_J \simeq 1/2$, $m_I \simeq -1/2$) and β_0 ($m_J \simeq -1/2$, $m_I \simeq 1/2$), in the vicinity of the β-e level crossing (Fig. 2). This experiment is therefore of special interest since radiative decay does not actually occur, thus displaying the key role played by the radiative reaction. The transition is induced by the mixing of the β_0 ($m_F = 0$) and e_{+1} ($m_F = +1$) states with a small dc electric field $\vec{E}$ perpendicular to the Zeeman field $\hat{B}$. The microwave electric field $\vec{\varepsilon}$ has both an axial component $\varepsilon_{\|}$ parallel to $\hat{B}$, which does not play an important role in this experiment, and a perpendicular component $\varepsilon_{\perp}$ which induces the Stark-transition. For the configuration of fields shown in Fig. 2, the invariant analysis of the scalar terms in the transition rate gives

$$R = a\,\vec{\varepsilon}_{\perp}^{\,2}\vec{E}^2 + b\,(\vec{\varepsilon}\cdot\vec{E})^2 + c\,(\vec{\varepsilon}\cdot\vec{E})(\vec{\varepsilon}\cdot\vec{E}\times\hat{B})$$

$$= \varepsilon_{\perp}^2 E^2\,[(a+b/2) + b/2\cos 2\phi + c/2\sin 2\phi]\,, \tag{V.1}$$

where ϕ is the angle of the electric field with respect to the $\vec{\varepsilon},\hat{B}$ plane. The last term has a θ_y-odd symmetry and has the same structure as the invariants discussed in the last section. In particular, it is proportional to the Stark-matrix elements of electric fields applied in the x and y directions, respectively. These matrix elements are

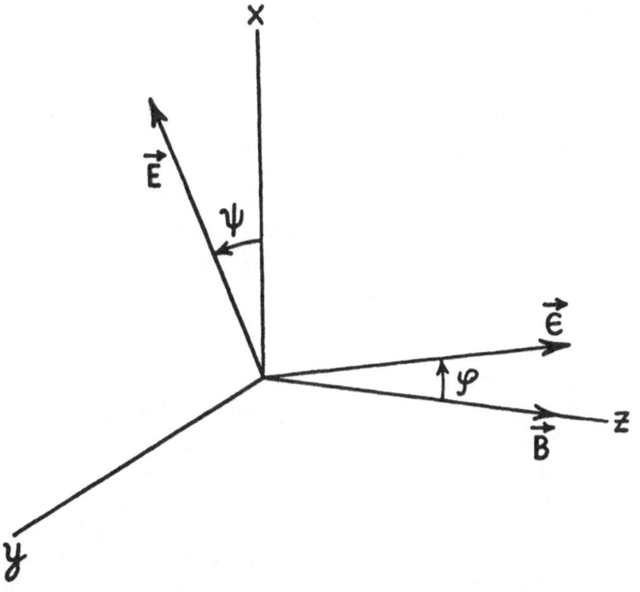

Fig. 2 Field configuration in the interaction region where the α_0-β_0 transition is driven.

relatively imaginary as a consequence of rotational invariance and T-symmetry. In order to express the invariant coefficients a,b,c in terms of the physical quantities which describe the atomic structure we expand to first order the perturbed β_0' state, using (II.9),

$$|\beta_0'\rangle = |\beta_0\rangle + \frac{V\cos\theta\, e^{-i\phi}}{\Delta E(\beta_0-e_{+1})+i\hbar\Gamma/2}\,|e_{+1}\rangle - \frac{V\sin\theta\, e^{i\phi}}{\Delta E(\beta_0-f_{-1})+i\hbar\Gamma/2}\,|f_{-1}\rangle, \tag{V.2}$$

where $|\ \rangle$ denotes the right eigenvector of H_{eff}, $V = \sqrt{3}\, ea_0 E$, and θ is the hyperfine mixing angle (θ 1/20 in the vicinity of the β-e level crossing). The corresponding left eigenvector is

$$\langle\bar{\beta}_0'| = \langle\bar{\beta}_0| + \frac{V\cos\theta\, e^{i\phi}}{\Delta E(\beta_0-e_{+1})+i\hbar\Gamma/2}\,\langle\bar{e}_{+1}| - \frac{V\sin\theta\, e^{-i\phi}}{\Delta E(\beta_0-f_{-1})+i\hbar\Gamma/2}\,\langle\bar{f}_{-1}|, \tag{V.3}$$

according to the expansion (II.11) The transition rate induced by the microwave field is $R(\alpha_0 \to \beta_0) = 1/4\ L(\omega,\tau)\ |\langle\beta_0'|e\varepsilon\cdot r|\alpha_0'\rangle|^2$, where $L(\omega,\tau)$ describes the transition lineshape. Evaluation of the rate and comparison with (V.1) yields

$$a = \kappa/4|\Delta_1^{-1} + \Delta_2^{-1}|^2,\quad b = -\kappa\,\mathrm{Re}(\Delta_1\Delta_2^*),\quad c = \kappa\,\mathrm{Im}(\Delta_1\Delta_2^*), \tag{V.4}$$

where

$$\begin{aligned}\Delta_1^{-1} &= [\Delta E(\beta_0-e_{+1})+i\hbar\Gamma/2]^{-1} - [\Delta E(\alpha_0-f_{-1})+i\hbar\Gamma/2]^{-1}\\ \Delta_2^{-1} &= [\Delta E(\beta_0-f_{-1})+i\hbar\Gamma/2]^{-1} - [\Delta E(\alpha_0-e_{+1})+i\hbar\Gamma/2]^{-1},\end{aligned} \tag{V.5}$$

and $\kappa = 9/4\ (ea_0)^4\sin^2 2\theta$, and a_0 is the Bohr radius. The invariant c is proportional to Γ, the lifetime of the 2P state which, as we saw before, explicitly exhibits that θ_y terms are possible only through the radiative reaction. The sign of the θ_y-odd invariant changes for $\vec{B}\to-\vec{B}$, $E_x \to -E_x$ ($\phi\to\pi-\phi$) or $E_y\to -E_y$ ($\phi\to\pi-\phi$), providing characteristic signatures for the presence of this term. Each reversal induces an asymmetry in the transition rate which is maximum for $\phi \approx \pi/4$:

$$A_1 = c/(2a + b). \tag{V.6}$$

Measurement of A_1 provides a determination of c. The sign of c depends much on the way damping is treated in the presence of external fields. The measurement of the magnetic field dependence of A_1 through the β-e level crossing is compared in Fig. 3 with the predicted asymmetry obtained from the non-Hermitian Hamiltonian formulation previously discussed. Because of the high degeneracy obtained at the level crossing, this measurement represents a distinct characterization of the interaction of an atom and its radiation field in the presence of external perturbation. The negative sign of the asymmetry explicitly confirms the phenomenological description of damping interferences given in sections II and III. Alternatively, the assumption that the use of an effective Hamiltonian is correct

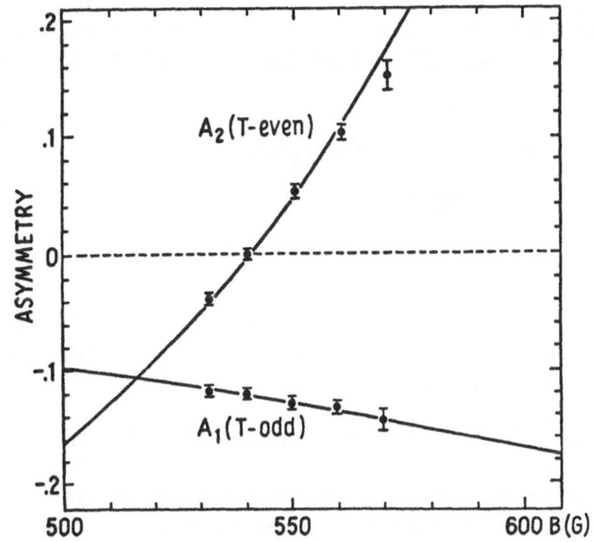

Fig. 3 Comparison of the experimental (solid circles) and theoretical (solid lines) values of the asymmetries A_1 and A_2. The errors are one standard deviation.

allows us to obtain from the measurement of A_1 an experimental value of the decay rate of the $2P_{1/2}$ state $\Gamma = 99.3 \pm 1.0$ MHz. The integrity of the measurement of A_1 was checked by measurement of the magnetic field-dependent asymmetry in the α_0-β_0 transition rate resulting from the sign change of the second (θ_y even) term of eq. (V.1) when $E = E_x$ ($\phi = 0$) $\rightarrow$ $E = E_y$ ($\phi = \pi/2$),

$$A_2 = b/(2a + b). \qquad (V.7)$$

The agreement of the experimental data with the theoretical prediction is also shown in Fig. 3.

B. Quenching anisotropy of $^4He^+$

In this experiment[6] a metastable (2S) beam of $^4He^+$ is polarized in a given m_J sublevel. The beam is subsequently quenched by the application of an electric field perpendicular to the polarization axis ($\vec{P}$), in zero magnetic field (Fig. 4). The angular distribution of the emitted radiation contains to lowest order three invariants,

$$I(k) = a\,\vec{E}^2 + b\,(\hat{k}\cdot\vec{E})^2 + c\,(\hat{k}\cdot\vec{E})(\hat{k}\cdot\vec{E}\times\vec{P}) \qquad (V.8)$$

where the relativistically allowed 2S-1S M1 rate has been ignored. The last invariant has a characteristic θ_y odd symmetry. The invariant coefficient may be expressed in term of the $^4He^+$ atomic structure[11],

$$\begin{aligned} a &= \kappa[|\Delta_1|^{-2} + 5/2|\Delta_2|^{-2} + \mathrm{Re}(\Delta_1{}^*\Delta_2)^{-1}], \\ b &= -\kappa[3/2|\Delta_2|^{-2} + 3\mathrm{Re}(\Delta_1{}^*\Delta_2)^{-1}], \\ c &= -\kappa\,\mathrm{Im}(\Delta_1{}^*\Delta_2)^{-1}. \end{aligned} \qquad (V.9)$$

The coefficient κ is $3\pi/4\ (ea_0/Z)^2\Gamma$, where Z (=2) is the nuclear charge and Γ is the decay rate of the 2P state. The energy differences Δ_1 and Δ_2 are respectively,

$$\Delta_1 = E(2S\text{-}2P_{1/2}) + i\hbar\Gamma/2, \quad \Delta_2 = E(2S\text{-}2P_{3/2}) + i\hbar\Gamma/2. \qquad (V.10)$$

From (V.8) we see that there is an anisotropy of the emitted radiation with respect to the ($\vec{E}$,$\vec{P}$) plane. More precisely, one can measure an asymmetry in the detected Lyman-α flux with two detectors located at

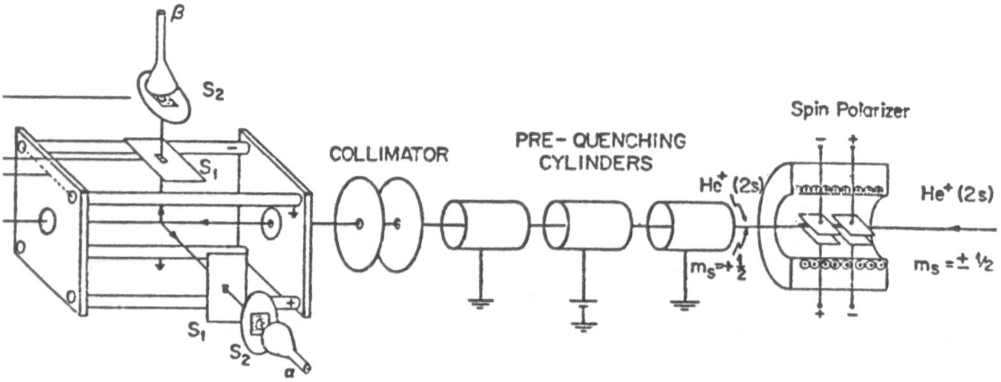

Fig. 4 Schematic diagram of the apparatus used in ref. 6 to measure the anisotropy of the Lyman-α radiation emitted in the quenching of $^4He^+$.

45° with respect to this plane,

$$A = (I(\pi/4) - I(-\pi/4))/(I(\pi/4) + I(-\pi/4)) = c/(2a+b), \quad \text{(V.11)}$$

which has the same form as in the previous experiment. This asymmetry can be calculated in terms of the respective energy differences and one finds[11]

$$A = \frac{3\hbar\Gamma[\Delta E(2P_{3/2}-2P_{1/2})]}{4(\Delta_2^r)^2 - 2\Delta_1^r\Delta_2^r + 7(\Delta_1^r)^2 + 9(\hbar\Gamma)^2/4} \quad \text{(V.12)}$$

which is proportional to the lifetime of the 2P state. The quantities Δ_1^r and Δ_2^r are the real part of Δ_1 and Δ_2 respectively. This asymmetry was accurately measured using the odd signature of the last invariant as $(\vec{k}\cdot\vec{E})$ or $\vec{P}$ are reversed. The measured value $A = 0.00760 \pm 0.0001$ is in good agreement with the predicted theoretical value of 0.007618 derived from eq. (V.12).[6]

VI. CONCLUSION

We have shown how the damping of quantum states could give rise to dynamical effects associated with the phase lags induced by the reaction of the quantum system to the radiation field. By analyzing the role of time reversal symmetry for damped systems, we concluded that such dynamical interferences were allowed in the transition moments between the damped eigenstates of the system, but were excluded in the probability to remain in a given quantum state. The resulting interferences always involve the width of some excited states and cannot be obtained from the associated classical system by a judicious application of the correspondence principle. More precisely, the interferences which are quantum mechanically observable are determined by the symmetries of the Hamiltonian. Because of the intrinsic quantum nature of parity inversion symmetry, the interference structure cannot be obtained classically. The two experiments briefly described represent therefore a unique manifestation of the quantum nature of the dynamical interferences induced by damping of quantum excitations. At present damping interferences have only been observed in hydrogenic system, because the higher symmetry of the pure Coulomb field provides additional degeneracy which enhances considerably the effects of damping. However, the effects discussed in this paper are universal and should be observable in all systems where level-crossing configurations occur. This could, for example, be the crossing of collective modes in condensed systems, such as the zero sound-spin wave level crossing of liquid ^{3}He recently analyzed.[12]

The author has greatly benefited from discussions with Drs. W. L. Williams and R. R. Lewis. Fellowship support was provided by the General Motors Company. This work was supported by the National Science Foundation under Grant DMR-8211508.

REFERENCES

1. D. Bohm, "Quantum Theory," Prentice-Hall, Inc., Englewood Cliffs, N.J. (1951).
2. D. M. Greenberger and A. W. Overhauser, Rev. Mod. Phys. 51:43 (1979).
3. A. van Wijngaarden, E. Goh, G. W. F. Drake, P. S. Farago, J. Phys. B9:201 (1976).
4. L. P. Lévy, Final State Interferences in the Radiative Decay of H(2S), to appear in Phys. Rev. A.
5. L. P. Lévy and W. L. Williams, Phys. Rev. Lett. 48:1011 (1982). L. P. Lévy, Ph.D. dissertation, Univ. of Michigan (Univ. Microfilms, Inc.).
6. A. van Wijngaarden, R. Helbing, J. Patel and G. W. F. Drake, Phys. Rev. A25:862 (1972).
7. See, for example, S. H. Patel, Y. Tomozowa and Y. P. Yao, Phys. Rev. 142: 1041 (1965).
8. The quantity <i'|d|j'> may in some circumstances be the observable of interest. J. S. Bell, in Proceedings of the Intnl. Workshop on Neutral Current Interactions in Atoms, p. 288, Cargèse, 1979, edited by W. L. Williams (Univ. of Michigan, Ann Arbor, 1980). J. S. Bell and G. Karl, Nuovo Cimento 41A:487 (1977).
9. E. P. Wigner, "Group Theory," p. 347, Academic Press, New York (1959).
10. M. Hillery and P. G. Mohr, Phys. Rev. A21:24 (1980).
11. A. van Wijngaarden and G. W. F. Drake, Phys. Rev. A25:400 (1982)
12. J. B. Ketterson, Phys. Rev. Lett. 50:259 (1983).

NONPERTURBATIVE TREATMENT OF DECAYING STATES

H.M. Nussenzveig

Departamento de Física
Pontifícia Universidade Católica
Rio de Janeiro, Brazil

1. INTRODUCTION

Decaying states appeared in quantum mechanics in 1928, in Gamow's treatment[1] of α-decay. However, they were already employed in classical physics by J.J. Thomson[2] almost exactly one century ago. Although our aim will be the application to quantum electrodynamics, it is useful to begin with simple examples, to show where problems arise and how one can deal with them.

Thomson's work dealt with the free modes of oscillation of the electromagnetic field around a perfectly conducting sphere. Since incoming radiation would provide a forcing term, "free" means that the radiation must be purely outgoing. The scattering matrix is defined in terms of the ratio between outgoing and incoming amplitudes. Thus, free modes of oscillation are associated with poles of the S-matrix. In Thomson's problem, these poles correspond to "complex eigenfrequencies"

$$\omega_n = \omega'_n - i\gamma_n \quad , \quad \gamma_n > 0 \quad , \tag{1.1}$$

which at first looks very nice, because

$$\exp(-i\omega_n t) = \exp(-i\omega'_n t)\exp(-\gamma_n t) \tag{1.2}$$

decays exponentially as $t\to\infty$, as one would expect (the field must get damped by radiation).

However, free outgoing spherical electromagnetic waves depend on t through $t-r/c$, so that (1.2) leads to an r-dependence of the form

$$\exp(i\omega_n r/c) = \exp(i\omega'_n r/c)\exp(\gamma_n r/c) \;, \qquad (1.3)$$

which blows up exponentially as $r \to \infty$. Respectable solutions of Maxwell's equations are not supposed to do this!

The root of the trouble is easily found. The field far away was emitted long ago, and (1.2) blows up as $t \to -\infty$. It is obviously unphysical to assume that the free modes existed at arbitrarily remote times in the past: they must have been excited somehow by supplying a finite amount of energy, after which they could decay freely.

Taking into account the excitation process provides a natural cutoff to the "exponential catastrophe". We conclude that, in order to make sense of "decaying states", the decay should be considered together with the excitation. One way to do this is to look for the general solution of the initial-value problem[3]: the excitation conditions are reflected in the choice of initial data.

Long-lived decaying states, associated with sharp resonances, are found in all areas of physics. They can often be regarded as perturbed stationary states, in the sense that they would become stationary if some interaction, responsible for the decay, were switched off: the stationary states of a free atom, uncoupled from the radiation field, are a good example. Since the stationary states provide a basis, in terms of which an arbitrary state can be expanded, it is natural to ask whether a similar expansion is possible in terms of decaying states.

Such questions can be answered in the framework of simple models, e.g., in nonrelativistic potential scattering.[4,5] Gamow's model of α-decay (fig.1) was a problem of this type.

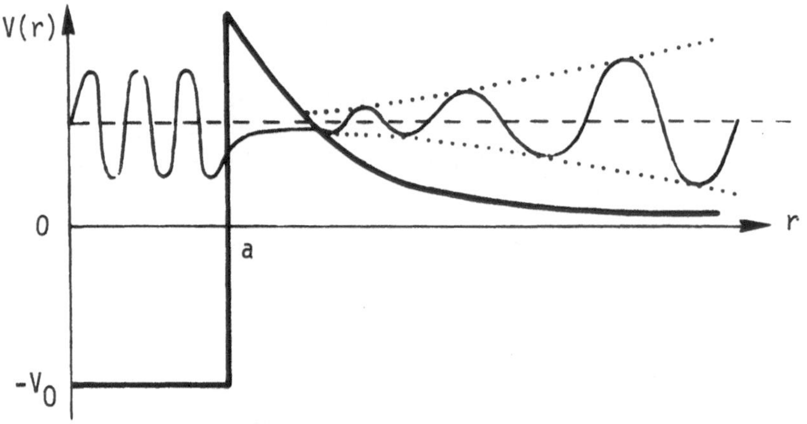

Fig.1. Wave packet associated with a decaying state in Gamow's model of α-decay. The exponential growth with r in the outside region is cut off due to formation at a finite time in the past.

A very simple model that can be explicitly solved is the delta-shell potential

$$V(r) = (A/2a)\delta(r-a) , \tag{1.4}$$

where, in units $\hbar=m=1$, A is a dimensionless parameter that measures the opacity of the shell. We consider only s waves. When $A\to\infty$, we get an impenetrable barrier at $r=a$ and the well-known stationary states of a particle in a spherical box, corresponding to the energy levels

$$E_n = \frac{1}{2}k_n^2 , \quad k_n = n\pi/a \quad (n=1,2,\ldots) . \tag{1.5}$$

For $A\gg1$, we get poles of the s-wave S-matrix $S(k)$ close to these points, in the lower half-plane,

$$k_n a \approx n\pi - i(n\pi/A)^2 \quad (n\pi\ll A) ,$$

$$E_n = \frac{1}{2}k_n^2 = E'_n - \frac{i}{2}\Gamma_n , \tag{1.6}$$

where E_n are the "complex energies" of the associated Gamow decaying states. Obviously, these complex E_n cannot be eigenvalues of the Hamiltonian; a "complex energy eigenfunction" is not normalizable, because of the exponential catastrophe.

Let $\Psi_j(r,t) = \phi_j(r,t)/r$ $(j=1,2)$ be the Schrödinger wave function in region j, where $j=1$ is the internal region $(0\le r<a)$ and $j=2$ is the external region $(r>a)$; the initial values are given by $\phi_j(r,0)$. If $\phi_2(r,0) = 0$, we have a decay problem of the Gamow type; if $\phi_1(r,0) = 0$, we have a scattering problem for the initial wave packet in region 2: in particular, we can describe resonance scattering, by concentrating its energy spectrum near one of the unperturbed eigenstates (1.5).

The general solution of the initial-value problem is given by

$$\phi_j(r,t) = \int_0^a G_{j1}(r,\rho,t)\phi_1(\rho,0)d\rho + \int_a^\infty G_{j2}(r,\rho,t)\phi_2(\rho,0)d\rho , \tag{1.7}$$

where $G_{jk}(r,\rho,t)$ are the propagators (Green's functions). Specifically, G_{11} describes the evolution of the internal excitation, G_{22} describes the scattering process, and $G_{12}(r,\rho,t) = G_{21}(\rho,r,t)$ represents the transmission between the internal and the external region.

The exact treatment[4] depends crucially on detailed knowledge about the analytic behavior of the S-matrix in the complex k plane. The main result is that $G_{jk}(r,\rho,t)$ can be expanded in terms of decaying state propagators associated with the poles k_n of $S(k)$. As $A\to\infty$, this expansion goes over into the usual expansion in terms of the stationary states (1.5), so that it represents the desired generalization of the stationary state expansion.

The basic decaying state propagator is given by

$$M(x,k_n,t) = -\frac{1}{2\pi i}\int_{-\infty}^{\infty}\exp\left[i(kx-Et)\right]\frac{dk}{k-k_n} \quad (\mathrm{Im}k_n<0) \ , \qquad (1.8)$$

where $E = \frac{1}{2}k^2$, so that M is the solution of the free-particle Schrödinger equation that, for t=0, reduces to

$$M(x,k_n,0) = \theta(-x)\exp(ik_n x) \ , \qquad (1.9)$$

where $\theta(x)=1$ for $x>0$, $\theta(x)=0$ for $x<0$. Thus, M is a cutoff Gamow decaying state. The cutoff contained in the Heaviside step function, which removes the exponential catastrophe, arises from turning on the excitation process through initial conditions at t=0.

For a long-lived decaying state, M will propagate much like a Gamow eigenfunction during many lifetimes, decaying exponentially. However, like any free-particle Schrödinger wave packet, it must eventually spread for sufficiently large times. This leads to a fundamental limitation of the exponential decay law.

The spreading law for a wave packet of free massive particles is easily found. If the initial wave packet is confined to a region of spatial extent Δx (of order $1/|\mathrm{Im}k_n|$ in the case of (1.9)), the associated momentum uncertainty $\Delta p \sim 1/\Delta x$ increases the volume of this region like $(\Delta p\ t)^3$. Normalization therefore requires that $|\Psi|^2$ must decay like t^{-3}, so that $\Psi(r,t)$ behaves like $t^{-3/2}$ as $t\to\infty$. This asymptotic behavior is indeed found for (1.8), superseding exponential decay for large enough t. For long-lived decaying states, the deviations appear after so many lifetimes that the particle is practically certain to have decayed, so that they would not be detected in practice. Deviations also occur for very short times, reflecting the details of the excitation process.

Why do we find free-particle decay for $t\to\infty$? The long-time behavior depends on the threshold behavior in momentum space as $p\to 0$. The associated slow-particle components have very long de Broglie wavelengths, so that they are almost unaffected by the confining potential, which has a short range.

Resonance scattering is described by G_{22}. The expansion in decaying state propagators allows one to treat in detail[4] the scattering of an initial wave packet, characterized by its spectral width and detuning, so as to discuss the dependence of the decay on the excitation process. Humblet and Rosenfeld[6] extended this method to the time-dependent theory of nuclear reactions.

The above formulation allows a completely explicit treatment of decay and resonance scattering, showing the role of decaying states and the possibility of expanding the general solution in terms of them, staying entirely within the normal framework of quantum mechanics (usual Hilbert space).

2. THE WEISSKOPF-WIGNER PROBLEM

It would be hard to improve on Weisskopf's own description[7] of how (in 1930) he came to work on this problem:

"When I came to my doctoral thesis I was thinking about quantum electrodynamics. Dirac had published his first paper on quantum electrodynamics, on the interaction of light with particles. I was especially interested in the question of radiation damping, the natural width of spectral lines. I dabbled around alone and tried to find exponential solutions of electrodynamics: I did not get very far because I was too young and ignorant. I asked the great Wigner for help; he was a few years older than I, but he already was very famous at that time. Of course, he helped me, right away; together we wrote a paper on the natural width of spectral lines, a paper that contains for the first time a divergent integral. I tried to convince Wigner that the integral could be made to vanish. Wigner said: "No, no, it is infinite." I didn't believe him, but he was right, of course. This paper, part of which later became my thesis, was the first paper in which divergent integrals appeared. They have not yet been resolved; they are still there after 40 years. One ought to be ashamed of it."

The problem treated by WW was the spontaneous decay of an excited atom interacting with the quantized radiation field. They assumed that the atomic state vector initially corresponds to an excited bound state (unperturbed by the interaction with radiation) and that the initial state vector for the field is the photon vacuum. In order to treat the time evolution of this state, nonperturbative methods are required, since ordinary time-dependent perturbation theory would at best be applicable only for times short enough that the probability of decay is negligible.

The procedure applied by WW was to make a guess; this is made to look respectable by referring to it in German as an "Ansatz". They replaced the system by what would now be called a two-level atom. The interaction with the field was treated in the dipole approximation. Only two types of states were taken into account: those where the atom is excited and there are no photons, and those where the atom is in the ground state and there is one emitted photon, assuming that only these states would be important. Finally, WW looked for a "decaying state" solution with exponential decay at all times. These assumptions led to the expected value for the lifetime of the excited state, together with a frequency shift given by Weisskopf's divergent integral.

No discussion was given of the validity of all these assumptions (including the exponential law) nor of how to improve on the approximation. Note that the WW initial state looks very unphysical: the atom would have to prepared in a bare excited state, with no photons in the field. It would seem preferrable to discuss a more realistic excitation process, e.g., through resonance scattering.

During the early 50's, Heitler and coworkers[8] tried to improve on this situation. Instead of starting from a bare excited state, they looked for a canonical transformation that would lead to dressed states. However, this concept is ambiguous, specially for excited states, and their treatment, which was very complicated, left many of the above questions unanswered. Low's covariant S-matrix treatment[9] did not address them either.

3. THE MODEL

In order to extend to QED the nonperturbative treatment of decaying states mentioned in Sect.1, we would like to have a model for decay and resonance scattering that incorporates as many realistic features as possible while still remaining exactly soluble.

An early version of such a model was treated by L. Davidovich,[10] and the present improved version was briefly discussed later.[11] It might be called "multilevel H atom in generalized rotating wave approximation". We start from the exact nonrelativistic H atom Hamiltonian (neglecting nuclear recoil)

$$\hat{H} = \underbrace{\hat{H}_A + \hat{H}_F}_{\hat{H}_0} + \underbrace{\hat{H}_{I,1} + \hat{H}_{I,2}}_{\hat{H}_I} \quad , \tag{3.1}$$

$$\hat{H}_A = \frac{\hat{\underset{\sim}{p}}^2}{2m} - \frac{e^2}{r} \quad , \tag{3.2}$$

$$\hat{H}_F = \frac{1}{8\pi}\int :\left[(\underset{\sim}{\hat{E}}^{\perp})^2 + (\underset{\sim}{\nabla}\times\underset{\sim}{\hat{A}})^2\right]: d^3r \quad , \tag{3.3}$$

$$\hat{H}_{I,1} = -\frac{e}{m}\,\hat{\underset{\sim}{p}}\cdot\hat{\underset{\sim}{A}}(\underset{\sim}{r}) \quad , \tag{3.4}$$

$$\hat{H}_{I,2} = \frac{e^2}{2m}\,\hat{\underset{\sim}{A}}^2(\underset{\sim}{r}) \quad , \tag{3.5}$$

together with the canonical commutation relations. We have set $\hbar=c=1$ throughout. We work in Coulomb gauge, in the Schrödinger picture and with the minimal-coupling interaction.

We take as basis states the eigenstates of $\hat{H}_0$. The atomic basis states are the eigenstates of $\hat{H}_A$,

$$|j\rangle = |n_j\, \ell_j\, m_j\rangle \tag{3.6}$$

for bound states (n,ℓ,m = principal, angular momentum and magnetic quantum numbers, respectively); continuum states will not be included in the model. For photons, we use Fock space in the angular momentum (multipole) representation. One-photon states are labelled by

$$|\beta k) = |JM\tau,k) \quad , \tag{3.7}$$

where 2^J is the multipole order ($J=1,2,3,\ldots$), M ranges from $-J$ to J, $\tau=0$ for electric multipoles and $\tau=1$ for magnetic multipoles, and k is the photon wave number (magnitude of photon momentum).

The multipole fields are given by[12]

$$\underset{\sim}{A}_{JM1}(k,\underset{\sim}{r}) = j_J(kr)\underset{\sim}{T}_{J,J,M}(\hat{\underset{\sim}{r}}) \quad , \tag{3.8}$$

$$\underset{\sim}{A}_{JM0}(k,\underset{\sim}{r}) = -(J/J+1)^{1/2} j_{J+1}(kr)\underset{\sim}{T}_{J,J+1,M}(\hat{\underset{\sim}{r}}) + (J+1/2J+1)^{1/2} j_{J-1}(kr)\underset{\sim}{T}_{J,J-1,M}(\hat{\underset{\sim}{r}}) \quad , \tag{3.9}$$

where j_J is the spherical Bessel function of order J and

$$\underset{\sim}{T}_{JLM}(\hat{\underset{\sim}{r}}) = \sum_{M'} C(L1J;M-M',M')Y_{L,M-M'}(\hat{\underset{\sim}{r}})\hat{\underset{\sim}{\varepsilon}}_{M'} \tag{3.10}$$

are the vector spherical harmonics (eigenfunctions of $\underset{\sim}{J}^2, J_z, \underset{\sim}{L}^2, \underset{\sim}{S}^2$); the coefficients C are Clebsch-Gordan coefficients and $\hat{\underset{\sim}{\varepsilon}}_M$ are the spin eigenvectors

$$\hat{\underset{\sim}{\varepsilon}}_{\pm 1} = \mp(\hat{\underset{\sim}{x}}\pm i\hat{\underset{\sim}{y}})/\sqrt{2} \quad , \quad \hat{\underset{\sim}{\varepsilon}}_0 = \hat{\underset{\sim}{z}} \quad . \tag{3.11}$$

The vector potential operator is expanded as

$$\hat{\underset{\sim}{A}}(\underset{\sim}{r}) = 2\sum_{\tau=0}^{1}\sum_{J=1}^{\infty}\sum_{M=-J}^{J}\int_0^\infty dk\sqrt{k}\, \hat{a}_{JM\tau}(k)\underset{\sim}{A}_{JM\tau}(k,\underset{\sim}{r}) + \text{h.c.} \quad , \tag{3.12}$$

where h.c. stands for hermitean conjugate and $\hat{a}_\beta(k)$ is the annihilation operator for photons with the set of quantum numbers (3.7), satisfying the canonical commutation relations

$$[\hat{a}_\beta(k),\hat{a}_{\beta'}^{+}(k')] = \delta_{\beta\beta'}\delta(k-k') \quad , \tag{3.13}$$

$$[\hat{a}_\beta(k),\hat{a}_{\beta'}(k')] = [\hat{a}_\beta^{+}(k),\hat{a}_{\beta'}^{+}(k')] = 0 \quad . \tag{3.14}$$

The free Hamiltonians (3.2), (3.3) become

$$\hat{H}_A = \sum_j E_{oj}|j\rangle\langle j| \quad , \tag{3.15}$$

$$\hat{H}_F = \sum_\beta \int_0^\infty dk\, k\, \hat{a}_\beta^{+}(k)\hat{a}_\beta(k) \quad , \tag{3.16}$$

where E_{0j} is the energy of the unperturbed bound state (3.6), given by Balmer's formula. The interaction Hamiltonian (3.4) becomes

$$\hat{H}_{I,1} = \frac{e}{m} \sum_{nm\beta} \int_0^\infty dk\sqrt{k}\, f_{nm\beta}(k)\hat{a}_\beta(k)|n\rangle\langle m| + \text{h.c.} \quad , \tag{3.17}$$

where

$$f_{nm\beta}(k) = -2\langle n|\hat{\underset{\sim}{p}}\cdot\underset{\sim}{A}_\beta(k,\underset{\sim}{r})|m\rangle$$
$$= -2\int d^3r\langle n|\underset{\sim}{r}\rangle\hat{\underset{\sim}{p}}\cdot\underset{\sim}{A}_\beta(k,\underset{\sim}{r})\langle\underset{\sim}{r}|m\rangle \tag{3.18}$$

and $\langle\underset{\sim}{r}|m\rangle = \psi_m(\underset{\sim}{r})$ is the H atom Schrödinger eigenfunction. The dyadic operator $|n\rangle\langle m|$ gives rise to transitions $m\rightarrow n$. The quadratic part (3.5) of the interaction Hamiltonian does not depend on atomic operators.

4. THE RESOLVENT OPERATOR METHOD

The basic ideas of this method go back to Heitler and coworkers,[8] but the most convenient formulation is found in a classic paper by Van Hove.[13] The resolvent operator is defined by

$$\hat{\mathcal{G}}(z) \equiv (z-\hat{H})^{-1} \quad , \tag{4.1}$$

where z is a complex variable. It exists as a bounded operator for $z \in \rho(\hat{H})$, the resolvent set of $\hat{H}$; we must exclude $\sigma(\hat{H})$, the spectrum of $\hat{H}$. In the present problem, σ is contained in the real axis: it will include poles associated with bound states and a cut along the continuous energy spectrum.

Thus, for Im $z \neq 0$ (without crossing the cut), $\hat{\mathcal{G}}(z)$ is a holomorphic operator function; this defines the physical sheet. The time evolution operator is given by

$$\exp(-i\hat{H}t) = -\frac{1}{2\pi i}\int_C e^{-izt}\hat{\mathcal{G}}(z)dz \;, \quad t>0 \quad , \tag{4.2}$$

where C is any straight line Imz = constant>0 parallel to the real axis in the upper half-plane.

Let $|\alpha\rangle$ be the eigenstates of $\hat{H}_o(=\hat{H}-\hat{H}_I)$,

$$\hat{H}_o|\alpha\rangle = E_{o\alpha}|\alpha\rangle \;, \quad \langle\alpha|\alpha'\rangle = \delta(\alpha-\alpha') \quad , \tag{4.3}$$

where $E_{o\alpha}$ are the unperturbed energies. The matrix elements of a general operator $\hat{\mathcal{O}}$ will be of the form

$$\langle\alpha|\hat{\mathcal{O}}|\alpha'\rangle = \delta(\alpha-\alpha')\mathcal{O}_d(\alpha) + \mathcal{O}_{nd}(\alpha,\alpha') \quad , \tag{4.4}$$

where $\mathcal{O}_{nd}$ does not contain a $\delta(\alpha-\alpha')$ singularity. Correspondingly, $\hat{\mathcal{O}}$

can be split into $\hat{\mathcal{O}}_d$, its diagonal part with respect to $\hat{H}_o$ (that gives rise to the first term in (4.4)) and $\hat{\mathcal{O}}_{nd}$, its nondiagonal part. Let us perform this decomposition for $\hat{\mathcal{G}}(z)$:

$$\hat{\mathcal{G}}(z) = \hat{\mathcal{D}}(z) + \hat{\mathcal{N}}(z) \quad , \tag{4.5}$$

where $\hat{\mathcal{D}}$ is the diagonal part of $\hat{\mathcal{G}}$ (with respect to $\hat{H}_o$) and $\hat{\mathcal{N}}$ is the nondiagonal part.

If the system is initially in an eigenstate $|a\rangle$ of $\hat{H}_o$, at t=0, the persistence amplitude of this state after a time t is given by

$$\langle a|\exp(-i\hat{H}t)|a\rangle = -\frac{1}{2\pi i}\int_C e^{-izt}\mathcal{D}_a(z)dz \quad , \tag{4.6}$$

where

$$\mathcal{D}_a(z) = \langle a|\hat{\mathcal{D}}(z)|a\rangle \quad . \tag{4.7}$$

The behavior of the persistence amplitude therefore depends crucially on the analytic behavior of $\mathcal{D}_a(z)$.

The analogue $\hat{\Sigma}(z)$ of Dyson's mass operator[14] is defined by

$$\hat{\mathcal{D}}(z) = \hat{\mathcal{G}}_o(z) + \hat{\mathcal{D}}(z)\hat{\Sigma}(z)\hat{\mathcal{G}}_o(z) \quad , \tag{4.8}$$

where

$$\hat{\mathcal{G}}_o(z) = (z-\hat{H}_o)^{-1} \tag{4.9}$$

is the unperturbed resolvent. We have

$$\hat{\mathcal{D}}(z) = \left[z - \hat{H}_o - \hat{\Sigma}(z)\right]^{-1} \quad , \tag{4.10}$$

so that $\hat{\Sigma}(z)$ represents the modification of the free propagator due to the interaction.

As z approaches the real axis from above or below, $\hat{\Sigma}(z)$ can approach different limits:

$$\lim_{z\to E\pm i0} \hat{\Sigma}(z) = \hat{\Delta}(E) \mp \frac{i}{2}\hat{\Gamma}(E) \quad , \tag{4.11}$$

where $\hat{\Delta}$ and $\hat{\Gamma}$ are diagonal operators with real eigenvalues. Denoting by an index α the eigenvalues of diagonal operators in the state $|\alpha\rangle$, (4.10) and (4.11) yield

$$\lim_{z\to E\pm i0} \mathcal{D}_\alpha(z) = \left[E-E_{o\alpha}-\Delta_\alpha(E) \pm \frac{i}{2}\Gamma_\alpha(E)\right]^{-1} \quad . \tag{4.12}$$

Thus, $\hat{\Delta}(E)$ plays the role of level shift operator, and $\hat{\Gamma}(E)$ that of level width operator. The latter is a positive semidefinite operator,

$$\hat{\Gamma}(E) \geq 0 \quad , \tag{4.13}$$

which means that $\Gamma_\alpha(E) \geq 0$ for all $|\alpha>$.

Van Hove's Classification of States:

Let $E_{o\alpha}$ be given by (4.3), and assume that

$$E - E_{o\alpha} - \Delta_\alpha(E) = 0 \quad , \tag{4.14}$$

as an implicit equation for E, has exactly one real root $E=E_\alpha$ (this will be true for the cases considered here). Then, there are only three possibilities:

$$\text{(i) } \Gamma_\alpha(E_\alpha) \neq 0 \quad . \tag{4.15}$$

By (4.12), this means that E_α lies on a cut of $\mathcal{D}_\alpha(z)$. Van Hove calls this a dissipative state: the interaction leads to some damping or diffusion process for such a state. An example is provided by the Weisskopf-Wigner excited initial state (cf. Sect. 10 below). The Weisskopf-Wigner approximation corresponds to

$$\mathcal{D}_\alpha(z) = (z-E_\alpha + \frac{i}{2}\Gamma_\alpha)^{-1} \quad , \tag{4.16}$$

which has a pole at $z=E_\alpha - \frac{i}{2}\Gamma_\alpha$ on the physical sheet. This "complex eigenvalue" violates the analyticity of $\mathcal{D}_\alpha(z)$ on this sheet, so that it is unacceptable.

$$\text{ii) } \Gamma_\alpha(E) = 0 \text{ for all real E} \; . \tag{4.17}$$

In this case, by (4.12), $\mathcal{D}_\alpha(z)$ has a simple pole at $z=E_\alpha$ and no cut. Van Hove calls this an asymptotically stationary state: its time evolution for $t\to\infty$ is not affected by the interaction $\hat{H}_I$, which produces only transient perturbation effects. An example is provided by $|\alpha> = |\psi(\text{out})>$, an outgoing state in scattering by a short range potential (asymptotic free-particle state extrapolated back to t=0). In the present case, a one-photon wave packet scattered from the atom in the ground state defines an asymptotically stationary state (Sect. 12). The interaction produces transitions between asymptotically stationary states, but is does not "dress" such a state.

Transitions are produced by the nondiagonal part $\hat{\mathcal{N}}(z)$ of $\hat{\mathcal{G}}(z)$ in (4.5). The transition operator $\hat{\mathcal{U}}(z)$ is defined by

$$\hat{\mathcal{N}}(z) = \hat{\mathcal{D}}(z)\hat{\mathcal{U}}(z)\hat{\mathcal{D}}(z) \quad . \tag{4.18}$$

If $|\alpha>, |\alpha'>$ are asymptotically stationary, the S-matrix element $<\alpha|\hat{S}|\alpha'>$ exists (without any need of adiabatic switching), and it is given by

$$<\alpha|\hat{S}|\alpha'> = \delta(\alpha-\alpha')-2i\pi\delta(E_\alpha-E_{\alpha'})\mathcal{U}_{\alpha\alpha'}(E_\alpha) \quad , \tag{4.19}$$

so that

$$\mathcal{U}_{\alpha\alpha'} = <\alpha|\hat{\mathcal{U}}|\alpha'> \tag{4.20}$$

is the transition amplitude.

$$\text{(iii)}\ \Gamma_\alpha(E_\alpha) = 0 \ , \text{ but } \Gamma_\alpha(E)\neq 0 \text{ for some real } E \ . \tag{4.21}$$

In this case, by (4.12), $\mathcal{D}_\alpha(E)$ has a cut on the real axis (where $\Gamma_\alpha(E)\neq 0$) and a simple pole at $E=E_\alpha$ outside of the cut. For such a state, the interaction produces persistent perturbation effects, which include both renormalization (in QED, self-energy, charge and wave function renormalization) and cloud effects, that give rise to dressing. One must then look for a new set of asymptotically stationary states, $|\alpha>_{as}$, that are dressed states of the coupled system. This is the most difficult situation to handle, and in QED it has only been treated perturbatively. However, we will see that one can go a long way in the nonperturbative treatment of decaying states without including persistent perturbation effects.

5. POSSIBLE APPROXIMATIONS

So far, everything has been exact. However, the prospects of exact solution for the full Hamiltonian look dim, so that we must introduce approximations. Ordinary perturbation theory does not work here: we need a nonperturbative approach. The basic philosophy will be to approximate the Hamiltonian, i.e., to find an exactly soluble model Hamiltonian, trying to retain the largest possible piece of the full $\hat{H}$ for which we can find an exact solution. We also try to make it as realistic as possible, e.g., by avoiding approximations in the matrix elements. Having solved a model Hamiltonian, one can then add back terms omitted from $\hat{H}$ and try to find out their effect, at least by perturbative methods. A somewhat analogous approach has been advocated by Grimm and Ernst.[15]

We employ two basic approximations for reducing the H atom Hamiltonian (3.1) to our model Hamiltonian:

(i) Finite Number of Atomic States:

This means cutting off the spectrum of $\hat{H}_A$ beyond some discrete level N, so that (3.15) is replaced by

$$\hat{H}_A = \sum_{n=1}^{N} E_{on}|n><n| \quad . \tag{5.1}$$

This eliminates the continuous spectrum, as well as the accumulation point of discrete levels at the ionization threshold. A very special case would be N=2, a two-level atom.

(ii) Generalized RWA for Resonant Interaction:

We are interested in processes where one particular excited atomic state is singled out, either as initial state or as one to be selectively excited by resonance scattering. Let $|r\rangle$ (for "resonant") denote this state. We now throw away the following pieces of the full interaction Hamiltonian $\hat{H}_I$: (A) The quadratic part $\hat{H}_{I,2}$, given by (3.5); (B) All nonresonant terms in $\hat{H}_{I,1}$, i.e., those terms in (3.17) that do not contain $|r\rangle$; (C) All "counter-rotating" terms in (3.17), of the type $\hat{a}_\beta(k)|n\rangle\langle r|$+h.c. . The remaining piece of the interaction defines the "generalized RWA" interaction Hamiltonian:

$$\hat{H}_I^{RWA} = \frac{e}{m}\sum_{n\beta}{}' \int_0^\infty dk\sqrt{k}\, f_{rn\beta}(k)\hat{a}_\beta(k)|r\rangle\langle n| + \text{h.c.} \quad , \tag{5.2}$$

where the prime in the summation sign means that n=r is to be excluded.

The terms retained in (5.2) produce two types of transitions:

(a) Photon absorption, always associated with transitions from other levels to $|r\rangle$.

(b) Photon emission, always associated with transitions from $|r\rangle$ to other levels.

For N=2, this reduces to the ordinary RWA for the two-level atom, with only "energy-conserving" transitions included. For N>2 and levels above $|r\rangle$, however, it is just the "energy-nonconserving" transitions that are retained. The reason for this choice will become apparent in Sect.6. Note also that, while all the other levels interact with $|r\rangle$, they do not undergo spontaneous decay to the ground state. The states $|n\rangle|0)$ ($n\neq r$), where $|0)$ denotes the photon vacuum, are annihilated by $\hat{H}_I^{RWA}$, so that they are not only asymptotically stationary, but actually stationary eigenstates of the full model Hamiltonian. This is of course an unrealistic feature of the model. It also implies that $|r\rangle$ can decay to the ground state only by one-photon processes: cascades are not allowed.

The following approximations will not be employed, for reasons discussed below:

(iii) Dipole Approximation:

This would correspond to setting, in the interaction Hamiltonian,

$$\hat{\mathbf{A}}(\mathbf{r}) = \hat{\mathbf{A}}(0) \quad ,$$

which is equivalent, in the interaction matrix elements, to neglecting retardation within the atom. In this approximation, only electric dipole waves couple to the atom, so that only one set of values of β contributes:

$$\beta \equiv (J=1;\ M=0,\pm1;\ \tau=0) \quad . \tag{5.4}$$

The corresponding electric dipole field in (3.9) is

$$\underset{\sim}{A}_{1M0}(\underset{\sim}{k},\underset{\sim}{0}) = \hat{\underset{\sim}{\varepsilon}}_M/\sqrt{3\pi} \quad , \tag{5.5}$$

which is a constant (independent of k). Thus, (3.18) reduces to the constant transition dipole matrix element, and this leads to ultraviolet divergent integrals (i.e., as $k\to\infty$) when the treatment is pursued further. Of course, the dipole approximation is inconsistent with the limit $k\to\infty$, so that, for consistency, as well as to eliminate divergences (Weisskopf's divergent integrals!), this approximation is usually coupled with a cutoff.

(iv) Cutoff:

To introduce a sharp cutoff, we multiply the matrix element $f_{rn\beta}(k)$ in (5.2) by

$$\rho(k) = \theta(K-k) \quad , \tag{5.6}$$

where θ is the Heaviside step function. This cuts off the integral at k=K. For consistency, K should be taken as $\sim a_B^{-1}$, where a_B is the Bohr radius. However, K is often taken at the electron Compton wave number, corresponding to a cutoff energy of the order of the electron mass, where relativistic effects begin to appear.

Approximations (iii) and (iv) are often employed in the two-level version of this model, which was probably first treated by Dirac as a model for resonance scattering[16]. In quantum field theory, it became known as the Lee model.[17] While the sharp cutoff renders the divergent integrals finite, it gives rise to a spurious pole of the resolvent on the real axis, that would contribute an unphysical nondecaying term to the time evolution;[10] this corresponds to the "ghost state" in the Lee model.

We will not employ approximations (iii) and (iv) in our model Hamiltonian. Fortunately, as will be seen in Sect.8, the exact H atom matrix elements, including retardation, can be explicitly evaluated. Furthermore, the exact atomic form factors provide a natural cutoff around the Bohr radius, and they have extremely nice analytic properties, which are very helpful in obtaining the explicit solution of the problem.

6. EXACTLY SOLUBLE MODEL. RESOLVENT MATRIX ELEMENTS

Applying approximations (i) and (ii), the Hamiltonian of our model becomes

$$\hat{H} = \sum_{n=1}^{N} E_{on}|n\rangle\langle n| + \sum_{\beta}\int_0^{\infty} dk\ k\ \hat{a}_\beta^{+}(k)\hat{a}_\beta(k)$$

$$+ \lambda \sum_{n\beta}{}' \int_0^{\omega} dk\sqrt{k}\left[f_{rn\beta}(k)\hat{a}_\beta(k)|r\rangle\langle n| + f^{*}_{rn\beta}(k)\hat{a}_\beta^{+}(k)|n\rangle\langle r|\right]$$

$$= \hat{H}_A + \hat{H}_F + \hat{H}_I \quad , \tag{6.1}$$

where $\lambda=e/m$, and the summation in $\hat{H}_I$ ranges not only over a finite set of values of n, but also over a finite set of values of β; for the H atom, this follows from the selection rules, as will be seen in Sect.8

It will be very instructive to discuss the structural properties of the Hamiltonian thus defined in a more general context, without specifying as yet the matrix elements $f_{rn\beta}(k)$. The crucial property of $\hat{H}$ that renders the model exactly soluble follows from the generalized RWA, that yields an additional conservation law.

This law corresponds to the conservation of the total number of excitations (comprehending both atomic excitation and photons), which is associated with the eigenvalues of the following operator:

$$\mathfrak{E} = \frac{1}{2}|r\rangle\langle r| - \frac{1}{2}\sum_{n}{}' |n\rangle\langle n| + \sum_{\beta}\int_0^{\infty} dk\ \hat{a}_\beta^{+}(k)\hat{a}_\beta(k) \quad . \tag{6.2}$$

With $\hat{H}_I$ defined by (6.1), each time the photon occupation number increases (decreases) by 1, corresponding to emission (absorption), the occupation number of state r decreases (increases) by 1, and the occupation number of some other state $n\neq r$ increases (decreases) by 1, so that $\mathfrak{E}$ must be conserved.

Indeed, it is readily verified that

$$[\mathfrak{E}, \hat{H}] = 0 \quad . \tag{6.3}$$

This leads to a splitting of the Hilbert space into disjoint sectors, generalizing what happens in the Lee model. The sector $\mathfrak{E}=-1/2$ is trivial; the atom sits in some state $|n\rangle\neq|r\rangle$ (an artificial feature of the model is that such states do not decay, as was noted in Sect. 5), and no photons are present: nothing happens! The next sector, however, $\mathfrak{E}=1/2$, is nontrivial, containing both decay and resonance scattering. From now on, we restrict our consideration to this sector, setting

$$\mathfrak{E} = 1/2 \quad . \tag{6.4}$$

The most general state vector in this sector is of the form

$$|\Psi\rangle = u|r;0\rangle + |\Phi\rangle \quad , \tag{6.5}$$

where u is a complex amplitude,

$$|r;0\rangle = |r\rangle|0) \quad , \tag{6.6}$$

$$|\Phi\rangle = \sum_{n\beta}{}' \int_0^\infty dk \ \phi_{n\beta}(k)\,|n;\beta,k\rangle \quad , \tag{6.7}$$

$$|n;\beta,k\rangle = |n\rangle\hat{a}_\beta^+(k)\,|0) = |n\rangle|\beta,k) \quad . \tag{6.8}$$

As in (3.6), (3.7), atomic states are denoted by angular brackets and photon states by round ones. Normalization requires

$$\langle\Psi|\Psi\rangle = |u|^2 + \sum_{n\beta}{}' \int_0^\infty dk\,|\phi_{n\beta}(k)|^2 = 1 \quad , \tag{6.9}$$

where the various terms have the standard probability interpretation. The probability amplitudes u and $\phi_{n\beta}$ are in general time-dependent.

From (4.9), (6.1) and (6.5), we get

$$\hat{\mathcal{G}}_0(z)\,|r;0\rangle = (z-E_{0r})^{-1}\,|r;0\rangle \quad , \tag{6.10}$$

$$\hat{\mathcal{G}}_0(z)\,|\Phi\rangle = \sum_{n\beta}{}' \int_0^\infty dk\,\frac{\phi_{n\beta}(k)}{z-E_{0n}-k}\,|n;\beta,k\rangle \quad , \tag{6.11}$$

$$\hat{H}_I\,|r;0\rangle = \lambda \sum_{n\beta}{}' \int_0^\infty dk\sqrt{k}\; f^*_{rn\beta}(k)\,|n;\beta,k\rangle \quad , \tag{6.12}$$

$$\hat{H}_I\,|\Phi\rangle = \left[\lambda \sum_{n\beta}{}' \int_0^\infty dk\sqrt{k}\; f_{rn\beta}(k)\phi_{n\beta}(k)\right]|r;0\rangle \quad . \tag{6.13}$$

The results always stay within the sector $\mathcal{E}=1/2$, as they must.

Exact Matrix Elements of the Resolvent

The exact matrix elements of $\hat{\mathcal{G}}(z)$ within the sector (6.4) follow from (6.10) to (6.13), together with the identity

$$\hat{\mathcal{G}}(z) = \hat{\mathcal{G}}_0(z) + \hat{\mathcal{G}}(z)\hat{H}_I\,\hat{\mathcal{G}}_0(z) \quad , \tag{6.14}$$

which follows from $\hat{1} = (z-\hat{H}+\hat{H}_I)(z-\hat{H}_0)^{-1} = (z-\hat{H})\hat{\mathcal{G}}_0(z) + \hat{H}_I\hat{\mathcal{G}}_0(z)$, by left multiplication of both sides with $\hat{\mathcal{G}}(z)$. Thus,

$$\langle r;0|\hat{\mathcal{G}}(z)|r;0\rangle = (z-E_{or})^{-1}\left[1+\langle r;0|\hat{\mathcal{G}}(z)\hat{H}_I|r;0\rangle\right]$$

$$= (z-E_{0r})^{-1}\left[1+\lambda \sum_{n\beta}{}' \int_0^\infty dk\sqrt{k}\; f^*_{rn\beta}(k)\langle r;0|\hat{\mathcal{G}}(z)|n;\beta,k\rangle\right] \quad . \tag{6.15}$$

Again applying (6.14), together with (6.13),

$$<r;0|\hat{\mathcal{G}}(z)|n;\beta,k> = <r;0|\hat{\mathcal{G}}(z)\hat{H}_I|n;\beta,k>(z-E_{0n}-k)^{-1}$$
$$= \lambda\sqrt{k}(z-E_{0n}-k)^{-1} f_{rn\beta}(k)<r;0|\hat{\mathcal{G}}(z)|r;0> \quad . \tag{6.16}$$

Substituting this result in (6.15) and solving with respect to $<r;0|\hat{\mathcal{G}}(z)|r;0>$, we find (cf.(4.5))

$$<r;0|\hat{\mathcal{G}}(z)|r;0> = <r;0|\hat{\mathcal{B}}(z)|r;0>$$
$$=\mathcal{D}_{r;0}(z) = 1/D(z) \quad , \tag{6.17}$$

where

$$D(z) = z-E_{0r}-\lambda^2 \sum_{n\beta}{}' \int_0^\infty \frac{k|f_{rn\beta}(k)|^2}{z-E_{0n}-k} dk \quad . \tag{6.18}$$

The resolvent matrix element (6.17) is the analogue of the propagator G_{11} in (1.7): by (4.6), it will yield the persistence amplitude of the state $|r;0>$ (cf. Sect.10). From (4.10), we see that the last term of (6.18) is the expectation value of $\hat{\Sigma}(z)$ in the state $|r;0>$, corresponding to the only irreducible self-energy diagram in this sector: emission and reabsorption of a photon from this state.

From (6.7), (6.16) and (6.17), we get

$$<r;0|\hat{\mathcal{G}}(z)|\Phi> = \frac{\lambda}{D(z)} \sum_{n\beta}{}' \int_0^\infty \frac{\sqrt{k}\, f_{rn\beta}(k)\phi_{n\beta}(k)}{z-E_{0n}-k} dk \quad , \tag{6.19}$$

which is the analogue of the propagator G_{21} in (1.7). Taking the hermitean conjugate and employing

$$\hat{\mathcal{G}}^+(z) = \hat{\mathcal{G}}(z^*), \; D^*(z) = D(z^*) \quad , \tag{6.20}$$

we get the analogue of G_{12} in (1.7),

$$<\Phi'|\hat{\mathcal{G}}(z)|r;0> = \frac{\lambda}{D(z)} \sum_{n\beta}{}' \int_0^\infty \frac{\sqrt{k}\, f^*_{rn\beta}(k)\phi'^*_{n\beta}(k)}{z-E_{0n}-k} dk \quad . \tag{6.21}$$

Finally, the analogue of G_{22} in (1.7) is

$$<\Phi'|\hat{\mathcal{G}}(z)|\Phi> = \sum_{n\beta}{}' \int_0^\infty \frac{\phi'^*_{n\beta}(k)\phi_{n\beta}(k)}{z-E_{0n}-k} dk$$
$$+ \frac{\lambda^2}{D(z)} \sum_{m\gamma}{}' \int_0^\infty \frac{\sqrt{k}\, f^*_{rm\gamma}(k)\phi'^*_{m\gamma}(k)}{z-E_{0m}-k} dk \sum_{n\beta}{}' \int_0^\infty \frac{\sqrt{k}\, f_{rn\beta}(k)\phi_{n\beta}(k)}{z-E_{0n}-k} dk \quad . \tag{6.22}$$

Expressions (6.17) to (6.22) give all matrix elements of the resolvent in the sector of interest, regardless of the choice of the interaction matrix elements $f_{rn\beta}(k)$. They may therefore be useful to cover a variety of different physical situations. The analytic properties of the resolvent, and in particular those of $1/D(z)$, depend crucially on the form of $f_{rn\beta}(k)$ for the specific interaction considered.

These results may be regarded as a mold, into which we pour the content of the physical problem to be treated, by specifying $f_{rn\beta}(k)$. In particular, we still do not know what will be the physical states of the theory: the matrix elements of the resolvent have been computed with respect to the unperturbed basis of (6.5). Depending on the choice of $f_{rn\beta}(k)$, the spectrum of the exact Hamiltonian may be very different from the unperturbed one, so that the physics is strongly affected. In fact, if we interpret the continuum (with different labels) as a continuum of atomic states rather than of the photon field, taking the latter to be that of a monochromatic laser, we can get models of ionization and autoionization similar to those treated here by other lecturers.

7. DISCUSSION

We are interested in the fate of the resonant unperturbed state $|r;0\rangle$ in the presence of interaction. According to Van Hove's classification of states (cf. (4.12) and (6.17)), this depends on the behavior of

$$\lim_{z\to E\pm i0} \mathcal{D}_{r;0}(z) = \frac{1}{D(E\pm i0)} \quad .$$

Since

$$\frac{1}{E-\mathcal{E}\pm i0} = \frac{P}{E-\mathcal{E}} \mp i\pi\delta(E-\mathcal{E}) \quad , \tag{7.2}$$

where P denotes the Cauchy principal value, we have, from (6.18),

$$\mathcal{D}_{r;0}(E\pm i0) = \left[E-E_{0r}-\Delta_{r;0}(E)\pm \frac{i}{2}\Gamma_{r;0}(E)\right]^{-1} \quad , \tag{7.3}$$

where

$$\Delta_{r;0}(E) = \lambda^2 \sum_{n\beta}' P\int_0^\infty \frac{k|f_{rn\beta}(k)|^2}{E-E_{0n}-k}dk \quad , \tag{7.4}$$

$$\begin{aligned}\Gamma_{r;0}(E) &= 2\pi\lambda^2 \sum_{n\beta}' \int_0^\infty k|f_{rn\beta}(k)|^2\delta(E-E_{0n}-k)dk \\ &= 2\pi\lambda^2 \sum_{n\beta}' \theta(E-E_{0n})(E-E_{0n})|f_{rn\beta}(E-E_{0n})|^2 \quad ,\end{aligned} \tag{7.5}$$

and θ denotes the Heaviside step function.

According to (7.3) and (7.5), $\mathcal{D}_{r,0}(z)$ has a sequence of cuts that "pile up", starting at the thresholds $E_{01}<E_{02}<\ldots<E_{0,r-1}<E_{0,r+1}<\ldots<E_{0N}<0$ associated with the unperturbed bound state energies, excluding E_{0r}. The corresponding continuous spectra arise from the presence of a photon with energy $k\geq 0$ together with the atom in a bound state $n\neq r$. In view of the existence of these cuts, $|r;0\rangle$ cannot be an asymptotically stationary state: either it is dissipative or there are persistent perturbation effects (cf. Sect. 4).

Which of these possibilities is realized depends on whether or not there exists a root of (cf.(4.14))

$$E-E_{0r}-\Delta_{r;0}(E) = 0 \tag{7.6}$$

for E values below the cut, i.e., for $E<E_{01} = \min E_{0n}$. If such a real root exists, it will be a real pole of $\mathcal{G}_{r;0}(E)$, i.e., a bound state of the interacting atom-photon system. Otherwise, $|r;0\rangle$ will be dissipative. Substituting (7.4) in (7.6), we must consider whether or not

$$E-E_{0r}-\lambda^2{\sum_{n\beta}}' \, P\int_0^\infty \frac{k|f_{rn\beta}(k)|^2}{E-E_{0n}-k}\,dk = 0 \quad , \quad E<E_{01} \quad , \tag{7.7}$$

has a solution. Since the denominators of the integrals in (7.7) are negative and increase in modulus as E decreases below E_{01}, there will be no solution if

$$\lambda^2{\sum_{n\beta}}' \int_0^\infty \frac{k|f_{rn\beta}(k)|^2}{(E_{0n}-E_{01})+k}\,dk < E_{0r}-E_{01} \tag{7.8}$$

and there will be exactly one solution in the opposite case: the interaction is then so strong that the upper level E_{0r} is "Lamb--shifted" below the unperturbed ground state E_{01}, giving rise to a new ground state.

For our H atom model, due to the smallness of the coupling constant λ, we are in the weak interaction regime: (7.8) will be satisfied; $|r;0\rangle$ is dissipative, leading to spontaneous decay (Sect. 10). For a stronger interaction or small enough $E_{0r}-E_{01}$ (nearly degenerate excited state), however, one might have the reverse inequality, leading to a bound "atom-photon" state. This would be a dressed state, with energy renormalized by the level shift (7.7) and a "wave function renormalization constant" given by the residue at the pole,

$$"Z_2" = [D'(E)]^{-1} = \left[1+\lambda^2{\sum_{n\beta}}' \int_0^\infty \frac{k|f_{rn\beta}(k)|^2}{(k+E_{0n}-E)^2}\,dk\right]^{-1} \quad , \tag{7.9}$$

where E is the solution of (7.7). These are persistent perturbation effects, expected according to Van Hove's classification.

The possibility of such a bound "atom-photon" state within a model of this class was pointed out by Grimm and Ernst.[15] Cohen-Tannoudji and Avan[18] discussed a model with a discrete state coupled to a continuum, showing a transition between exponential decay for weak coupling and Rabi-type oscillation, with the emergence of a new discrete state below the continuum, for strong coupling. They suggested, as a possible application, laser excitation from a discrete bound state to a narrow autoionizing level near the ionization limit of an autoionizing continuum. Similar effects indeed occur in the model of autoionization resonances discussed here by Rzazewski (with the photon continuum replaced by the electronic continuous spectrum), for a laser photon frequency below the ionization threshold.

Another analogous model that was treated by Messina and Persico[19] is a system of N spins $\frac{1}{2}$ in a static magnetic field, interacting with a phonon mode. For strong spin-phonon interaction, the ground state differs from the state with no spin-phonon excitations: it contains coherent phonons, coupled with correlated spin waves. It was suggested that this situation may be found in Jahn-Teller phase transitions.

A sort of photonic Jahn-Teller effect has been proposed by Pfeifer[20] to explain the stability of optical isomers: chiral molecules are not found in an eigenstate of the Coulomb Hamiltonian, which is parity-conserving, but rather in left-handed or right-handed form. These chiral forms are stable under perturbations (e.g., collisions or external fields); transitions from one isomer to the other do not occur. The two lowest energy levels of the Coulomb Hamiltonian, E_{01} and E_{0r}, associated with even and odd parity, respectively, would be nearly degenerate. Pfeifer considers a model of the present type with only these two levels (N=2) and condition (7.8) not satisfied, due to the near-degeneracy. This signals the occurrence of a spontaneous symmetry breaking, leading to degenerate chiral ground states of opposite handedness, dressed by a cloud of soft (infrared) coherent photons, and separated by a superselection rule, which would explain their stability under external perturbations.

To determine whether or not such models are realistic, one would have to investigate whether the predicted effects persist when the full Hamiltonian is taken into account. However, they demonstrate our point that the present model can describe, in principle, a variety of different physical situations.

8. EXACT RETARDED H ATOM MATRIX ELEMENTS

We now come back to our specific model, by identifying $f_{r\eta\beta}(k)$ in (6.1) with the exact H atom matrix elements (including retardation), given by (3.18), where $\langle \underset{\sim}{r}|n\rangle = \Psi_n(\underset{\sim}{r})$ is the bound-state eigenfunction associated with a set of quantum numbers (3.6), and $\underset{\sim}{A}_\beta$ is given by (3.8), (3.9). More explicitly,

$$f_{nn'\beta}(k) \doteq -2 \int \Psi^*_{n\ell m}(\underset{\sim}{r})(-i\underset{\sim}{\nabla})\cdot \underset{\sim}{A}_{JM\tau}(k,\underset{\sim}{r})\Psi_{n'\ell'm'}(\underset{\sim}{r})d^3r \quad , \tag{8.1}$$

where

$$n \equiv (n\ell m) \quad , n' \equiv (n'\ell'm') \quad , \beta \equiv (JM\tau) \tag{8.2}$$

and $\hbar=c=1$, as before. This still allows us to set the unit of length. We choose the atomic unit, which is the Bohr radius, setting

$$a_B = (me^2)^{-1} = (\alpha m)^{-1} = 1 \quad , \tag{8.3}$$

where α is the fine structure constant. The unit of momentum is $a_B^{-1} = \alpha m = 1$.

In these units, Balmer's formula for the unperturbed H atom energy levels becomes

$$E_{0n} = -\alpha/(2n^2) \quad (n=1,2,3,\ldots) \quad , \tag{8.4}$$

and the coupling constant in (6.1) (cf.(5.2)) is

$$\lambda = e/m = \alpha^{3/2} \quad . \tag{8.5}$$

Furthermore,

$$\Psi_{n\ell m}(\underset{\sim}{r}) = R_{n\ell}(r)Y_{\ell m}(\theta,\phi) \quad , \tag{8.6}$$

where ℓ ranges from 0 to n-1 and m from $-\ell$ to ℓ. The radial functions are

$$R_{n\ell}(r) = N_{n\ell}\, e^{-r/n}(2r/n)^{\ell}\, L^{2\ell+1}_{n+\ell}(2r/n) \quad , \tag{8.7}$$

where $N_{n\ell}$ is a normalization factor and L_p^q are Laguerre polynomials.

The angular momentum algebra allows the angular integrations in (8.1) to be performed explicitly, with the following results:[21]

$$f_{nn'\beta}(k) = \frac{i}{k}\, a_{nn'\beta}\{\left[J(J+1)+\Delta(W+1)\right] \cdot$$
$$\times \int_0^\infty j_J(kr)R_{n\ell}\, \frac{\partial R_{n'\ell'}}{\partial r}\, rdr + \left[-J(J+1)+\Delta(W+1)\right] \cdot$$
$$\times \int_0^\infty j_J(kr)\, \frac{\partial R_{n\ell}}{\partial r}\, R_{n'\ell'}\, rdr\} \qquad (\tau=0) \quad , \tag{8.8}$$

$$f_{nn'\beta}(k) = i\, b_{nn'\beta} \int_0^\infty j_J(kr)R_{n\ell}\, R_{n'\ell'}\, rdr \qquad (\tau=1) \quad , \tag{8.9}$$

where

$$\Delta = \ell-\ell' \quad , \quad W=\ell+\ell' \quad , \tag{8.10}$$

and $a_{nn'\beta}$ and $b_{nn'\beta}$ are real coefficients depending only on angular momentum quantum numbers (e.g., products of Wigner 3j-symbols).

The following exact selection rules must be fulfilled:

$$|\Delta| = |\ell-\ell'| \leq J \leq \ell+\ell' = W \quad , \quad M=m-m' \quad , \tag{8.11}$$

from angular momentum conservation, and

$$J+\ell+\ell' \equiv \tau(\text{mod } 2) \tag{8.12}$$

from parity conservation. Since $\ell+\ell' \leq (n-1)+(n'-1)$ and n and n' in (6.1) are bounded, (8.11) implies that J also is bounded, and therefore (cf.(3.7)) the summation over β in (6.1) indeed ranges over a finite set of values.

The k dependence of $f_{nn'\beta}(k)$ is given by the radial matrix elements in (8.8) and (8.9). The evaluation of H atom transition matrix elements by group theoretical methods has been discussed by Barut and Wilson,[22] who refer also to previous calculations. Here we apply a direct approach to obtain the form of these matrix elements.

By (8.7), $R_{n\ell}$ and $\partial R_{n\ell}/\partial r$ are sums of terms of the form $r^j \exp(-r/n)$, where j is an integer. The basic integral employed for the evaluation is[23]

$$\int_0^\infty e^{-\alpha x} J_\nu(\beta x) x^{\mu-1}\, dx$$

$$= \frac{(\beta/2\alpha)^\nu \Gamma(\nu+\mu)}{\alpha^\mu \Gamma(\nu+1)} \left(1+\frac{\beta^2}{\alpha^2}\right)^{1/2-\mu} F\left(\frac{\nu-\mu+1}{2}, \frac{\nu-\mu}{2}+1; \nu+1; -\frac{\beta^2}{\alpha^2}\right), \tag{8.13}$$

where

$$F(a,b;c;z) = 1+\frac{ab}{c}z+\frac{a(a+1)b(b+1)}{2!c(c+1)}z^2+\ldots$$

$$= F(b,a;c;z) \qquad (|z|<1) \tag{8.14}$$

is Gauss's hypergeometric function. If a or b is a negative integer or zero, F reduces to a polyomial:

$$F(-j,b;c;z) = P_j(z) \quad , \quad \deg P_j = j \quad , \tag{8.15}$$

i.e., P_j is a polynomial of degree j.

Substituting (8.7) in (8.8) and introducing the notation

$$\kappa \equiv \kappa_{nn'} = \frac{1}{n}+\frac{1}{n'} \tag{8.16}$$

for the "inverse Bohr transition radius", we find

$$f_{(n\ell m)(n'\ell'm')(JM0)}(k) = a_{\ell m\ell'm'JM}\left(\frac{k}{\kappa}\right)^{J-1}\left(\frac{\kappa}{k^2+\kappa^2}\right)^{W} \cdot$$

$$\times \sum_{i+j=0}^{n+n'+W-2} \left\{ b_{n\ell n'\ell'Jij}\left(\frac{\kappa}{k^2+\kappa^2}\right)^{i+j+1} \cdot \right.$$

$$\times F\left(\frac{J-W-i-j}{2}, \frac{J-W-i-j+1}{2}; J+\frac{3}{2}; -\frac{k^2}{\kappa^2}\right)$$

$$+ c_{n\ell n'\ell'Jij}\left(\left[J(J+1)+\Delta(W+1)\right](\ell'+j)+\left[-J(J+1)+\Delta(W+1)\right](\ell+j)\right) \cdot$$

$$\left. \times \left(\frac{\kappa}{k^2+\kappa^2}\right)^{i+j} F\left(\frac{J-W-i-j+1}{2}, \frac{J-W-i-j+2}{2}; J+\frac{3}{2}; -\frac{k^2}{\kappa^2}\right)\right\}, \quad (8.17)$$

where a,b,c are numerical (k-independent) coefficients. The selection rules (8.11) (8.12) imply that every F function in (8.17) is of the form (8.15), i.e., reduces to a polynomial. The only "dangerous" case in which this could fail, for the last term of (8.17), would be if $J=W$, $i=j=0$, but the coefficient of this term then vanishes. A similar expression is found from (8.9) for $\tau=1$.

Finally, we get

$$f_{(n\ell m)(n'\ell'm')(JM\tau)}(k) = \frac{k^{J+\tau-1}\mathcal{P}_s(k^2)}{(k^2+\kappa^2_{nn'})^{n+n'-1}}, \quad (8.18)$$

where $\mathcal{P}_s$ is a polynomial of degree

$$\deg \mathcal{P}_s = s = n+n'-2-\frac{1}{2}(J+\ell+\ell'+\tau) \quad . \quad (8.19)$$

In particular, we have

$$f_{(n\ell m)(n'\ell'm')(JM\tau)}(k) = \begin{cases} \mathcal{O}(k^{J+\tau-1}) & (k\to 0) \quad , \\ \mathcal{O}(k^{-\ell-\ell'-3}) & (k\to\infty) \quad . \end{cases} \quad (8.20)$$

Since $\ell+\ell'\geq J$ by the selection rules (8.11), and $J\geq 1$, this implies that the matrix elements $\to 0$ at least as fast as k^{-4} as $k\to\infty$.

Explicit examples (remember that $f_{nm\beta}$ is associated with absorption) are, for the Lyman α and β transitions,

$$f_{(21m)(100)(1,-m,0)}(k) = \text{constant}/\left[k^2+(\tfrac{3}{2})^2\right]^2 , \quad (8.21)$$

$$f_{(31m)(100)(1,-m,0)}(k) = \text{constant} \times \left[2k^2+(\tfrac{4}{3})^2\right]/\left[k^2+(\tfrac{4}{3})^2\right]^3 , \quad (8.22)$$

both of which satisfy (8.20). Since $\kappa_{nn'}\sim 1$ always, we see that the exact atomic form factors introduce a smooth (but very fast) cutoff at momenta of the order of the inverse Bohr transition radius, so that $k\to\infty$ may be replaced by $k\gg 1$ in (8.20).

The explicit form (8.18) of the transition matrix elements has extremely simple analytic properties: it is a rational function of k. Its denominator is a power of $(k^2+\kappa_{nn'}^2)$: this is the momentum space cutoff associated with the exponential behavior of the radial wave functions (8.7).

9. THE FUNCTION D(z)

Let

$$r \equiv (rLm_r) \quad , \quad n \equiv (n\ell m) \quad . \tag{9.1}$$

Then, by (6.18) and (7.5),

$$D(z) = z - E_{0r} - \frac{1}{2\pi} \sum_{n\beta}{}' \int_0^\infty \frac{\Gamma_{rn\beta}(k)}{z-E_{0n}-k} \, dk \quad , \tag{9.2}$$

where

$$\Gamma_{rn\beta}(k) \equiv 2\pi\lambda^2 k |f_{rn\beta}(k)|^2 \quad , \tag{9.3}$$

and, by (8.5),

$$\lambda^2 = \alpha^3 (\approx 4\times10^{-7}) \quad . \tag{9.4}$$

From (9.3) and (8.18),

$$\Gamma_{rn\beta}(k) = 2\pi\lambda^2 k \, P_{rn\beta}(k^2)/(k^2+\kappa_{rn}^2)^d \tag{9.5}$$

where $P_{rn\beta}$ is a polynomial of degree D, with

$$d = 2(r+n-1) \quad , \quad D \equiv \deg P_{rn\beta} = r+n-3- \tfrac{1}{2}(L+\ell-J-\tau) \quad , \tag{9.6}$$

$$\Gamma_{rn\beta}(k) = \begin{cases} \mathcal{O}(k^{2J+2\tau-1}) & (k\to0), \\ \mathcal{O}(k^{-2L-2\ell-5}) = \mathcal{O}(k^{-7}) & (k>>1). \end{cases} \tag{9.7}$$

Substituting (9.5) in (9.2), we find that the integrals can be explicitly computed, with the following result:

$$D(z) = z - E_{0r} - \frac{1}{2\pi} \sum_{n\beta}{}' \Gamma_{rn\beta}(z-E_{0n}) \ell n \left[- \frac{(z-E_{0n})}{\kappa_{rn}} \right] + \lambda^2 \sum_{n\beta}{}' \frac{Q_{rn\beta}(z-E_{0n})}{[(z-E_{0n})^2+\kappa_{rn}^2]^d} \quad , \tag{9.8}$$

where, for each $\xi = z-E_{0n}$,

$\ell n(-\xi) = \ell n|\xi|+i(\arg \xi-\pi)$,

$$0 \leq \arg\xi \leq 2\pi \text{ on the physical sheet,} \tag{9.9}$$

and $Q_{rn\beta}(\xi)$ is again a polynomial, with

$$\deg Q_{rn\beta} = 2d-1 \quad , \tag{9.10}$$

so that $Q_{rn\beta}$ has 2d coefficients. These coefficients can be explicitly determined from 2d conditions, namely, that D(z) is regular on the physical sheet at the points

$$z = E_{0n} \pm i\kappa_{rn} \quad , \tag{9.11}$$

even though, from (9.8) and (9.5), it might appear that each of these points is a pole of order d.

It follows from (9.8) that

$$D(z) = z-E_{0r}+\lambda^2\mathcal{O}(z^{-1}) \text{ as } |z|\to\infty \; . \tag{9.12}$$

Note also that, according to (9.9), we have, on the physical sheet,

$$\ell n\left[-(E\pm i0-E_{0n})/\kappa_{rn}\right] = \ell n|(E-E_{0n})/\kappa_{rn}| \mp i\pi\theta(E-E_{0n}) \quad , \tag{9.13}$$

so that (7.3) and (7.5) are verified. Note that D(z) has a logarithmic branch point at each unperturbed bound state energy E_{0n}, except the resonant one, n=r.

10. DECAY OF A WEISSKOPF-WIGNER INITIAL STATE

Due to the smallness of the coupling constant (9.4), condition (7.8) is satisfied for our H atom model: the l.h.s. is $\mathcal{O}(\alpha^3)$, whereas the r.h.s., by (8.4), is $\mathcal{O}(\alpha)$. Therefore, the Weisskopf-Wigner initial state of the system,

$$|\Psi(t=0)\rangle = |r;0\rangle \tag{10.1}$$

is dissipative, and its persistence amplitude at time t (cf.(4.6), (6.17)) is

$$\begin{aligned} u(t) &= \langle r;0|\; \exp(-i\hat{H}t)|r;0\rangle \\ &= -\frac{1}{2\pi i}\int_{C}\frac{e^{-izt}}{D(z)}\,dz \qquad (t>0) \quad . \end{aligned} \tag{10.2}$$

We consider the closed path $\mathcal{C}$ shown in Fig.2, where all straight lines C_j are parallel to the bisector of the fourth quadrant. Each path C_j winds around the branch point E_{0j}, so that the l.h.s. of C_j and the r.h.s. of C_j are located on different Riemann sheets, but the r.h.s. of C_j and the l.h.s. of C_{j+1} are on the same sheet. Since

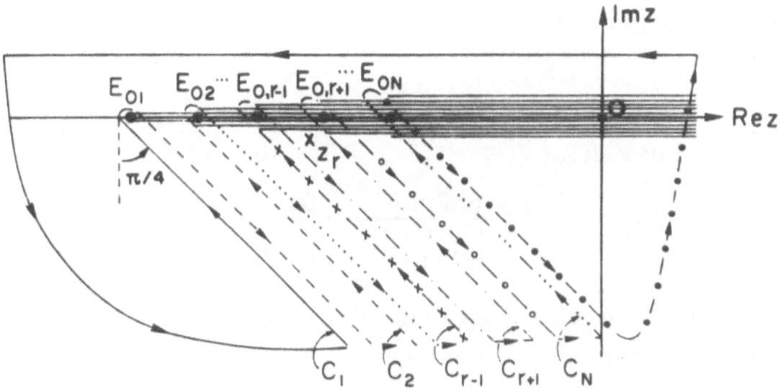

Fig. 2. The path $\mathcal{C}$; portions belonging to the same Riemann sheet are similarly represented. The pole z_r is indicated.

$\exp(-izt)\to 0$ as $|z|\to\infty$, Imz<0, and $D(z) = \mathcal{O}(|z|)$ as $|z|\to\infty$ (cf. (9.12)), the arcs of circles at infinity do not contribute. Therefore,

$$u(t) = -\frac{1}{2\pi i}\sum_{j=1}^{N}{}' \int_{C_j} \frac{e^{-izt}}{D(z)}\, dz + \sum \mathrm{Res}\left[\frac{e^{-izt}}{D(z)}\right], \tag{10.3}$$

where the residues are taken at all poles of $1/D(z)$ within the path $\mathcal{C}$. Where are these poles?

<u>The Poles of the Resolvent</u>

Let $D_{(j)}(z)$ be the branch of $D(z)$ reached along path $\mathcal{C}$ after going around E_{0j} clockwise on C_j, with $D_{(0)}(z) = D(z)$, which is <u>regular</u> (physical sheet). Since equivalent points after and before winding are related by

$$(z-E_{0j})_{(j)} = (z-E_{0j})_{(j-1)}\exp(-2i\pi) \quad , \tag{10.4}$$

it follows from (9.8) and (9.9) that

$$D_{(j)}(z) = D_{(j-1)}(z) + i\sum_{\beta}\Gamma_{rj\beta}(z-E_{0j}) \quad , \tag{10.5}$$

so that

$$D_{(j)}(z) = D(z) + i\sum_{n\leq j,\beta}{}' \Gamma_{rn\beta}(z-E_{0n}) \quad . \tag{10.6}$$

In particular, for $j=r-1$, the poles of $1/D_{(r-1)}(z)$ are the roots of (cf.(9.8))

$$z = E_{0r} - i \sum_{n<r,\beta} \Gamma_{rn\beta}(z-E_{0n}) + \frac{1}{2\pi} \sum_{n\beta}' \Gamma_{rn\beta}(z-E_{0n}) \cdot$$

$$\times \ln\left[-(z-E_{0n})/\kappa_{rn}\right] - \lambda^2 \sum_{n\beta}' \frac{Q_{rn\beta}(z-E_{0n})}{\left[(z-E_{0n})^2+\kappa_{rn}^2\right]^d} \quad . \tag{10.7}$$

One root is located in the lower half-plane, close to E_{0r}. It can be obtained by iteration, starting with

$$z_r^{(0)} = E_{0r} - i0 = E_{0r} \exp(2i\pi) \quad , \tag{10.8}$$

so that, by (9.9),

$$\ln|-(z_r^{(0)} - E_{0n})/\kappa_{rn}| = \ln|(E_{0r}-E_{0n})/\kappa_{rn}| + i\pi\theta(r-n) \quad , \tag{10.9}$$

and the iteration yields

$$z_r^{(1)} = E_{0r} + \Delta_{r;0}(E_{0r}) - \frac{i}{2} \Gamma_r(E_{0r}) \quad , \tag{10.10}$$

where (cf.(7.4))

$$\Delta_{r;0}(E_{0r}) = \frac{1}{2\pi} \sum_{n\beta}' P \int_0^\infty \frac{\Gamma_{rn\beta}(k)}{E_{0r}-E_{0n}-k} dk$$

$$= \lambda^2 \sum_{n\beta}' \frac{(E_{0r}-E_{0n})P_{rn\beta}((E_{0r}-E_{0n})^2)}{\left[(E_{0r}-E_{0n})^2+\kappa_{rn}^2\right]^d} \ln \left|\frac{E_{0r}-E_{0n}}{\kappa_{rn}}\right|$$

$$-\lambda^2 \sum_{n\beta}' \frac{Q_{rn\beta}(E_{0r}-E_{0n})}{\left[(E_{0r}-E_{0n})^2+\kappa_{rn}^2\right]^d} \quad , \tag{10.11}$$

and

$$\Gamma_r(E_{0r}) = \sum_{n<r,\beta} \Gamma_{rn\beta}(E_{0r}-E_{0n}) \quad . \tag{10.12}$$

Further corrections from other iterations are small, so that we can stop here, taking $z_r = z_r^{(1)}$ as the position of the pole shown in Fig.2.

What about other poles? Looking at (10.7), we see that, while $E_{0r}=\mathcal{O}(\alpha)$, all other terms are multiplied by $\lambda^2=\mathcal{O}(\alpha^3)$. Therefore, the only other possible roots must be located close to one of the poles (9.11) of the rational functions. One finds d poles clustered around $E_{0n}-i\kappa_{rn}$,

$$z_p \simeq E_{0n}-i\kappa_{rn} + c\,\lambda^{2/d} \exp\left(\frac{2i\pi p}{d}\right) \qquad (p=0,1,\dots d-1) \quad , \tag{10.13}$$

where c is a constant, and d similar poles in the upper half-plane, clustered around $E_{0n}+i\kappa_{rn}$. All these poles fall outside path $\mathcal{C}$, which, on this sheet, is bounded by C_{r-1} and C_{r+1} in Fig.2 (remember that $E_{0n}=\mathcal{O}(\alpha)$, $\kappa_{rn}=\mathcal{O}(1)$); this is the reason why these paths were taken at an angle of $-\pi/4$.

The argument principle shows that $1/D(z)$, on this sheet, has precisely the (finite) number of poles given by these results, so that all poles have been found, and only z_r contributes. The same discussion applies to the poles on all other sheets, so that, as happened in the potential scattering model of Sect.1, we get the complete pole distribution on all relevant unphysical sheets. None of the other poles falls within $\mathcal{C}$, so that only the resonant pole z_r contributes to the residue in (10.3). In this sense, we can compare u(t) with the "decaying state propagator" (1.8) found in potential scattering.

11. THE ASYMPTOTIC DECAY LAW

The residue contribution in (10.3), according to the above discussion, is

$$u_{res}(t) = \frac{\exp(-iz_r t)}{D'(z_r)} = \left[1+\mathcal{O}(\alpha^4 \ell n\alpha)\right]\exp(-iE_r t - \frac{1}{2}\Gamma_r t), \qquad (11.1)$$

where (cf.(10.10)) $E_r = E_{0r} + \Delta_{r;0}(E_{0r})$. Therefore, the lifetime of level r is $\tau_r = 1/\Gamma_r$. Note that, by (10.12), the width of level r is the sum of the partial widths associated with the transitions from r to all lower levels, as it should be; analytically, this arises from the threshold crossings.

The radiative level shift ("Lamb shift") of level r in this model is given by (10.11). Note that it is related to the "absorptive parts" Γ_{rn} by dispersion relations, as it should (cf. Cohen-Tannoudji's lectures here). The logarithmic terms are reminiscent of Bethe logarithms with cutoffs κ_{rn}, but they are nonperturbative and (apart from higher iterations) exact within this model.

The effects of retardation, at least for transitions involving low-lying states, are numerically negligible, although they have played a crucial role in the solution. Thus, for the 2P→1S Lyman α transition (cf.(8.21)),

$$\left[(E_{0r}-E_{0n})^2+\kappa_{rn}^2\right]^{-4} = \left[(3\alpha/8)^2+(3/2)^2\right]^{-4} = (2/3)^8\left[1+(\alpha/4)^2\right]^{-4},$$

the retardation correction $(1+\alpha^2/16)^{-4}$ differs from unity in the fourth significant figure. Therefore, (10.11) and (10.12) give the usual results for the level shift, $\mathcal{O}(\alpha^4\ell n\alpha)$, and for the linewidth, $\mathcal{O}(\alpha^4)$, as compared with $\mathcal{O}(\alpha)$ for the energy levels in our units.

DEVIATIONS FROM EXPONENTIAL DECAY

Deviations arise from the cut contributions in (10.3). Using the notations (10.5), we have

$$\int_{C_j} \frac{e^{-izt}}{D(z)}\, dz = \int_{E_{0j}}^{E_{0j}+e^{-i\pi/4}\infty} \left[\frac{1}{D_{(j)}(z)} - \frac{1}{D_{(j-1)}(z)}\right] e^{-izt} dz$$

$$= i\exp(-iE_{0j}t)\int_0^{\infty} \frac{\exp(-e^{i\pi/4}\rho t)\,\sum_\beta \Gamma_{rj\beta}(e^{-i\pi/4}\rho)}{D_{(j)}(E_{0j}+e^{-i\pi/4}\rho)D_{(j-1)}(E_{0j}+e^{-i\pi/4}\rho)}\, d\rho, \quad (11.2)$$

where we have employed (10.5).

The asymptotic behavior of such a Fourier-Laplace integral as $t\to\infty$ is determined (Tauberian theorems) by the threshold behavior of $\Gamma_{rj\beta}(z)$ as $z\to 0$. Taking into account (9.7), we finally get

$$u(t) \approx \exp(-iE_r t - \frac{1}{2}\Gamma_r t) + \sum_n{}' A_{rn}\, t^{-2(J_{rn}+\tau_{rn})} \exp(-iE_{0n}t)$$
$$(t\to\infty) \quad , \quad (11.3)$$

where A_{rn} are constants. The slowest-decaying transitions are electric dipole ones ($J=1$, $\tau=0$), which will therefore dominate for $t\to\infty$, leading to $\mathcal{O}(t^{-2})$ corrections to exponential decay. The main corrections arise from interference between the first and second term of (11.3) in $|u(t)|^2$.

By summing over all electric dipole transitions, it is found[10] that, for the Lyman-α line, the corrections to exponential decay are negligible for $t \lesssim t_0$, where

$$\exp(-\frac{1}{2}\Gamma_r t_0) \sim \frac{2\alpha}{3\pi} \langle r|(\underset{\sim}{x}/t_0)^2|r\rangle \ . \quad (11.4)$$

The matrix element in (11.4) is the mean square radius of the excited state divided by the square of the distance travelled by light during time t_0 ($c=1$). This yields a value for t_0 of the order of 10^2 lifetimes, so that the atom has by then decayed for all practical purposes.

The t^{-2} asymptotic decay law is similar to that of a free photon wave packet when account is taken of the causal character of light propagation (this is connected with the positive semidefiniteness of the photon energy). Indeed, for fixed $\underset{\sim}{x}$, the causal propagator[24]

$$D_0{}^c(\underset{\sim}{x},t) = \frac{1}{4\pi}\,\delta_+(\underset{\sim}{x}^2-t^2) = \frac{1}{4\pi}\left[\delta(\underset{\sim}{x}^2-t^2) + \frac{i}{\pi}\frac{P}{(\underset{\sim}{x}^2-t^2)}\right] \quad (11.5)$$

behaves like t^{-2} as $t\to\infty$.

There is a strong analogy between these results and the discussion of the decaying state propagator $M(x,k_n,t)$, given by (1.8), in potential scattering. Just as that propagator resolves the problems posed by a Gamow decaying state associated with a complex pole k_n, the present solution resolves the Weisskopf-Wigner decaying state problem associated with the "complex energy" $z_r = E_r - i\Gamma_r/2$.

The analogue of the propagator G_{12} in (1.7) is the matrix element (6.21), which yields the probability amplitude for observing photons of various frequencies from the decay of the initial state (10.1), giving the spectrum of the emitted radiation. The expected Lorentzian spectrum is found[10] to be a very good approximation.

12. RESONANCE SCATTERING OF A WAVE PACKET

As mentioned in Sect.2, the Weisskopf-Wigner initial state (10.1), corresponding to a bare excited state, does not provide a very realistic description of the excitation process. A more realistic model is obtained by considering the resonance scattering of an incident photon wave packet by the atom (resonance fluorescence). Resonance with the excited state $|r\rangle$ is required for the consistency of the model (Sect.5,(ii)).

It follows from the generalized RWA that the state $|1;\beta,k\rangle = |1\rangle|\beta,k)$, where $|1\rangle$ is the atomic ground state, is asymptotically stationary, so that we can take an incident one-photon wave packet

$$|\Phi\rangle = \int_0^\infty \phi(k)\,|1;\beta,k\rangle \quad , \tag{12.1}$$

where $|\phi(k)|^2$ should be peaked around the resonant transition energy, having small overlap with other transitions. A suitable choice for this purpose is to take $\phi(k)$ proportional to $(k-\bar{E}+i\gamma/2)^{-1}$, which yields a Lorentzian for $|\phi(k)|^2$. The incident spectral width is given by γ, and variation of $\bar{E}$ changes the detuning.

The atomic excitation amplitude is given by (6.19), the analogue of G_{21} in (1.7):

$$\langle r;0|\exp(-i\hat{H}t)|\Phi\rangle = -\frac{1}{2\pi i}\int_C e^{-izt}\langle r;0|\hat{\mathcal{G}}(z)|\Phi\rangle dz \quad , \tag{12.2}$$

which can be computed in the same way as $u(t)$. One finds[10] that the excitation probability has a characteristic <u>rise time</u> T_r and a <u>decay time</u> T_d given by

$$T_r \sim \min(\gamma^{-1},\Gamma_r^{-1}) \ , \ T_d \sim \max(\gamma^{-1},\Gamma_r^{-1}) \quad , \tag{12.3}$$

as ought to be expected.

With the help of (6.22), one can also obtain the scattered wave packet, as well as the resonance fluorescence line shape. The fluorescence spectrum reflects the overlap between the spectrum of the incident wave packet, which plays the role of a filter, and the natural line shape found in the Weisskopf-Wigner case. For excitation by a broad line ($\gamma >> \Gamma_r$), corresponding to a sharp pulse in time, we approach the Weisskopf-Wigner excitation conditions: the resonant state is excited with rather small probability $(\Gamma_r/\gamma)^2$, but it decays with the natural line shape.

13. CONCLUSION

What happens when one adds back the terms that were omitted from the full Hamitonian? The counter-rotating terms induce virtual transitions from the ground state to excited states, with emission of a photon, as well as reverse transitions. This corresponds to a "self-energy diagram" for the ground state. As a consequence, the interacting ground state differs from that of the unperturbed system: it undergoes persistent perturbation effects.

In order to restore the previous situation, one must first construct asymptotically stationary states, after which one may proceed as before. For this purpose, one may look for a dressing transformation. This has been done, but only with the help of perturbative techniques, so that it falls outside of the scope of the present lectures. Therefore, we confine ourselves to stating the results.[10,11]

To second order in perturbation theory, the dressing transformation accomplishes its purpose: when coupled with mass renormalization, it yields an interacting ground state that is asymptotically stationary to second order, and whose energy is corrected, to this order, by the nonrelativistic ground-state Lamb shift. The Lamb shift of the excited states, as it should, arises in an entirely different way. All excited states, as was found for $|r\rangle$ in our soluble model, give rise to complex poles in the analytic continuation of the resolvent, and the Lamb shift appears as a displacement in the real part of each pole.

One can also discuss the line shape in resonance fluorescence (Kramers-Heisenberg dispersion formula), comparing the minimal coupling interaction adopted here with the $-\tilde{d}.\hat{\tilde{E}}$ version. The results are found to be identical, though distributed in different ways between resonant and background contributions. If one wants to approximate the answer by the resonant term alone, one choice may be more convenient than the other, but which one yields a better approximation depends on the specific transition that is considered.

REFERENCES

1. G. Gamow, Z.Phys. 51: 204 (1928).
2. J.J. Thomson, Proc.Lond.Math.Soc.15(1): 197 (1884).
3. G. Beck and H.M. Nussenzveig, Nuovo Cimento 16: 416 (1960).
4. H.M. Nussenzveig, Nuovo Cimento 20: 694 (1961).
5. H.M. Nussenzveig, "Causality and Dispersion Relations", Academic Press, New York (1972).
6. J. Humblet and L. Rosenfeld, Nucl.Phys. 26: 529 (1961).
7. V. Weisskopf, "Physics in the Twentieth Century", p.4, MIT Press, Cambridge (1972).
8. E. Arnous and W. Heitler, Proc.Roy.Soc. (London) A 220: 290 (1953); W. Heitler, "The Quantum Theory of Radiation", 3rd ed., Oxford University Press, London (1954).
9. F.E. Low, Phys.Rev. 88: 53 (1952).
10. L. Davidovich, Ph.D.thesis, University of Rochester (1975).
11. L. Davidovich and H.M. Nussenzveig, in "Foundations of Radiation Theory and Quantum Electrodynamics", A.O. Barut, ed., Plenum Publ.Corp., New York (1980).
12. M.E. Rose, "Elementary Theory of Angular Momentum", John Wiley, New York (1957).
13. L. Van Hove, Physica 21: 901 (1955).
14. F.J. Dyson, Phys.Rev. 75: 486, 1736 (1949).
15. E. Grimm and V. Ernst, J.Phys.A 7: 1664 (1974); Z.Phys.A 274: 293 (1975).
16. P.A.M. Dirac, "The Principles of Quantum Mechanics", 4th ed., Oxford University Press, London (1958).
17. T.D. Lee, Phys.Rev. 95: 1329 (1954); G. Källén and W. Pauli, Dan.Mat.Fys.Medd. 30: Nº7 (1955).
18. C. Cohen-Tannoudji and P. Avan, Colloques Internat. du C.N.R.S. 273: 93 (1977).
19. A. Messina and F. Persico, J.Phys.C 6: 3557 (1973).
20. P. Pfeifer, Phys.Rev.A 26: 701 (1982) and other references therein; J.Phys. A 14: L129 (1981).
21. H.E. Moses, Phys.Rev. A 8: 1710 (1973).
22. A.O. Barut and R. Wilson, Phys.Rev. A 13: 918 (1976).
23. G.N. Watson, "A Treatise on the Theory of Bessel Functions", 2nd ed., Cambridge University Press (1944).
24. E.C.G. Stueckelberg and D. Rivier, Helv.Phys.Acta 23: 215 (1950); M. Fierz, Helv.Phys.Acta 23: 731 (1950).

$\vec{A}\cdot\vec{p}$ OR $\vec{r}\cdot\vec{E}$? MINIMAL COUPLING AND MULTIPOLAR HAMILTONIANS IN THE QUANTUM THEORY OF RADIATION

Kurt Haller

Department of Physics
University of Connecticut
Storrs, CT 06268

I. INTRODUCTION

From the earliest beginnings of the quantum theory, the canonical procedure for coupling matter to the electromagnetic field has been based on a local interaction between charge-current densities and the four-component electromagnetic vector potential.[1] This locally gauge-invariant interaction, known by the sobriquet "minimal coupling," is generated by replacing space-time derivatives of charged matter fields with gauge-covariant derivatives, as shown by

$$\partial_\mu \phi \to (\partial_\mu - ie\, A_\mu)\phi \tag{I-1a}$$

$$\partial_\mu \phi^\dagger \to (\partial_\mu + ie\, A_\mu)\phi^\dagger \,. \tag{I-1b}$$

The resulting form for the Schroedinger equation, for spinless charged particles subject to electromagnetic forces, is[†]

$$[-\frac{1}{2m}(\vec{\nabla} - ie\,\vec{A})\cdot(\vec{\nabla} - ie\,\vec{A}) + e\,A_0 - i\frac{\partial}{\partial t}]\,\psi = 0\,. \tag{I-2}$$

Local gauge invariance has become the underlying principle in formulating interactions between matter and electromagnetic radiation,[2] and similarly, local gauge-covariance has come to determine the coupling rules for non-Abelian fields used in the Weinberg-Salam unified electro-weak theory[3,4] and in quantum chromodynamics.[5]

[†] we adopt the natural system of units, with $\hbar = c = 1$, and the Minkowski metric $g_{\mu\nu} = \delta_{\mu,\nu}$ with $x_4 = i\,t$, and $A_4 = i\,A_0$.

Although the minimal coupling rule does not inevitably fix the coupling of matter and radiation fields unambiguously (because of an ambiguity of Lagrangians by a four-divergence),[6] minimal coupling has become standard for charged particle dynamics in quantum mechanics. The special role that this coupling mode assigns to local interactions between matter fields and the gauge-independent part of the vector potential finds experimental support in the Aharanov-Bohm effect.[7,8†]

Equation (I-2) can be rewritten in the form $(H - i\,\partial/\partial t)\psi = 0$, where $H = H_0 + H_1$, and H_0 is the limit of H as $e \to 0$. This leads to

$$H_0 = (p^2/2m) \tag{I-3a}$$

and

$$H_1 = -\frac{e}{2m}(\vec{p}\cdot\vec{A}+\vec{A}\cdot\vec{p}) + \frac{e^2}{2m}|\vec{A}|^2 + e\,A_0\,, \tag{I-3b}$$

where $\vec{p}$ is understood to act on everything to its right, including the wave function whose presence is implicit.

Alternatively to the gauge-invariant minimal coupling, many workers in atomic and molecular physics, and in quantum optics, prefer to use an interaction Hamiltonian that couples charges and currents to electric and magnetic field strengths rather than to the four components of the vector potential.[9,10] The resulting interaction Hamiltonian is an infinite series in multipole order, and is therefore useful mostly when electrons are in bound state orbitals. To leading order this interaction features the electronic dipole term $-e\vec{r}\cdot\vec{E}$; the next order contains the magnetic dipole interaction $-(e/2m)\vec{B}\cdot\vec{L}$ (where $\vec{L}$ is the orbital angular momentum and $\vec{B}$ the magnetic field) and an electric quadropole term.

An early suggestion that the electric dipole interaction can be used in radiation theory was made by Fermi.[11] His analysis, which still has relevance to the discussion underway today, can be summarized as follows:

To leading order, (i.e. neglecting the "seagull" term, $(e^2/2m)|\vec{A}|^2$, in eq. (I-3b)), the matrix element for photon emission

---

† The interpretation of the Aharanov-Bohm effect as a phenomenon beyond description by a theory with local interactions between electrons and electromagnetic field strengths $\vec{E}$ and $\vec{B}$ has its critics; but these critics generally accept the minimal coupling rule for coupled Schroedinger and electromagnetic fields (see, for example, Ref. 31).

(or absorption) is given by:

$$<f|H_1|i> = -\frac{e}{m}\frac{\hat{\varepsilon}(\vec{k})}{(2k)^{\frac{1}{2}}}\cdot<f_A\,|\exp[\pm i\vec{k}\cdot\vec{r}]\,\vec{p}\,|i_A> \tag{I-4}$$

where $|i_A>$ and $|f_A>$ designate electrons in bound atomic orbitals, $\hat{\varepsilon}(\vec{k})$ represents the polarization of the emitted (absorbed) photon, and the $\pm$ in the exponential assigns + for photon absorption, - for emission. If the photon wavelength is much larger than the atom, so that $kr \ll 1$ in the region in which $|i_A>$, and $|f_A>$ differ significantly from zero, then we can approximate $\exp[\pm i\vec{k}\cdot\vec{r}]$ by 1. Use† of $(\vec{p}/m) = i[H_0,\vec{r}]$ allows us to represent $<f|H_1|i>$ by

$$<f|H|i> = \frac{ie(E_i-E_f)}{(2k)^{\frac{1}{2}}}\,\hat{\varepsilon}(\vec{k})\cdot<f_A|\vec{r}|i_A> \tag{I-5}$$

where E_i and E_f represent the H_0 eigenvalues of $|i_A>$ and $|f_A>$. To the precision allowed by $\delta E\delta t \approx \hbar$ the transitions are limited to those for which initial and final energies are identical. We therefore set $E_i - E_f = \pm k$ and obtain

$$<f|H_1|i> = \pm\, ie\left(\frac{k}{2}\right)^{\frac{1}{2}}\hat{\varepsilon}(\vec{k})\cdot<f_A|\vec{r}|i_A>\,, \tag{I-6}$$

where + is associated with photon emission, - with absorption. Equation (I-6) is identical to the expression that would have been obtained by using $-e\vec{r}\cdot\vec{E}$ under the same assumptions.

Later authors have suggested that the Hamiltonian (eqs. I-3a,b) itself may be changed,[12-26] so that in the new, transformed Hamiltonian a multipole series, with $-e\vec{r}\cdot\vec{E}$ the leading term, replaces the $-(e/m)\,\vec{A}\cdot\vec{p} + (e^2/2m)|\vec{A}|^2$ that one finds in Coulomb gauge formulations. Some have argued that this new Hamiltonian can be derived from the original one by making a gauge transformation, and that the new Hamiltonian may be used freely because gauge transformations never affect observable quantities. Others have implied that this particular gauge transformation (i.e., the one that ostensibly transforms from the old to the new Hamiltonian) violates generally accepted theorems that observable quantities are invariant to gauge transformations. Some authors appear to take the position that it is so difficult to distinguish between observable and unphysical, gauge-dependent parts of the vector potential, that the only safe way to avoid error is to gauge-transform the minimal coupling Hamiltonian into an expression that replaces potentials with electric and magnetic field strengths.[15,26] Still other authors have eliminated

†In this case H_0 also includes the part of A_0 due to static charges on the nucleus.

A_μ from the Hamiltonian, in favor of the multipole series whose initial term is $-e\vec{r}\cdot\vec{E}$, by making transformations that they do not relate to any change of gauge. In these latter cases questions have arisen about the justification for replacing the original Hamiltonian with the transformed one, and about the circumstances under which the replacement "multipole" Hamiltonian may safely be used. In one treatment even the original statement of the problem avoids the electromagnetic potentials A_μ, in favor of the field strengths $\vec{E}$, $\vec{B}$, $\vec{D}$ and $\vec{H}$. Assumptious about the commutation rules among these operators; and between them and the charged particle velocity operator, define the dynamics so as to generate Maxwell's equations.[27]

The discussion in this paper will address and hopefully clarify the relation between the minimal coupling and the multipolar forms of the Hamiltonian for the system of nonrelativistic spinless charged particles and an electromagnetic field.

II. CAN GAUGE TRANSFORMATIONS HAVE DYNAMICAL CONSEQUENCES?

In this section we will discuss the question raised in its heading, and answer it in the negative. We will restrict the discussion for the present by postponing, to the next section, consideration of the technical issues that arise when the electromagnetic potential is treated as a quantized gauge field.

Let us carry out a gauge transformation on eq. (I-2). We replace $\vec{A}$ by $\vec{A} + \vec{\nabla}\chi$, A_0 by $A_0 - \partial\chi/\partial t$, and ψ by $\exp[ie\chi]\psi$ for an arbitrary χ. The transformed equation reads

$$\{-\frac{1}{2m}(\vec{\nabla}-ie\vec{A}-ie\vec{\nabla}\chi)\cdot(\vec{\nabla}-ie\vec{A}-ie\vec{\nabla}\chi) + eA_0 - e(\frac{\partial\chi}{\partial t}) - i\frac{\partial}{\partial t}\}\exp[ie\chi]\psi \tag{II-1}$$

It is trivial to see that $\exp[ie\chi]$ can be shifted to the extreme left-hand side of eq. (II-1) and that, in the process, derivatives $\vec{\nabla}\chi$ and $(\partial\chi/\partial t)$ are generated in just the appropriate places to subtract the $\vec{\nabla}\chi$ and $(\partial\chi/\partial t)$ that appear between the brackets { } in eq. (II-1). The result of that process is simply to regenerate eq. (I-2). It might appear superficially that the effect of a gauge transformation is to replace the original $\vec{A}$ and A_0 by $\vec{A}' = \vec{A} + \vec{\nabla}\chi$ and $A_0' = A_0 - \partial\chi/\partial t$ respectively, and that the transformed vector potential, A_μ', might possibly lead to at least marginally different dynamical consequences. But that view fails to account for the very crucial replacement of ψ by $\psi' = \exp[ie\chi]\psi$. It does not do to claim that ψ' is another wavefunction, different from ψ, and that the gauge-transformed Schroedinger equation (eq. II-1) does have a different set of vector potentials, albeit for the new ψ', which obeys a different wave equation. We must interpret the results of calculations by using operators that correspond to observable dynamical variables. The expectation values of observable quantities are gauge-invariant, and in evaluating them the implicit difference between ψ and ψ' (by the factor $\exp[ie\chi]$) must ultimately be taken into account. For example, the gauge-invariant charge-current density is given by

$$\rho = e\psi^{\dagger}\Psi \tag{II-2a}$$

and

$$\vec{j} = \frac{ie}{2m}\left[(\vec{\nabla}\psi^{\dagger})\psi - \psi^{\dagger}(\vec{\nabla}\psi)\right] - \frac{e}{m}\rho\vec{A} \,. \tag{II-2b}$$

The gauge-transformed $\vec{j}$ has $\vec{A}$ replaced by $\vec{A}+\vec{\nabla}\chi$, but the extra terms from $\vec{\nabla}(\exp ie\chi\ \psi)$ and from $\vec{\nabla}(\exp -ie\chi\ \psi^{\dagger})$ exactly compensate for the substitution of $\vec{A}+\vec{\nabla}\chi$ for $\vec{A}$, so that the gauge-transformed and original $\vec{j}$ are identical. The fact that ψ and ψ' differ by e-dependent terms must not be ignored in this case. Another example, more germane to the kinds of physical predictions for which this theory is actually used, deals with the transition amplitude, $T_{f,i}$ from an initial state $|i>$ to a final state $|f>$. $T_{f,i}$ is given by

$$T_{f,i} = <f|T|i> \tag{II-3a}$$

with

$$T = H_1 + H_1\ (E_i - H + i\varepsilon)^{-1}H_1 \ . \tag{II-3b}$$

Since H_I and H_0 are not separately gauge invariant,† gauge transformations of H_1 alone cannot be expected to leave $T_{f,i}$ gauge-invariant. It should be noted that eqs. (II-3a,b) give $T_{f,i}$ correctly only when $|i>$ and $|f>$ are eigenstates of H_0 (in fact, H_0 is exactly that part of H for which $|i>$ and $|f>$ are eigenstates). Since $\exp[ie\chi]|i>$ and $\exp[ie\chi]|f>$ are not generally eigenstates of the same H_0 as $|i>$ and $|f>$, a more generally applicable form of $T_{f,i}$ is necessary to investigate the gauge-dependence of $T_{f,i}$. Such a form is given by

$$T_{f,i} = <f|(\overleftarrow{H} - H_f)[1 - (E_i - H + i\varepsilon)^{-1}(\vec{H} - E_i)]|i> \tag{II-4}$$

where the arrows indicate that $\overleftarrow{H}$ acts on $<f|$ and $\vec{H}$ on $|i>$ before any further calculations are performed. In the next section we will demonstrate that $\vec{H}|i>$ and $<f|\overleftarrow{H}$ are gauge-invariant, even though H is not. Given that fact, eq. (II-4) makes it easy to see that $T_{f,i}$ has the same value, on and off the energy shell, in any gauge.

---

† H is also not gauge-invariant. The fact that this feature causes no difficulty, will be discussed in the next section.

The fact that H is gauge-dependent is one of a number of topics that can be quite vexing in semiclassical radiation theory, but that are easily resolved in the fully quantized theory. Another such topic deals with the fact that gauge transformations turn electron wave functions, that are independent of the electronic charge e, into e-dependent states. We want, somehow, to ask "In which gauge are the electron wave functions really e-independent?" but the question neither makes sense nor is answerable in semiclassical radiation theory. The gauge-dependence of H makes it appear necessary to invent a gauge-independent "energy operator" that differs from H. In fact, the fully quantized theory makes it clear that such a move, which is so contrary to the spirit of the quantum theory, is entirely unnecessary.

Before turning to the next section, we focus the foregoing discussion with the following simple illustrative example: Consider H_1, given in eq. (I-3b), applied to a one-photon absorption or emission process, so that the "seagull" term, $(e^2/2m)|\vec{A}|^2$ can be neglected. We will also limit our discussion to lowest order multipole moment, and choose the Coulomb gauge so that H_1 reduces to $-(e/m)\vec{A}^T\cdot\vec{p}$ (A^T is the transverse part of $\vec{A}$). Next we carry out the transformation $\vec{A} \rightarrow \vec{A} + \vec{\nabla}\chi$ and $A_0 \rightarrow A_0 - \partial_0\chi$ with $\chi = \pm\, \vec{r}\cdot\vec{A}^T$ but make no effort to transform state vectors. Then $\vec{\nabla}\chi = \pm\, e\vec{A}^T$ + higher order multipole contributions (which we neglect) and $\partial_0\chi = \pm\, e\vec{r}\cdot\vec{E}^T$. If we choose $-\, e\vec{r}\cdot\vec{A}^T$ for χ we obtain the transformed Hamiltonian $H_1' = -e\vec{r}\cdot\vec{E}^T$; if we choose the opposite sign for χ we obtain $H_1'' = -2(e/m)\vec{A}^T\cdot\vec{p}+e\vec{r}\cdot\vec{E}^T$. This transformation resembles a gauge transformation closely enough so that a number of authors have mistaken it for such. But because no provisions have been made to include $\psi \rightarrow \exp[ie\chi]\psi$, this is not a gauge transformation, and H_1', as well as H_1'' determine dynamical laws different from those obtained with H_1. By applying Fermi's argument, summarized in section I, we see however that H_1, H_1' and H_1'' all have identical on-energy-shell transition matrix elements. That is too little to demand of gauge-equivalent interactions. But since $H_1 \rightarrow H_1' \rightarrow H_1''$ are not gauge transformations, we need to understand why even on-energy-shell transition amplitudes are the same, to this approximation, in these cases. We will deal with this question again in sections IV and V.

III. THE QUANTUM ELECTRODYNAMICS OF THE SCHROEDINGER FIELD

This section deals with the theory of the quantized Schroedinger field ψ interacting with the quantized electromagnetic vector potential A_μ.[28] We will begin our discussion with a Coulomb gauge formulation in which A_μ includes a transverse part $\vec{A}^T$, and a time-like component

$$A_0^c\,(\vec{x}) = \int \frac{d\vec{y}\,\rho(\vec{y})}{4\pi|\vec{x}-\vec{y}|} \quad . \tag{III-1}$$

The gauge-fixing condition that defines the Coulomb gauge, $\vec{\nabla}\cdot\vec{A} = 0$, eliminates the longitudinal component of A_μ. $-\vec{E}^T$, where $\vec{E}^T$ refers to the transverse electric field, is the canonically conjugate momentum to A^T, so that the commutation rules for A_i^T and E_j^T are

$$[A_i^T(\vec{r}),\ E_j^T(\vec{r}\,')] = -i\ (\delta_{i,j}\delta(\vec{r}-\vec{r}\,') + \frac{\partial}{\partial r_i}\frac{\partial}{\partial r_j}\frac{1}{4\pi|\vec{r}-\vec{r}\,'|})\ . \qquad \text{(III-2a)}$$

ψ and ψ^+ obey the anticommutation rule (we take ψ to describe spinless electrons)

$$\{\psi^+(\vec{r}),\ \psi(\vec{r}\,')\} = \delta(\vec{r}-\vec{r}\,') \qquad \text{(III-2b)}$$

The charge-current densities are as given in eqs. (II-2a,b) except that in this case $\vec{A}$ takes on the special value $\vec{A}^T$ appropriate for the Coulomb gauge. The Hamiltonian is given by

$$H^c = H_0 + H_1 \qquad \text{(III-3a)}$$

with†

$$H_0 = \tfrac{1}{2}\int [|\vec{E}^T(\vec{r})|^2 + |\vec{B}(\vec{r})|^2]d\vec{r} + \int \psi^\dagger(\vec{r})[-(2m)^{-1}\nabla^2 + V(\vec{r})]\psi(\vec{r})d\vec{r} \qquad \text{(III-3b)}$$

and

$$H_1^c = -\int \vec{J}(\vec{r})\cdot\vec{A}^T(\vec{r})d\vec{r} + \frac{e}{2m}\int \rho(\vec{r})\vec{A}^T(\vec{r})\cdot\vec{A}^T(\vec{r})d\vec{r}$$

$$+ \int \frac{\rho(r)\rho(r')}{8\pi|\vec{r}-\vec{r}\,'|}\ d\vec{r}\ d\vec{r}\,'\ , \qquad \text{(III-3c)}$$

where $\vec{J}(\vec{r})$ denotes

$$\vec{J} = \frac{ie}{2m}\ [(\nabla\psi^\dagger)\psi - \psi^\dagger(\nabla\psi)]\ . \qquad \text{(III-3d)}$$

The Hamiltonian H^c describes the interactions of charged spinless electrons and transverse photons. There are no unphysical photons in this gauge.

---

†V(r) is either a non-electromagnetic potential or the A_0 due to an object other than one of the electrons described by the Schroedinger field, such as a static nucleus.

It is one of the features of the fully quantized theory that, in any other gauge, the time evolution of all state vectors that describe physically realizable states is governed by the very same Hamiltonian as in the Coulomb gauge! This is not because there is anything very special about the Coulomb gauge. The Coulomb gauge is only one of many gauges and all gauges are physically equivalent. It is because, in a formulation in which field excitations are defined the same way in every gauge, Hamiltonians have identical forms in all gauges, except for gauge-dependent parts that can trivially be shown to make no contribution to the time evolution of physically realizable state vectors.[28]

To clarify the preceding remarks we consider the theory in one of the Lorentz gauges, the socalled "Feynman gauge." In that gauge the Lagrangian has the form†

$$L = -(\tfrac{1}{2}m)\,[\partial j + ieA_j]\psi^\dagger[\partial_j - ieA_j]\psi - \psi^\dagger V\psi$$
$$+ i\psi^\dagger(\partial/\partial t + eA_4)\psi - \tfrac{1}{4}F_{\mu\nu}F_{\mu\nu} - G\partial_\mu A_\mu + \tfrac{1}{2}G^2\,. \qquad \text{(III-4)}$$

G represents a "gauge-fixing" field[29] necessary to preserve canonical commutation rules, and to avoid constraints on the momentum conjugate to A_4.[30] Without the gauge-fixing field, the momentum canonically conjugate to A_4 would vanish identically. The quantum electrodynamics of the Schroedinger field in the Feynman gauge is developed in ref. (28).

We will not repeat all of this work here, but we will quote the Hamiltonian, which is given by

$$H^f = \int [\tfrac{1}{2}|\vec{E}|^2 + \tfrac{1}{2}|\vec{B}|^2 - \tfrac{1}{2}G^2 - i\,E_j\,\partial_j A_4 + G\,\partial_j A_j$$
$$+ \frac{1}{2m}\partial_j\psi^\dagger\partial_j\psi + \psi^\dagger V\psi - A_j J_j + \frac{e}{2m}\rho A_j A_j$$
$$+ \rho A_0\,]\,d\vec{r} \qquad \text{(III-5)}$$

where iG is the momentum adjoint to A_4. This Hamiltonian can be expressed in a momentum space decomposition by expanding A_μ in terms of annihilation and creation operators for transverse, longitudinal and timelike excitations. The latter two, (longitudinal and timelike) correspond to annihilation and creation operators, $a_L(\vec{k})$ and $a_L^\dagger(\vec{k})$, respectively for longitudinal excitations, and $a_4(\vec{k})$ and $a_4^\dagger(\vec{k})$ respectively for timelike ones. The fact that the vector potential has a Minkowski metric makes it mandatory to quantize the

†Eq. (III-4) corrects some transcription errors in ref. (28).

theory in an indefinite metric space,[31] if inconsistencies are to be avoided. In our discussion we will use "ghost" operators, which correspond to excitations of zero-norm "ghost" particles. They are given by

$$a_Q(\vec{k}) = (2)^{-\frac{1}{2}} [a_L(\vec{k}) + ia_4(\vec{k})] \;, \tag{III-6a}$$

$$a_R(\vec{k}) = (2)^{-\frac{1}{2}} [a_L(\vec{k}) - ia_4(\vec{k})] \;, \tag{III-6b}$$

$$a^*_Q(\vec{k}) = (\;)^{-\frac{1}{2}} [a^\dagger_L(\vec{k}) + ia^\dagger_4(\vec{k})] \;, \tag{III-6c}$$

and

$$a^*_R(\vec{k}) = (\;)^{-\frac{1}{2}} [a^\dagger_L(\vec{k}) - ia^\dagger_4(\vec{k})] \;, \tag{III-6d}$$

The excitations for these "ghosts" have zero norm because $[a_Q(\vec{k}),\, a^*_Q(\vec{k}')] = [a_R(\vec{k}),\, a^*_R(\vec{k}')] = 0$. It is a_Q and a^*_R, as well as a_R and a^*_Q that have the usual commutation relations of conjugate annihilation and creation operators, although a_Q and a^*_Q, as well as a_R and a^*_R are each others respective adjoints in the indefinite metric space. G is given by $G = \partial_\mu A_\mu$, and the equations of motion for this theory are the conventional Schroedinger equation, and the electromagnetic equations

$$\vec{\nabla}\cdot\vec{E} - \rho = -\partial/\partial t(G) \tag{III-7a}$$

$$\partial\vec{E}/\partial t - \vec{\nabla}\times\vec{B} + \vec{j} = -\vec{\nabla}G \tag{III-7b}$$

$$\vec{\nabla}\cdot\vec{B} = 0 \tag{III-7c}$$

$$\partial\vec{B}/\partial t + \vec{\nabla}\times\vec{E} = 0 \tag{III-7d}$$

and

$$(\nabla^2 - \partial^2/\partial t^2)G = 0 \tag{III-7e}$$

The Hamiltonian given in eq. (III-5) appears to differ so profoundly from that in eqs. (III-3a→d) that the claim made at the beginning of this section, that time evolution in all gauges is mediated by the same Hamiltonian, appears to be contradicted. But that appearance is a superficial illusion. Inspection of eqs. (III-7a→e) demonstrates that we are not dealing with a quantized version of Maxwell theory unless we set G = 0. Since G is a quantized operator, that cannot be done as an operator identity without introducing inconsistencies.

The only consistent approach to this problem is to define a set of "physical" states by a constraint, in the form[32,33]

$$G^{(+)} \, |\nu\rangle = 0 \, , \tag{III-8a}$$

where $G^{(+)}$ is the positive frequency part of G. Equations (III-8a) and (III-7a) have the following immediate consequence: States that obey the constraint equations (eq. (III-8a)), also obey Gauss's Law. $G^{(+)}$ is easily evaluated, and leads to the explicit form of eq. (III-8A)

$$[a_Q(\vec{k}) + (2k^{3/2})^{-1} \, \rho(\vec{k})] \, |\nu\rangle = 0 \, . \tag{III-8b}$$

The Hamiltonian in eqs. (III-3a→d) can therefore only be used consistently with states that obey eqs. (III-8b), since it is only for these states that expectation values of $\vec{E}$ and $\vec{B}$ obey Maxwell's equations. But eq. (III-8b) is difficult to implement and has complicated solutions that are coherent superpositions of electrons, and photon ghosts. This fact helps us to understand why the Hamiltonians in eqs. (III-3a→d) and (III-5) look so different. The electron excitations have different meanings! In the Coulomb gauge the one-electron state $\varepsilon^{\dagger}(\vec{p})|0\rangle$ represents an electron accompanied by its Coulomb field. The corresponding state in the theory governed by H^f, given in eq. (III-5), is an eigenstate of the limiting form of H^f in the limit $e \to 0$. This state represents a completely "bare" electron devoid even of its Coulomb field. We cannot decide to what extent the Hamiltonians H^f and H^c resemble each other until this discrepancy in the definition of the electron field excitations has been remedied.[34]

It is possible to transform H^f with an operator U, unitary in the indefinite metric space, such that this objective is met. We find that[35]

$$U\left(a_Q(\vec{k}) + (2k^{3/2})^{-1} \rho(\vec{k})\right) U^{-1} = a_Q(\vec{k}) \, . \tag{III-9}$$

We apply the same transformation consistently to all operators (and states) so that we establish unitary equivalence between the original and the transformed representations. The results are:

$$\tilde{H}^f = U \, H^f \, U^{-1} = H^c + H_Q \tag{III-10}$$

and

$$\tilde{\vec{E}}(\vec{r}) = U\vec{E}(\vec{r})U^{-1} = \vec{E}^{T}(\vec{r}) - \vec{\nabla}_{\vec{r}} \int \frac{\rho(\vec{r}\,')}{4\pi|\vec{r}-\vec{r}\,'|} d\vec{r}\,'$$

$$+ \sum_{k} k^{-\frac{1}{2}} \{a_Q(\vec{k})\exp[i\vec{k}\cdot\vec{r}]$$

$$+ a_Q^*(\vec{k})\exp[-i\vec{k}\cdot\vec{r}]\} \qquad \text{(III-11)}$$

In this transformed representation the following apply: The new form of eq. (III-8a→b) is

$$a_Q(\vec{k})|n\rangle = 0 \qquad \text{(III-12)}$$

and its solutions, which can be found by inspection, constitute a Fock space of n-electron states, transverse photons, and $a_Q^*(\vec{k})$ ghosts, (but which excludes $a_R^*(\vec{k})$ ghosts). The expectation value of $\tilde{\vec{E}}$ for one-electron states is $-\vec{\nabla} \langle\varepsilon(\vec{p})| \int d\vec{r}\,' \rho(\vec{r}\,') [4\pi|\vec{r}-\vec{r}\,'|]^{-1}|\varepsilon(\vec{p})\rangle$, so that the electron field excitations in the transformed representation do represent electrons with their accompanying Coulomb field. H_Q in eq. (III-10) is a self adjoint operator of the form

$$H_Q = \sum_k h_Q(\vec{k})\, a_Q(\vec{k}) + h_Q^*(k)\, a_Q^*(k) \qquad \text{(III-13)}$$

with h_Q totally free of a_R and a_R^* operators. <u>States that obey the constraint equation (eq. (III-12)) at any one time obey it forever after; their time evolution, in the quotient space of observable states, is completely determined by H^c alone</u>. The $\sum_k h_Q(k)\, a_Q(\vec{k})$ part of H_Q vanishes when acting on a state that obeys eq. (III-12). The $\sum_k h_Q^*(\vec{k})\, a_Q^*(\vec{k})$ creates zero-norm ghosts which never can contribute through internal photon loops because $a_R(k)$ and $a_R^*(\vec{k})$ we never part of $\tilde{H}^f$, and because $a_Q^*(k)$ commutes with every operator except $a_R(\vec{k})$. In other Lorentz gauges (Landau gauge,[36] Fried-Yennie gauge,[37] etc.) the only difference is in the form of h_Q and h_Q^* in eq. (III-10) (they still are devoid of a_R and a_R^*). Gauges other than Lorentz gauges can be generated through operator gauge transformations in which electron and photon field operators are gauge transformed, but the Hamiltonian of the original Lorentz (or Coulomb) gauge kept intact in the form of $\tilde{H}^f$ or H^c.[39] In the new gauges the field equations for A_μ and the commutation rules are changed, but the time-evolution of state vectors in the physical subspace is still entirely controlled by H^c. The gauge-dependent parts of H are wholly irrelevant to the time-evolution of state vectors that obey the constraint equation; they also never can affect transition amplitudes (on or off the energy shell). In any gauge, radiative processes are governed by the $\int [-\vec{J}^T\cdot\vec{A}^T + (e/2m)\rho|A^T|^2]d\vec{r}$ common to H^c, H^f and $\tilde{H}^f$. Changing <u>that</u> part of the time evolution operator to a multipolar form is wholly beyond what a gauge transformation can effect.

IV. UNITARY TRANSFORMATION OF THE MINIMAL COUPLING HAMILTONIAN

In this section we will examine the changes that can be made in the Hamiltonian H^c, given in eq. (III-3a-d), under unitary transformations not related to any change of gauge. A number of authors have given such transformations.[12-14,20-23] In ref. (28) we have used $W = \exp[i\alpha]$ with

$$\alpha = - \int r_i \, \rho(\vec{r}) \, \underset{r}{F} \, A_i^T(\vec{r}) \, d\vec{r} \tag{IV-1a}$$

and

$$F_r = (1 - \tfrac{1}{2} r_i \frac{\partial}{\partial r_i} + \frac{1}{3} r_i r_j \frac{\partial}{\partial r_i} \frac{\partial}{\partial r_j} \cdots) \ . \tag{IV-1b}$$

We obtain $\overline{H} = W \, H W^{-1}$ with

$$\overline{H} = H_0 + H_1 \tag{IV-2a}$$

and

$$\begin{aligned} H_1 = & - \int r_i \, \rho(\vec{r}) \, F_r \, E_i^T(\vec{r}) d\vec{r} - \int [\vec{r} \times \vec{J}(\vec{r})]_i \, G_r \, B_i(\vec{r}) \, d\vec{r} \\ & + \int \rho(\vec{r}) \, \rho(\vec{r}) \, [8\pi |\vec{r}-\vec{r}\,'|]^{-1} \, d\vec{r} \, d\vec{r}\,' + (e/2m) \int \rho(\vec{r}) |\vec{r} \times G_r \vec{B}(\vec{r})|^2 d\vec{r} \\ & + \int r_i \, r_j' \, \rho(\vec{r}) \rho(\vec{r}\,') \, F_r F_{r'} \, [\frac{\partial}{\partial r_i} \frac{\partial}{\partial r_j} - \delta_{i,j} \nabla^2] \, \frac{1}{8\pi |\vec{r}-\vec{r}\,'|} \, d\vec{r} \, d\vec{r}\,' \\ & + \text{surface terms} \end{aligned} \tag{IV-2b}$$

where

$$G_r = \left(\tfrac{1}{2} - \frac{1}{3!} r_i \frac{\partial}{\partial r_i} + \frac{1}{4!} r_i r_j \frac{\partial}{\partial r_i} \frac{\partial}{\partial r_j} \cdots\right) \ . \tag{IV-2c}$$

The process of carrying out this transformation is straightforward, though tedious, and will not be repeated here. Other authors have made similar transformations that appear to differ from this one in detail only.[22] It is this kind of unitary transformation that produces the changes in the interaction Hamiltonians which were discussed in the illustrative example at the end of section II.

What is very pertinent to our concern is the theoretical rationale for substituting $\overline{H}$ for H in the theory of radiative processes. Some authors view the unitary transformation that leads from H^c to $\overline{H}$ (we use $\overline{H}$ to designate the multipolar Hamiltonian) as part of a general similarity transformation in which unitary equivalence is established between two representations.[23,40,41] Such a transformation requires that all operators (and states) be subject to the same transformation and, in turn, guarantees that all matrix elements and equations of motion are invariant to the transformation. The transformed operators, under the transformation $\xi = W\xi W^{-1}$, are $\overline{B}_i = B$, $\overline{\rho} = \rho$, and

$$\overline{E}_i^T(\vec{r}) = E_i^T(\vec{r}) - r_i\,\rho(\vec{r}) - \partial/\partial r_i\,\partial/\partial r_j \int r_j'\,\rho(\vec{r}\,')\,[4\pi|\vec{r}-\vec{r}\,'|]^{-1}\,d\vec{r}\,'$$

$$+ \text{ higher order multipoles .} \qquad \text{(IV-3)}$$

$\overline{J}_i$ also differs from J_i but will not be given here for lack of space. Some authors who interpret the multipolar Hamiltonian as being the minimal coupling Hamiltonian, in a unitarily equivalent representation, have chosen new names for the transformed fields (such as designating $\overline{E}_i^T$ as an electric displacement vector in the multipolar representation)[23] but in our view this practice has the danger of obscuring the fact that $\overline{E}_i^T$ is just the electric field in the new representation. It is also necessary to transform the states if one wants to maintain unitary equivalence, and in that case the transformed one-electron state is a coherent superposition of a "bare" electron and transverse photons, given by $\overline{\varepsilon}^\dagger(\vec{p})|0\rangle = W\,\varepsilon\,(\vec{p})|0\rangle$. $\overline{\varepsilon}^\dagger(p)|0\rangle$ can be formally expanded as a non-terminating series as shown by $\overline{\varepsilon}^\dagger(p)|0\rangle =$

$$\varepsilon^\dagger(p)|0\rangle - ie\sum_{q,\vec{k}}\int u_q^\dagger(\vec{r})\,u_p(\vec{r})\,(2k)^{-\frac{1}{2}}\vec{r}\cdot\hat{\varepsilon}(k)\,F_r(\exp[-i\vec{k}\cdot\vec{r}])\;d\vec{r}\times$$

$$\times\;\varepsilon^\dagger(q)\,a_\varepsilon^\dagger(\vec{k})|0\rangle$$

$$-\frac{e^2}{2}\sum_{q,\vec{k}}\int[u_q^\dagger(\vec{r})\,u_p(\vec{r})\,(k)^{-1}|\vec{r}\cdot\hat{\varepsilon}(\vec{k})|^2|F_r(\exp -i\vec{k}\cdot\vec{r}])|^2]\;d\vec{r}\times$$

$$\times\;\varepsilon^\dagger(q)|0\rangle$$

$$-\frac{e^2}{2}\sum_{q,\vec{k},\vec{k}}\int[u_q^\dagger(\vec{r})u_p(\vec{r})(kk')^{-\frac{1}{2}}\,\vec{r}\cdot\hat{\varepsilon}(\vec{k})\vec{r}\cdot\hat{\varepsilon}'(\vec{k}')\;F_r(\exp[-i\vec{k}\cdot\vec{r}])\times$$

$$\times\;F_r(\exp[-i\vec{k}'\vec{r}])\;d\vec{r}\;\varepsilon^\dagger(q)a_{\varepsilon'}^\dagger(\vec{k}')a_\varepsilon^\dagger(\vec{k})|0\rangle$$

$$+\cdots\text{ etc} \qquad \text{(IV-4)}$$

where u_p and $u_{p'}$ designate electron orbitals. The similarity transformation outlined here can be viewed in the passive mode, in which operators are unchanged and represented in terms of new variables; or it can be viewed in the active mode in which all operators are changed. In either case the need to implement the similarity transformation imposes essentially the same formal requirements whether we adopt the active or passive interpretations.

These formal requirements are a source of difficulty when one tries to interpret the multipolar Hamiltonian as the time evolution operator in a representation in which every operator and every state are the unitarily transformed versions of a corresponding form in a representation in which the minimal coupling Hamiltonian determines time evolution. To justify that interpretation we would have to use the transformed electron states to represent electrons in atomic orbitals (or make equivalent changes in relabeling the old states). The enormous complexity of these coherent superpositions would be a severe burden. Use of such electron wave functions is never adopted in actual calculations and obviously is not necessary in actual practice for calculating radiative transition rates. In the next section we will explore another justification for using the multipolar Hamiltonian that does not involve such burdensome requirements.

V. HYBRID TRANSFORMATIONS

It is possible to support the use of the multipolar Hamiltonian in a formalism in which the states and other operators of the theory remain untransformed. In that case one can not maintain unitary equivalence between two representations, and the change from the minimal coupling to the multipolar Hamiltonian is not part of a similarity transformation. We have called such transformations "hybrid transformations."[28] We have shown that when the initial and final states are left untransformed, substitution of the multipolar for the minimal coupling Hamiltonian still leaves scattering amplitudes unchanged.

We consider the scattering amplitudes determined by two different Hamiltonians: The Coulomb gauge Hamiltonian H^c (eqs. III-3a→d) and $\overline{H}$ (IV-2a→c), for which $\overline{H} = W H^c W^-$. The scattering wave functions for these two cases are

$$\psi_i> = |i> + (E_i - H^c + i\varepsilon)^{-1} H_1^c \; |i> \qquad \text{(V-1)}$$

and

$$\overline{\psi}_i> = |i> + (E_i - \overline{H} + i\varepsilon)^{-1} \mathcal{H}_1 \; |i> \qquad \text{(V-2)}$$

with $(H_0 - E_i)|i\rangle = 0$. We also give the transition amplitudes from an initial state $|i\rangle$ to a final state $|f\rangle$ for both the minimal coupling (Coulomb gauge) case, and the multipolar case. This leads to

$$T_{f,i} = \langle f|\, H_1^c \,|\psi_i\rangle \tag{V-3}$$

for the Coulomb gauge case and

$$\overline{T}_{f,i} = \langle f|\, \mathcal{H}_1 \,|\overline{\psi}_i\rangle \tag{V-4}$$

for the multipolar case. It is easy to show that $T_{f,i}$ and $\overline{T}_{f,i}$ are related by†

$$\overline{T}_{f,i} = T_{f,i} + (E_f - E_i)\, \langle f|(1-W)|\psi_i\rangle$$
$$+ i\varepsilon \langle f|\{(W-1)(E_i - H^c + i\varepsilon)^{-1}\, H_1^c -$$
$$- \mathcal{H}_1\, (E_i - \overline{H} + i\varepsilon)^{-1}\, (W-1)\}|i\rangle \tag{V-5}$$

and $|\overline{\psi}_i\rangle$ is related to $|\psi_i\rangle$ by

$$|\overline{\psi}_i\rangle = W|\psi_i\rangle - i\varepsilon(E_i - \overline{H} + i\varepsilon)^{-1}\, (W-1)|i\rangle. \tag{V-6}$$

To determine the relation between $T_{f,i}$ and $\overline{T}_{f,i}$ we examine the term $\langle f|(1-W)|\psi_i\rangle$. If this term has an implicit $(E_f - E_i)^{-1}$ singularity, then $T_{f,i}$ and $\overline{T}_{f,i}$ might differ on the energy shell by $(E_f - E_i)\langle f|(1-W)|\psi_i\rangle$. In our case however that cannot occur.

We use eq. (V-1) to write

$$\langle f|(1-W)|\psi_i\rangle = \langle f|(1-W)|i\rangle + \langle f|(1-W)(E_i - H_0 + i\varepsilon)^{-1}H_1^c|\psi_i\rangle \tag{V-7}$$

and we will designate the second term on the right hand side of eq. (V-7) as $s_{f,i}$. $s_{f,i}$ is potentially "dangerous" since, if $(1-W)$ commutes with H_0, we will obtain a singular $(E_i - E_f + i\varepsilon)^{-1}$ coefficient from $s_{f,i}$, and with it an observable discrepancy between $T_{f,i}$ and $\overline{T}_{f,i}$. In this particular case, however, that cannot occur.

---

†Proofs of eqs. (V-5) and (V-6) are given in refs. (28) and (34).

We find that $(1-W) = \beta = -i\alpha - \frac{1}{2}(i\alpha)^2 + \cdots$, and $s_{f,i}$ reduces to $\Sigma_n(E_i - E_n + i\varepsilon)^{-1} \langle f|\beta|n\rangle\langle n|H_1^c|\psi_i\rangle$. It is clear that $\langle f|\beta|n\rangle$ does not generate a $\delta_{f,n}$ singular contribution, and that $s_{f,i}$ is an improper integral that has no $(E_i - E_f + i\varepsilon)^{-1}$ singularity. $(E_i - E_f)s_{f,i}$ therefore vanishes when $E_i = E_f$.† We next turn to the term in eq. (V-5) proportional to $i\varepsilon$. Here we need to be concerned about $(i\varepsilon)^{-1}$, or even $(i\varepsilon)^{-n}$ contributions from the coefficient of $i\varepsilon$. In earlier work we pointed out that in properly mass-renormalized theories $(i\varepsilon)^{-1}$ is the most singular contribution possible,[34,35] and such singular terms only occur in self energy modifications to external lines, which, in turn, only affect wave function renormalization constants. The term proportional to $i\varepsilon$ in eq. (V-5) does contribute in that case. In a different context this effect accounts for the fact that the renormalized rather than the unrenormalized S-matrix elements are identical for different gauges in quantum electrodynamics,‡ a fact first demonstrated, by a different argument, by Bialynicki-Birula.[42]

The fact that $(E_f - E_i)s_{f,i}$ vanishes for $E_f = E_i$, and the fact that the term in eq. (V-5) proportional to $(i\varepsilon)^{-1}$ contributes at most to wave function renormalization graphs, allows us to substitute $\overline{H}$ for H^c in calculating S-matrix elements for elastic and inelastic scattering, radiative absorption and decay processes, and, in short, for any process determined by an on-shell transition amplitude. It is this, the natural extension of Fermi's argument,[11] that justifies the substitution of the multipolar for the minimal coupling Hamiltonian in radiation theory, and that explains the illustrative example at the end of section II.

It is interesting to note that not every unitary operator gives the same license to use a substitute Hamiltonian as does W.‡ For example, were we to substitute R for W, where R is a finite rotation, and were we to let H^c stand for a Hamiltonian with a short-range potential that lacks rotational symmetry then we would obtain a quite different result. In that case we find

---

† $s_{f,i}$ may have infrared or ultraviolet infinities that require regularization and/or renormalization. We will not discuss that problem here.

‡ The Feynman rules in the manifestly coveriant Lorentz gauges imply that the Gupta-Bleuler subsidiary condition has not been implemented. Otherwise the unrenormalized as well as the renormalized S-matrix elements would be identical in all gauges (see Ref. 34).

‡ Aharanov and Au (Ref. 43) have coined the very apt phrase "pseudo-perturbations" to describe the case in which $T_{f,i} = \overline{T}_{f,i}$ on the energy shell.

$s_{f,i} = \langle f|(1-R)(E_i - H_0 + i\varepsilon)^{-1} H_1^c|\psi_i\rangle$, and since $(1-R)$ and H_0 commute,

$$s_{f,i} = \frac{1}{E_i - E_f + i\varepsilon} \langle f|(1-R)H_1|\psi_i\rangle . \tag{V-6}$$

In this case, instead of demonstrating that $\overline{T}_{f,i} = T_{f,i}$ on the energy shell, the argument leads to a trivial identity which supplies no information about the relation between $T_{f,i}$ and $\overline{T}_{f,i}$.

In earlier work we have used a hybrid transformation to account for the following fact:[34,35] Even though we are not entitled to use H^f, in eq. (III-5), without implementing the constraint in eqs. (III-8a,b), H^f is regularly used by itself, without regard to any constraint, to develop Feynman rules (for relativistic electron theories) from the Dyson-Ward expansion of the S-matrix. In this case we also found that, because H^f and $\tilde{H}_f$, given in eq. (III-10), are related by a hybrid transformation, this apparent paradox can be resolved.

A final caveat will constitute the concluding remark of this section. When we interpret the relation between the minimal coupling and multipolar Hamiltonians as defined by a hybrid rather than a general similarity transformation, and use the multipolar Hamiltonian $\overline{H}$ to generate equations of motion for the untransformed fields through $\partial E/\partial t = i[\overline{H},\vec{E}]$ and $\partial\vec{B}/\partial t = i[\overline{H},\vec{B}]$, then we fail to obtain Maxwell's equations.[44-47] Under these circumstances $\overline{H}$ may be substituted for H^c in calculating S-matrix elements for radiative processes, but not for effecting time-evolution through finite time intervals.

VI. SUMMARY AND ACKNOWLEDGMENTS

The following remarks represent the most important conclusions of this article:

A) The minimal coupling Hamiltonian is consistent with Maxwell's equations and with the Schroedinger equation. When A_μ is quantized, and the necessary constraints are properly implemented, the gauge-dependent parts of this Hamiltonian are manifestly unable to affect any of its dynamical consequences. It is the minimal coupling Hamiltonian, not the multipolar Hamiltonian, whose validity in time-evolution equations may be taken for granted.

B) The multipolar Hamiltonian cannot be obtained from the minimal coupling Hamiltonian by making a gauge transformation.

C) The multipolar Hamiltonian can be obtained from the minimal coupling Hamiltonian by a unitary transformation.

If one is content to use the multipolar Hamiltonian for calculating radiative transition rates (which involve on-shell transition amplitudes) one may do so without transforming anything else.

If one insists on implementing a transformation that protects off-shell as well as on-shell transition amplitudes, and that preserves Maxwell's equations, it is necessary to carry out complicated transformations on the electron states, as well as on all the other operators in the theory. It is questionable, in this author's opinion, whether any useful purpose is ever served by such a complicated and tedious formal maneuver. Finally, caution must be used in substituting the multipolar for the minimal coupling Hamiltonian in calculations that require renormalization of electron self energy divergences.

Support is acknowledged from Department of Energy Grant No. DE-AC02-79ER 10336. A.

REFERENCES

1. P.A.M. Dirac, Proc. Roy. Soc. London A117, 610 (1928).
2. S. Weinberg, "Dynamic and Algebraic Symmetries," in Lectures on Elementary Particles and Quantum Field Theory, S. Deser, M. Grisaru, and H. Pendleton, eds., M.I.T. Press, Cambridge, (1970).
3. S. Weinberg, Phys. Rev. Lett. 19, 1264 (1967).
4. A. Salam, Proceedings of the Eighth Nobel Symposium, N. Svartholm, ed., Wiley-Interscience, N.Y. (1968).
5. W. Marciano and H. Pagels, "Quantum Chromodynamics," Phys. Reports 36C, No. 3 (1978).
6. S. Gasiorowicz, "Elementary Particle Theory," Wiley, N.Y. (1966), Chapter 5.
7. Y. Aharanov and D. Bohm, Phys. Rev. 115, 485 (1959).
8. J. J. Sakurai, "Advanced Quantum Mechanics," Addison-Wesley, Reading (1967), section 1-5.
9. R. Loudon, "The Quantum Theory of Light," Clarendon Press, Oxford (1973), Chapter 8.
10. M. Sargent, M. O. Scully, and W. E. Lamb Jr., "Laser Physics," Addison-Wesley, Reading (1974), section 2-1.
11. E. Fermi, Rev. Mod. Phys. 4 (1932), 87.
12. J. Fiutak, Can. J. Phys. 41, 12 (1963).
13. W. P. Healy, J. Phys. A10, 279 (1977).
14. W. P. Healy, Phys. Rev. A16, 1568 (1977).
15. D. H. Kobe, Phys. Rev. Lett. 40, 538 (1978).

16. D. H. Kobe, Am. J. Phys. 46, 342 (1978).
17. D. H. Kobe and A. L. Smirl, Am. J. Phys. 46, 624 (1978).
18. D. H. Kobe, Am. J. Phys. 49, 581 (1981).
19. D. H. Kobe, Am. J. Phys. 50, 128 (1982).
20. E. A. Power and S. Zienau, Nuovo Cim. 6, 7 (1957).
21. E. A. Power and S. Zienau, Phil. Truns. R. Soc. London A251, 427 (1959).
22. M. Babiker, E. A. Power and M. Thirunamachandran, Proc. Roy. Soc. (London) A338, 235 (1974).
23. E. A. Power and M. Thirunamachandran, Am. J. Phys. 46, 370 (1978).
24. M. Babiker and R. Loudon, Proc. Roy. Soc. (London) A385, 439 (1983).
25. R. G. Woolley, Proc. Roy. Soc. (London) A321, 557
26. K. H. Yang, Ann. Phys. (N.Y.) 101, 62 (1976).
27. K. Rzazewski and K. Wodkiewicz, Ann. Phys. (N.Y.) 130, 1 (1980).
28. K. Haller and R. B. Sohn, Phys. Rev. A20, 1541 (1979).
29. B. Lautrup, Kgl. Dan. Mat. Pys. Medd. 35 (1967), No. 11.
30. P.A.M. Dirac, "Lectures on Quantum Mechanics," Belfer Graduate School of Science N.Y. (1964).
31. F. Strocchi and A. S. Wightman, J. Math. Phys. 15, 2198 (1974); err. 17, 1930 (1976).
32. S. N. Gupta, Proc. Phys. Soc. (London) 63, 681 (1950).
33. K. Bleuler, Helv. Phys. Acta 23, 567 (1950).
34. K. Haller, Acta Phys. Austriaca 42, 103 (1976).
35. K. Haller and L. F. Landovitz, Phys. Rev. D2, 1498 (1970).
36. L. D. Landau and I. M. Khalatnikov, Sov. Phys. JETP 2, 69 (1956).
37. H. M. Fried and D. R. Yennie, Phys. Rev. 112, 1391 (1958).
38. S. P. Tomczak and K. Haller, Nuovo Cimento B8, 1 (1972).
39. K. Haller, Nucl. Phys. B57, 589 (1973).
40. E. A. Power and T. Thuranamachandran, Phys. Rev. A22, 2894 (1980).
41. W. P. Healy, Phys. Rev. A22, 2891 (1980).
42. I. Bialynicki-Birula, Phys. Rev. D2, 2877 (1970).
43. Y. Aharanov and C. K. Au, Phys. Rev. A20, 1553 (1979).
44. L. Mandel, Phys. Rev. A20, 1590 (1979).
45. K. Haller, Phys. Rev. A26, 1796 (1982).
46. W. P. Healy, Phys. Rev. A26, 1798 (1982).
47. E. A. Power and T. Thirunamachandran, Phys. Rev. A26, 1800 (1982).

MANIFEST GAUGE INVARIANCE IN NONRELATIVISTIC QUANTUM MECHANICS WITH CLASSICAL ELECTROMAGNETIC FIELDS

Donald H. Kobe

Department of Physics
North Texas State University
Denton, Texas 76203

Since gauge invariance is such a fundamental symmetry principle, it is surprising that a manifestly gauge-invariant formulation of nonrelativistic quantum mechanics did not emerge until a few years ago. In 1976 Yang[1] gave such a formulation in which he took as physical observables only Hermitian, gauge-invariant operators.[2,3] For a charged particle in an external classical electromagnetic field, the Hamiltonian is not in general the energy operator. An energy operator is defined as the sum of the kinetic and potential energies, which is the Hamiltonian minus the scalar potential of the external time-dependent electromagnetic field.[4-6] The eigenvalue equation for the energy operator gives the energy eigenvalues and energy eigenstates at any time. The probability amplitude for finding the particle in an energy eigenstate is the inner product of the eigenstate of the energy operator and the wave function. This amplitude is gauge invariant, and therefore can be used to obtain a valid probability. The energy eigenstates are coupled by the matrix elements of the quantum-mechanical power operator.[7]

The manifestly gauge-invariant formulation can be contrasted with the conventional approach to the interaction of a charged particle and a classical electromagnetic radiation field.[8] The conventional approach uses the interaction $\vec{A}\cdot\vec{p}$ plus $\vec{p}\cdot\vec{A}$ plus A^2 plus ϕ (constants omitted), where $\vec{A}$ is the vector potential, ϕ is the scalar potential of the time-dependent electromagnetic field, and $\vec{p} = -i\hbar\nabla$ is the canonical momentum operator. For scattering cross sections this interaction gives the same results as the manifestly gauge-invariant approach, because the potentials are switched on and off adiabatically. Nevertheless, if the electromagnetic field remains on during a measurement, different state probabilities are obtained.[9]

GAUGE INVARIANCE OF THE SCHRÖDINGER EQUATION

For the sake of simplicity a single nonrelativistic quantum mechanical charged particle in a classical external electromagnetic field is considered. The magnetic field $\vec{B}$ and the electric field $\vec{E}$ can be expressed in terms of a vector potential $\vec{A}$ and a scalar potential ϕ as

$$\vec{B} = \nabla\times\vec{A} , \qquad \vec{E} = -\nabla\phi - \partial\vec{A}/\partial tc . \tag{1}$$

The Hamiltonian for the particle of mass m and charge q in the external classical electromagnetic field is

$$H(\vec{A},\phi) = (2m)^{-1}(\vec{p} - q\vec{A}/c)^2 + V(\vec{r}) + q\phi . \tag{2}$$

The potential energy $V(\vec{r})$, a nonrelativistic concept, is defined such that its negative gradient $-\nabla V(\vec{r})$ is a conservative force. The negative gradient of the scalar potential ϕ of the time-dependent external electromagnetic field is $-\nabla q\phi = q\vec{E} + q\partial\vec{A}/\partial tc$, which is not a conservative force. The scalar potential ϕ transforms as the zero component of a four vector under Lorentz transformations. The Schrödinger equation with the Hamiltonian in Eq. (2) is

$$H(\vec{A},\phi)\psi = i\hbar\partial\psi/\partial t . \tag{3}$$

A gauge transformation to a new vector potential $\vec{A}'$ and a new scalar potential ϕ' can be made,

$$\vec{A}' = \vec{A} + \nabla\Lambda , \qquad \phi' = \phi - \partial\Lambda/\partial tc , \tag{4}$$

where $\Lambda(\vec{r},t)$ is an arbitrary differentiable function. Under these transformations on the potentials, the electric and magnetic fields in Eq. (1) are unchanged. A gauge transformation on the wave function is

$$\psi' = \exp(iq\Lambda/\hbar c)\psi . \tag{5}$$

Under Eqs. (4) and (5) the Schrödinger equation in Eq. (3) is form invariant:

$$H(\vec{A}',\phi')\psi' = i\hbar\partial\psi'/\partial t . \tag{6}$$

The active view of gauge transformations is used here in which the canonical momentum operator $\vec{p} = -i\hbar\nabla$ remains unchanged under gauge transformations.

GAUGE INVARIANCE OF OPERATORS

For a quantum mechanical operator to correspond to a physical observable it must be both Hermitian and gauge invariant.[2,3] An operator $\theta(\vec{p},\vec{A})$, which depends on the canonical momentum $\vec{p}$ and the vector potential $\vec{A}$, is gauge invariant if its expectation value in the old gauge is equal to its expectation value in the new gauge,

$$<\psi|\theta(\vec{p},\vec{A})\psi> = <\psi'|\theta(\vec{p},\vec{A}')\psi'> \ . \tag{7}$$

If Eq. (5) is used in the left-hand side of Eq. (7), it can be written as

$$<\psi|\theta(\vec{p},\vec{A})\psi> = <\psi'|\theta'(\vec{p},\vec{A})\psi'> \ , \tag{8}$$

where the unitarily transformed operator θ' is defined as

$$\theta'(\vec{p},\vec{A}) = \exp(iq\Lambda/\hbar c)\theta(\vec{p},\vec{A})\exp(-iq\Lambda/\hbar c) \ . \tag{9}$$

If Eqs. (7) and (8) are compared, the operator corresponding to a physical observable must satisfy

$$\theta'(\vec{p},\vec{A}) = \theta(\vec{p},\vec{A}') \ . \tag{10}$$

In other words, for an operator to be gauge invariant, a unitary transformation on the operator must induce a gauge transformation on the vector potential on which the operator depends.

Since a function of operators is defined by its Taylor series expansion and since $\vec{A}$ is a function, Eq. (10) can be written as $\theta(\vec{p}',\vec{A}) = \theta(\vec{p},\vec{A}')$, where $\vec{p}'$ is defined as in Eq. (9). Since $\vec{p}' = \vec{p} - q\nabla\Lambda/c$ and $\vec{A}'$ is given by Eq. (4), Eq. (10) gives a functional equation

$$\theta(\vec{p} - q\nabla\Lambda/c,\vec{A}) = \theta(\vec{p},\vec{A} + \nabla\Lambda) \ . \tag{11}$$

A solution to this functional equation is

$$\theta(\vec{p},\vec{A}) = \Omega(\vec{p} - q\vec{A}/c) \ , \tag{12}$$

where Ω is a function of a single (vector) argument. Therefore an operator corresponding to a physical observable depends on the kinetic (or mechanical) momentum $\vec{p} - q\vec{A}/c$, if it depends on the canonical momentum $\vec{p}$ or on the vector potential $\vec{A}$.

The Hamiltonian is not a physical observable because it is not gauge invariant. When the unitary transformation in Eq. (9) is applied to the Hamiltonian in Eq. (2), the result is

$$H'(\vec{A},\phi) = H(\vec{A}',\phi) = H(\vec{A}',\phi') + q\partial\Lambda/\partial tc \ . \tag{13}$$

Therefore the Hamiltonian is not a gauge-invariant operator and its expectation value can have any arbitrary value because Λ is an arbitrary function. The Hamiltonian in this time-dependent problem is therefore not the energy, which is a physical observable.[1,4]

An energy operator can be defined as

$$\mathcal{E}(\vec{A}) = (2m)^{-1}(\vec{p} - q\vec{A}/c)^2 + V(\vec{r}) \quad , \tag{14}$$

which is the sum of the kinetic and potential energies.[1,4] Equation (14) differs from the Hamiltonian by the scalar potential term $q\phi$ which is not a conservative potential energy, so $\mathcal{E}(\vec{A}) = H(\vec{A},\phi) - q\phi$. A unitary transformation as in Eq. (9) on Eq. (14) gives $\mathcal{E}'(\vec{A}) = \mathcal{E}(\vec{A}')$, so the operator is gauge invariant.

An Ehrenfest theorem for the energy operator $\mathcal{E}(\vec{A})$ in Eq. (14) is obtained[1] by differentiating its expectation value with respect to the time,

$$d\langle\psi|\mathcal{E}(\vec{A})\psi\rangle/dt = \langle\psi|P\psi\rangle = \int d^3r \; \vec{J}\cdot\vec{E} \quad . \tag{15}$$

The quantum mechanical power operator P in Eq. (15) is[7]

$$P = (q/2)(\vec{E}\cdot\vec{v} + \vec{v}\cdot\vec{E}) \quad , \tag{16}$$

where $\vec{E}$ is the time-dependent electric field in Eq. (1) and the velocity operator $\vec{v} = (\vec{p} - q\vec{A}/c)/m$. The electric current density in Eq. (15) is $\vec{J} = q\mathcal{R}e\,\psi^*\vec{v}\psi$. Since the electromagnetic field loses energy at a rate $\vec{J}\cdot\vec{E}$ per unit volume by Poynting's theorem, conservation of energy demands that the particle gains energy at a rate $\vec{J}\cdot\vec{E}$ per unit volume. Therefore, the operator $\mathcal{E}(\vec{A})$ can be identified as the energy operator for the particle because it is gauge invariant and satisfies the appropriate Ehrenfest theorem in Eq. (15).

PROBABILITY AMPLITUDES

If an experiment is designed to measure the energy, then the eigenstates of the energy operator should be used.[1,4] The energy operator eigenvalue equation is

$$\mathcal{E}(\vec{A})\psi_n = \varepsilon_n\psi_n \quad . \tag{17}$$

Since the energy operator in general depends on the time as a parameter, the energy eigenvalue ε_n and energy eigenstate ψ_n also depend on the time in general. The probability amplitude for finding the particle in an energy eigenstate at time t is

$$c_n(t) = \langle\psi_n(t)|\psi(t)\rangle \quad . \tag{18}$$

If the time derivative of this amplitude is taken and the Schrödinger equation in Eq. (3) is used, the equation of motion is

$$i\hbar\dot{c}_n - \varepsilon_n c_n = \sum_m \langle\psi_n|(q\phi - i\hbar\partial/\partial t)\psi_m\rangle c_m \ . \qquad (19)$$

The gauge-invariant operator $q\phi - i\hbar\partial/\partial t$ induces transitions between the energy eigenstates.

For the nondegenerate case where $\varepsilon_n \neq \varepsilon_m$ if $n \neq m$, Eq. (19) can be rewritten as[7]

$$i\hbar\dot{c}_n - \tilde{\varepsilon}_n c_n = \sum_m{}' \, i\tau_{nm}\langle\psi_n|P\psi_m\rangle c_m \ , \qquad (20)$$

where the prime on the summation indicates that $m \neq n$ and $\tilde{\varepsilon}_n = \varepsilon_n + \langle\psi_n|(q\phi - i\hbar\partial/\partial t)\psi_n\rangle$ is a dressed energy. The characteristic time associated with the transition from state n to m is $\tau_{nm} = \hbar/(\varepsilon_n - \varepsilon_m)$. Equation (20) shows that the states are coupled by matrix elements of the power operator P in Eq. (16). If the electric field vanishes, the power operator also vanishes and there is no coupling between the energy eigenstates.[10] Spontaneous emission is neglected here because the field is classical.

All of the equations in this section are either form invariant (gauge covariant) or numerically unchanged under gauge transformations. The energy eigenstate ψ_n transforms as ψ does in Eq. (5) under a gauge transformation. The energy operator is form invariant under unitary transformation, so the energy eigenvalue ε_n in Eq. (17) is gauge invariant. The probability amplitude in Eq. (18) is also gauge invariant. In the new gauge, $c_n'(t) \equiv \langle\psi_n'(t)|\psi'(t)\rangle = c_n(t)$, since the phase factor on each side of the inner product cancel. Equations (19) and (20) are also both gauge invariant.

CONCLUSION

In this paper a brief review of the manifestly gauge-invariant formulation of quantum mechanics is given.[1,4] No discussion of approximations, like the electric dipole approximation, is given.[5,11] No comparison of the gauge-invariant formulation with the conventional approach is given.[12] No applications of the gauge-invariant formulation are given.[13-15] These important topics have been discussed elsewhere. Only the novel features of the gauge-invariant formulation like the energy operator, the power operator,[16] and gauge-invariant probability amplitudes[17] are discussed here.

Even though the electromagnetic field is treated classically here, there are many practical problems for which this treatment is adequate. The structure of quantum field theory has been guided in the past by the corresponding classical field theory. The manifestly gauge-invariant semiclassical theory may give a pattern for the development of a manifestly gauge-invariant quantum electrodynamics.

ACKNOWLEDGMENT

I should like to express my gratitude to Dr. Kuo-Ho Yang for his many helpful discussions, for his encouragement, and for his friendship during the last seven years.

REFERENCES

1. K.-H. Yang, Ann. Phys. (N.Y.) 101:62 (1976).
2. D. H. Kobe and K.-H. Yang, J. Phys. A: Math. Gen. 13:3171 (1980).
3. C. Cohen-Tannoudji, B. Diu, F. Laloë, "Quantum Mechanics," Vol. 1, Wiley, New York (1977), pp. 315-328.
4. D. H. Kobe and A. L. Smirl, Am. J. Phys. 46:624 (1978).
5. D. H. Kobe, Int. J. Quant. Chem. S 12:73 (1978).
6. D. H. Kobe, Phys. Rev. A 19:205 (1979); 19:1876 (1979).
7. D. H. Kobe, E. C.-T. Wen, and K.-H. Yang, Phys. Rev. D 26:1927 (1982).
8. L. I. Schiff, "Quantum Mechanics," 3rd. edn., McGraw-Hill, New York (1968), pp. 398-403.
9. C. Leubner and P. Zoller, J. Phys. B: At. Mol. Phys. 13:3613 (1980).
10. D. H. Kobe and K.-H. Yang, Am. J. Phys. 51:163 (1983).
11. D. H. Kobe, Am. J. Phys. 50:128 (1982).
12. D. H. Kobe and E. C.-T. Wen, J. Phys. A: Math. Gen. 15:787 (1982).
13. D. H. Kobe, J. Phys. A: Math. Gen. 16:737 (1983).
14. P. K. Kennedy and D. H. Kobe, J. Phys. A: Math. Gen. 16:521 (1983).
15. D. H. Kobe, J. Phys. B: At. Mol. Phys. 16:1159 (1983).
16. K.-H. Yang, Phys. Lett. 92A:71 (1982).
17. K.-H. Yang, J. Phys. A: Math. Gen. 15:437 (1982).

CONSEQUENCES OF HIGH FIELD INTENSITY IN SEMICLASSICAL ELECTRODYNAMICS

Howard R. Reiss

Arizona Research Laboratories
University of Arizona
Tucson, Arizona

and

Physics Department
The American University
Washington, D.C.

INTRODUCTION

Within the context of non-relativistic quantum mechanics and plane-wave classical electromagnetic fields, the specific consequences of very high field intensity are explored. Four principal departures from conventional behavior are identified: (i) The $-e\vec{E}\cdot\vec{r}$ interaction term frequently employed in electrodynamics is not adequate. (ii) The long-wavelength approximation (LWA) and the electric-dipole approximation (EDA) are not equivalent. (iii) In a two-body bound system, the equations of motion cannot be separated into independent equations describing center-of-mass motion and relative motion. (iv) A multipole treatment of interaction terms loses its utility because new, unfamiliar terms appear, and because the relative orders of magnitude of conventional multipole terms are altered.

The investigation of intense-field phenomena is done in two different gauges to correspond to two gauges in common use. One is the standard Coulomb (C) gauge. The other is the less-familiar "electric-field" or EF gauge, which is an extension of the standard gauge with $-e\vec{E}\cdot\vec{r}$ interaction term to a gauge with a full four-component potential for the electromagnetic field.

EF GAUGE

The Göppert-Mayer[1] gauge commonly employed in atomic and molecular physics, in which a plane-wave electromagnetic field is represented by a scalar potential $-\vec{r}\cdot\vec{E}(t)$, is useful and correct only as long as the magnetic component of the plane wave is of no importance. Since this presumption cannot be made a priori for very intense fields, it is necessary to provide a complete specification of the field potentials. The simplest such extension which reduces to the scalar $-\vec{E}\cdot\vec{r}$ potential in the long-wavelength and low-intensity case is the EF gauge.[2]

Let $A^\mu(x)$ be a four-vector potential in a "simple" gauge, which is defined to be any gauge connected to Coulomb gauge by a gauge transformation of the form

$$A^\mu \to \tilde{A}^\mu = A^\mu + k^\mu \Lambda \ . \tag{1}$$

In Eq. (1), k^μ is the propagation four-vector of the plane-wave field, Λ is a scalar function, and the relativistic notation employs a time-favoring real metric so that, for example,

$$k\cdot x = k_\mu x^\mu = k^\mu x_\mu = \omega t - \vec{k}\cdot\vec{r} \ . \tag{2}$$

Units with $\hbar=c=1$ are used in this paper. The initial $A^\mu(x)$ is to be a function of $k\cdot x$ only. EF gauge is generated from this four-potential by the transformation function

$$\chi = -\ x\cdot A \ . \tag{3}$$

This gives the EF-gauge four-potential $\tilde{A}^\mu$ as

$$\tilde{A}^\mu = A^\mu + \partial^\mu \chi = -k^\mu x\cdot A' \ , \tag{4}$$

where the prime on A denotes a derivative with respect to $k\cdot x$. If the initial four-potential A^μ is in C gauge, then $x\cdot A' = -\vec{r}\cdot\vec{A}'$, and so the EF-gauge four-potential is

$$\tilde{A}^\mu = -(k^\mu/\omega)\ \vec{r}\cdot\vec{E} \ . \tag{5}$$

In Eq. (5), ω is the time part of k^μ, $\vec{r}$ is the space part of x^μ, and $\vec{E}$ is the electric field vector $\vec{E}(\vec{r},t)$. The scalar component of the EF-gauge potential in (5) is

$$\tilde{A}^o = -\vec{r}\cdot\vec{E}(\vec{r},t) \ , \tag{6}$$

which yields the Göppert-Mayer potential

$$\phi = -\vec{r}\cdot\vec{E}(t) \tag{7}$$

in the LWA limit. The EF potential is a "completed" Göppert-Mayer potential which fully describes electric and magnetic fields.

EF gauge is a Lorentz gauge (i.e., it satisfies $\partial_\mu \tilde{A}^\mu=0$), but it is not a simple gauge, since it does not satisfy Eq. (1). Thus, the usual gauge-invariance investigations of QED do not include EF gauge. Furthermore, EF gauge is not spacelike, as all simple gauges are. Rather, it is lightlike as Eq. (4) or (5) shows.

SINGLE-PARTICLE SCHRÖDINGER EQUATION

The single-particle Schrödinger equation in EF gauge is

$$i\partial_t\psi = \{-e\vec{r}\cdot\vec{E}(\vec{r},t)+(1/2m)[\vec{p}+e\hat{k}\vec{r}\cdot\vec{E}(\vec{r},t)]^2+V(\vec{r})\}\psi \, , \tag{8}$$

where $\hat{k} \equiv \vec{k}/\omega$. In the LWA, Eq. (8) becomes

$$i\partial_t\psi = \{-e\vec{r}\cdot\vec{E}(t)+(1/2m)[\vec{p}+e\hat{k}\vec{r}\cdot\vec{E}(t)]^2 + V(\vec{r})\}\psi \, . \tag{9}$$

Notice that a vector potential, and hence a magnetic field, remains in Eq. (9). Specifically, there is an interaction term in Eq. (9)

$$\begin{aligned} -e\vec{A}\cdot\vec{p}/m &= (e/m\omega)(\vec{r}\cdot\vec{E})(\vec{k}\cdot\vec{p}) \\ &= (e/2m\omega)[(\vec{k}\cdot\vec{p})(\vec{r}\cdot\vec{E})+(\vec{E}\cdot\vec{p})(\vec{k}\cdot\vec{r})]-(e/2m\omega)\vec{B}\cdot\vec{L} \\ &= \text{electric quadrupole + magnetic dipole} \, , \end{aligned} \tag{10}$$

where $\vec{B}$ is the magnetic induction and $\vec{L}$ is the angular momentum. An immediate consequence of Eq. (10) is that the LWA equation of motion, Eq. (9), contains multipole interaction terms beyond the electric dipole term. That is, the LWA and the EDA are not the same in EF gauge.

In C gauge in the LWA, the interaction term $-e\vec{A}(t)\cdot\vec{p}/m$ is purely electric dipole. Therefore, LWA and EDA are the same in C gauge, and consequently C gauge and EF gauge become inequivalent at the level of the LWA. Although different gauges must be entirely equivalent when considered in full, they need not be equivalent when any approximation is introduced, such as the LWA.

TWO-BODY SCHRÖDINGER EQUATION

The two-body EF-gauge Schrödinger equation in the LWA for particles of masses m_1, m_2, charges e_1, e_2, and position vectors $\vec{r}_1$, $\vec{r}_2$ is

$$i\partial_t\psi(\vec{r},\vec{R}) = [-e_t\vec{R}\cdot\vec{E}-e_r\vec{r}\cdot\vec{E}+(1/2m_t)(\vec{p}_R+e_t\hat{k}\vec{R}\cdot\vec{E}+e_r\hat{k}\vec{r}\cdot\vec{E})^2$$

$$+ (1/2m_r)(\vec{p}_r+e_r\hat{k}\vec{R}\cdot\vec{E}+e_e\hat{k}\vec{r}\cdot\vec{E})^2 + V(r)]\psi(\vec{r},\vec{R}) \,. \tag{13}$$

In Eqs. (12) and (13), the masses and charges are

$$m_t = m_1 + m_2 \,, \qquad m_r = m_1\, m_2/m_t$$
$$e_t = e_1 + e_2 \,, \quad e_r = (e_1m_2-e_2m_1)/m_t \,, \quad e_e = (e_1{m_2}^2+e_2{m_1}^2)/{m_t}^2. \tag{14}$$

The LWA two-body equations in C gauge separate in c.m. and relative coordinates to the two independent equations

$$i\partial_t\psi_R = (1/2m_t)(\vec{p}_R-e_t\vec{A})^2\psi_R$$
$$i\partial\psi_r = [(1/2m_r)(\vec{p}_r-e_r\vec{A})^2 + V(r)]\psi_r \,. \tag{15}$$

In Eqs. (13) and (15), the LWA is interpreted to mean

$$|\vec{k}\cdot\vec{r}| << 1 \,, \qquad |\vec{k}\cdot\vec{R}| << 1 \,. \tag{16}$$

Whereas the C-gauge equations in (15) are uncoupled, the EF-gauge equation in (13) has crossed $\vec{r},\vec{R}$ terms. These crossed terms, when compared to leading terms in the usual uncoupled equations, are of order of magnitude[2]

$$\frac{|e(\vec{E}\cdot\vec{R})(\vec{k}\cdot\vec{p}_r)/\omega m_r|}{|e\vec{E}\cdot\vec{r}|} = O(z^{1/2})$$
$$\frac{|e(\vec{E}\cdot\vec{r})(\vec{k}\cdot\vec{p}_R)/\omega m_t|}{{p_R}^2/2m_t} = O(z^{1/2}), \tag{17}$$

where z is the field-intensity parameter

$$z = \frac{e^2\langle\vec{E}^2\rangle}{m^2\omega^2} \,. \tag{18}$$

The angular bracket in Eq. (18) refers to a time average over a wave period. Although it requires $2.4\text{x}10^{18}$ W/cm^2 of 1.06 μm radiation to have a value of unity for z, the square-root dependence on z in Eq. (17) means that crossed terms will become quantitatively important at about $z \simeq 10^{-4}$, which occurs at an energy flux of $6\text{x}10^{13}$ W/cm^2 at 1.06 μm. When crossed terms are important, the EF-gauge equations of motion (13) cannot be separated. This loss of separability would not appear if only the

Göppert-Mayer $-\vec{r}\cdot\vec{E}$ potential were used. Since the EF potentials give a complete statement of the fields, and $-\vec{r}\cdot\vec{E}$ does not, this failure to find intensity-induced coupling between c.m. and relative coordinates is a defect of the $-\vec{r}\cdot\vec{E}$ potential.

The LWA equations of motion in C gauge given in Eq. (15) have been found to be separable, seemingly without reference to the field intensity. However, if the potentials are extended to the next multipole order of approximation beyond the LWA, then the two-body equation of motion in C gauge is

$$i\partial_t\psi(\vec{r},\vec{R}) = \{(1/2m_t)[\vec{p}_R-e_t\vec{A}-e_t(\vec{k}\cdot\vec{R})\vec{E}/\omega-e_r(\vec{k}\cdot\vec{r})\vec{E}/\omega]^2$$
$$+ (1/2m)[\vec{p}_r-e_r\vec{A}-e_r(\vec{k}\cdot\vec{R})\vec{E}/\omega-e_e(\vec{k}\cdot\vec{r})\vec{E}/\omega]^2 + V\}\psi(\vec{r},\vec{R}) \,. \quad (19)$$

The crossed terms between $\vec{r}$ and $\vec{R}$ which appear in Eq. (19) are not the same as those evaluated in Eq. (17). Nevertheless, the separation-blocking terms in Eq. (19) have the same orders of magnitude given in Eq. (17), and so the same statements apply to the supplemented C-gauge Eq. (19) as to the LWA EF-gauge Eq. (13).

INTENSITY EFFECTS IN THE MULTIPOLE EXPANSION

The terms which mix c.m. and relative coordinates in the equations of motion (13) and (19) have been remarked upon as blocking the separation of variables into independent equations. These mixed $\vec{r},\vec{R}$ terms have another important consequence. They constitute terms which have no conventional multipole meaning. Yet, as demonstrated, for example, by the first line in Eq. (17), these terms can be very significant as compared to the usually dominant electric dipole term. This fact greatly limits the customary utility of the multipole expansion. Another limiting feature exists as well which has not been exhibited yet.

The exact classical solution for charged particle motion in an electromagnetic plane wave shows that,[3] for large field intensity,

$$\lim_{z\to\infty} |\vec{k}\cdot\vec{R}| = \frac{1}{4} \quad (20)$$

in the frame in which the particle is, on average, at rest. The approach to the limit is such that z=1 corresponds to about half the limiting value of Eq. (20). Equation (20) is in violation of the LWA condition in Eq. (16). The multipole expansion must therefore be modified. In particular, the electric field at one particle in a two-body system can be written as

$$\vec{E}(\vec{r},t) = \vec{E}_o \sin(\omega t-\vec{k}\cdot\vec{r}_1)$$
$$= \vec{E}_o \sin(\omega t-\vec{k}\cdot\vec{R}-\vec{k}\cdot\vec{r}m_2/m_t)$$
$$\approx \vec{E}_o \cos\vec{k}\cdot\vec{R} \sin\omega t - \vec{E}_o \sin\vec{k}\cdot\vec{R} \cos\omega t$$
$$- \vec{E}_o\, \vec{k}\cdot\vec{r}(m_2/m_t)\sin\vec{k}\cdot\vec{R} \sin\omega t$$
$$- \vec{E}_o\, \vec{k}\cdot\vec{r}(m_2/m_t)\cos\vec{k}\cdot\vec{R} \cos\omega t \ , \qquad (21)$$

with a similar expression for $\vec{E}(\vec{r}_2,t)$.

If Eq. (21) and its analogue for $\vec{E}(\vec{r}_2,t)$ are employed in place of the expressions obtained from Eq. (16), then the equations of motion replacing Eqs. (13) and (19) become far more complicated than those equations. There are new mixed $\vec{r},\vec{R}$ terms beyond those in (13) and (19), and some of these new terms are as large as those previously identified. It is clear from Eq. (21) that these new terms which will appear in the equations of motion have no conventional multipole meaning.

REFERENCES

1. M. Göppert-Mayer, Ann. Phys. (Leipzig) 9, 273 (1931).
2. H. R. Reiss, Phys. Rev. A 19, 1140 (1979).
3. E. S. Sarachik and G. T. Schappert, Phys. Rev. D 1, 2738 (1970).

INTERACTION HAMILTONIAN IN QUANTUM OPTICS

OR: $\vec{p} \cdot \vec{A}$ VS. $\vec{E} \cdot \vec{r}$ REVISITED

R. R. Schlicher[‡], W. Becker[†], J. Bergou[‡*], and M. O. Scully[†‡]

† Institute for Modern Optics
Department of Physics and Astronomy,
University of New Mexico
Albuquerque, NM 87131, USA

‡ Max-Planck Institut für Quantenoptik
D-8046 Garching bei München
West Germany

ABSTRACT

The question of the equivalence of the two matter-field interaction Hamiltonians which are commonly used in quantum optics, i.e. $e\vec{E} \cdot \vec{r}$ and $(e/m)\vec{p} \cdot \vec{A}$, is considered in the context of nonrelativistic quantum mechanics. We restrict our attention to classical fields in the dipole approximation. After a review of the concept of gauge invariance in quantum mechanics we describe a general procedure for obtaining gauge invariant transition probabilities in an arbitrary gauge. This is based on using eigenstates of the gauge invariant energy operator as the states between which transitions occur. We compare this procedure with the frequently employed "hybrid procedure" in which gauge transformations are applied to the operators, but not to the states, and discuss the conditions under which the hybrid procedure yields correct results. All general results are illustrated with the example of the two-level atom subject to a weak field which can be dealt with by perturbation theory. We discuss the so called momentum translation (MTA) approximation as an example of the

* Alexander von Humboldt Fellow on leave from the Central Research Institute for Physics, H-1525 Budapest, P. O. Box 49, Hungary.

dangers associated with misconceptions on gauge invariance. In addition to examples in which the MTA is known to fail, e.g., the harmonic oscillator and the two-level atom, we apply it to the case of a Rydberg wave function which is strongly localized at a particular radius, and show that the MTA approach fails in this case as well.

I. INTRODUCTION

In these lectures we will treat the interaction of a quantum mechanical atomic system with an external radiation field. We will focus only on classical fields in dipole approximation and on non-relativistic systems. This particular situation is of current interest since it describes most forms of laser interaction with atoms and molecules. One central problem is the calculation of the probability for transitions induced by radiation. Since we have to know the Hamiltonian of the system for the Schrödinger equation, the form of the matter - radiation interaction depends on the choice of gauge. Of course all physical results must be gauge-independent. In practice, however, even on a very elementary level there is much confusion in the literature concerning the gauge in which the probability of transitions induced by optical radiation should be calculated. This might be due to the fact that until recently most textbooks on quantum mechanics and time-dependent perturbation theory in quantum mechanics barely mention the subject of gauge transformations and invariance and its application to the matter-radiation interaction. It is the object of this paper to present these fundamental questions on a simple tutorial level.

In most practical calculations, two forms of the Hamiltonian are used for the interaction of matter with optical radiation: the interaction $-e\vec{r}\cdot\vec{E}(t)$ or the interaction $-(e/m)\vec{p}\cdot\vec{A}(t)$. As will be seen, these two interactions correspond to two different gauges of the total Hamiltonian of the system. In 1931 Maria Göppert-Mayer [1] used the $\vec{E}\cdot\vec{r}$-Hamiltonian for the first time in a multiphoton calculation. Since Lamb's pioneering work on the spectrum of atomic hydrogen [2], it is known that in certain situations it is more convenient to choose the $\vec{E}\cdot\vec{r}$ interaction. However, the results of Ref. 2 have often been ill-interpreted in that it is argued that the two perturbations would yield physically different results and that only the $\vec{E}\cdot\vec{r}$ interaction is the "correct" interaction, in contradiction to the known gauge invariance of quantum mechanics. Power and Zienau [3] gave the general unitary transformation connecting the minimal coupling Hamiltonian $\vec{p}\cdot\vec{A}$ to the Hamiltonian corresponding to the multipole expansion of the fields. In the electric dipole approximation the latter form leads to the $\vec{E}\cdot\vec{r}$ interaction.

But the general attitude for practical calculations has been to use one or the other forms of the interaction with the same set of unperturbed basis vectors [4]. In some cases this procedure leads to different results for the two interactions. This has led to some confusion in the literature concerning the question of obtaining gauge-independent transition probabilities. In 1976 Jaynes [5] commented:

> "... a whole generation of physicists has stumbled on this problem and lived, not only under the shadow of the immediate difficulty: 'How can I ever know whether a practical calculation has been done right?,' but the deeper mystery: 'How is it possible that a theory, for which formal gauge invariance is proved easily once and for all, can lead to grossly non-invariant results as soon as we try to apply it to the simplest real problem?'"

Since the mid-seventies the problem attracted even more attention, due to the rapid development of quantum optics. The fundamentals of gauge invariance and gauge transformations in quantum mechanics are explained in the textbook by Cohen-Tannoudji et al. [6]. Yang [7] obtained gauge invariant transition amplitudes by calculating transitions between eigenstates of the energy operator. The energy operator is to be distinguished from the unperturbed Hamiltonian, as we shall outline in detail. Only in the gauge with the $\vec{E} \cdot \vec{r}$ interaction do the energy operator and the unperturbed Hamiltonian coincide. Therefore, transition amplitudes calculated with the $\vec{E} \cdot \vec{r}$ interaction are gauge invariant. If the transition amplitude is calculated with the $\vec{p} \cdot \vec{A}$-interaction using eigenstates of the same unperturbed Hamiltonian, which is what has usually been done in practical calculations, then these results are gauge-dependent. Forney et al. [8] proved that in the limit of large times $t \to \infty$ the transition amplitudes calculated between eigenstates of the energy operator and between eigenstates of the unperturbed Hamiltonian nevertheless coincide in the $\vec{p} \cdot \vec{A}$ interaction.

With these three basic publications the subject could almost have been exhausted. However, the way Ref. 7 presented its results and the recent applications of these results [9,10] has raised critical comments in the literature [11-14]. Furthermore, even nowadays one can find many examples in the literature claiming different results with the different gauges, e.g. for the off-resonant two-photon absorption [15] as well as for other examples [16,17]. This, in turn, raised even more controversy [18-22]. Thus, we face over the last five years a rapidly increasing number of publications on this fundamental yet elementary topic.

The careful distinction between gauge invariant and gauge dependent quantities is not only necessary for the interpretation of transition probabilities. It is also essential for the interpretation of wavefunctions. In 1970 Reiss [23] published the so-called "momentum translation approximation" (MTA) for the solution of the Schrödinger equation with the $\vec{p} \cdot \vec{A}$ interaction. The MTA wavefunction is supposed to include approximately the interaction of a bound system with an external field to an arbitrary order. With a simple gauge transformation one can show that this interpretation is wrong [24]. Actually, the MTA wave function turns out to be an eigenfunction of the energy operator and, consequently, the correct noninteracting bound state wave function in the Coulomb gauge. Hence multiphoton transition probabilities derived by using the MTA method are unreliable and likely to be spurious. The failure of the MTA is a good example of the pitfalls one may encounter if one is not aware of the restrictions imposed by gauge invariance on the physical interpretation of wave functions. Also, the MTA wave function was recently invoked to approximate the nuclear wave functions in a calculation of forbidden nuclear beta-decay in the presence of a strong external field [25], again with the misguided intention of approximating the nucleus-field interaction. The fact that there is no enhancement of forbidden nuclear beta decay due to a (moderately) strong external field [26] should not come as a surprise if one knows the correct interpretation of the MTA wave function. This is yet another example of the dangers of misinterpretations of the gauge question in quantum mechanics.

In this paper we first describe in Section 2 the situation existing until the mid-seventies. We introduce the two interactions $\vec{E} \cdot \vec{r}$ and $\vec{p} \cdot \vec{A}$ and show how they yield different probabilities for transitions between eigenstates of the unperturbed Hamiltonian. We refer to this procedure as the "hybrid" procedure [19]. In Section 3 we summarize some basics of gauge invariance in quantum mechanics, following Ref. 6, and apply them to the two gauges usually dealt with in quantum optics. In Section 4 we outline the formalism of Refs. 7 and 8, which yields the required gauge invariant probabilities for transitions between eigenstates of the energy-operator. We refer to this procedure as the gauge-invariant procedure. We see that the hybrid procedure yields gauge dependent results since it neglects the gauge transformation on the wavefunctions. Comparing the hybrid and the gauge invariant procedures in the R-gauge, we find that for a wide class of physical situations, notably whenever the vector potential of the field vanishes at the time of the preparation of the system and at the time of measurement, both procedures yield identical results [8,27]. This applies to all fields with finite frequency bandwidth if the system is prepared at $t_o \to -\infty$ and if the measurement takes place at $t \to +\infty$. Special care has to be taken in the case of a monochromatic wave which does not have a turning on or off process. The equivalence of both procedures for larger times also holds true for a monochromatic wave if the field is resonant with the transition energy [28,29,18-21]. In Section 5 we discuss the MTA-wavefunction and compare it with a non-interacting state. In Section 6 we summarize our revision of the $\vec{E} \cdot \vec{r}$ versus $\vec{p} \cdot \vec{A}$ problem.

Since it is the intention of these lecture notes to demonstrate the material on a textbook level, we shall try to apply all the general results to an example. The number of examples is limited since only a few quantum systems interacting with an electromagnetic field can be discussed in closed form. Instructive examples are the two-level atom in the strong signal regime [2,9], the charged harmonic oscillator [see, for example, Refs. 10,16,24, 30], or the free electron in a spatially uniform time-varying field [31]. Further examples can be found in Ref. 10. We choose in our paper the example of a two-level atom under the impact of a weak field which can be treated as a perturbation. Thus we can restrict these lecture notes to single-photon processes and first order perturbation theory. The results for two-photon and multi-photon processes will only be mentioned briefly and the relevant literature will be quoted.

2. FORMULATION OF THE $\vec{E}\cdot\vec{r}$ VERSUS $\vec{p}\cdot\vec{A}$ PROBLEM

2.1. The $\vec{E}\cdot\vec{r}$ interaction Hamiltonian

The $\vec{E}\cdot\vec{r}$-Hamiltonian can be derived intuitively from classical electrodynamics. The interaction between a dipole and an external field provides a simple but instructive physical picture of the interaction of radiation with matter. If a dipole with dipole moment $\vec{p} = e\vec{r}$ is placed in an external spatially uniform field $\vec{E}_o$, the electrostatic interaction energy between the dipole and the field is

$$V = -\vec{p} \cdot \vec{E}_0 \ . \tag{2.1}$$

For a time dependent field $\vec{E}(t)$, Eq. (2.1) describes the instantaneous energy $V(t)$. If the external field interacts, for instance, with an atom of charge density $\rho(\vec{r})$, the electric dipole moment is given by

$$\vec{p} = \int \vec{r}\rho(\vec{r})d^3r \ . \tag{2.2}$$

In quantum mechanics the density $\rho(\vec{r})$ is given by the overlap of the initial and final state of the system, ψ_i and ψ_f,

$$\rho(\vec{r},t) = \Psi_f^*(\vec{r},t)\Psi_i(\vec{r},t) \tag{2.3}$$

and Eq. (2.1) represents the matrix element of the interaction Hamiltonian. The dipole moment (2.2) is different from zero if ψ_i and ψ_f have opposite parities.

In this paper we shall often deal with monochromatic electromagnetic fields $\vec{E}(\omega t-\vec{k}\vec{r})$. For the interaction of an atom with such a plane wave of optical or even lower frequency, the atomic radius is much smaller than the wavelength of the external field. Atomic radii are of the order of the Bohr radius $a \sim 5 \cdot 10^{-11}$ m, the wavelength of visible light is about $\lambda \sim 5 \cdot 10^{-7}$ m. Hence it is a good approximation to ignore the spatial variation $\vec{k}\vec{r}$ of such an electromagnetic field over the dimensions of the atom and to describe it by a uniform field $\vec{E}(t)$. If we choose the nucleus as the origin of the coordinate system, we can set $\vec{k}\vec{r} \cong a/\lambda \cong 0$. With this restriction the Hamiltonian can only give rise to electric-dipole interaction. We neglect the electric-quadrupole and magnetic-dipole interactions and higher terms. This approximation is known as the electric dipole approximation (EDA). We will use it throughout this paper.

The previous arguments suggest the Hamiltonian to be identified with

$$H_1(r) = -e\vec{E}(t)\cdot\vec{r}\ . \tag{2.4}$$

This Hamiltonian has a simple physical interpretation and depends only on a physical quantity, the electric field $\vec{E}$. The Schrödinger equation for a particle with charge e interacting with an external field $\vec{E}(t)$ then reads

$$i\hbar\frac{\partial}{\partial t}\Psi(\vec{r},t) = (H_0 + H_1(t))\Psi(\vec{r},t)\ . \tag{2.5}$$

The unperturbed Hamiltonian of the particle, bound in a static potential $V(\vec{r})$, is

$$H_0 = \frac{\vec{p}^2}{2m} + V(\vec{r})\ . \tag{2.6}$$

The eigenvalues and eigenstates of this Hamiltonian are denoted by

$$H_0\phi_n^0(\vec{r}) = E_n\phi_n^0(\vec{r}) \tag{2.7}$$

and the solution of the field-free Schrödinger equation

$$i\hbar \frac{\partial}{\partial t} \Psi^0(\vec{r},t) = H_0\Psi^0(\vec{r},t) \tag{2.8}$$

has the form

$$\Psi_n^0(\vec{r},t) = e^{-iE_n t/\hbar} \phi_n^0(\vec{r}) . \tag{2.9}$$

Eq. (2.5) does not include the energy of the radiation field since we only consider the semiclassical theory of classical fields and quantum-mechanical atoms in this paper. Furthermore, it should be mentioned that the Hamiltonian (2.4) in the dipole approximation can be justified from classical electrodynamics in a much more rigorous way with the help of a Hamiltonian formalism and a multipole expansion of the electric field [3,32]. A part of this procedure will be outlined in Sections 2.2 and 3.2 for the particular physical situation considered in this paper.

2.2. The $\vec{p}\cdot\vec{A}$ interaction Hamiltonian

We consider now another approach to the matter-field interaction. The rigorous way to derive the interaction is to start from the classical Hamiltonian. We can include the interaction of a particle of charge e with an external electromagnetic field into the free particle Hamiltonian $H = \vec{p}^2/2m$ by replacing the canonical momentum $\vec{p}$ and the energy E of the particle by

$$\vec{p} \rightarrow \vec{p} - e\vec{A}^g(\vec{r},t), \; E \rightarrow E - eU^g(\vec{r},t) . \tag{2.10}$$

The classical Hamiltonian for the motion of a free particle then reads

$$H^g(\vec{r},\vec{p};t) = \frac{1}{2m} [\vec{p} - e\vec{A}^g(\vec{r},t)]^2 + eU^g(\vec{r},t) . \tag{2.11}$$

This Hamiltonian is expressed in terms of the canonical variables $\vec{r},\vec{p}$ and the electromagnetic potentials $\vec{A}^g(\vec{r},t)$, $U^g(\vec{r},t)$. The superscript g indicates that the potentials and therefore also the Hamiltonian depend on the gauge chosen. Common gauges are the Coulomb gauges (C-gauges) $(\vec{A}^g, U^g)$ which satisfy the condition

$$\vec{\nabla} \cdot \vec{A}^g(\vec{r},t) = 0 \tag{2.12}$$

so that in quantum mechanics $[\vec{p}, \vec{A}^g] = 0$. We use a long-wavelength approximation (LWA) for the vector potential $\vec{A}^g(\vec{r} = 0,t) = \vec{A}^g(t)$, similar to the electric dipole approximation in Section 2.1. If we add the static potential $V(\vec{r})$ to the free particle Hamiltonian (2.11) (if $V(\vec{r})$ is a Coulomb potential this could again be derived more rigorously from Maxwell's equations in a Hamiltonian formalism), we obtain the Schrödinger equation

$$i\hbar \frac{\partial}{\partial t} \Phi^g(\vec{r},t) = (H_0 + H_2^g(t))\Phi^g(\vec{r},t) . \tag{2.13}$$

The unperturbed Hamiltonian H_o is given by Eq. (2.6) and the interaction by

$$H_2^g(t) = - \frac{e}{m} \vec{A}^g(t) \cdot \vec{p} + \frac{e^2}{2m} (\vec{A}^g(t))^2 + eU^g(\vec{r},t) . \tag{2.14}$$

Since there are no sources of the electromagnetic field present in the problem considered it is always possible to find a Coulomb gauge with vanishing scalar potential

$$U(\vec{r},t) = 0 . \tag{2.15}$$

The gauge $(\vec{A},0)$ which satisfies the conditions (2.12) and (2.15) is usually called radiation gauge (R-gauge). Quantities in this gauge will be denoted without a subscript. The vector potential $\vec{A}$ is related to the electric and magnetic fields by

$$\vec{E} = - \frac{\partial}{\partial t} \vec{A} , \quad \vec{B} = \vec{\nabla} \times \vec{A} . \tag{2.16}$$

Hence the LWA implies neglecting the magnetic field.

The $\vec{p} \cdot \vec{A}$-interaction Hamiltonian takes in the R-gauge the form

$$H_2(t) = -\frac{e}{m}\vec{A}(t) \cdot \vec{p} + \frac{e^2}{2m}\vec{A}^2(t) \qquad (2.17)$$

and the Schrödinger equation reads

$$i\hbar \frac{\partial}{\partial t} \Phi(\vec{r},t) = (H_0 + H_2(t))\Phi(\vec{r},t) \ . \qquad (2.18)$$

In the LWA the $\vec{A}^2$ - term cannot give rise to transitions between different levels. It can be eliminated with the help of the unitary transformation

$$\Phi'(\vec{r},t) = \exp\{i \frac{e^2}{2m\hbar} \int_{t_0}^{t} d\tau \vec{A}^2(\tau)\} \ \Phi(\vec{r},t) \ . \qquad (2.19)$$

The wavefunction Φ' then obeys the equation of motion

$$i\hbar \frac{\partial}{\partial t} \Phi'(\vec{r},t) = \left(H_0 - \frac{e}{m}\vec{A}(t) \cdot \vec{p}\right)\Phi'(\vec{r},t) \ . \qquad (2.20)$$

2.3. The $\vec{p}\cdot\vec{A}$ versus $\vec{E}\cdot\vec{r}$ controversy

The two different interaction Hamiltonians $H_1(t)$ (2.4) and $H_2(t)$ (2.17) seem to give different physical results since the matrix elements of these two Hamiltonians, calculated between eigenstates of the "unperturbed" Hamiltonian H_o (2.6), do not agree. We shall restrict the discussion in this section to monochromatic plane wave fields. Furthermore, for the sake of simplicity, we only consider linearly polarized fields. The electric field then takes the form

$$\vec{E}(t) = \vec{E}_0 \sin\omega t \qquad (2.21)$$

and the corresponding vector potential in the R-gauge reads

$$\vec{A}(t) = \vec{A}_0 \cos\omega t, \quad \vec{A}_0 = \frac{1}{\omega}\vec{E}_0 \ . \qquad (2.22)$$

It is convenient to separate the time dependence from the Hamiltonians H_1 and H_2 and to denote their amplitudes by

$$W_1 = - ie\vec{E}_0 \cdot \vec{r}, \quad W_2 = - \frac{e}{m} \vec{A}_0 \cdot \vec{p} . \tag{2.23}$$

By using the commutator relation

$$[\vec{r},H_0] = i\hbar \frac{\vec{p}}{m} \tag{2.24}$$

we find for the matrix elements of W_1 and W_2, calculated between an initial eigenstate $|i\rangle$ of H_o and a final eigenstate $|f\rangle$, the ratio

$$\frac{\langle f|W_2|i\rangle}{\langle f|W_1|i\rangle} = \frac{- \frac{e}{m} \vec{A}_0\langle f|\vec{p}|i\rangle}{- ie\vec{E}_0\langle f|\vec{r}|i\rangle} = \frac{E_f - E_i}{\hbar\omega} . \tag{2.25}$$

Hence the matrix elements of the two interactions H_1 and H_2 differ by the ratio of the transition energy over the field frequency.

It seems to be straightforward that this difference will also show up in measurable quantities like transition rates. Let us consider time dependent perturbation theory. We expand the solution $\psi(\vec{r},t)$ of the Schrödinger equations (2.5) or (2.18) with the Hamiltonians $H_0 + H_1(t)$ or $H_0 + H_2(t)$, respectively, in a complete set $\{\psi_n^o(\vec{r},t)\}$ (2.9) of eigenstates of H_o

$$\Psi(\vec{r},t) = \sum_n c_n(t)e^{-iE_nt/\hbar} \phi_n^0(\vec{r}) . \tag{2.26}$$

The time dependent expansion coefficients $c_n(t) = \langle\psi_n^o(t)|\psi(t)\rangle$ are usually called probability amplitudes. By inserting this ansatz for ψ into Eq. (2.5) or (2.18), we obtain the equations of motion of the coefficients c_n:

$$\dot{c}_n(t) = - \frac{i}{\hbar} \sum_m c_m(t)e^{-i(E_m-E_n)t/\hbar} \langle\phi_n^0|H_{1/2}(t)|\phi_m^0\rangle . \tag{2.27}$$

Usually $|c_n(t)|^2$ is interpreted as the probability of finding the system in the eigenstate $|n\rangle$ at time t. We see from Eq. (2.27) that this probability is proportional to $|\langle f|H_{1/2}|i\rangle|^2$. Hence using the two different interactions H_1 and H_2 would lead to transition probabilities which differ by a factor $[(E_f-E_i)/\hbar\omega]^2$ (2.25).

Let us demonstrate this difference with a slightly different version of Lamb's original example [2] that gave rise to so much confusion: a two level atom with an excited level b which has a finite lifetime $1/\Gamma$. At time $t_0 = 0$ the atom is in its ground state a. It is then driven by a weak external plane wave field of frequency ω, which is supposed to be almost in resonance with the atomic transition energy $\hbar\omega \cong E_f-E_i$ (Fig. 1). We are interested in the probability for transitions from a to b. In the weak field limit we can use the first order of perturbation theory. Furthermore, since we are only interested in the resonance regime $\hbar\omega \cong E_f-E_i$, we can use the rotating wave approximation (RWA) and write the interactions H_1 and H_2 in the form

$$H_{1/2}(t) = \frac{1}{2} W_{1/2}\, e^{-i\omega t} \quad . \tag{2.28}$$

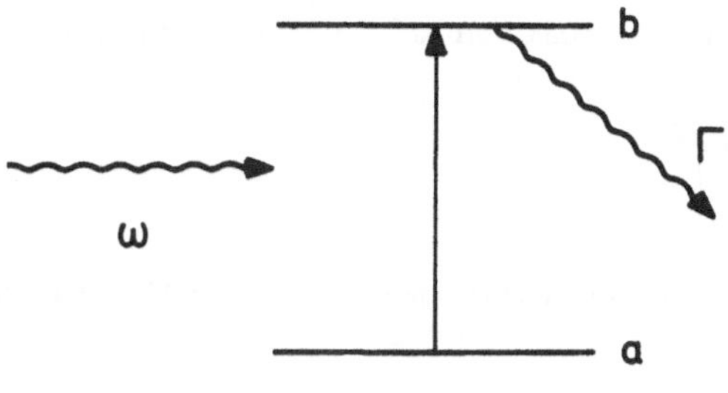

Fig. 1 Energy-level diagram of a two-level atom with spontaneous decay rate Γ.

The decay of the level b is treated phenomenologically by the introduction of a damping term. The equations of motion for the two levels then read

$$\dot{a} = 0$$
$$\dot{b} = -\frac{1}{2}\Gamma b - \frac{i}{2\hbar} \langle b|W_{1/2}|a\rangle e^{-i(E_a-E_b+\hbar\omega)t/\hbar} a \tag{2.29}$$

with the initial condition a(0) = 1, b(0) = 0. These equations can easily be solved with the help of the substitution

$$\tilde{b}(t) = e^{\frac{1}{2}\Gamma t}\, b(t)\ . \tag{2.30}$$

We find for the "transition probability"

$$|b(t)|^2 = \left(\frac{\langle b|W_{1/2}|a\rangle}{2\hbar}\right)^2 \frac{1 + e^{-\Gamma t} - 2e^{-\frac{1}{2}\Gamma t}\cos[(E_b - E_a - \hbar\omega)t/\hbar]}{\frac{1}{\hbar^2}(E_b - E_a - \hbar\omega)^2 + (\Gamma/2)^2}\ . \tag{2.31}$$

In the limit of infinite lifetime $\Gamma=0$ of level b this equation reduces to

$$|b_0(t)|^2 = \left(\frac{\langle b|W_{1/2}|a\rangle}{2\hbar}\right)^2 \left\{\frac{\sin[(E_b - E_a - \hbar\omega)t/2\hbar]}{\frac{1}{2\hbar}(E_b - E_a - \hbar\omega)}\right\}^2 . \tag{2.32}$$

Obviously the two interactions H_1 and H_2 lead to "transition probabilities" which differ by the square of the ratio (2.25).

In an experiment one is usually interested in the limit $t \to \infty$. By using the representation of the δ-function

$$\lim_{t\to\infty} \frac{1}{t}\left(\frac{\sin\omega t}{\omega}\right)^2 = \pi\delta(\omega) \tag{2.33}$$

we find the transition rate per unit time for an atom without spontaneous decay

$$\lim_{t\to\infty} \frac{1}{t}|b_0(t)|^2 = \frac{2\pi}{\hbar}|\langle b|\frac{1}{2}W_{1/2}|a\rangle|^2\delta(E_b - E_a - \hbar\omega)\ . \tag{2.34}$$

Due to the energy conserving δ-function in Eq. (2.34) we obtain in the limit of large times only a non-vanishing transition rate if the field is on resonance with the atomic energy difference $\hbar\omega = E_b - E_a$. Under this resonance condition the ratio (2.25) becomes unity and the two interactions H_1 and H_2 yield the same transition rates per time.

The situation is different for the two-level atom with a spontaneously decaying excited level. The probability (2.31) for finding the atom in level b becomes constant for large times

$$\lim_{t\to\infty} |b(t)|^2 = |\langle b|\frac{1}{2\hbar} W_{1/2}|a\rangle|^2 \frac{1}{\frac{1}{\hbar^2}(E_b - E_a - \hbar\omega)^2 + (\Gamma/2)^2} . \tag{2.35}$$

Since the populated level b decays with the rate Γ, the rate of spontaneous decays of b in an atom, which is originally in its ground state, is

$$\Gamma \cdot \lim_{t\to\infty} |b(t)|^2 = |\langle b|\frac{1}{2\hbar} W_{1/2}|a\rangle|^2 \frac{\Gamma}{\frac{1}{\hbar^2}(E_b - E_a - \hbar\omega)^2 + (\Gamma/2)^2} . \tag{2.36}$$

Since the population (2.35) of level b is constant, Eq. (2.36) also describes the rate per unit time for transitions from a to b. In contrast to Eq. (2.34), Eq. (2.36) does not exhibit an energy-conserving δ-function. Therefore also non-resonant transitions $\hbar\omega \neq E_b - E_a$ are allowed and the two interactions H_1 and H_2 then yield different rates (2.36).

Lamb commented in 1952 on this result: " Of course the difference between the perturbations H_1 and H_2 just corresponds to a gauge transformation under which the theory is known to be invariant, so that both perturbations must lead to the same physical predictions. Nevertheless, a closer examination shows that the usual interpretation of probability amplitudes is valid only in the former gauge, and no additional factor $[(E_f - E_i)/\hbar\omega]^2$ actually occurs." It is the purpose of the following two chapters to outline in detail this "closer examination" in order to show how we can obtain the measurable transition rates with the two different interactions H_1 and H_2 and how they then lead to the same physical predictions.

The procedure outlined in this section, namely to expand the solution of the Schrödinger equations (2.5) and (2.18) in terms of the same set of eigenstates of H_0 (2.6) and to interpret the expansion coefficients c_n in (2.26) as probability amplitudes, is sometimes misleadingly called the "conventional procedure" [7,9,10]. Due to reasons which will become clear in the next Section, we prefer the name "hybrid procedure", adapted from Ref. 19.

3. GAUGE TRANSFORMATIONS IN QUANTUM MECHANICS

3.1. Gauge Transformations

In this Section we briefly summarize the basic concept of gauge invariance to the extent needed in Section 4. For the derivation and justification of the arguments used in the following we refer the reader to Ref. 6.

The vector potential $\vec{A}^g(\vec{r},t)$ and the scalar potential $U^g(\vec{r},t)$ are related to the electric field $\vec{E}(\vec{r},t)$ and the magnetic field $\vec{B}(\vec{r},t)$ by

$$\vec{E}(\vec{r},t) = -\vec{\nabla}U^g(\vec{r},t) - \frac{\partial}{\partial t}\vec{A}^g(\vec{r},t) ,$$

$$\vec{B}(\vec{r},t) = \vec{\nabla}\times\vec{A}^g(\vec{r},t) . \tag{3.1}$$

$\vec{A}^g$ and U^g are not uniquely determined by Maxwell's equations. $\vec{E}$ and $\vec{B}$ remain unchanged if we transform from a gauge ($\vec{A}^g$, U^g) to a gauge ($\vec{A}^{g'}$, $U^{g'}$) according to

$$\vec{A}^g(\vec{r},t) \rightarrow \vec{A}^{g'}(\vec{r},t) = \vec{A}^g(\vec{r},t) + \vec{\nabla}\chi(\vec{r},t) ,$$

$$U^g(\vec{r},t) \rightarrow U^{g'}(\vec{r},t) = U^g(\vec{r},t) - \frac{\partial}{\partial t}\chi(\vec{r},t) . \tag{3.2}$$

$\chi(\vec{r},t)$ is an arbitrary function of $\vec{r}$ and t. Also the state vector $|\psi_g\rangle$ of a physical system depends on the gauge. The wavefunction in a gauge g is transformed to another gauge g' by a unitary transformation:

$$\Psi^g(\vec{r},t) \rightarrow \Psi^{g'}(\vec{r},t) = T(\vec{r},t)\Psi^g(\vec{r},t) ,$$

$$T(\vec{r},t) = e^{ie\chi(r,t)/\hbar} . \tag{3.3}$$

Hence for a gauge transformation one has to transform simultaneously the wavefunction, according to Eq. (3.3), and all the potentials in the operators, according to Eq. (3.2).

The Schrödinger equation is forminvariant under gauge transformations. In order to see this, we start from the Schrödinger equation in gauge g with the Hamiltonian H^g (2.11) and a static potential $V(\vec{r})$

$$i\hbar \frac{\partial}{\partial t} \Psi^g(\vec{r},t) = \{\frac{1}{2m} (\vec{p} - eA^g(\vec{r},t))^2 + eU^g(\vec{r},t) + V(\vec{r})\}\Psi^g(\vec{r},t) \ . \tag{3.4}$$

By inserting $\Psi^g = T^+\Psi^{g'}$ in Eq. (3.4) we find the equation of motion for $\psi^{g'}$

$$i\hbar \frac{\partial}{\partial t} \Psi^{g'} = \{\frac{1}{2m} (\vec{p} - e[\vec{A}^g + \vec{\nabla}\chi])^2 + e[U^g - \frac{\partial}{\partial t}\chi] + V\}\Psi^{g'} \ . \tag{3.5}$$

If we use the expressions (3.2) for $A^{g'}$ and $U^{g'}$, Eq. (3.5) becomes

$$i\hbar \frac{\partial}{\partial t} \Psi^{g'}(\vec{r},t) = \{\frac{1}{2m} (\vec{p} - e\vec{A}^{g'}(\vec{r},t))^2 + eU^{g'}(\vec{r},t) + V(\vec{r})\}\Psi^{g'}(\vec{r},t) \ .$$

Hence the Schrödinger equation has the same form in any gauge chosen.

3.2. Electric Field Gauge

We now consider a special gauge transformation. If we apply to the R-gauge in the LWA $(\vec{A}(t),0)$ the gauge transformation

$$\chi_{RE}(\vec{r},t) = - \vec{A}(t) \cdot \vec{r} \tag{3.7}$$

where $\vec{A}$ is the vector-potential in the R-gauge, we obtain with the help of Eqs. (3.2) and (3.1) the gauge $(0, - \vec{E}(t) \cdot \vec{r})$. By inserting these two gauges in the Schrödinger equation (3.4) we see that the transformation (3.7) leads from Eq. (2.18) to Eq. (2.5). Hence the Hamiltonian $H_0 + H_1(t)$, which we found intuitively in Section 2.1., is related to the Hamiltonian $H_0 + H_2(t)$, which was derived rigorously in Section 2.2., by the gauge transformation (3.7). This transformation was applied for the first time by M. Göppert-Mayer in 1931 [1]. We call the gauge $(0, -\vec{E} \cdot \vec{r})$, which leads to the electric-dipole interaction $H_1(t)$, the electric field gauge (E-gauge). This gauge is also referred to in the literature as the electric-dipole, length, Göppert-Mayer, or Lamb gauge.

3.3. Hybrid transformations

A gauge transformation requires a transformation of the potentials according to Eq. (3.2) and of the wavefunctions according to Eq. (3.3). We call it a hybrid transformation if only one of these two tranformations is carried out.

The first types of hybrid transformations are the unitary transformations (3.3) on the wavefunctions without a modification of the potentials. If we consider the Schrödinger equation (3.4), the transformation (3.3) of the wavefunction leads to Eq. (3.5) as the equation of motion for the transformed wavefunction $\psi^{g'}$. But now the potentials in Eq. (3.5) are still given by $\vec{A}^g$ and U^g. This hybrid transformation is an essential step of the so-called "momentum translation approximation" method [23] which will be discussed in Section 5.

The second type of hybrid transformations consists of a transformation of the potentials without a transformation of the wavefunctions. We have seen in Section 3.2 that the Hamiltonians $H_o + H_1(t)$ and $H_0 + H_2(t)$ are related by such a transformation of the potentials. Nevertheless we assumed in Section 2.3 the same solution for the Schrödinger equations (2.5) and (2.18), and we expanded the solution in the same set $\{\phi_n^o\}$ of eigenstates of H_0. Obviously one cannot in general expect gauge invariant physical predictions from this hybrid procedure. We shall see in Section 4 how we have to handle the wavefunctions in the two different gauges in order to obtain gauge invariant physical predictions.

3.4 Physical Quantities

It is useful to distinguish between physical and non-physical quantities. Physical quantities are defined as those quantities whose corresponding operator $G^g \equiv G(\vec{A}^g, U^g)$ is forminvariant under a unitary transformation (3.3)

$$G^{g'} = T\, G^g\, T^+ \quad . \tag{3.8}$$

We obtain these observables in the gauge g' either by the substitution (3.2) of the potentials or by the unitary transformation (3.3). The difference between physical and non-physical quantities lies in the gauge invariance of the eigenvalues. The eigenvalues of an observable are identical in all gauges, whereas the eigenvalues of non-physical quantities depend on the chosen gauge. Let us denote the eigenvalues and eigenstates of G^g by g_n and $|\zeta_n^g\rangle$, respectively:

$$G^g |\xi_n^g\rangle = g_n |\xi_n^g\rangle \quad . \tag{3.9}$$

Only for physical quantities the eigenvalues g_n are gauge invariant:

$$G^{g'}|\xi_n^{g'}> = TG^{g}T^{+}T|\xi_n^{g}> = Tg_n|\xi_n^{g}> = g_n|\xi_n^{g'}> . \tag{3.10}$$

Hence, non-physical quantities are only a calculation tool, rather than an actual observable quantity.

Let us consider some examples which will become important in the next section. The starting point for all considerations is the fact that the operators $\vec{r}$ and $\vec{p}$, associated with the position and the canonical momentum of the particle, are the same in all gauges. In the position representation they have the gauge-independent form

$$\hat{\vec{r}} = \vec{r}, \quad \hat{\vec{p}} = - i\hbar\vec{\nabla} . \tag{3.11}$$

With this rule the operator of the mechanical momentum

$$\vec{\pi}^{g} = \vec{p} - e\vec{A}^{g}(\vec{r},t) \tag{3.12}$$

is a physical, measurable quantity since

$$T\vec{\pi}^{g}T^{+} = T(\vec{p} - e\vec{A}^{g})T^{+} = \vec{p} - e\vec{A}^{g} - e\vec{\nabla}\chi = \vec{\pi}^{g'} . \tag{3.13}$$

Similarly, the instantaneous energy operator of the system, consisting of the kinetic energy and the static potential

$$\varepsilon^{g} = \frac{1}{2m}(\vec{p} - e\vec{A}^{g}(\vec{r},t))^2 + V(\vec{r}) \tag{3.14}$$

represents a physical quantity as well as any other operator which is only a function of other physical quantities like $\vec{\pi}^{g}$.

Comparing Eqs. (3.4) and (3.14), and using the fact that the Schrödinger equation is gauge invariant, the operator

$$F^{g} = eU^{g}(\vec{r},t) - i\hbar \frac{\partial}{\partial t} \tag{3.15}$$

must also be gauge invariant. Correspondingly, this expression has units of energy. The gauge invariance of F^g can easily be demonstrated by

$$TF^gT^+ = eU^g - i\hbar\left(T\frac{\partial}{\partial t}T^+\right) - i\hbar\frac{\partial}{\partial t} = F^{g'} \ . \tag{3.16}$$

In general, the explicit form of an operator corresponding to a physical quantity is gauge dependent.

On the other hand, the canonical momentum $\vec{p}$ is not a physical quantity, since $\vec{p}$ has the same form in all gauges $\vec{p}^g = \vec{p}^{g'}$ whereas

$$T\vec{p}\,T^+ = \vec{p} - e\vec{\nabla}\chi \neq \vec{p} \ . \tag{3.17}$$

In a similar way the operator H_o (2.6) does not depend on the potentials, $H_o^g = H_o^{g'}$, and is therefore a non-physical quantity

$$TH_0T^+ = H_0 - \frac{e}{2m}(\vec{p}\vec{\nabla}\chi + \vec{\nabla}\chi\vec{p}) + \frac{e^2}{2m}(\vec{\nabla}\chi)^2 \neq H_0 \ . \tag{3.18}$$

In general, any operator which is a function of non-physical quantities alone, like the canonical momentum $\vec{p}$ or the vector or the scalar potentials $\vec{A}^g$ or U^g, represents a non-physical quantity. Hence also the total Hamitonian H^g, which governs the Schrödinger equations (3.4),

$$H^g = \frac{1}{2m}(\vec{p} - e\vec{A}^g(\vec{r},t))^2 + eU^g(\vec{r},t) + V(\vec{r}) \tag{3.19}$$

is not a physical quantity, viz.

$$TH^gT^+ = H^{g'} + e\frac{\partial}{\partial t}\chi \neq H^{g'} \ . \tag{3.20}$$

H^g differs from the observable energy (3.14) by the potential energy eU^g. It should be noticed that the property (3.20) is essential for the forminvariance of the Schrödinger equation as shown in Eqs. (3.4) - (3.6).

We conclude that the time evolution of a physical system is determined by Hamiltonians like H^g or H_0, which in general are not observable quantities. True physical quantities are, for example, the mechanical momentum and the energy of the system.

Since the energy operator ε^g is time-dependent, its eigenstates ϕ_n^g and its eigenvalues E_n are also in general time-dependent

$$\varepsilon^g(t)\phi_n^g(\vec{r},t) = E_n\phi_n^g(\vec{r},t) \ . \tag{3.21}$$

Only in the LWA, i.e. if one neglects the magnetic field, the eigenvalues of ε^g are time-independent. This can be seen with the help of the gauge transformation (3.7). In the LWA the operator relation holds

$$(\hat{\vec{p}} - e\vec{A}(t))^n \, e^{ie\vec{A}(t)\vec{r}/\hbar} = e^{ie\vec{A}(t)\vec{r}/\hbar} \, \hat{\vec{p}}^n \ . \tag{3.22}$$

Hence, applying the transformation (3.7) to the energy operator ε^g (3.14) and to Eq. (3.21) leads to the unperturbed Hamiltonian H_0 (2.6) and to Eq. (2.7). The eigenstates ϕ_n^g are then related to the eigenstates ϕ_n^o of H_o by

$$\phi_n^g(\vec{r},t) = e^{ie\vec{A}^g(t)\vec{r}/\hbar} \, \phi_n^0(\vec{r},t) \tag{3.23}$$

and the eigenvalues E_n of ε^g coincide with the time-independent eigenvalues E_n of H_0 since the eigenvalues of physical quantities are gauge independent.

3.5 Comparison of Hamiltonian/energy and canonical/kinetic momentum in the E- and R-gauges

The discussion of this section is restricted to the LWA. We first summarize the relations between the Hamiltonian and the energy operator ε^g (3.14) in the different gauges. In the E-gauge the unperturbed energy operator ε^E is equal to the unperturbed Hamiltonian H_o. Hence the eigenstates of H_o are also the eigenstates of ε^E.

By switching to the R gauge, the energy operator transforms according to Eq. (3.8), whereas the total Hamiltonian $H_0 + H_1(t)$ (2.4) in the E-gauge transforms according to Eq. (3.2o). The unperturbed energy operator ε^R then coincides in the R-gauge with the total Hamiltonian $H_0 + H_2(t)$ (2.17). In the R-gauge the total Hamiltonian does not coincide with the total energy operator ε^g_{tot}, which is the sum of kinetic and potential energy (ε^g) plus the interaction energy $-e\vec{E}\vec{r}$ (in the LWA)

$$\varepsilon^g_{tot}(t) = \frac{1}{2m}(\vec{p} - e\vec{A}^g(t))^2 + V(\vec{r}) - e\vec{E}(t) \cdot \vec{r} \,. \qquad (3.24)$$

In Eq. (3.24) the term $-e\vec{E}(t) \cdot \vec{r}$ is considered as the gauge-independent interaction energy, whereas in the Hamiltonian $H_o + H_1(t)$ it represents the gauge-dependent scalar potential.

The eigenvalues of $_R\varepsilon^R$ are the same as the eigenvalues of H_0, but the eigenstates of ε^R and H_o are different. The eigenfunctions ϕ^R_n of ε^R are related to the eigenfunctions ϕ^o_n of H_0 via Eq. (3.23). Notice that although the wavefunctions ϕ^R_n are eigenstates of the total Hamiltonian $H_0 + H_2(t)$, they are not solutions of the Schrödinger equation (2.18) since they are time dependent due to the time dependent Hamiltonian.

A quantity is a constant of motion, if for any solution of the Schrödinger equation the expectation value of the corresponding operator A is a constant of time

$$\frac{d}{dt}\langle A\rangle = \langle\frac{1}{i\hbar}[A_1 H_{tot}(t)] + \frac{\partial}{\partial t}A\rangle \,. \qquad (3.25)$$

We see that the energy ε^g is neither in the E gauge

$$\langle\frac{1}{i\hbar}[\varepsilon^E, H_0 + H_1(t)] + \frac{\partial}{\partial t}\varepsilon^E\rangle = \langle\frac{e}{m}\vec{E}_0 \cdot \vec{p}\rangle \qquad (3.26)$$

nor in the R gauge

$$\langle\frac{1}{i\hbar}[\varepsilon^R, H_0 + H_2(t)] + \frac{\partial}{\partial t}\varepsilon^R\rangle = \langle\frac{e}{m}\vec{E}_0 \cdot (\vec{p} - eA(\vec{t}))\rangle \qquad (3.27)$$

a constant of motion.

We conclude this section with an example of the gauge invariance of the matrix elements of a physical quantity. Let us consider the mechanical momentum $m\vec{v}^g = \vec{\pi}^g$ (3.12) in the E gauge and in the R gauge. In the E gauge $\vec{\pi}^g$ coincides with the operator of the canonical momentum $\vec{p} = -i\hbar\vec{\nabla}$. If the system is in some state ψ, the expectation value of the mechanical momentum reads in the E gauge

$$< \Psi|m\vec{v}|\Psi> = <\Psi|\vec{p}|\Psi> = \frac{\hbar}{i} <\Psi|\vec{\nabla}|\Psi> . \tag{3.28}$$

In order to obtain the expectation value in the R gauge, we have to transform the wavefunction ψ according to (3.3) and (3.7) and the operator of the mechanical momentum according to (3.2) and (3.7):

$$<e^{ie\vec{A}\vec{r}/\hbar}\Psi|-i\hbar\vec{\nabla} - e\vec{A}|e^{ie\vec{A}\vec{r}/\hbar}\Psi> = \frac{\hbar}{i} <\Psi|\vec{\nabla}|\Psi> . \tag{3.29}$$

As expected, the two matrix elements (3.28) and (3.29) are identical.

4. GAUGE INVARIANT TRANSITION AMPLITUDES

4.1. Transition probabilities

In order to find the transition probability we must carefully study the question: what is the probability we are actually looking for and how is this probability defined? This might sound trivial, but the careless handling of this question is the origin of the entire $\vec{E}\cdot\vec{r}$ versus $\vec{p}\cdot\vec{A}$ controversy.

Only the measurable physical quantities are meaningful in the description of the state of a quantum mechanical system. Let us determine this state, denoted by $\psi^g>$, by a measurement of a physical quantity whose operator in a gauge g is G^g (3.10). It is one of the basic postulates of quantum mechanics that the probability of obtaining the eigenvalue g_n in a measurement is

$$P_n = |<\xi_n^g|\Psi^g>|^2 . \tag{4.1}$$

The probability (4.1) is gauge invariant:

$$<\xi_n^{g'}|\Psi^{g'}> = <\xi_n^g|T^+T|\Psi^g> = <\xi_n^g|\Psi^g> . \tag{4.2}$$

The choice of the observable G^g depends on the physical situation we are interested in. In this paper we consider bound particles which are perturbed by an external field. Before the perturbation is switched on, the system is characterized by an energy eigenvalue E_i. When the perturbation is switched on we are interested in the probability of finding the system in an eigenstate with eigen-energy E_n. In this situation we have to choose the energy operator (3.14) for G^g. In order to obtain the probability (4.1) of finding the system in an eigenstate with energy E_n, we need the eigenstates ϕ_n^g (3.21) of ε^g.

The central importance of the LWA, i.e. the neglect of effects of the magnetic fields, for the choice of the energy operator (3.14) must be stressed. First of all, the energy eigenvalues E_n of ε^g are only time independent in the LWA, as discussed in Section 3.4. Hence only in the LWA the eigenvalues of ε^g coincide with the observable energy spectrum of the system. Furthermore, only in the LWA the eigenstates ϕ_n^g of ε^g are linked to the eigenstates ϕ_n^o of H_0 via Eq. (3.23). In practical calculations, the eigenstates ϕ_n^o of the energy operator in the absence of the external field H_0 are usually known, and not the eigenstates ϕ_n^g of ε^g.

We have seen that it is meaningful only to define the probability (4.1) for a system to be in an eigenstate of a <u>physical</u> quantity. In contrast, the hybrid procedure of Section 2.3 expands the wavefunction (2.26) in terms of eigenstates of H_0 and interprets the expansion coefficients as probability amplitudes. Only in the E-gauge does the unperturbed Hamiltonian H_0 coincide with the energy operator and the eigenstates ϕ_n^o are the energy eigenstates, on which we have to project the wavefunction ψ in order to obtain the desired probability (4.1). Therefore, only in the E-gauge the wavefunction (2.26) is expanded in terms of energy-eigenstates, and the interpretation of the expansion coefficients c_n as probability amplitudes for finding the system in an eigenstate of the observable energy is correct. This explains Lamb's statement about the preference for the E-gauge.

In any other gauge H_0 is a non-physical quantity and its eigenstates are not the energy-eigenstates of the system. The expansion coefficients c_n in Eq. (2.26) are then the "probability amplitudes" for finding the system in an eigenstate of H_0. However, if H_0 is a non-physical quantity, this "probability" is gauge dependent and has to be distinguished from the measurable, gauge invariant probability of finding the system in an energy-eigenstate. We can formulate the reason for the gauge dependence of the "probabilities", calculated in the hybrid procedure, also in another way: the splitting of the total Hamiltonian H^g into a basis-defining unperturbed part H_0 and a perturbation is not gauge invariant, if we do not work in the E-gauge. Most textbooks which describe time-dependent perturbation theory are not careful enough with this point. Of course the expansion (2.26) is mathematically correct. But we have to determine the relationship between the measurable gauge invariant probability amplitudes and the gauge dependent, purely mathematical expansion coefficients c_n in (2.26).

4.2. Equations of Motion

The discussion in the last section showed that it is useful to expand the wavefunction of our system in terms of eigenstates of the energy operator ε^g

$$\Psi^g(\vec{r},t) = \sum_n d_n(t) e^{-iE_n t/\hbar} \phi_n^g(\vec{r},t) . \tag{4.3}$$

The expansion coefficients d_n then coincide with the probability amplitudes for transitions of the system to an eigenstate of the energy operator with energy E_n

$$d_n(t) = \langle \phi_n^g(t) | \Psi^g(t) \rangle e^{iE_n t/\hbar} . \tag{4.4}$$

It is convenient to work in the interaction picture, i.e. to extract the rapidly oscillating factor $\exp(-iE_n t/\hbar)$ from the time dependence of the coefficients d_n. Inserting the wavefunction (4.3) into the Schrödinger equation

$$\{\varepsilon^g(t) + F^g(t)\}\Psi^g(\vec{r},t) = 0 \tag{4.5}$$

and projecting with $\langle \phi_m^g |$ on this equation yields the equations of motion of the amplitudes d_n

$$\dot{d}_n(t) = -\frac{i}{\hbar} \sum_m d_m(t) e^{-i(E_m-E_n)t/\hbar} \langle \phi_n^g(t) | F^g(t) | \phi_m^g(t) \rangle . \tag{4.6}$$

F^g is defined in Eq. (3.15). Instead of inserting the different perturbations H_1 and H_2 in Eq. (2.27), the different interactions enter Eq. (4.6) via the potentials in the different gauges. Therefore, the difference (2.25) between the matrix elements of the "perturbations" H_1 and H_2, which is not changed by the transformation of the states in the R gauge, does not play any role in the probability amplitudes. The matrix elements on the r.h.s. of Eq. (4.6) must have the same form in any gauge, since the amplitudes d_n are gauge invariant by construction. This gauge invariance can be seen from Eq. (3.16).

Let us compare the E-gauge and the R-gauge in the LWA. The formalism is now so constructed that both gauges must yield the same probability. We can see this explicitly by inserting both gauges in Eq. (4.6). In any case, Eq. (4.6) in the LWA becomes

$$\dot{d}_n(t) = \frac{i}{\hbar} e \sum_m d_m(t) e^{-i(E_m - E_n)t/\hbar} \vec{E}(t) \langle \phi_n^0 | \vec{r} | \phi_m^0 \rangle . \quad (4.7)$$

We shall refer to this procedure as the gauge invariant procedure.

The gauge dependent expansion coefficients c_n^g of the hybrid procedure (2.26)

$$c_n^g(t) = \langle \phi_n^0 | \psi^g(t) \rangle \, e^{iE_n t/\hbar} \quad (4.8)$$

are related to the gauge invariant probability amplitudes d_n of the gauge invariant procedure in a simple way. Comparison of Eqs. (2.26) and (4.3) and projection on one eigenstate yields

$$c_n^g(t) = \sum_m d_m(t) e^{-i(E_m - E_n)t/\hbar} \langle \phi_n^0 | \phi_m^g(t) \rangle . \quad (4.9)$$

In the same way, d_n can be expressed in terms of the c_n^g. In the LWA Eq. (3.23) can be used for the overlap of the two wavefunctions on the r.h.s. of Eq. (4.9):

$$c_n^g(t) = \sum_m d_m(t) e^{-i(E_m - E_n)t/\hbar} \langle \phi_n^0 | e^{ie\vec{A}^g(t)\vec{r}/\hbar} | \phi_m^0 \rangle . \quad (4.10)$$

It should be pointed out that the expansion of the wavefunction ψ^g (4.3) in terms of eigenstates of the energy operator ε^g is only a question of convenience in the particular physical situation we consider. We are interested here in the probability for transitions of a bound particle to eigenstates of the energy operator ε^g characterized by a gauge invariant energy eigenvalue E_n. If we were interested in t ransitions to an eigenstate of some other observable G^g (3.8), it would be more convenient to expand the wavefunction ψ^g in terms of the eigenstates $|\xi_n^g\rangle$ (3.10) of G^g, because the expansion coefficients then coincide automatically with the probability amplitudes. But this procedure is by no means the only correct one. It only makes the calculations more economic. In principle, we can expand the wavefunction ψ^g in any basis and relate the expansion amplitudes afterwards to the wanted transition probabilities [11]. For example, for a free charged particle in a spatially uniform external field it is more natural to prepare the system to be in an eigenstate of the mechanical momentum rather than in an energy eigenstate. It is then more convenient to choose for G^g the kinetic momentum $\vec{\pi}^g$ (3.12) (which then commutes with ε^g). This example is carried out in Ref. 31.

4.3. Two-level atom

A somewhat more elegant approach to time-dependent perturbation theory is the formalism with time-evolution operators. In order to separate the unperturbed Hamiltonian H_o from the perturbation $V(t)$ in the Schrödinger equation

$$i\hbar \frac{\partial}{\partial t} |\Psi(t)\rangle = (H_0 + V(t))|\Psi(t)\rangle \tag{4.11}$$

we choose the interaction picture. The wavefunction $|\psi_I\rangle$ in the interaction picture is related to $|\psi\rangle$ via the transformation

$$|\Psi(t)\rangle = U_0(t)|\Psi_I(t)\rangle, \quad U_0(t) = e^{-iH_0t/\hbar} \quad . \tag{4.12}$$

$|\psi_I\rangle$ satisfies the equation

$$i\hbar \frac{\partial}{\partial t} |\Psi_I(t)\rangle = U_0^+(t)V(t)U_0(t)|\Psi_I(t)\rangle. \tag{4.13}$$

The formal solution of Eq. (4.13) can be written as

$$|\Psi_I(t)\rangle = U_I(t)|\Psi(0)\rangle, \; U_I(t) = P \exp\{-\frac{i}{\hbar}\int_0^t d\tau U_0^+(\tau)V(\tau)U_0(\tau)\} \; . \tag{4.14}$$

Here we used $|\psi_I(0)\rangle = |\psi(0)\rangle$ according to Eq. (4.12). P denotes the time ordering operator. Transforming back to the Schrödinger picture, we obtain as a formal solution of Eq. (4.11)

$$|\Psi(t)\rangle = U_0(t)U_I(t)|\Psi(0)\rangle \; . \tag{4.15}$$

The gauge invariant probability amplitudes d_n (4.4) are then given in the E-gauge by

$$d_n^E(t) = \langle\phi_n^0|U_0(t)U_I^{(1)}(t)|\phi_i^0|\rangle e^{iE_nt/\hbar} \tag{4.16}$$

and in the R-gauge by

$$d_n^R(t) = \langle\phi_n^0| e^{-\frac{i}{\hbar} e\vec{A}(t)\vec{r}} U_0(t) U_I^{(2)}(t) e^{\frac{i}{\hbar} e\vec{A}(0)\vec{r}} |\phi_i^0\rangle e^{\frac{i}{\hbar} E_n t} . \quad (4.17)$$

The superscript on U_I denotes the interaction $H_1(t)$ or $H_2(t)$, respectively, used in Eq. (4.14) for the perturbation $V(t)$.

We demonstrate the equivalence of Eqs. (4.16) and (4.17) with the example of a two-level atom, as introduced in Section 2.3. In first order of perturbation theory the time-evolution operator U_I (4.14) becomes

$$U_I = 1 - \frac{i}{\hbar} \int_0^t d\tau U_0^+(\tau) V(\tau) U_0(\tau) . \quad (4.18)$$

Using again the RWA for the electric field and assuming the atom to be in its ground state a at t = 0, the probability amplitude of the excited state b takes in the E-gauge the form

$$\begin{aligned} d_b^E(t) &= -\frac{i}{\hbar} \langle\phi_b^0| \int_0^t d\tau e^{iH_0\tau/\hbar} H_1(\tau) e^{-iH_0\tau/\hbar} |\phi_a^0\rangle \\ &= -\frac{i}{\hbar} \tfrac{1}{2} \langle\phi_b^0|W_1|\phi_a^0\rangle \int_0^t d\tau e^{i(E_b-E_a-\hbar\omega)\tau/\hbar} \\ &= \tfrac{1}{2} ie\vec{E}_0 \langle\phi_b^0|\vec{r}|\phi_a^0\rangle \frac{e^{i(E_b-E_a-\hbar\omega)t/\hbar} - 1}{E_b - E_a - \hbar\omega} . \end{aligned} \quad (4.19)$$

This result has to be compared to the corresponding calculation in the R-gauge. Let us first calculate the transition rate with the hybrid procuedure, i.e. we only replace H_1 by H_2 in Eq. (4.19) and do not take the phases $\exp(ie\vec{A}\vec{r}/\hbar)$ in Eq. (4.17) into account. As shown in Section 2.3, especially with Eq. (2.25), we then obtain

$$\tilde{d}_b^R(t) = \tfrac{1}{2} ie\vec{E}_0 \langle\phi_b^0|\vec{r}|\phi_a^0\rangle \frac{E_b - E_a}{\hbar\omega} \frac{e^{i(E_b-E_a-\hbar\omega)t/\hbar} - 1}{E_b - E_a - \hbar\omega} \quad (4.20)$$

The amplitude $\tilde{d}_b^R$ differs from d_b^E by the characteristic factor $(E_b-E_a)/\hbar\omega$. Carrying out the calculation for the correct gauge invariant amplitude d_b^R (4.17), we see that we obtain additionally to the term of the hybrid procedure two more terms linear in the coupling constant e, due to the phase factors $\exp(ie\vec{A}\vec{r}/\hbar)$ of the wavefunctions

$$d_b^R(t) = \langle\phi_b^0|(1 - \frac{i}{\hbar} e\vec{A}(t)\vec{r})e^{-\frac{i}{\hbar}H_0 t}\left(1 - \frac{i}{\hbar}\int_0^t d\tau e^{\frac{i}{\hbar}H_0\tau} H_2(\tau) e^{\frac{i}{\hbar}H_0\tau}\right)$$

$$\times \left(1 + \frac{i}{\hbar} e\vec{A}(0)\vec{r}\right)|\phi_a^0\rangle e^{\frac{i}{\hbar}E_b t}$$

$$= -\frac{i}{\hbar} e\langle\phi_b^0|\vec{r}|\phi_a^0\rangle\left(\vec{A}(t)e^{\frac{i}{\hbar}(E_b - E_a)t} - \vec{A}(0)\right) + \gamma_b^R(t) \tag{4.21}$$

$$= \frac{1}{2} ie\vec{E}_0\langle\phi_b^0|\vec{r}|\phi_a^0\rangle\left(e^{\frac{i}{\hbar}(E_b - E_a - \hbar\omega)} - 1\right) / (E_b - E_a - \hbar\omega)$$

Here we used the vector potential (2.22) in the RWA

$$\vec{A}(t) = \frac{1}{2\omega}\vec{E}_0 e^{-i\omega t} . \tag{4.22}$$

As expected, the amplitudes d_b^R (4.21) and d_b^E (4.19) are identical.

4.4 Discussion of the hybrid procedure in the R-gauge

As we will show now, the hybrid procedure in the R-gauge yields the correct gauge invariant result in many situations of experimental interest, although it differs from the correct gauge invariant procedure by a unitary transformation of the wavefunctions.

Usually the field is turned off at the moment t_0 of the preparation of the system, $\vec{E}(t_0) = 0$, and at the moment t of the measurement, $\vec{E}(t) = 0$. We shall prove now that under these conditions the hybrid procedure in the R-gauge yields the correct transition probability for a wavepacket with finite frequency bandwidth $\Delta\omega$

$$\vec{E}(t) = -\int_{\omega_0 - \frac{\Delta\omega}{2}}^{\omega_0 + \frac{\Delta\omega}{2}} d\omega\, \vec{E}(\omega) \cos\omega t . \tag{4.23}$$

This describes as well an electromagnetic field which is suddenly turned on and off at t_0 and t as a field which is adiabatically switched on and off at large times. In the R gauge the corresponding vector potential reads

$$\vec{A}(t) = \int d\omega\, \vec{E}(\omega) \frac{\sin\omega t}{\omega} . \tag{4.24}$$

By using the representation of the δ-function

$$\lim_{t\to\infty} \frac{\sin\omega t}{\omega} = \pi\delta(\omega) \tag{4.25}$$

one can see that $\vec{A}(t)$ vanishes in the limit $t\to\pm\infty$, assuming $\vec{E}(\omega=0) = 0$:

$$\lim_{t\to\pm\infty} \vec{A}(t) = \pm\pi\vec{E}(0) = 0 \ . \tag{4.26}$$

Therefore also the transformation operator T_{RE}^{-1} for the wavefunctions becomes unity for $t_o \to -\infty$ and $t\to\infty$:

$$\lim_{t\to\pm\infty} T_{RE}^{-1}(\vec{r},t) = \lim_{t\to\pm\infty} e^{\frac{i}{\hbar} e\vec{A}(t)\vec{r}} = 1 \ . \tag{4.27}$$

As Eqs. (4.10) or (4.17) show, the amplitudes d_n and c_n^g of the gauge invariant and the hybrid procedure in the R-gauge coincide for $t\to\pm\infty$. The coincidence at the time of preparation $t_0\to-\infty$ is necessary in order to have equal initial conditions in the different procedures or gauges [27]. For this class of experiments, i.e. wavepackets with $t_0\to-\infty$ and $t\to+\infty$, the hybrid procedure is justified for any multiphoton process.

Only the condition

$$\vec{A}(t_0) = \vec{A}(t) = 0 \tag{4.28}$$

is necessary for the agreement between hybrid and gauge invariant procedure. Whenever the vector potential would vanish at finite times t_0 and t, no matter what way, the hybrid procedure would already be justified for measurements at finite times. But one has to be careful with this kind of mathematical arguments since they usually do not apply to a physical situation. If we assume in the R gauge a vector potential which is suddenly switched off with a step function at time t, the corresponding electric field exhibits a δ-function peak at time t [22]. Only the physical electric fields can be turned off, suddenly or adiabatically. And only the E-gauge is the "preferential" gauge [8] where the potentials vanish whenever the field does. In the R-gauge the vector potential does not necessarily vanish when the field is turned off.

We now demonstrate the gauge independence of the transition amplitudes in our example of a two-level atom in first order perturbation theory. We consider an arbitrary field $\vec{E}(t)$. According to Eq. (2.27) the amplitude of the excited level b reads in the hybrid procedure

$$b^g(t) = b^g(t_0) - \frac{i}{\hbar}\int_{t_0}^{t} d\tau\, e^{\frac{-i}{\hbar}(E_b-E_a)\tau} \langle b|H^g_{int}(\tau)|a\rangle . \qquad (4.29)$$

In the E gauge the interaction Hamiltonian is given by Eq. (2.4) and we find

$$b^E(t) = b^E(t_0) + \frac{i}{\hbar}\, e\langle b|\vec{r}|a\rangle \int_{t_0}^{t} d\tau e^{\frac{-i}{\hbar}(E_b-E_a)\tau}\vec{E}(\tau) . \qquad (4.30)$$

For the R gauge the perturbation is given by Eq. (2.17). With the help of Eq. (2.24), integration by parts and Eq. (2.16), the expansion coefficient in the R-gauge takes the form

$$b^R(t)=b^R(t_0) + \frac{i}{\hbar}\, e\langle b|\vec{r}|a\rangle \times \left\{ e^{\frac{-i}{\hbar}(E_b-E_a)\tau}\vec{A}(\tau)\Big|_{t_0}^{t} + \int_{t_0}^{t} d\tau e^{\frac{-i}{\hbar}(E_a-E_b)\tau}\vec{E}(\tau)\right\} . \qquad (4.31)$$

If the field is switched on at time t_0 and switched off at time t, the first gauge dependent term on the r.h.s. of Eq. (4.31) vanishes and the amplitudes (4.30) and (4.31) coincide.

The strictly monochromatic field $\vec{E}(\omega') = \vec{E}_o\delta(\omega-\omega')$ is an exception. It does not have a turning on or off process. For a field with vanishing frequency bandwidth the limits (4.26) and therefore also (4.27) do not exist in the classical sense, but only as a distribution:

$$\lim_{t\to\pm\infty} \vec{A}(t) = \pm\pi\vec{E}_0\delta(\omega) . \qquad (4.32)$$

The usual procedure consists of replacing the monochromatic field $\Delta\omega = 0$ by a wavepacket with infinitesimally small frequency bandwidth $\Delta\omega \to 0$. If the monochromatic field is adiabatically turned on and off in this way, Eq. (4.26) holds again and we obtain the same transition probability in the hybrid procedure and in the gauge invariant procedure, for any order of the perturbation series [3,8].

The equivalence of the hybrid and the gauge invariant procedure in the limit $t_o \to -\infty$, $t \to \infty$ can also be shown explicitly for a strictly monocromatic field without a turning on or off process. For this equivalence it is necessary that the field frequency $\hbar\omega$ is in resonance with the transition energy E_f-E_i, i.e. that $\hbar\omega$ is equal to or an integer part of E_f-E_i. We have seen this result already in section 2.3 for first order transitions $\hbar\omega = E_f - E_i$. (For single photon transitions both procedures actually coincide on resonance at any time, since the ratio (2.25) becomes unity). The equivalence was also proven for second order perturbation theory [28,29] and for arbitrary order [21].

On the other hand, for off-resonant processes in a monochromatic field the hybrid procedure leads also in the limit $t_0 \rightarrow -\infty$, $t \rightarrow \infty$ to a wrong answer. We have seen this in Section 2.3 for a single photon process. A similar example for the two-photon absorption was constructed in Ref. 15.

5. THE MOMENTUM TRANSLATION APPROXIMATION

5.1. Review of the MTA-method

The MTA method is supposed to provide an approximate solution of the Schrödinger equation (2.18) in the R-gauge [23]. It is a non-perturbative approach, which is supposed to describe multiphoton-processes of any order. The method is claimed to consistently work once and for all in the R-gauge. The formal solution of Eq. (2.18) can be written as

$$\Phi(\vec{r},t) = e^{-iH_0(t-t_0)/\hbar}\Phi(\vec{r},t_0) - \frac{i}{\hbar}\int_{t_0}^{t} d\tau e^{-iH_0(t-\tau)/\hbar} H_2(\tau)\Phi(\vec{r},\tau) \ . \tag{5.1}$$

The first basic step of the MTA-method is the unitary transformation on the wavefunction Φ

$$\Phi(\vec{r},t) = e^{ie\vec{A}(t)\vec{r}/\hbar}\Psi(\vec{r},t) \ . \tag{5.2}$$

We can now write Eq. (5.1) as (we assume $\vec{A}(t_0) = \vec{A}(t) = 0$)

$$\Phi(\vec{r},t) = e^{-iH_0(t-t_0)/\hbar}\Psi(\vec{r},t_0) - \frac{i}{\hbar}\int_{t_0}^{t} d\tau e^{-iH_0(t-\tau)/\hbar} H_2(\tau)\, e^{ie\vec{A}(\tau)\vec{r}/\hbar}\Psi(\vec{r},\tau) \ . \tag{5.3}$$

The transformation (5.2) is just the same as the transformation (3.3) and (3.7) discussed in Section 3.2. Therefore ψ satisfies Eq. (2.5). But the interpretation of Eq. (2.5) is now different from Section 3.2: the potentials are still given by $(\vec{A},0)$ and not by $(0,-\vec{E}\cdot\vec{r})$. Thus the transformation (5.2) is not a gauge transformation but a hybrid transformation in the sense of Section 3.3.

The second basic step of the MTA procedure argues that in situations where the field frequency ω is much smaller than the considered transition energy E_f-E_i (i.e. for multiphoton processes of high order) the perturbation $H_1^i(t)$ in Eq. (2.5) is much smaller than the perturbation $H_2(t)$ in Eq. (2.18) according to Eq. (2.25). Therefore, if $H_2(t)$ is too large to be treated as a small perturbation with respect to H_0 in Eq. (2.18), $H_1(t)$ might still appear as a small perturbation compared to H_0 in Eq. (2.5). The error is then supposed to be small if one neglects the interaction $-e\vec{E}(t)\cdot\vec{r}$ in Eq. (2.5). Thus the solution of Eq. (2.5) can be approximated by

$$\Psi_0(\vec{r},t) = e^{-iH_0(t-t_0)/\hbar}\,\Psi(\vec{r},t_0) \ . \tag{5.4}$$

This gives the approximate solution for Φ

$$\begin{aligned}\Phi(\vec{r},t) &= e^{-iH_0(t-t_0)/\hbar}\,\Psi(\vec{r},t_0)\\ &- \frac{i}{\hbar}\int_{t_0}^{t} d\tau\, e^{-\frac{i}{\hbar}H_0(t-\tau)}\,H_2(\tau)\,e^{\frac{i}{\hbar}e\vec{A}(\tau)\vec{r}}\,e^{-\frac{i}{\hbar}H_0(\tau-t_0)}\,\Psi(\vec{r},t_0) \ .\end{aligned} \tag{5.5}$$

The exponential $\exp(ie\vec{A}\vec{r}/h)$ can be expanded in a sum over Bessel functions. Calculating the transition rate with the wavefunction (6.5), we obtain an infinite sum over multiphoton terms

$$c_n = \sum_{\ell} T_{ni}^{(\ell)}\,\delta(E_n - E_i - (\ell+1)\hbar\omega) \ . \tag{5.6}$$

5.2 Counterexamples

Our example in Section 4.3 showed that approximations, based on the difference (2.25) of the matrix elements of the two interactions $\vec{p}\cdot\vec{A}$ and $\vec{E}\cdot\vec{r}$, cannot be trusted. Although the matrix element of the $\vec{p}\cdot\vec{A}$ interaction is much larger, both interactions yield the same transition rate.

As shown in Ref. 24 the MTA is an inconsistent approximation.

The first-order correction to Ψ, neglected in Eq. (5.4), leads to terms of the same order of magnitude as those given by the MTA.

A good example, where the MTA fails, is the harmonic oscillator. This problem can be solved exactly and does not exhibit any multiphoton transitions [24]. In contrast, the MTA predicts multiphoton transitions of any order for the harmonic oscillator.

In order to save the MTA it was claimed that this discrepancy is only a peculiar feature of the harmonic oscillator, since "the harmonic oscillator appears to be the only problem in which multiphoton transitions between adjacent levels are not possible" [33]. But there are further examples where the MTA fails. Let us consider an example where one can approximately include the interaction $-e\vec{E}\vec{r}$ in Eq. (2.5). Such an example is a Rydberg wavefunction which is localized between r_o and $r_o + \delta r$ with a "delta-like" shape.

Assuming the particle to be bound in a s-state $\Psi(n,l=0,m=0)$ in the spherically symmetric potential $V(|\vec{r}|)$ and the electric field $\vec{E}(t)$ to be linearly polarized in z-direction, the Schrödinger equation for Ψ reads

$$i\hbar \frac{\partial}{\partial t} \Psi(r,t) = \{\frac{1}{2m} \vec{\nabla}_r^2 + V(r) - eE(t)r \cos\theta\}\Psi(r,t) \ . \tag{5.7}$$

For a well localized wavefunction it is a good approximation to replace the radial coordinate r in the perturbation $-eEr\cos\vartheta$ by r_o. This leads to the approximate solution of Eq. (5.7)

$$\begin{aligned}\Psi(r,t) &\approx e^{\frac{i}{\hbar} e \int_{t_0}^{t} d\tau E(\tau) r_0 \cos\theta} \Psi_0(r,t) \\ &= e^{-ie\vec{A}(t)\vec{r}_0/\hbar} e^{-iH_0(t-t_0)/\hbar} \Psi(r,t_0) \ .\end{aligned} \tag{5.8}$$

If we insert this improved solution into Eq. (5.3)

$$\begin{aligned}\Phi(\vec{r},t) &= e^{-iH_0(t-t_0)/\hbar} \Phi(\vec{r},t_0) \\ &- \frac{i}{\hbar} \int_{t_0}^{t} d\tau e^{\frac{-i}{\hbar} H_0(t-\tau)} H_2(\tau) e^{\frac{i}{\hbar} e\vec{A}(\tau)(\vec{r}-\vec{r}_0)} e^{\frac{-i}{\hbar} H_0(\tau-t_0)} \Psi(\vec{r},t_0)\end{aligned} \tag{5.9}$$

we find that the MTA factor $\exp(ie\vec{A}(t)\vec{r}/\hbar)$ cancels for a wavefunction sharply peaked about r_0. In particular, if we expand the exponential $\exp\{ie\vec{A}(\vec{r}-\vec{r}_o)/\hbar\}$ in terms of Besselfunctions $J_n(ie\vec{A}_0(\vec{r}-\vec{r}_0)/\hbar)$ we see that for sharply localized wavefunctions no multiphoton transitions can occur anymore, since $J_n(0) = \delta_{no}$.

5.3. Noninteracting bound-state wavefunction

We call a bound state non-interacting if it is an eigenstate of the energy operator. This state has the same constant energy eigenvalue E, no matter which gauge we use and whether the field is on or off. This non-interacting state has the character of a usual stationary state. Since wavefunctions are gauge dependent, we cannot expect the non-interacting state to be represented by the same wavefunction in different gauges.

Physical interpretations are straightforward in the E-gauge, since the vector potential vanishes and the Hamiltonian H_0 coincides with the energy operator. The non-interacting state is then given by one of the eigenstates $\phi_n^o(\vec{r})$ (2.7) of H_o.

In any other gauge the non-interacting state is represented by an eigenstate $\phi_n^g(\vec{r},t)$ (3.21) of the energy operator ε^g. In the dipole approximation the ϕ_n^g are related to the ϕ_n^o by Eq. (3.23). Thus in the R-gauge the noninteracting state is

$$\phi_n^R(\vec{r},t) = e^{ie\vec{A}(t)\vec{r}/\hbar}\phi_n^0(\vec{r}) \ . \tag{5.10}$$

The question might arise why we do not obtain the non-interacting state in the R-gauge by simply setting $\vec{A} = 0$ in the Schrödinger equation (2.18). This would lead us to Eq. (2.8), which has the solution ψ_n^o (2.9). This means that we would have the same non-interacting states ϕ_n^o in any gauge, since whenever we set the electromagnetic potentials equal to zero, we end up with the unperturbed Hamiltonian H_0. The wavefunctions ϕ_n^o, however, do not conserve energy in the R-gauge

$$\begin{aligned}\frac{d}{dt}\langle\phi_n^0|\varepsilon^R|\phi_n^0\rangle &= -\frac{i}{\hbar}\langle\phi_n^0|[\varepsilon^R,H_0 + H_2] + \frac{\partial}{\partial t}\varepsilon^R|\phi_n^0\rangle \\ &= \langle\phi_n^0|\frac{e}{m}(\vec{p} - e\vec{A})\vec{E}|\phi_n^0\rangle = -\frac{e^2}{m}\vec{A}(t)\vec{E}(t) \neq 0 \ .\end{aligned} \tag{5.11}$$

Let us now compare the energy-eigenstate (5.10) in the R-gauge with the so-called MTA wavefunction. In the MTA procedure the solution of Eq. (2.5) is approximated by the field-free solution Ψ^o_n (2.9) and then inserted into Eq. (5.2), as described in Section 5.1:

$$\psi^{MTA}_n(\vec{r},t) = e^{ie\vec{A}(t)\vec{r}/\hbar}\psi^0_n(\vec{r},t) = e^{-iE_n t/\hbar}\phi^R_n(\vec{r},t) . \qquad (5.12)$$

We see that the MTA wavefunction is an eigenstate of the energy operator ε^R with energy eigenvalue E_n [26]. Reiss [33], however, claims that the wavefunction (5.12) approximately includes the interaction of the particle with the field to arbitrary order due to the "integrating factor" exp $(ie\vec{A}(t)\vec{r}/h)$. He states [34] that "by construction ϕ^o_n is the non-interacting solution in Coulomb-gauge, and so ϕ^R_n contains interaction with the field".

This controversy can be summarized in one question: are the eigenstates of the unperturbed Hamiltonian H_0 or the eigenstates of the energy operator ε^g the non-interacting states? Reiss uses the eigenstates ϕ^o_n of H_0, which have the same analytical form in any gauge, as the non-interacting states. This is wrong. They might be called non-perturbated states since they have the same form as in the case of vanishing potentials, but they are not non-interacting states since they do not correspond to constant energy eigenvalues as shown in Eq. (5.11). For the definition of a non-interacting state we have to consider the real physical quantities, the gauge invariant eigenvalues of the observable energy. The non-interacting state must be an eigenstate of the energy operator with the same time independent energy eigenvalue for any value of the electromagnetic field and for any gauge. The wavefunction (5.12) satisfies this condition. The central misunderstanding is the conclusion that the functional dependence of the wavefunction (5.12) on the vector potential $\vec{A}$ does imply an interaction with the field. This wavefunction is altered by the external field, but it does not interact with it. Actually, the non-interacting state has to, in general, be altered by the external field in order to keep the same energy.

Having the correct interpretation of the MTA-wave function (5.12) as a noninteracting bound state wave function in mind we can now conclude that Eq. (5.3) can only be trusted to first order in the coupling constant e. Multiphoton transitions ($n > 1$) which are generated by the wave function (5.3) as exhibited in Eqn. (5.6) are unreliable and can be spurious as demonstrated by the case of the harmonic oscillator and the example of Section 5.2. By adopting the approximation (5.4) which entirely neglects the field, Eqn. (5.3) has become meaningless beyond first order.

6. SUMMARY

One usually calculated the probability for transitions between eigenstates of the unperturbed Hamiltonian H_0 without worrying much about the physical interpretation of H_0. But H_o is not a physical gauge invariant quantity and we cannot measure the system in an eigenstate of this purely mathematical operator. This "transition probability" is gauge dependent. In a real physical situation one actually has to consider transitions between energy eigenstates, not between eigenstates of H_0. The energy operator is by definition a physical quantity and the amplitudes of transitions between its eigenstates must be gauge invariant.

Only in the E-gauge does H_0 coincide with the energy operator. Thus, the use of the E-gauge is foolproof since the calculated amplitudes are the gauge invariant amplitudes for transitions between energy eigenstates. In any other gauge one should use the gauge invariant procedure which tells us the proper choice of basis states to ensure gauge independent probabilities. It is convenient to expand the wavefunction in terms of a complete set of energy eigenstates because the expansion coefficients are then automatically the desired probability amplitudes. One then ends up again with the same set of equations of motion for the transition amplitudes as with the E-gauge.

We have seen that the gauge invariance of quantum mechanics requires that a particular <u>physical</u> state and the operator corresponding to a <u>physical</u> quantity have different representations in different gauges. (But it should be stressed that their physical interpretation is the same in any gauge). Therefore the energy operator ε^g and its eigenstates ϕ_n^g transform under a gauge transformation according to Eqs. (3.8) and (3.3), respectively, whereas H_0 and its eigenstates ϕ_n^o have the same form in any gauge. This unitary transformation of the basis states is the essential difference between the hybrid procedure and the gauge procedure. The use of the energy operator ε^g (3.14) as the basis defining operator is restricted to fields in dipole approximation. Only in the LWA the gauge invariant and time independent eigenvalues of ε^g coincide with the energy spectrum of the system in the absence of the external field.

On the other hand, for a wide class of physical situations omitting the transformation on the wave functions does not affect the transition probabilities. The hybrid procedure is justified in the R-gauge for pulses, if the preparation of the system takes place at $t_o \to -\infty$ and the measurement at $t \to +\infty$, since the vector potential vanishes for wavepackets at $t = \pm\infty$. This holds also true for resonant monochromatic fields, no matter whether or not they are adiabatically switched on and off. Only for non-resonant transitions induced by a monochromatic field does the hybrid procedure yield contradictory results in different gauges. This is just the example which Lamb investigated and which brought up the $\vec{p} \cdot \vec{A}$ vs. $\vec{E} \cdot \vec{r}$ problem.

Acknowledgments

We enjoyed and benefitted from discussions with D. Kobe, W. E. Lamb, Jr., P. Meystre and H. Walther. This work was supported in part by the Air Force Office of Scientific Research under Contract No. AFOSR-81-0128.

References

1. M. Göppert-Mayer, Ann. der Phys. 9 (1931) 273.
2. W.E. Lamb, Jr. and R.C. Retherford, Phys. Rev. 79 (1950) 549; W.E. Lamb, Jr., Phys. Rev. 85 (1952) 259.
3. E.A. Power and S. Zienau, Nuovo Cimento 6 (1957) 7; Phil. Trans.R.Soc. 251 (1959) 427.
4. see, for example, J.M. Worlock, Laser Handbook, ed. F.T. Arecchi and E.O. Schulz-DuBois (North-Holland, Amsterdam, 1972) Vol. 2, p. 1323; and references therein.
5. E.T. Jaynes, 1976, cited from Ref. 9.
6. C. Cohen-Tannoudji, B. Diu and F. Laloe, Quantum Mechanics (Hermann/Wiley, Paris, 1977).
7. K.-H. Yang, Ann.Phys. (N.Y.) 101 (1976) 62.
8. J.J. Forney, A. Quattropani, and F. Bassani, Nuovo Cimento 37B (1977) 78.
9. D.H. Kobe and A.L. Smirl, Am.J.Phys. 46 (1978) 624.
10. K.-H. Yang, Phys. Lett. 64A (1977) 276; 84A (1981) 165; 92A (1982) 71; 94A (1983) 259; J. Phys. A 15 (1982) 437; 15 (1982) 1201; 16 (1983) 919; 16 (1983) 935;
    D.H. Kobe, Int. J. Quantum Chem., Symp. 12 (1978) 73; Phys. Rev. A19 (1979) 205; A19 (1979) 1876: Am.J.Phys. 49 (1981) 581; 50 (1982) 128; 51 (1983) 163; J.Phys. A 16 (1983) 737; J. Phys. B 16 (1983) 1159;
    D.H. Kobe and K.-H. Yang, J.Phys. A 13 (1980) 3171, Am.J.Phys. 51 (1983) 163;

D.H. Kobe and E.C.-T. Wen, Phys. Lett. 80A (1980) 121; J. Phys. A 15 (1982) 787;
D.H. Kobe, E.C.-T. Wen and K.-H. Yang, Phys. Rev. D26 (1982) 1927;
P.K. Kennedy and D.H. Kobe, J. Phys. A 16 (1983) 521;
D.H. Kobe and P.K. Kennedy, J. Phys. B 16 (1983) L443;
D. Lee and A.C. Albrecht, J. Chem. Phys. 78 (1983) 3382;
D. Lee, A.C. Albrecht, K.-H. Yang and D.H. Kobe, Phys. Lett. 96A (1983) 393.

11. C. Leubner, Phys. Lett. 82A (1981) 223.
12. Y. Aharonov and C.K. Au, Phys. Lett. 86A (1981) 269, 95A (1983) 412.
13. T.E. Feuchtwang, E. Kazes, H. Grotch and P.H. Cutler, Phys. Lett. 93A (1982) 4;
E. Kazes, T.E. Feuchtwang, H. Grotch and P. Cutler, Phys. Rev. D27 (1983) 1388.
14. S.T. Epstein, Phys. Lett. 97A (1983) 29.
15. D.H. Kobe, Phys. Rev. Lett. 40 (1978) 538.
16. E.A. Power, Multiphoton Processes, ed. J.H. Eberly and P. Lambropoulos (Wiley, New York, 1978) p. 11.
17. S. Geltman, J. Phys. B 10 (1977) 831.
18. G. Grynberg and E. Giacobino, J. Phys. B 12 (1979) L93.
19. K. Haller and R.B. Sohn, Phys. Rev. A20 (1979) 1541.
20. Y. Aharonov and C.K. Au, Phys. Rev. A20 (1979) 1553.
21. S. Olariu, I. Popescu and C.B. Collins, Phys. Rev. D20 (1979) 3095.
22. C. Leubner and P. Zoller, J. Phys. B 13 (1980) 3613.
23. H.R. Reiss, Phys. Rev. A1 (1970) 803.
24. C. Cohen-Tannoudji, J. Dupont-Roc, C. Fabre and G. Grynberg, Phys. Rev. A8 (1973) 2747.
25. H.R. Reiss, Phys. Rev. C27 (1983) 1199; C27 (1983) 1229.
26. W. Becker, R.R. Schlicher and M.O. Scully, Phys. Rev. C, to be published.
27. J. Bergou, J.Phys. B 16 (1983) L647.
28. W.L. Peticolas, R. Norris and K.E. Rieckhoff, J.Chem.Phys. 42 (1968) 4164.
29. F. Bassani, J.J. Forney, and A. Quattropani, Phys. Rev. Lett. 39 (1977) 1070; A. Quattropani, F. Bassani and S. Carillo, Phys. Rev. A25 (1982) 3079.
30. I.I. Gol'dman and V.D. Krivchenkov, Problems in Quantum Mechanics (Pergamon, New York, 1961) problems 3-13; G.W. Parker, Am.J.Phys. 40 (1972) 120; J. Bergou and F. Ehlotzky, J.Phys.B 15 (1982) L185.
31. J. Bergou, J. Phys. A 13 (1980) 2817; C. Leubner, Am. J. Phys. 49 (1981) 738; J. Bergou and S. Varro, J.Phys. B 15 (1982) L179.
32. R. Loudon, The Quantum Theory of Light (Clarendon, Oxford, 1973)
33. H.R. Reiss Phys. Rev. A23 (1981) 3019.
34. H.R. Reiss, Phys. Rev. C, to be published.

IS REALITY REALLY REAL? - AN INTRODUCTION TO BELL'S INEQUALITIES

P. Meystre

Max-Planck-Institut für Quantenoptik
D-8046 Garching
Fed. Rep. Germany

1. INTRODUCTION

Ever since its development in the 1920's, quantum mechanics has been the object of numerous discussions, which are still going on, and will probably keep going on for some time. At the onset, one should agree on one point, namely, that quantum mechanics works extremely well, and allows us to predict the most minute aspects of, say, atomic spectra, with incredible accuracy. The problem is not there, but rather, lies in its interpretation. What is the meaning of the wave function, what is performed in a measurement, etc. ..., are questions which have fascinated, and still fascinate, many physicists. Some people make a living out of discussing these problems, but for most of us, this is a hobby, that we talk about during coffee breaks or in the evening, around a pitcher of beer. I am certainly not an expert on the foundations of quantum mechanics. But over the last few years, I have read a substantial amount of papers on this topic, and have realized that during my studies I had been "brain-washed" into accepting things which I should not have ... at least, not readily. I have come to understand that we live in a very strange world indeed, where the most trivial, self-evident truths don't apply. In this lecture, I would like to explain why it is so. What I will say is not new, it is just my way of understanding and summarizing the work of very clever people such as Einstein, Bohr, Bohm, Bell, Wigner, and many others. Of course, I may well misunderstand and misquote them at some point or the other, and apologize in advance for this. I hope nevertheless that these notes may be useful to some other "operational" physicists, who spend their life doing very concrete calculations or experiments, but ask themselves now and then "what on earth it all means".

In Section 2, I briefly review the Einstein-Podolski-Rosen "paradox". This is the starting point of most discussions of the foundations of quantum mechanics, puts the whole problem into a proper frame, and allows to introduce the central concepts of "locality" and "reality". In Section 3, I then open a parenthesis to discuss a variation on Bertlmann's socks adapted to the present summer school. This example, introduced by J.S. Bell to illustrate the strangeness of quantum mechanics, sparked my understanding of the problem, and maybe, it will help somebody else, too.

In Section 4, I then derive Bell's inequalities in a simple form, trying to explicitly show at which point each assumption enters -this is not necessarily obvious to see in the published literature. In Section 5, I review the most recent experiments, by Aspect and coworkers, using photon cascade in atomic transitions to test this inequality. In Section 6, I mention a loophole in Aspect's experiments, which leads to a brief discussion of delayed choice experiments. Finally, in Section 7, I ponder about the implications of these results. Is our world not local, or not real? ... I must say at the onset that any reader expecting to find an answer to these questions in this paper is going to be bitterly disappointed.

2. THE EINSTEIN-PODOLSKI-ROSEN "PARADOX"

It is well-known that Einstein, although he did not deny its operational success, never quite accepted quantum mechanics. In a famous paper[1], Einstein, Podolsky and Rosen (EPR) proposed a Gedankenexperiment aimed at proving that quantum mechanics is not a complete theory. We discuss here a variation of this experiment proposed by Bohm[2].

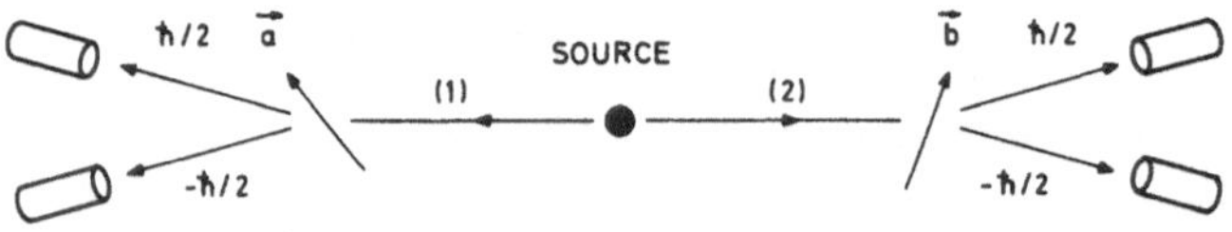

Fig. 1: Experimental set-up of the EPR-Bohm Gedanken experiment. The second analyzer $\vec{b}$ is not relevant here, but plays a central role in Bell's theorem (see Section 4).

Consider a source in which pairs of identical spin 1/2 particles are produced by, say, the decay of a diatomic molecule in the singlet state. Upon emerging from the source, these two particles fly towards two space-like separated analyzers and detectors, such as Stern-Gerlach magnets. (see Fig. 1).

Long after the particles are emitted, an observer sets the analyzer (magnet) 1, in order to measure the spin component $S_a = \vec{a} \cdot \vec{S}_1$ of particle 1 along $\vec{a}$. For a spin-1/2 particle, the result is $\pm \hbar/2$. Because the total spin of the system is zero, we know for sure, without having to perform a measurement, that the spin component S_a of the second particle along this same direction is then $\mp \hbar/2$.

At this point, EPR introduce the concept of "reality"[1]: "If, without in any way disturbing a system, we can predict with certainty (i.e. with probability equal to unity) the value of a physical quantity, then there is an element of physical reality corresponding to this physical quantity." EPR further require that "every element of the physical reality must have a counterpart in the physical theory".

According to this criterion, we can then attribute an element of physical reality to the spin component S_a. However, our observer might have chosen to set the detector 1 in direction $\vec{a}'$, thus measuring the spin component $S_{a'}$, and inferring, without in any way disturbing particle 2, its spin component $S_{a'}$. Thus, there is also an element of physical reality attached to $S_{a'}$.

According to quantum mechanics, however, one cannot predict precise values for non-commuting observables. If quantum mechanics is complete, the two observables S_a and $S_{a'}$ cannot have simultaneous reality (unless $\vec{a} = \vec{a}'$), in contradiction with the preceeding argument. Thus, EPR conclude that quantum mechanics is not complete[3].

Of course, many of the founders of quantum mechanics, and in particular Bohr[4], have refuted this argument, maintaining that the specification of the experimental procedure plays a central role in quantum mechanics, but for many years, no real progress was made. In fact, a majority of physicists believed that this whole discussion belonged more to the realm of philosophy that to physics, since it looked like no experiment was able to determine which was the correct attitude.

The situation was changed drastically since the early 1960's, due in particular to the work of J.S. Bell, who studied an extention of the original EPR experiment, where the analyzers 1 and 2 (see Fig. 1) are set at different angles $\vec{a}$ and $\vec{b}$. One then measures the joint probability of obtaining a given value (say, + $\hbar/2$) for the spin components S_a and S_b .

Let us call $P_{++}(a,b)$ the joint probability of measuring $S_a = \hbar/2$ and $S_b = \hbar/2$, $P_{+-}(a,b)$ the joint probability of measuring $S_a = \hbar/2$, and $S_b = -\hbar/2$, etc... According to quantum mechanics, we have

$$
\begin{aligned}
P_{++}(a,b) &= P_{--}(a,b) = \frac{1}{2}\sin^2 \theta/2, \\
P_{+-}(a,b) &= P_{-+}(a,b) = \frac{1}{2}\cos^2 \theta/2,
\end{aligned}
\qquad (1)
$$

where θ is the angle between $\vec{a}$ and $\vec{b}$. This indicates that quantum mechanics predicts strong correlations between the measurements at detectors 1 and 2. For instance, for $\theta = 0$, we get $P_{++}(a,b) = P_{--}(a,b) = 0$, and $P_{+-}(a,b) = 1/2$.

Of course, strong correlations are well-known in every-day life, too. Think for instance of two friends taking blindly one ball each out of a bag containing one white ball and one black ball, putting them into their pocket and then travelling one to the moon, and the other to Boulder. The two friends agree to look at the color of the balls at a given, prearranged time. When the traveler to the moon sees that he has, say, the black ball, he immediately knows that his friend has the white one, without need to even check. There is a strong correlation due to a common cause. (It is also legitimate for the moon traveler to attribute an element of physical reality to his friends's white ball - or is it? EPR would say yes.)

Are the correlations observed on the EPR-type experiments also due simply to common causes (the two spins are after all prepared by the decay of a single, common, molecule)? In the original EPR experiment, the answer could have been, yes! But the surprise is that in Bell's more elaborate version, the answer is, to a large degree of certainty, no (that is, provided that Aspect's results still hold in a "delayed choice" version of his experiment). And to understand better what this means, it is time to turn to Bertlmann's socks.

3. DR. BERTLMANN GOES TO BOULDER

In a recent paper [5], J.S. Bell discussed the case of Dr. Bertlmann's socks. I found this example, directly inspired by d'Espagnat[6], very illuminating, and tried to adapt it to the context of this school, which takes place in the foothills of the beautiful Rocky Mountains. In a series of experiments, I have asked participants to this Institute to perform two of three tests, consisting of hikes on trails on flat ground, on slopes of 45°, and on vertical cliffs, respectively.

Now, for a fresh bunch of participants, one has that

(those who can hike at 0° but not at 45°)

+

(those who can hike at 45° but not at 90°) (2)

is not less than

(those who can hike at 0° but not at 90°)

which is trivially correct, and not very deep, since members of the last group can either hike at 45°, and belong also to the second group, or not, in which case they belong to the first group. (Note that we do not assume that if somebody can hike at some angle, he can hike at lower angles, too.). However, it is hard to perform such an experiment, because if we bring somebody incompetent on a steep cliff, he may not be available for the next test! Also, after one test, a hiker is not fresh anymore, and may not pass another test!

But I have noticed that people always go in pairs of equivalently trained hikers, so that if one member passes a test, the other one would pass it, too.

Thus, I can rewrite relation (2) as

(the number of pairs where one member can hike at 0° and the other not at 45°)

+

(the number of pairs where one member can hike at 45° and the other not at 90°) (3)

is not less than

(the number of pairs where one member can hike at 0° and the other not at 90°).

Assuming that the number of participants at the school is so large that one can go from single events to probabilities, this may be rewritten as

(the probability that one hiker can make it at 0° and the other not at 45°)

+

(the probability that one hiker can make it at 45° and the other not at 90°) (4)

is not less than

(the probability that one hiker can make it at 0° and the other not at 90°).

Now, spins in the EPR experiment are very much like hikers -or socks - except that they are anticorrelated: if one spin passes a test, the other one will not - if the spin component S_a is equal to $+\hbar/2$, then S_b is $-\hbar/2$. Thus, for spins, we must reexpress relation (4) as[24]

Prob (one spin having $\frac{\hbar}{2}$ at 0° and the other not $(-\frac{\hbar}{2})$ at 45°)

+

Prob (one spin having $\frac{\hbar}{2}$ at 45° and the other not $(-\frac{\hbar}{2})$ at 90°)

is not less than (5)

Prob (one spin having $\frac{\hbar}{2}$ at 0° and the other not $(-\frac{\hbar}{2})$ at 90°)

This is, in essence, Bell's inequality. From Eq. (1), we can also compute the predictions of quantum mechanics for such an experiment. Taking by convention that +h/2 corresponds to passing the test, we have

$$\frac{1}{2}\sin^2(22.5) + \frac{1}{2}\sin^2(22.5) = 0.1464,$$

and

$$\frac{1}{2}\sin^2(45) = 0.25.$$

Thus, quantum mechanics clearly violates Relation (5)! As we shall see later on, experiments up to now agree with quantum mechanics, and are in violation of Bell's inequalities. This indicates that at the microscopic level, things don't behave like socks or hikers anymore!

To understand better what is so peculiar about the microscopic world, let us now derive somewhat more formally Bell's inequalities, and try to isolate the hypotheses leading to them.

4. BELL'S INEQUALITIES

In 1964, Bell[8] showed that by performing correlation experiments of the type just discussed, one can distinguish between the predictions of quantum mechanics and those of so-called "local-realistic hidden-variable theories". Later on, his work was extended in particular by Clauser, Holt, Horne and Shimony [9]. Here, I limit my discussion to a simple case, using the derivation that I personally like best [10,11].

The first point is that one should not be afraid by the term "local-realistic hidden-variable theories". What hides behind it is the idea that, following EPR, quantum mechanics is not complete. At some level, not yet understood, in an ultimate theory, there must be more, or different, variables providing a complete description of the system under study. Since these are not yet known, or measurable, they are hidden - thus hidden variables.

"Local-realistic" is a fancy word which, basically, means that one would like to have a "reasonable" ultimate theory. By reasonable, one means that three "self-evident truths" should hold, which d'Espagnat calls [6]

- realism,
- locality,
- free use of inductive inference.

I will try to explain these as we go along.

Consider an arrangement like that of Figure 1, with a source emitting two correlated particles (1) and (2). A property of these particles is measured by detectors 1 and 2, with settings $\vec{a}$, resp. $\vec{b}$, of the analyzers. (The analyzers could be Stern-Gerlach magnets for spins, polarizers for photons, etc ...) Using the principle of <u>inductive inference</u> - legitimate conclusions can be drawn from regularities in the results of experiments - we are entitled to speak about probabilities, rather than single events.

Let us denote $p_1(a)$ and $p_2(b)$ the probabilities of detecting particle (1), resp. (2) for settings $\vec{a}$ and $\vec{b}$ of the analyzers. (For instance, in the EPR-Bohm experiment, we could measure if a spin component is up along direction $\vec{a}$ and down along $\vec{b}$. If we had a complete theory at hand, $p_1(a)$ would be a function of all parameters $\{\lambda\}$ describing completely the emission process in the source. But at the present stage of physics (1983) we have no way to know, or measure, or even guess what these parameters might be. They are hidden, out or our control. What we detect in a series of measurements is some average over them:

$$p_1(a) = \int d\lambda \rho(\lambda) p_1(a,\lambda), \tag{6}$$

where $d\lambda$ is a - unknown - measure over the space of hidden variables, and $\rho(\lambda)$ some weight function. For simplicity, we write λ instead of $\{\lambda\}$. Similarly,

$$p_2(b) = \int d\lambda \rho(\lambda) p_2(b,\lambda). \tag{7}$$

We may, but don't have to require

$$\int d\lambda \rho(\lambda) = 1. \tag{8}$$

Suppose for a moment that we actualy could control the hidden parameters $\{\lambda\}$, and know precisely what their value is. We could then ask the joint probability $p_{12}(a,b,\lambda)$ of detecting both particles for settings $\vec{a}$ and $\vec{b}$ of the analyzers. If the detectors are space-like separated, and the settings chosen long after the emission process, the result at one detector should be unaffected by the setting of the other. This is the principle of locality: no influence of any kind can travel faster than the speed of light -there can be no cross-talk between detectors 1 and 2. Thus, the counting-rates at detectors 1 and 2 should be uncorrelated, so that

$$p_{12}(a,b,\lambda) = p_1(a,\lambda) \cdot p_2(b,\lambda). \tag{9}$$

Note, however, that the actually measured joint probability needs not be uncorrelated: Integrating over the hidden variables, we get with (9)

$$p_{12}(a,b) = \int d\lambda \rho(\lambda) p_1(a,\lambda) p_2(b,\lambda). \tag{10}$$

The weight function $\rho(\lambda)$, which contains all informations about the correlations between hidden variables in the source, leads in general to a nonfactorizable joined probability distribution $p_{12}(a,b)$ (correlation through common cause). We present some examples of functions $\rho(\lambda)$ in Section 7.

A simple theorem [10] states that for any four numbers x,x',y, and y' between 0 and 1, the following inequalities hold:

$$-1 \leqq xy - xy' + x'y + x'y' - x' - y \leqq 0. \tag{11}$$

Noting that probabilities lie between 0 and 1, and chosing two possible directions $\vec{a}$ and $\vec{a}'$, respectively $\vec{b}$ and $\vec{b}'$, for the analyzers 1 and 2, we obtain

$$\begin{aligned} -1 \leqq\ & p_1(a,\lambda)p_2(b,\lambda) - p_1(a,\lambda)p_2(b',\lambda) \\ & + p_1(a',\lambda)p_2(b,\lambda) + p_1(a',\lambda)p_2(b',\lambda) \\ & - p_1(a',\lambda) - p_2(b,\lambda) \leqq 0, \end{aligned}$$

or, with Eq. (9)

$$- 1 \leq p_{12}(a,b,\lambda) - p_{12}(a,b',\lambda) + p_{12}(a',b,\lambda) + p_{12}(a',b',\lambda) - p_1(a',\lambda) - p_2(b,\lambda) \leq 0. \quad (12)$$

Integrating Eq. (12) over the hidden variables yields finally

$$- \int d\lambda \rho(\lambda) \leq p_{12}(a,b) - p_{12}(a,b') + p_{12}(a',b) + p_{12}(a',b') - p_1(a') - p_2(b) \leq 0 \quad (13)$$

The left-hand side of this inequality is equal to -1 if condition (8) holds, but we actually don't need it. Keeping the right-hand side only yields:

$$\frac{p_{12}(a,b) - p_{12}(a,b') + p_{12}(a',b) + p_{12}(a',b')}{p_1(a') + p_2(b)} \leq 1, \quad (14)$$

which is one form of Bell's inequalities, as derived by Clauser and Horne [10].

It is important to realize that when introducing two possible directions $\vec{a}$ and $\vec{a}'$, resp. $\vec{b}$ and $\vec{b}'$ for the analyzers, we implicitly assume that we can speak about the outset of measurements even if we don't actually perform them. (We cannot make simultaneous measurements of particle (1) along both $\vec{a}$ and $\vec{a}'$ directions!) This is the hypothesis of <u>reality</u> introduced by EPR [1], according to which there is an objective physical reality independent of whether we make an observation or not. Thus, three assumptions were indeed needed to derive Bell's inequalities: inductive inference, locality, and reality[6].

In practice, it is not the form (14) of Bell's theorem which is tested. Rather, one makes supplementary hypotheses, some of which can at least in principle be tested experimentally. In particular, if the joint probability $p_{12}(a,b)$ depends only on the angle θ between $\vec{a}$ and $\vec{b}$, and the probability $p_1(\vec{a})$ is independent of the direction of the analyzer, we can chose $\vec{a}$, $\vec{a}'$, $\vec{b}$, and $\vec{b}'$ as in Fig. 2.

Equation (14), becomes then

$$\frac{3p_{12}(\theta) - p_{12}(3\theta)}{p_1 + p_2} \leq 1. \quad (15)$$

For the spin-1/2 case discussed in Section 1, quantum mechanics predicts $p_1 = p_2 = 1/2$, and with Eq. (1) and $\theta = 45°$, we obtain [12]

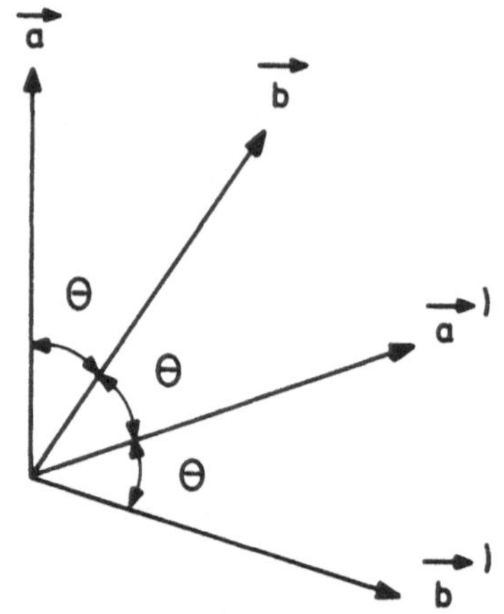

Fig. 2 Geometry used to derive the form (15) of Bell's theorem.

$$\frac{3p_{12}(\theta) - p_{12}(3\theta)}{p_1 + p_2} = 1.207,$$

in violation of Bell's theorem. Thus, quantum-mechanical correlations cannot be fully accounted for by any "local-realistic" hidden-variables theory. The question, then, is to determine which "self-evident truth" is violated by quantum mechanics.

Of course, the situation would be easiest if Bell's inequalities were not violated in experiments, in which case, our world would be "normal", after all. However, the best evidence today, although not definitive, clearly favors quantum mechanics.

5. EXPERIMENTS

The experimental efforts until 1978 to check Bell's theorem are summarized in the review by Clauser and Shimony[9]. Here, we briefly discuss the best experiments to-date, recently performed at Orsay by Aspect and coworkers [13,14]. Instead of spins, as in the original EPR-Bohm Gedankenexperiment, the system used here consists of pairs of optical photons emitted in an atomic radiative cascade (see Fig. 3) [15].

The $4p^2\ {}^1S_0$ level of calcium is populated by two-photon excitation, and decays back to the $4s^2\ {}^1S_0$ state over the $4s4p\ {}^1P_1$ level, emitting two photons of wavelengths $\lambda_1 = 5513$ Å and $\lambda_2 = 4227$ Å.

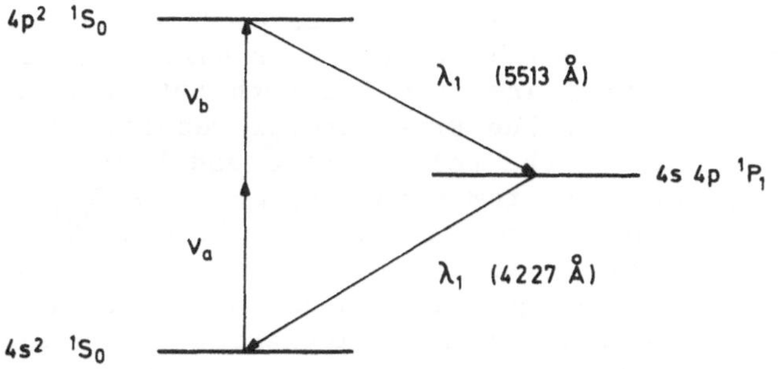

Fig. 3 Radiative cascade scheme used in the Aspect experimental test of Bell's theorem.

Because the change of angular momentum in the transition is $J = 0 \to J=1 \to J=0$, no net angular momentum is carried by the photons. For emitted photons counterpropagating in the ± z-directions, the state of polarization of the total system must therefore be of the form

$$\psi = \frac{1}{\sqrt{2}} \quad \{\binom{1}{o}\binom{1}{o} + \binom{o}{1}\binom{o}{1}\}, \tag{16}$$

where $\binom{1}{o}$ represents the polarization along the $\hat{x}$-axis and $\binom{o}{1}$ along the $\hat{y}$-axis. The first column-vector describes the λ_1-photon and the second the λ_2-photon. Thus, in such a cascade, the polarization states are completely correlated, as the spins are completely anticorrelated in the Bohm set-up. But the difference between complete correlations and complete anticorrelation is irrelevant, as already discussed within the example of Section 3. Thus, photon cascades are appropriate systems to test Bell's inequalities.

The experimental difficulties are, however, considerable [9,16]. Most of the painstaking work consists in improving detectors, polarizers, sources, etc... The best experiment so far [14], however, has attained a high degree of perfection and is becoming agonizingly close to the idealized EPR-Bohm-Bell scheme. (There are, however, a couple of loopholes left. The most severe is discussed briefly in Section 6).

In photon-cascade experiments, one does not measure directly the quantitiy Eq. (14), but rather, another combination of correlation functions which satisfies the inequality [14]:

$$-2 \leqq S \leqq 2.$$

The detailed form of S in irrelevant here. It is sufficiant to note that this combination is more direct - or basic - in the sense that no single probabilities are used and both output channels of each analyzer are monitored. The experimental results, for an angle of 22.5° or 67.5° between the polarizers $\vec{a}$ and $\vec{b}$ is, however, S_{exp} = 2.697 ± 0.015, while quantum mechanics predicts S_{QM} = 2.70 ± 0.05. (The theoretical uncertainty takes into account imperfections in the detection system). Thus, the experiment shows a spectacular violation of Bell's inequalities and is in excellent agreement with the predictions of quantum mechanics. Assuming that these results remain valid when the last two experimental loopholes are eliminated, which is most likely, we must therefore conclude that local-realistic hidden variable theories are wrong.

6. DELAYED-CHOICE EXPERIMENTS

As recognized already by Aspect et al., the experiments of Ref. 14 still have a couple of loopholes left. The first one, related to the low efficiencies of the detectors, does not appear to be very severe. The other is, however, of a conceptual nature and must be eliminated in future experiments. Specifically, the difficulty arises from the fact that the setting of the analyzers was fixed <u>before</u> the emission process has taken place. Thus, in principle, the analyzers could "tell" the source how to emit before hand, and the experiments do not actually test the locality or unlocality of quantum mechanics.

To eliminate this loophole, "delayed choice" experiments must be performed [16,17]. Aspect et al. [18] have recently gone one step in this direction by performing a series of experiments using variable polarizers which jump between two orientations in a time short compared to the photon transit time. The results are still in violation of Bell's inequalities. However, the experimental arrangement is not yet ideal: the change of analyzer direction is not random, but rather quasiperiodic, although the switches of the polarizers are driven by different generators at different frequencies. These recent experiments make "local-realistic" hidden-variable theories more and more improbable, but truly delayed-choice experiments, where the analyzers are set <u>randomly long after the emission process</u>, are nevertheless still absolutely required. Nobody really doubts that in this case, Bell's inequalities will still be violated. And nobody really knows what this means. We now know that the microscopic world is very strange, but <u>what</u> is strange in it is not yet clear.

7. INTERPRETATION ?

In Section 4, we derived Bell's inequalities using only three "self-evident truths": locality, reality, and the use of inductive inference. Their violation in photon cascade and other experiments indicates that at least one of these assumptions is wrong. Which one? Nobody actually knows, but many people seem to believe that locality is the bad guy. However, there is no proof of that.

I would like to suggest that, maybe, none of these "self-evident truths" is wrong. In the derivation of Bell's theorem, there was a fourth hypothesis, which one mostly doesn't pay much attention to, namely the fact that probabilities are positive and bounded by 1. However, if one abandons this requirement, one can easily build a local-realistic hidden variables model violating Bell's inequalities and reproducing the quantum mechanical results. In the following, I give a heuristic example of how to do this for the case of photon cascades. Using classical-looking representations such as the Wigner distributon, Scully has developed a systematic way to build such "hidden-variable theories"[19].

I start from Eqs. (6) and (10), and assume that the source is completely described by two sets of hidden variables α and β such that [20]

$$p_1(a) = \int_o d\alpha \int_o d\beta \rho(\alpha,\beta)\cos^2(a-\alpha), \tag{17}$$

where a is the angle of the λ_1-photon polarizer with respect to the origin, and α varies between 0 and 2π. Similarly

$$p_2(b) = \int_o d\alpha \int_o d\beta \rho(\alpha,\beta)\cos^2(b-\beta) \tag{18}$$

and

$$p_{12}(a,b) = \int_o d\alpha \int_o d\beta \rho(\alpha,\beta)\cos^2(a-\alpha)\cos^2(b-\beta). \tag{19}$$

Belifante[20] has shown that the condition of "maximum source correlation",

$$\rho(\alpha,\beta) = \frac{1}{2\pi}\delta(\alpha-\beta) \tag{20}$$

yields $p_1(a) = p_2(b) = 1/2$, as it should, but

$$p_{12}(a,b) = \frac{1}{4}[\cos^2(a-b) + \frac{1}{2}], \tag{21}$$

which is within the bounds allowed by Bell's inequalities. The "problem" of quantum mechanics is that it produces more correlations than allowed by the common causes involved in local-realistic theories. How can one then produce more correlations in the source? For instance, by using a distribution sharper than a delta-function.

Taking as an example

$$\rho\ (\alpha,\beta) = \frac{1}{2\pi}\,\delta(\alpha-\beta) - \frac{1}{8\pi}\,\delta''(\alpha-\beta) \qquad (20)$$

$$\equiv \rho(\alpha,\beta)_{classical} + \rho(\alpha,\beta)_{QM}$$

one finds readily

$$p_1\ (a) = p_2\ (b) = 1/2 \qquad (22)$$

and

$$p_{12}\ (a,b) = \frac{1}{2}\cos^2\ (ab), \qquad (23)$$

which is the correct (quantum mechanical) correlation function in the case of photon cascades [9]. (The half-angle appearing in Eq. (1) is characteristic of spin-1/2 particles). In such a model, one could interpret $\rho(\alpha,\beta)/2\pi$ as the "classical" source correlation, while $\rho''(\alpha-\beta)/8\pi$ appears as a singular, "quantum" correction.

The model described here is not unique. We have built a number of them [21], all doing the job. They are local, but involve negative probabilities. A fascinating point is that when trying to make quantum mechanics look classical, using e.g. Wigner distribution functions, it is usually not locality, but positive probability, which is sacrificed [22]. Could it be that our concept of probabilities is too naive, that they can, indeed, become negative, and that the quantum-mechanical wave-function is a way to handle them, very much like imaginary numbers are the way to handle the square-root of negative numbers? To stop these speculations, what is needed is somebody with a stroke of genius comparable to that of J.S. Bell, who comes up with a way to test separately the various "self-evident truths" used in local-realistic theories.

ACKNOWLEDGEMENTS

I have benefited from numerous discussions with A. Barut and M.O.Scully. I am greatful to A. Aspect for discussing his experiments in detail with me, for sharing his views of their implications, and for a careful reading of this manuscript. The fact that nonlocality can be avoided when introducting negative probabilities has been recognized by a number of people, including F. Laloe and Ph. Grangier[23]. That they did not publish their findings may be an indication of the general uneasiness about this concept. J.S. Bell has written [5], "... many physicists came to hold not only that it is difficult to find a coherent picture, but that it is wrong to look for one - if not actually immoral then certainly unprofessional." I am greatful to the Max-Planck Institute for Quantum Optics, and in particular to Prof. H. Walther, for allowing me to spend some of my time on such unprofessional ventures.

REFERENCES:

1. A. Einstein, B. Podolsky, and N. Rosen, Phys. Rev. 47, 777 (1935).
2. D. Bohm, Quantum Theory, Prentice-Hall, Englewood Cliffs, N.J. 1951, p. 614-619.
3. "...one must consider the description given by quantum mechanics as an incomplete and indirect description of reality, destined to be later replaced by an exhaustive and direct description. In my opinion one should, when seeking a unitary basis to the ensemble of physics, avoid in any case to dogmatically stick to the scheme of the present theory"
   Letter from A. Einstein to M. Born, 5. April 1948, in "Correspondance 1916-1955, Editions du Seuil, Paris, 1972, p. 186 (translation by P.M.).
4. N. Bohr, Phys. Rev. 48, 696 (1935).
5. J.S. Bell, J. de Physique (France), Colloque C2, Suppl. au n°3, Tome 42 (1981), p. C2-41.
6. B. d'Espagnat, Scientific American, Nov. 1979, p. 158.
7. N.D. Mermin, Am. J. Physics 49, 940 (1981).
8. J.S. Bell, Physica 1, 195 (1964).
9. A beautiful review is given in J.F. Clauser and A. Shimony, Rep. Prog. Phys. 41, 1881 (1978).
10. J.F. Clauser and M.A. Horne, Phys. Rev. D10, 526 (1974).
11. There are a number of ways to derive Bell's theorem. A particularly tutorial one was given by E.P. Wigner, Am. J. Phys. 38, 1005 (1970). the shortest proof I know of is by A. Peres and W.H. Zurek, Am. J. Phys. 50, 807 (1982). See also J.S. Bell, Comments on Atom. Mol. Phys. 9, 121 (1980). A process to construct further generalizations has been given by M. Froissart, Il Nuovo Cimento 64B, 241 (1981).
12. Since in the EPR set-up, the spins are anticorrelated, the joint-probability for detection $p_{12}(a,b)$ is interpreted as the probability of detecting spin 1 with component + h/2 and spin 2 with component - h/2.
13. A. Aspect, Ph. Grangier, and G. Roger, Phys. Rev. Letters 47, 460 (1981).
14. A. Aspect, Ph. Grangier, and G. Roger, Phys. Rev. Letters 49, 91 (1982).
15. Previous experiments using similar arrangements include those of S.J. Freeman and J.F. Clauser, Phys. Rev. Letters 28, 938 (1972); R.A. Holt and F.N. Pipkin, see R.A. Holt, Ph.D. Thesis, Harvard University (1973), unpublished; E.S. Fry and R.C. Thompson, Phys. Rev. Letters 37, 465 (1976); J.F. Clauser, Phys. Rev. Letters. 36, 1223 (1976). Except for the experiment of Holt and Pipkin, all results agree with quantum mechanics and violate Bell's inequalities.
    A critical discussion of these experiments, as well as of further measurements involving γ-rays and correlated proton pairs, is given in Ref. 9.

16. A. Aspect, Phys. Rev. D14, 1944 (1976).
17. J.A. Wheeler, see e.g. "Quantum Optics, General Relativity, and Measurement Theory, Ed. by P. Meystre and M.O. Scully, Plenum Publishing Co., 1983.
18. A. Aspect, J. Dalibard, and G. Roger, Phys. Rev. Letters 49, 1804 (1982).
19. M.O. Scully, private communication (1983).
20. F.J. Belifante, A Survey of Hidden-Variables Theories, Pergamon Press, Oxford (1973).
21. A. Barut, P. Meystre, and M.O. Scully, unpublished.
22. Measured probabilities are positive, though. It is only at the hidden-variables level that these strange things happen.
23. Ph. Grangier, These de Troisième Cycle, Université de Paris - Sud (1982), unpublished.
24. Note added in proof: I am thankful to Dr. W. Mückenheim for pointing out an inconsistency in an earlier form of this statement.

BELL'S THEOREM AND QUASICLASSICAL QUANTUM DISTRIBUTION THEORY:

A CALCULATIONAL EXAMPLE

L. M. Pedrotti and Marlan O. Scully*

Institute for Modern Optics
University of New Mexico
Albuquerque, New Mexico 87131
and
Max-Planck Institut fur Quantenoptik
D-8046 Garching bei Munchen
West Germany

ABSTRACT

A simple proof of Bell's Theorem is developed by considering the Einstein, Rosen, Podolski and Bohm spin singlet gedankenexperiment. The quantum mechanical violation of Bell's Theorem for this set-up is developed using the formalism of quasiclassical quantum distribution theory. This formalism allows one to easily note the difference, in this simple case, between quantum mechanics and the sort of theories which satisfy the Bell inequality.

I. INTRODUCTION

Bell's theorem has come to occupy a central position in the ongoing attempt by physicists to better understand the foundations and mathematical framework of quantum mechanics. Bell's theorem and its relationship to quantum mechanics is a subtle subject. In this paper, we hope to provide a background which, for the non-expert, will serve as a doorway into the subject. We do this by considering a specific and well known gedankenexperiment - the

* One of us (MOS) wishes to acknowledge enjoyable and stimulating discussions with A. Barut and P. Meystre who know how to make physics fun!

so-called Einstein, Rosen, Podolski and Bohm (EPRB) set-up. We first prove Bell's theorem in the context of this gedankenexperiment and then proceed to demonstrate the ways in which quantum mechanical predictions contradict Bell's theorem. We note here that the discussion of Section II has been motivated by and is closely related to the lectures of Prof. Meystre contained in this volume.

It is generally excepted that experimental reality contradicts Bell's theorem and supports quantum mechanics. In hopes of shedding some light on the subject, we redo the quantum mechanical calculation using the formalism of quantum distribution theory [1] in Section III. In this way we hope to clarify the difference between quantum mechanics and the more intuitive theoretical frameworks which satisfy Bell's theorem (e.g. hidden variable mechanics).

II. PROOF OF BELL'S THEOREM

We consider in this paper the EPRB gedankenexperiment illustrated in Fig. 1. A spin zero system "splits" into two spin 1/2 particles which then have anticorrelated values of spin projection along any given axis. For the purpose of proving Bell's theorem we are interested in the probability that particle one will pass a Stern-Gerlach apparatus (SGA_1 in Fig. 1) which is oriented at an angle θ with the vertical ($+\hat{z}$) direction and that particle two will pass a Stern-Gerlach apparatus (SGA_2) which is oriented at an angle ϕ to the vertical. We denote this joint passage probability by $P(\phi,\theta)$. A simple statement of Bell's theorem in this context is given by Eq. 1.

$$P(\theta,\phi) + P(\phi,\gamma) \geqslant P(\theta,\gamma) \; . \tag{1}$$

For simplicity and concreteness, in the proof of Bell's theorem which follows, we make the arbitrary assignments $\theta = 0^\circ$, $\phi = 45^\circ$, and $\gamma = 90^\circ$. To proceed with the proof some notational difficulties need first be overcome. To establish our notation, consider the expression,

$$P(0,45) = P\left(\begin{array}{ccc|ccc} \multicolumn{3}{c|}{\text{particle 1}} & \multicolumn{3}{c}{\text{particle 2}} \\ + & - & \square & - & + & \square \\ 0 & 45^\circ & 90^\circ & 0^\circ & 45^\circ & 90^\circ \end{array}\right) . \tag{2}$$

Here, the left side of the partition in the expanded notation refers to particle 1 and the right side particle 2. As shown in Eq. 2 there are three "slots" on each side of the partition in which we have put either a plus sign, a minus sign, or an empty box. The first, second, and third slots are reserved for

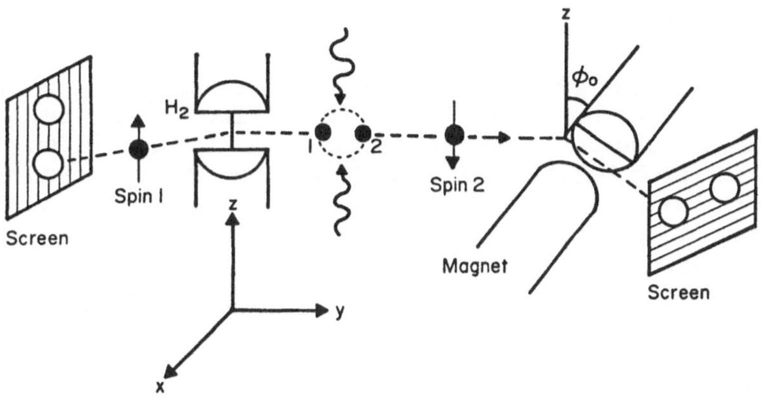

Fig. 1. EPRB set-up consists of a singlet spin system such as orthohydrogen, is split by an external radiation field and the corresponding spin-1/2 particles (protons) proceed to opposite ends of the laboratory where they are passed through Stern-Gerlach apparati oriented along the z-axis (θ = 0) in the case of particle 1 and at an angle ϕ_0 to the z-axis in the case of particle 2.

information concerning passage through an SGA oriented at the angles 0°, 45°, and 90° respectively. A plus sign refers to passage and a minus sign to blockage. An empty box means that the particular joint probability in question does not contain information about passage at that angle.

Now that we have explained the notation in general let us return to Eq. 2 to work a specific example. Recall that P(0,45) denotes the probability that particle 1 passes SGA_1 oriented along the z axis and particle 2 passes SGA_2 oriented at an angle of 45° to the vertical. So we put a plus sign in the first slot on particle one's side of the partition and in the 45° slot on particle two's side of the partition. Furthermore, since the spin projections of the two particles along a given axis are anti-correlated we must put a minus sign in the 45° slot of particle one and in the 0° slot of particle two. Empty boxes appear in the 90° slot of both particles since P(0,45) does not refer to the probability of passage of Stern-Gerlach apparati oriented at 90° to the vertical at all. Likewise we write,

$$P(45,90) = P(\square + - \mid \square - +), \tag{3a}$$

and

$$P(0,90) = P(+ \square - \mid - \square +) \; . \tag{3b}$$

The usefulness of this notation becomes apparent when we take the next step. Although the joint probability P(0,45) says nothing about passage at 90° we do know that for any given particle the probability that it will pass an SGA oriented at 90° to the vertical plus the probability that it will not pass such an apparatus must be equal to unity. Using this fact and the anticorrelation of the spin projections we write,

$$P(0,45) = P(+ - \square \mid - + \square) = P(+ - + \mid - + -) + P(+ - - \mid - + +) . \tag{4a}$$

Similarly,

$$P(45,90) = P(\square + - \mid \square - +) = P(+ + - \mid - - +) + P(- + - \mid + - +) . \tag{4b}$$

and,

$$P(0,90) = P(+ \square - \mid - \square +) + P(+ + - \mid - - +) + P(+ - - \mid - + +) . \tag{4c}$$

Given Eqs. 4, the proof of Bell's Theorem is completed easily. We add P(0,45) and P(45,90) to get,

$$P(0,45) + P(45,90) = P(+ - + \mid - + -) + P(+ - - \mid - + +) + P(+ + - \mid - - +) + P(- + - \mid + - +) . \tag{5}$$

We note that, using Eq. (4c), Eq. 5 can be written as,

$$P(0,45) + P(45,90) = P(0,90) + P(+ - + \mid - + -) + P(- + - \mid + - +) . \tag{6}$$

Classically, probabilities must be positive so that this implies,

$$P(0,45) + P(45,90) \geqslant P(0,90) . \tag{7}$$

This completes our "proof" of Bell's Theorem (Eq. 1) for the case $\theta = 0°$, $\phi = 45°$, and $\gamma = 90°$. A quick review of the preceeding argument will reveal that these choices in no way affect the validity of the proof and so it holds for the general case as well.

III. QUASI-CLASSICAL QUANTUM DISTRIBUTION TREATMENT OF SPIN SINGLET PROBLEM

We have shown the limits that Bell's theorem places on the possible values of the three joint probabilities of passage $P(\theta,\phi)$, $P(\phi,\gamma)$ and $P(\theta,\gamma)$. The Bell's theorem result is intuitively pleasing and so it is somewhat unsettling to find that the quantum mechanical predictions for these joint probabilities violate the Bell inequality for some values of the parameters. There has been much discussion of this point in the literature and several experiments have been devised and performed, to decide which is correct. The general consensus is that experimental reality supports quantum mechanics [2]. This has led many investigators to try to pinpoint where our intuition led us astray in the proof of Bell's theorem. In this section, we treat the EPRB spin singlet problem using the formalism of quantum distribution theory so that the two theoretical frameworks can be placed side by side for comparison. It is our belief that quantum distribution theory is more useful (in this regard) than the standard formulation of quantum mechanics since it more nearly approximates the methodology of classical physics. At any rate the calculation is "fun" and has implications beyond the scope of these notes.

Let us proceed then towards the quantum mechanical expression for $P(\theta,\phi)$. The Hilbert space for the EPRB spin singlet problem is spanned by the Pauli spin matrices $\hat{\sigma}_x^{(1)}$, $\hat{\sigma}_z^{(1)}$, $\hat{\sigma}_x^{(2)}$, and $\hat{\sigma}_z^{(2)}$. Note that we consider roations of the SGA in the x-z plane only (see Fig. 1) so that $\hat{\sigma}_y$ does not play a role. In this formalism the expectation value of an operator is given by

$$<\hat{Q}> = \int d\vec{m}^{(1)} d\vec{m}^{(2)}\, \mathcal{P}(\vec{m}^{(1)}, \vec{m}^{(2)}, t)\, Q(\vec{m}^{(1)}, \vec{m}^{(2)}) \ . \tag{8}$$

Here $(\vec{m}^{(1)}, \vec{m}^{(2)})$ is shorthand notation for $(m_x^{(1)}, m_z^{(1)}, m_x^{(2)}, m_z^{(2)})$ which are the classical variables associated with the Pauli spin operators. The function $\mathcal{P}(\vec{m}^{(1)}, \vec{m}^{(2)}, t)$ is called the probability density and is formally given by

$$\mathcal{P}(\vec{m}^{(1)},\vec{m}^{(2)},t) = \mathrm{Tr}[\rho(t)\,\delta(m_x^{(1)} - \hat{\sigma}_x^{(1)})\,\delta(m_z^{(1)} - \hat{\sigma}_z^{(1)})$$

$$\times\ \delta(m_x^{(2)} - \hat{\sigma}_x^{(2)})\,\delta(m_z^{(2)} - \hat{\sigma}_z^{(2)})]. \tag{9}$$

In Eq. (9), $\rho(t)$ is the density matrix describing our EPRB spin singlet problem, and the δ functions in Eq. (9) are defined as

$$\delta(m_i - \hat{\sigma}_i) = \int \frac{dz}{2\pi}\, e^{-iz(m_i - \hat{\sigma}_i)} \ . \tag{10}$$

Here we are interested in the expectation value of the operator $\hat{\pi}_\phi^{(2)}\hat{\pi}_\theta^{(1)}$ in the singlet state

$$|\psi\rangle = \frac{1}{\sqrt{2}} \quad |+_1 \; -_2\rangle - | -_1 \; +_2] \quad . \tag{11}$$

$\hat{\pi}_{\phi}^{(2)}$ and $\hat{\pi}_{\theta}^{(1)}$ are the projection operators corresponding to the passage of particle 2 through SGA_2 oriented at an angle ϕ to the vertical and particle 1 through SGA_1 oriented at an angle θ to the verical. The state $|+ ,- \rangle$ refers to particle one having spin up along the ϕ direction relative to the vertical and particle two having spin down along the same axis. In terms of the eigenstate corresponding to spin up along the z-axis, $|\uparrow\rangle$, these states may be written as

$$|+\rangle = e^{-i\phi\hat{\sigma}_y/2} \, |\uparrow\rangle$$

and

$$|-\rangle = e^{-i\phi\hat{\sigma}_y/2} \, |\downarrow\rangle \; . \tag{12}$$

In this notation the projection operators and their associated c-number representations are given by

$$\hat{\pi}_{\theta}^{(1)} = |\theta_1\rangle\langle\theta_1| = \frac{1}{2} \, (1 + \hat{\sigma}_z^{(1)} \cos\theta + \hat{\sigma}_x^{(1)} \sin\theta) \tag{13a}$$

$$\pi_{\theta}(\vec{m}^{(1)}) = \frac{1}{2} \, (1 + m_z^{(1)} \cos\theta + m_x^{(1)} \sin\theta) \tag{13b}$$

and

$$\hat{\pi}_{\phi}^{(2)} = |\phi_2\rangle\langle\phi_2| = \frac{1}{2} \, (1 + \hat{\sigma}_z^{(2)} \cos\phi + \hat{\sigma}_x^{(2)} \sin\phi) \tag{14a}$$

$$\pi_{\phi}(\vec{m}^{(2)}) = \frac{1}{2} \, (1 + m_z^{(2)} \cos\phi + m_x^{(2)} \sin\phi). \tag{14b}$$

The density matrix $\rho(t)$ is of course just $|\psi\rangle\langle\psi|$ with $|\psi\rangle$ given by Eq. (11). We now have assembled all of the expressions needed to evaluate $P(\theta,\phi) = \langle\psi| \hat{\pi}_{\phi}^{(2)}\hat{\pi}_{\theta}^{(1)}| \;\rangle$. Using Eqs. (13) and (14) in Eq. (8) gives

$$P(\theta,\phi) = \int d\vec{m}^{(1)} d\vec{m}^{(2)} \mathcal{P}(\vec{m}^{(1)},\vec{m}^{(2)})$$

$$\times \frac{1}{2} (1 + m_z^{(1)} \cos\theta + m_x^{(1)} \sin\theta)$$

$$\times \frac{1}{2} (1 + m_z^{(2)} \cos\phi + m_x^{(2)} \sin\phi) \quad . \tag{15}$$

Using Eq. (10) in Eq. (9) and performing some algebra leads to an expression for the probability density in the integrand of Eq. (15). This expression can be put into a simpler form is we make the change of variables,

$$m_x^{(1)} = m \sin \beta, \tag{16a}$$

$$m_z^{(1)} = m \cos\beta \ , \tag{16b}$$

$$m_x^{(2)} = m \sin\alpha \quad . \tag{16c}$$

and

$$m_z^{(2)} = m \cos\alpha \ . \tag{16d}$$

Then, in terms of α and β,

$$\mathcal{P}(\vec{m}^{(1)},\vec{m}^{(2)}) \equiv \mathcal{P}(\alpha,\beta)$$

$$= \delta(\alpha-\beta-\pi)\tfrac{1}{2}\{\tfrac{1}{2}[\delta(\alpha-\pi/4) + \delta(\alpha+\pi/4)]$$

$$+ \tfrac{1}{2}[\delta(\alpha - \frac{3\pi}{4}) + \delta(\alpha + \frac{3\pi}{4})]\} \quad . \tag{17}$$

Insertion of Eqs. (16) and (17) into Eq. (15) gives the integral form of $P(\theta,\phi)$. That is,

$$P(\theta,\phi) = \int \delta(\alpha-\beta-\pi)\tfrac{1}{2}\{\tfrac{1}{2}[\delta(\alpha-\pi/4)$$

$$+ \delta(\alpha+\pi/4)] + \tfrac{1}{2}[\delta(\alpha - \frac{3\pi}{4}) + \delta(\alpha + \frac{3\pi}{4})]\}$$

$$\times \tfrac{1}{2}[1 + m\cos(\theta-\beta)]\tfrac{1}{2}[1 + m\cos(\phi-\alpha)]d\alpha d\beta \ . \quad (18)$$

Normalization considerations require that $m = \sqrt{2}$. Eq. (18) is fairly easy to evaluate and gives the expected quantum mechanical result,

$$P(\theta,\phi) = \tfrac{1}{4}[1 - \cos(\theta-\phi)] \ . \quad (19)$$

We note in passing that the variables α and β appearing in Eq. (18) are related to certain "hidden" variable parameters in some works. This is discussed in Ref. 3.

We now may see that the quantum mechanical prediction for $P(\theta,\phi)$ does indeed violate Bell's theorem. As in Section I we use the values $\theta = 0°$, $\phi = 45°$, and $\gamma = 90°$ for the purposes of demonstration. Then, using Eq. (19) we find,

$$P(0,45) + P(45,90) \cong 0.1465 \leqslant P(0,90) = 0.250 \quad (20)$$

in direct disagreement with Bell's theorem. This result, of course, is not new.

Here we would like simply to see "why" it is that quantum mechanics gives a result different from that of Bell's theorem. The utility of the quasi-classical quantum distribution approach is that it is closely related to the language of statistical classical mechanics.

To see how the quantum mechanical description differs from the hidden variable logic, we write the general form of the classical description of $P(\theta,\phi)$ as

$$P(\theta,\phi)_{\text{classical}} = \int C(\alpha,\beta)P_\theta(\beta)\ P_\phi(\alpha)\ d\alpha d\beta \ . \quad (21)$$

In Eq. (21) $P_\phi(\alpha)$ and $P_\theta(\beta)$ represent the probabilities that particle 2 with the "hidden" variable α would pass SGA_2 oriented at an angle ϕ to the vertical and particle 1 with "hidden" variable β would pass SGA_1 oriented an angle θ to the vertical. Classically these probabilities are, of course, positive. The function $C(\theta,\phi)$ represents the correlation between the variables α

and β and so integration over these leads to the joint probability of passage, $P(\theta,\phi)|_{\text{classical}}$. The general form for $P(\theta,\phi)$ from quantum distribution theory closely resembles the classical form (Eq. 21) and is given by

$$P(\theta,\phi) = \int \mathcal{P}(\alpha,\beta)\ \pi_\theta(\beta)\ \pi_\phi(\alpha)\,d\alpha\,d\beta \qquad (22)$$

As in the classical case $\mathcal{P}(\alpha,\beta)$ plays the role of a correlation function between the variables α and β. However the projection operators $\pi_\theta(\beta)$ and $\pi_\phi(\alpha)$ given by Eqs. (13), (14) and (16) as

$$\pi_\theta(\beta) = \tfrac{1}{2}[1 + \sqrt{2}\cos(\theta - \beta)] \qquad (23a)$$

and

$$\pi_\phi(\alpha) = \tfrac{1}{2}[1 + \sqrt{2}\cos(\phi - \alpha)], \qquad (23b)$$

can be negative. Herein lies an essential difference between the classical and quantum descriptions of the EPRB spin singlet problem. The functions $\pi_\theta(\beta)$ and $\pi_\phi(\alpha)$, which are the closest quantum mechanical analogues to the classical probabilities $P_\theta(\beta)$ and $P_\phi(\alpha)$ (they enter into the corresponding expressions for $P(\theta,\phi)$ in the same way) can be negative. We recall that the proof of Bell's theorem does not hold if one allows negative probabilities.

We have shown that an attempt to make quantum mechanics look classical requires that we make correspondences between classical probabilities and quantum distribution theory functions which are not positive semi-definite. Thus we see (in this example) how it is that we have a quantum mechanical violation of Bell's theorem. A more detailed quasi-classical quantum distribution theory treatment of the EPRB spin singlet problem which emphasizes the "hidden variable" aspect of this formalism can be found in Ref. 3.

REFERENCES

1. This approach to the EPRB spin singlet problem was motivated by F. Belinfante, A Survey of Hidden Variable Theories (Pergammon, New York, 1973).
2. An excellent summary of the experimental tests up to 1978 is given in J. Clauser and A. Shimony, Rep. Prog. Phys., 41, 1881 (1978). For the most recent and convincing test see A. Aspect, P. H. Granger and G. Roger, Phys. Rev. Lett., 49, 91 (1982).
3. M. O. Scully, Phys. Rev., to be published.

INDEX